I0072035

Food Processing Technology

Food Processing Technology

Editor: Albert McCoy

www.callistoreference.com

Callisto Reference,
118-35 Queens Blvd., Suite 400,
Forest Hills, NY 11375, USA

Visit us on the World Wide Web at:
www.callistoreference.com

ISBN: 978-1-64116-170-1 (Hardback)

Cataloging-in-Publication Data

Food processing technology / edited by Albert McCoy.
p. cm.
Includes bibliographical references and index.
ISBN 978-1-64116-170-1
1. Food industry and trade--Technological innovations. 2. Food processing machinery.
3. Food processing plants--Equipment and supplies. I. McCoy, Albert.
TP370 .F66 2019
664--dc23

Table of Contents

Preface

Food processing is concerned with the transformation of raw food ingredients into marketable food products. It involves a variety of physical and chemical techniques to achieve this transformation. Principal techniques of this field are mincing, pickling, pasteurization, preservation, canning, packaging among many others. Some of the important parameters for food processing are hygiene, minimization of waste and energy efficiency. There are a number of advantages of food processing such as toxin removal, easy marketability, prevention of foodborne diseases, improvement of taste, etc. Extra nutrients can also be added to food in food processing. This book is a compilation of chapters that discuss the most vital concepts and emerging trends in the field of food processing. Most of the topics introduced herein cover new techniques of food processing and their applications. This book, with its detailed analyses and data, will prove immensely beneficial to professionals and students involved in this area at various levels.

Significant researches are present in this book. Intensive efforts have been employed by authors to make this book an outstanding discourse. This book contains the enlightening chapters which have been written on the basis of significant researches done by the experts.

Finally, I would also like to thank all the members involved in this book for being a team and meeting all the deadlines for the submission of their respective works. I would also like to thank my friends and family for being supportive in my efforts.

Editor

Effects of Different Cooking Methods on Heavy Metals Level in Fresh and Smoked Game Meat

Kobia Joyce[1], Emikpe BO[2*], Asare DA[1], Asenso TN[1], Yeboah Richmond[3], Jarikre TA[4] and Jagun-Jubril Afusat[4]

[1]*Department of Animal Science, Kwame Nkrumah University of Science and Technology, Kumasi, Ghana*
[2]*Department of Pathobiology, School of Veterinary Medicine, Kwame Nkrumah University of Science and Technology, Kumasi, Ghana*
[3]*Department of Biological Sciences, Kwame Nkrumah University of Science and Technology, Kumasi, Ghana*
[4]*Department of Veterinary Pathology, Faculty of Veterinary Medicine, University of Ibadan, Nigeria*

Abstract

This study investigated the effects of different cooking methods on heavy metals levels in fresh and smoked bush meat. Thirty-five fresh meat samples were washed thoroughly before employing the three different cooking methods (boiling, grilling and frying). Smoked meat samples obtained from central market were subjected to only boiling. Samples were protein digested and the digested solution was submitted to the Ghana Atomic Energy Commission for analysis. The Atomic Absorption Spectrometer was employed to detect the presence and concentration of heavy metal in game meats: (Iron (Fe), Copper (Cu), Cadmium (Cd), Lead (Pb), Manganese (Mn) and Zinc (Zn)). The data obtained was analyzed with a one- way analysis of variance. Results showed that different cooking methods had significant effect on heavy metals concentration with boiling increasing Fe, Pb, and Cu concentrations but reduced Zn (zinc) and Mn (manganese) concentration. Grilling increased Fe and Cu concentrations, but reduced Pb, Zn and Mn concentrations. Frying decreased Fe, Cu and Mn but increased Zn and Pb concentrations; lastly Cadmium (Cd) concentrations were within the recommended range and were not affected by any of the cooking methods. It is therefore recommended that consumers of game meat should be encouraged to subject fresh game meat to grilling which has the capacity to reduce the concentration of the most toxic heavy metal (Pb). These findings shall inform measures in consumer safety.

Keywords: Bush meat; Heavy metals; Frying; Grilling; Boiling

Introduction

Wildlife is identified as a potential source of animal protein and income for rural households. Most rural populace hunts for bush meat for consumption as well as market some to earn money aside domestic animals such as chickens, sheep or goats. Bush meat refers to carcass of animals hunted from the wild [1]. The most commonly hunted bush meat animals are bats, squirrels, grasscutter, giant rats, guinea fowls and snakes [1]. These animals are regularly consumed only on ceremonial and festive occasions and not a regular source of food [2,3].

Studies show that carcass of grass cutter is widely recognized as a source of animal protein in Africa especially West Africa which has a large market for the meat [2]. In Ghana, bush meat is preferred to fish and other domestic animal's meat by the people of high social class [4]. The cane rat meat is appreciated because of its culinary properties with demand consistently outstripping supply [3,5]. Bush meat has an important nutritional role in the diet of humans; it adds diversity and encourages people to consume greater quantities of staple foods [6]. The demand for grasscutter meat has been met through hunting from the wild. This has been done by the use of guns, cutlasses, chase dogs, baiting with chemicals, and by bush burning [7]. The method of capture can pose a health threat to consumers because of the possibility of heavy metal contamination of the meat [8]. More so, the method of hunting is becoming a growing concern due to bullets containing Pb and other metals [9-11]. The use of guns as well as chemicals has been reported to have harmful effects on consumers and other untargeted species [7]. According to Omojola et al. [12], meat has the potential of accumulating toxic minerals and represents a major source of heavy metals for human consumption which poses as danger to animal and human health [13,14].

The meat is often sold either as fresh (whole undressed) carcasses, or dressed and smoked [3,15,16]. Most people employ different cooking methods to process these meats and these methods include barbequing, boiling, grilling, pan frying, stir frying and roasting which is done to increase its flavor, taste, palatability and to make it tender [17]. Apart from the excellent taste, bush meat is nutritionally superior because of its higher protein and low fat level [18]. However, with the various cooking methods and their effects on palatability, there is a paucity of information on the effect on different cooking methods on heavy metals concentration in game animal meat. Therefore, this study attempts to evaluate the effect of different cooking methods on the heavy metals concentrations in fresh and smoked bush meat obtained in selected markets in Kumasi Ghana.

Materials and Methods

Study area

The study was carried out in Atwemunom Market and Central Market in the Kumasi Metropolis as well as parts of Sekyeredumase in the Ejura-Sekyedumase District within the Ashanti Region of Ghana. Sekyeredumase is located with latitude 6.747N and longitude -1.509W. The area is within the forest zone of Ghana characterized by high rainfall patterns and presence of forests and grasses which favor game animals such as grass cutter, giant rat, squirrels, and antelopes among others. The different kinds of bush meat available in these markets include both processed meat, fresh and smoked meat.

Study animals

Cane rat (*Thryonomys swinderianus*) and Giant rat were the study

***Corresponding author:** Emikpe BO, Department of Pathobiology, School of Veterinary Medicine, Kwame Nkrumah University of Science and Technology, Kumasi, Ghana, E-mail: banabis2001@yahoo.com

animals considered because these were the mostly consumed bush meat in the area of this study. A total of thirty-five (35) bush meat samples were obtained, each weighing between 80-130 grams of which five (5) were processed cane rat, fifteen (15) fresh cane rats and fifteen (15) fresh giant rat.

Sample collection and preparation

Fresh meat samples were analyzed and served as the standard. The muscles including the bones of the study cane rat and giant rat were distributed cautiously to three different cooking conditions of boiling, grilling, and frying. The meat was washed and boiled in a cooking pan with enough water to cover meat using gas stove. Again the processed meat of grass cutter was boiled; the processed meat was prepared as described by Omojola et al. [12]. All meat samples for grilling and frying were partially boiled for 5-8 minutes with water. No spices or food additives were added. Grilling of meat was done on a coal pot with charcoal fitted with metal grids for 10-15 minutes depending on the size of the meat. Frying of meat was carried out in a stainless frying pan containing vegetable cooking oil, it was fried for 3-5 minutes depending on this size and the temperature obtained was 170 degrees Celsius after frying. Fresh, boiled, fried and grilled meat samples were chopped into smaller parts and air-dried for 3-5 days to remove moisture. Air-dried meat samples were milled into powdery form using a blender.

Laboratory analysis

Digestion of samples and heavy metals detection: A gram (1 g) of each sample was weighed into a glass beaker for the digestion process. 10 ml of nitric acid was added to it. A minimum time frame of two to five minutes was used to boil the samples on the stove which was then allowed to cool down for five to ten minutes. The digested samples were filtered using 50 ml volumetric flasks, filter paper and a funnel. It was then topped up to 50 ml using distilled water. Heavy metals concentrations of iron (Fe), zinc (Zn), copper (Cu), manganese (Mn), lead (Pb) and cadmium (Cd) were determined using the Atomic Absorption Spectrometer.

Statistical analysis: The results obtained were statistically analysed by SPSS Version 20 using one- way analysis of variance (ANOVA) and Duncan Multiple Range Test to find out the significant differences between cooking methods effects on heavy metals concentration. The presence or absence of a significant difference was specified at 95% confidence interval and significance level of p=0.05.

Results

Cooking methods effects on heavy metals concentration

The result obtained for the effect of different cooking methods on the heavy metals concentration is represented in Table 1. Results obtained were compared to the accepted or safe limits levels of metal concentrations for human consumption. From Table 1, the average iron concentration in the fresh meat was 12.72 ppm; after employing the bush meat into the three cooking methods, boiling increased the iron concentration to 19.17 ppm, grilling increased the iron levels to 26.06 ppm. Frying bush meat reduced iron concentration level to 5.61 ppm. The average iron concentration for smoked bush meat, was 15.28 ppm but concentration increased after boiling to 16.41 ppm. The average fresh meat concentration of manganese was 5.16 ppm but after employing them into the three cooking methods, boiling reduced the manganese concentration to 0.64 ppm, frying reduced it 0.84 ppm and grilling reduced it to 0.41 ppm. Copper concentration in the fresh state of the bush meat was 0.96 ppm. However, after boiling, the concentration increased to 1.09 ppm, frying reduced it to 0.56 ppm and after grilling it increased to 1.99 ppm. Copper concentrations of smoked bush meat after boiling also increased from 0.94 ppm to 1.24 ppm. Results also showed that boiling, grilling as well as boiling of smoked bush meat reduced the zinc concentrations of bush meat comparing to the fresh bush meat, however only fried bush meat increased the zinc concentrations from 7.40 ppm to 7.64 ppm. Cadmium concentrations in the bush meat where not affected by the different cooking methods. The levels of cadmium concentration for the different cooking methods were the same as well as for the smoked and smoke-boiled bush meat (0.10). Concentration of lead in the fresh bush meat was 4.31 ppm. After boiling, it increased to 13.73 ppm and increased to 17.32 ppm after frying. The lead concentrations reduced drastically to 3.59 ppm after grilling. In the smoked bush meat, initial lead concentration was 0.07 ppm and it reduced to 0.05 ppm after the smoked bush meat was boiled.

Discussion

This study was carried out to determine the extent of the effects of different cooking methods on heavy metals levels in fresh and processed (smoked) bush meat. From the results obtained, the average iron (Fe) concentration for in the fresh meat was high above the recommended levels however after employing the three cooking methods, boiling increased the iron concentration, grilling increased the iron levels much higher ($p<0.05$) while frying reduced Fe concentration level. The average iron concentration for smoked bush meat was higher than the recommended level for human consumption but increased after boiling. This could be attributed to the interactions between the meat and the metal grid which is often made of iron. Therefore, during grilling, some of the Fe particles can get into the meat. This could have accounted for the higher Fe concentration in the grilled meat as well as for the smoked bush meat as they both employ the same processing procedure. In comparison of the results obtained to the accepted levels of iron (Fe) concentration for human consumption (4.49 ppm to 15.0 ppm), none

Different cooking methods	Heavy metals and their concentration levels (parts per million (ppm))					
	Fe	Mn	Cu	Zn	Cd	Pb
	Safe limits (4.49-15.0)	Safe limits (unknown)	Safe limits (0.87-5.0)	Safe limits (0.41-5.0)	Safe limits (0.33)	Safe limits (0.01-0.38)
Fresh	12.72	5.16	0.96	7.40	0.10	4.31
Boiled	19.17	0.64	1.09	7.20	0.10	13.93
Fried	5.61	0.84	0.56	7.64	0.10	17.32
Grilled	26.06	0.41	1.99	6.97	0.10	3.59
Processed (fresh)	15.28	5.33	0.94	7.39	0.10	0.07
Processed (Boiled)	16.41	1.25	1.24	6.66	0.10	0.05
Significant level (p<0.05)	0.00	0.11	0.26	0.39	1.00	0.48

Table 1: Effects of cooking methods on heavy metal concentrations in eat of bush meat.

of the cooking methods was able to reduce the iron concentrations and this may pose health hazards to humans.

The concentration of manganese in bush meat was reduced by the three cooking methods while smoked meat upon boiling reduced the manganese concentration. The reason for these findings has not been fully elucidated however this can be attributed to the possible effect of the heat treatments and water association of manganese.

Copper concentrations after boiling and grilling were increased while frying reduced it. Smoked bush meat upon boiling also increased the Cu concentrations from 0.94 ppm to 1.24 ppm however the concentration levels were within the recommended accepted levels of metal concentration for human consumption. This finding buttresses the report of Soewu et al. [19] who also reported that copper was detected in relative low quantities. Concentrations of lead in bush meat increased after boiling and after frying, however the concentrations reduced drastically when bush meat were grilled. Similar observation was noted in fresh and smoke-dried grasscutter (*Thryonomys swinderianus* Temminck) meat in Southern Nigeria [11]. In the smoked bush meat, the lead (Pb) concentration was lower than that of the fresh and the different cooking methods and it reduced the more when it was boiled ($p>0.05$). This can be attributed to the reduction of water in the meat as a result of the smoking and grilling of the meat which invariably could reduce the lead in the meat since lead is found mostly in the cytoplasm of the cell which reduced during smoking. The drop in concentration of heavy metal in the grilled samples were also corroborated in fish meat by Ahmed et al. [20], Eboh et al. [21], and Ajani et al. [22], attributed to heat effect on heavy metals and a possibility of the heavy metal being converted to other compounds. It was also reported that cooking methods (boiling, steaming, frying among others) can change the levels of toxic metals through various means, including the evaporation of water and volatile components, solubilization of the element and also by metal binding to other macronutrients present in the food item such as carbohydrates, lipids and proteins.

Conclusion

In conclusion, the different cooking methods had effects on the various heavy metals concentrations in bush meat. Iron (Fe), zinc (Zn) and manganese (Zn) concentration is affected by the different cooking methods. In relation to lead (Pb), frying and boiling increased the concentration level while grilling and smoking reduced the level of lead concentration in the meat to a considerable level. The concentration level of cadmium (Cd) was not affected by the different cooking methods. It can be recommended that bush meat should be smoked or grilled. These cooking methods have the capacity to reduce the concentration of lead (Pb) considering its risk implications. These findings will help to inform measures in consumer safety.

References

1. Adeola M, Decker E (1987) Wildlife utilization in rural Nigeria. International Symposium of Wildlife Management in Sub-Saharan Africa.
2. Asibey E (1974) (*Thryonomys Swinderianus*) Temminck, in Ghana. The grasscutter 34: 161-170.
3. Ajayi S (1971) Wildlife as a source of protein in Nigeria; some priorities for development. Niger Field 36: 115-127.
4. Adomah YJC (2009) Features and profitability of domestic grasscutter production in the brong Ahafo region. University of Science and Technology, Kumasi-Ghana.
5. National Research Council (NRC) (1991) Little known small animals with a promising economic future. Micro-livestock.
6. Annegers J (1973) Seasonal food shortages in West Africa. Ecol Food Nutri 2: 251-257.
7. Oduro W, Kankam B (2002) Environmental and public health hazards of traditional grasscutter (*Thryonomys Swinderianus*) hunting. Proceedings of a workshop on promoting grasscutter production for poverty reduction in Ghana Pp: 19-22.
8. Hunt WG, Watson RT, Oaks LJ, Parish CN, Burnham KK, et al. (2009) Lead bullet fragments in venison from rifle-killed deer: Potential for human dietary exposure. PLoS ONE 4: e5330- e5330.
9. Irschik I, Bauer F, Sager M, Paulsen P (2012) Copper residues in meat from wild artiodactyls hunted with two types of rifle bullets manufactured from copper. Eur J Wildlife Res Eur J Wildlife Res.
10. AESAN (2012) Scientific committee risk associated with the presence of lead in wild game meat in Spain. Report approved by the Scientific Committee on plenary session.
11. Okoro KI, Igene JO, Ebabhamiegbebho PA, Evivie SE (2015) Lead (Pb) and Cadmium (Cd) levels in fresh and smoke-dried grasscutter (*Thryonomys swinderianus* Temminck) meat. Africa J Agri Res 10: 3116-3122.
12. Omojola A, Hammed S, Attoh-Kotoku V, Wogar-Gogar G, Iyanda O, et al. (2014) Physicochemical and organoleptic characteristics of Muscovy drake meat as influenced by cooking methods. Africa J Food Sci 8: 184-189.
13. Aschner M (2002) Neurotoxic mechanism of fish-bone methylmetry. Environ Toxicol Phamacol 12: 101-102.
14. Aycicek M, Kaplan M, Yarman M (2008) Effect of cadmium on germination, seedling growth and metal contents of sunflower (*Helianthus annus* L). Asian J Chem 20: 2663-2672.
15. Den Hartog A, DeVos A (1973) The use of rodents as food in tropical Africa. FAO Nutr News 1: 1-4.
16. Barnes A (1994) A study FAO newsletter on traditional meat preservation technologies. Game animal meat drying in Ghana 5: 2-6.
17. Omojola AB, Ahmed SA, Attoh-Kotoku V, Wogar GS (2015) Effect of cooking methods on cholesterol, mineral composition and formation of total heterocyclic aromatic amines in Muscovy drake meat. J Sci Food Agric 95: 98-102.
18. Ledger H (1963) Notes on the relative body composition of wild and domesticated ruminant. Bull Epizootic Dis Africa 2: 163-165.
19. Soewu DA, Bakare OK, Ayodele IA (2012) Trade in the wild mammalian species for traditional medicine in Ogun state, Nigeria. Global J Med Res 12: 6-21.
20. Ahmed A, Dodo A, Bouba AM, Clement S, Dzudie T, et al. (2011) Influence of traditional drying and smoke-drying on the quality of three fish species (*Tilapia nilotica, Silurus glanis and Arius parkii*) from Lagdo lake, Cameroon. J Anim Vet Adv 10: 301-306.
21. Eboh L, Mepba HD, Ekpo MB (2006) Heavy metal contaminants and processing effects on the composition, storage stability and fatty acid profiles of five common commercially available fish species in Oron local government, Nigeria. Food Chem 97: 490-497.
22. Ajani F, Adetuaji VO, Oyedokun JO (2013) Biophysiochemical changes that occur in fish during different stages of traditional processing. University of Ibadan, Nigeria.

2

Monitoring of Change in Cantaloupe Fruit Quality under Pre-Cooling and Storage Treatments

Azam MM[1,2*], Eissa AHA[1,2] and Hassan AH[1]

[1]*Department of Agriculture Systems Engineering, College of Agricultural and Food Sciences, King Faisal University, Saudi Arabia*
[2]*Agriculture Engineering Department, Faculty of Agriculture, Minoufiya University, Shibin El-Kom, Egypt*

Abstract

During recent years, global concern for protection of the environment has led researchers to improving postharvest treatments such (precooling, handling, storage, etc.,). The present investigation was also directed to find out the precooling and storage alternatives for the extension of the storage life of cantaloupe fruits (*Cucumis melon*). Fruits were harvested at the mature stage and precooled by forcing cooling air with velocity of 1-2 m/s at three different air temperatures of 5°C, 10°C and 15°C. The fruits were analyzed for physiological characters such as loss in weight, fruit firmness, change in color surface of fruit and its flesh. Fruit temperature expectedly decreased from initial level of about 36-38°C to the desired storage temperature of 10°C by forcing cooling air at 5°C, for about 45 min. Cooling time increased to 105 and 165 min when cooling air temperature increased to 10°C and 15°C, respectively. During subsequent storage at 15°C with 90-95% relative humidity, precooled fruits were exhibit more desirable characteristics than that of non-precooled fruits (control), where precooling was retard softening. Non precooled fruits "control" turned ripe-soft after 15 days when firmness decreased to less than 10 N from initial value of 90 N at the unripe stage.

Keywords: Cantaloupe; Forced air cooling; Cold storage; Color; Firmness; Quality

Introduction

No doubt that fruits and vegetables considered a major component of vitamins, antioxidants, minerals, dietary fiber [1]. Both quantitative and qualitative postharvest losses in fresh fruits and vegetables are significant due to their high values reaching up to 50% in developing countries [2]. Therefore, postharvest conservation of the produce quality and quantity is an important measurement to enhance world food supplies in a largely effective manner [3]. Certainly improved technologies in postharvest can at some extent overcome these losses. Temperature and relative humidity are the most important environmental factors affecting quality of fresh produce and also the consumer acceptability for fruits and vegetables displayed in a produce department. Good temperature management is the simplest and easiest way of delaying produce deterioration and it can be applied through low storage temperatures which can depress physiological activity of tissues and activity of spoilage microorganisms, and, in general, the lower the storage temperature, the longer the produce postharvest life [4,5]. In a tropical environment, high temperature and high relative humidity were frequently occurred throughout the year and they will affect postharvest quality of fruit, include melon fruit quality. Several methods were applied for rapid cooling the product to the lowest safe storage temperature within hours of harvest, reducing the respiration rate and enzyme activity, slower ripening/senescence, maintenance of firmness, inhibition of pathogenic microbial growth and minimal water loss [6].

Temperature is the most single important factor which affects the storage life and the quality of fresh produce. The process of precooling is the removal of field heat as soon as possible after harvest since field heat arrest the deterioration and senescence process. The precooling process can be achieved via different methods. Forced air precooling is the most common technique and is adapted to many commodities [7].

Pre-cooling by removing field heat from freshly harvested fruits reduces microbial activity and respiration rates. Furthermore, the respiratory activity and senescence of fruit as well as ethylene production are temperature dependent. Due to the pre-cooling treatments, metabolic activity and consequently respiration rate and ethylene production of the fruits were reduced considerably. This also decreases the ripening rate, diminishes water loss and decay, and thus, helps preserving quality and prolongs shelf life of the fruits [8]. Thompson et al. [9] defined forced-air cooling as a technique of forcing cold air through containers, cooling individual pieces of product. It is commonly used for tree fruit, berries, melons and cut flowers. Forced-air cooling advantages include ease of use, rapid cooling and a dry product [10]. Color may be defined as the impact of the wavelengths of light in the visual spectrum, from 390 to 760 nm, on the human retina and is one of the major attributes which affect the consumer perception of quality [11]. Color is one of the main attributes that characterizes the freshness of most fresh-cut fruits and vegetables, as consumers take product appearance as a primary criterion in food choice and acceptability, and may even influence taste thresholds, sweetness perception and pleasantness [12].

The quality of the cantaloupe fruits are characterized by different factors that are associated with the characteristics of flesh as the soluble solids content, external and internal appearance of fruit, the thickness of the pulp and its aroma and flavor and all these characteristics determine the acceptance of the fruit by consumers and also are used as an index for grading and marketing [13].

The initial quality of the cantaloupe, based on physical appearance such as the size, net development, background colour, and being free

***Corresonding author:** Azam MM, Department of Agriculture Systems Engineering, College of Agricultural and Food Sciences, King Faisal University, Saudi Arabia, E-mail: mazam@kfu.edu.sa

from defects is a basic criterion for fruit to be selected for fresh-cut processing. These fruits all show rapid quality degradation when put into storage [14-16]. The quality reduction may be tolerated under optimum postharvest handling and a storage temperature of between 0°C and 1°C. These conditions have been previously recommended for delaying deterioration and for ripening. Interest in reducing the amount of fruit dumping at fresh fruit markets has led to an observation of how the cantaloupe undergoing a longer duration of postharvest storage may preserve and maintain the fresh quality after processing. The objective of this study is to determine the effect of forced air cooling, in combination with storage period to enhance the quality and characteristics of cantaloupe fruits.

Materials and Methods

System experimental design

The freshly harvested fruits sorted by size, color and absence of defects to see effect of fast heat removal on cantaloupe properties. Cooling air was forced in a refrigeration room at 5°C, 10°C and 15°C with air velocity of 1-2 m/s. During pre-cooling, two fruits were penetrated with thermocouple for temperature monitoring. Temperature was measured every 5 min. After attaining the required temperature 10°C, fruits were shifted after cooling to foam box which was set in cooled room maintained at 10°C. Some physical and mechanical properties of fruits were studied before and after pre-cooling processes to set the change in fruit quality properties. Fruits without forced air-precooled served as control. The study was done following procedures for completely randomized design experiments. Four replications were used, each replication having 4 fruits per observation period (Figure 1). Cantaloupes were considered cooled when they reached 7/8 of the cooling time, which is the time necessary for a reduction of 87.5% of the initial fruit temperature with respect to air temperature [9].

After pre-cooling process, samples were stored in storage unit for 15 days (10°C and 90 ± 5% Relative humidity) while sample without forced air-precooled (control) placed in room temperature (23°C and 78% Relative humidity). Through the storage period (0, 5, 10, 15 day), changing in fruit properties such as firmness and color were measured. Data were collected and incorporated into the computer to make the necessary statistical analysis at this stage to identify the most important results of this study (Figure 1).

Figure 1: Box for pre-cooling cantaloupe products.

Measurement of Fruit Responses

Color determination

Color was determined using the Hunter Lab System with a Minolta colorimeter CR320 model (Figure 2a). Chroma values were the means of three determinations for each fruit along the equatorial axis. The lightness L, a and b values were converted to standard L^*, a^* and b^* values as [17]. The L* (lightness), a* and b* (chromaticity) coordinates. The cantaloupe fruit (flesh, skin) color was measured for L*, a*[green (-) to red (+)], b* [blue (-) to yellow (+)] (Figure 2b): [18]

$$L^* = \frac{100L}{255}$$

$$a^* = \frac{240a}{255} - 120$$

$$b^* = \frac{240b}{255} - 120$$

The hue angle (H^o) and chroma (C^*) values and the total color difference (ΔE) were calculated as

$$H^o = \tan^{-1}\left(\frac{b^*}{a^*}\right)$$

$$C^* = \left(a^{*2} + b^{*2}\right)^{1/2}$$

$$\Delta E = \left[\left(L^* - L_o^*\right)^2 + \left(a^* - a_o^*\right)^2 + \left(b^* - b_o^*\right)^2\right]^{0.5}$$

Fruit firmness

Digital instrument (Effe-Gi, Ravenna, Italy) for measuring penetration resistance was used to measure firmness, which is used worldwide as a test of ripeness and maturity for many vegetables and fruits. Firmness for fruit was measured by two methods. First one measured the firmness for full fruit as showed in (Figure 3A) (beginning of the neck region), while the second method measured firmness for flesh fruit in four points (f1, f2, f3, f4) as showed in Figure 3.

Weight loss

It was calculated as the difference between the initial (prior to storage) weight and the final weight (after storage).

Results and Discussion

Figure 4, indicates the rate of decrease in pulp temperature during forcing cooling air at 5-15°C. Bringing down the fruit temperature from about 36-38°C to the desired storage temperature of 10°C was

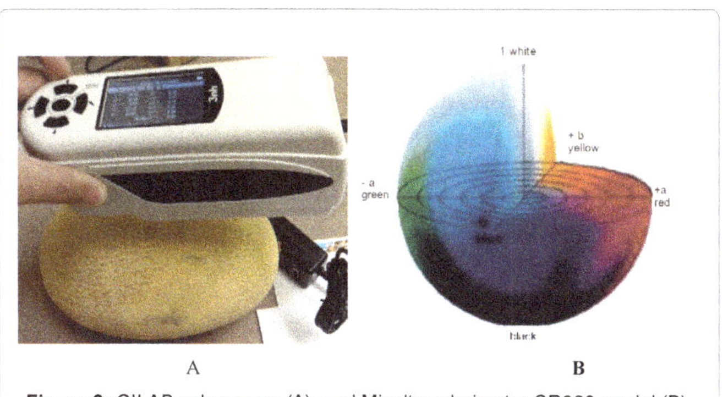

Figure 2: CILAB color space (A), and Minolta colorimeter CR320 model (B).

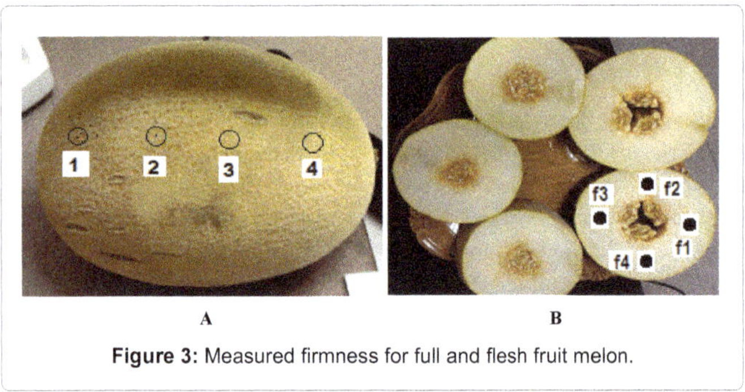

Figure 3: Measured firmness for full and flesh fruit melon.

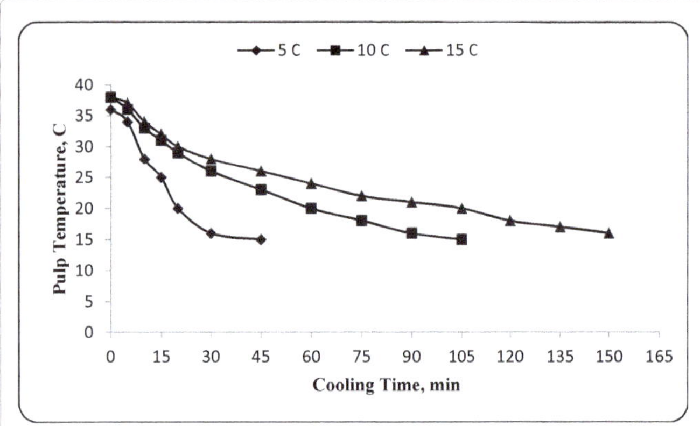

Figure 4: Changes in pulp temperature of cantaloupe during force air cooling at 15°C.

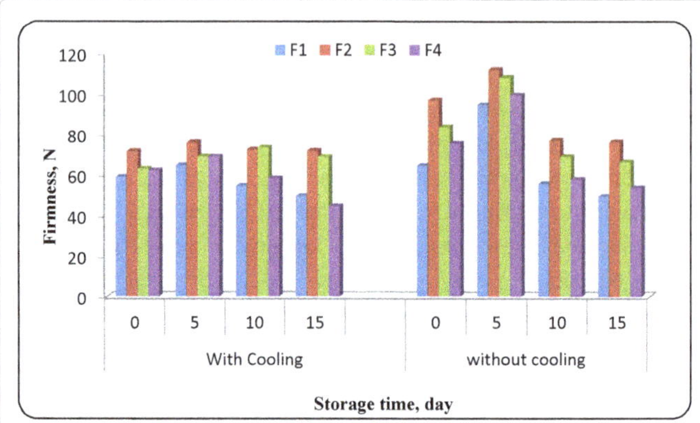

Figure 5: Effect of precooling treatment and storage time on firmness of full cantaloupe.

expectedly fastest at 5°C and slowest at 15°C. It took about 45 min for the pulp temperature to drop to 10°C by cooling air of 5°C and about 165 min by cooling air of 15°C. At cooling air of 10°C, the time spent to reduce the pulp temperature to 10°C was about 105 min. During subsequent 10°C storage, forced air-precooled fruits did not exhibit desirable characteristics indicative of more favorable treatment effects for quality maintenance and shelf life extension.

Firmness

Data in Figure 5 showed the changing in skin cantaloupe firmness under precooling and storage treatments. The values of skin firmness were varied from 44.65 to 76.65 N for cantaloupe cooling treatment, and from 49.50 to 112 N for cantaloupe without cooling treatment (control). Samples of all treatments showed decrease in flesh and full fruit firmness during all storage period.

Fruits which were stored without cooling (control) showed a skin softening at end of storage period. Mean values of treatments also revealed that fruit stored after cooling had lower pulp firmness. This indicates that the firmness of fruits was decreased as postharvest storage of the cantaloupe increased. It can be concluded that in comparison between treatment (with cooling and without cooling) the experimental results of measuring firmness in cantaloupe which pre-cooled were pretty much comparable with each other. From Table 1, average values of firmness of fresh cantaloupe decreased from 97.73 N to 67.33 N for stored precooled fruit and from 99.67 N to 62.63 N for stored non-cooled fruit as the storage period increased. And data in Figure 6 showed the effect of postharvest storage of cantaloupe on the flesh fruit firmness, which stored at 10°C and 90 ± 5% RH for 15 days. The slides that pre-cooled before storage recorded lowest values of firmness compared to slides stored without cooling. Differences in the percentage loss were affected by the fruits and storage duration. This indicates that the firmness of the flesh fruits were decreased as storage period of the cantaloupe increased as shown in Figure 6. Values of flesh fruit firmness which stored without cooling were increased at the second period (after 5 day) due to the high percentage of water loss while, it was decreased during the next periods from the harvesting day. But same values of flesh fruit firmness which stored after cooling recorded highest values in the end of storage period (15 day). This is may be due to some changing occurred in flesh fruit as shrinkage due to the high water loss percentage particularly during the period of first week from harvesting day (Table 1).

External and internal color

External (skin) color: Fruit which harvested at maturity stage showed an increase in yellow color development. There were differences between color parameters of skin fruit (L^*, a^*, b^*, h^*, ΔL^*, Δa^*, Δb^*, Δh^*, ΔE) under different treatments (Figure 7). Data showed correlation between different treatments and a* values. a* values were increased when skin fruit reached from immature to full matured stage. An increasing in a^* value means a decrease in the degree of greenness. Meanwhile, mean values of color parameters of skin fruit decreased over the postharvest storage time of the fruit. The decreases of color in the skin cantaloupe fruits were obviously associated with the ripeness stage of the fruit [19].

Maximum value of a^* (19.37) was found at treatment without cooling after 10 days storage, while maximum value of a^* at treatment with cooling (18.74) was found at the end of storage period. At the end of storage period values of a^* were reached to 18.74 and 13.84 for treatments with and without cooling, respectively. These changing during storage periods (loss of greenness) were occurred as a result of breakdown of chlorophyll pigments in the peel tissues. The change of peel color from green to yellow was due to accumulation of carotenoids and chlorophyll degradation at thylakoid membrane through chlorophyllase and oxidase enzymes.

Values of b^* were increased between treatments under this study. This values ranged from 7.71 to 9.14 in treatment with cooling, and ranged from 5.52 to 8.61) in treatment without cooling. Similarly, the luminosity (L^*) varied in same fashion (Figure 7). These results are in close agreement to those of earlier workers [20]. The luminosity (L*) of the skin fruit recorded similar values of 16.88 and 16.23 for treatment with-cooling and without cooling, respectively. But in end of the storage period, the value of L* of the skin fruit had a value of 16.44

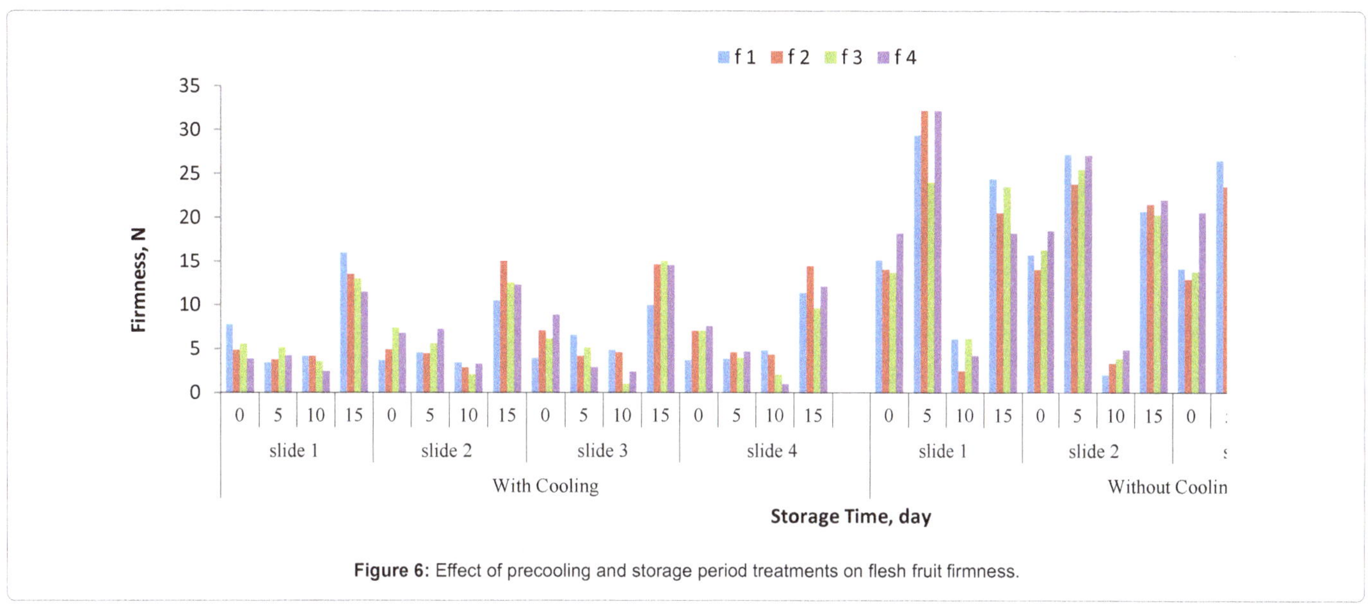

Figure 6: Effect of precooling and storage period treatments on flesh fruit firmness.

Treatment	Storage period	Firmness	L*	a*	b*	c*	h*	ΔL*	Δa*	Δb*	ΔC*	Δh*	ΔE
With Cooling	Period 1	70.11	16.88	18	9.14	20.19	26.91	0.31	0.21	0.41	0.38	0.71	1.34
	Period 2	97.73	15.85	18.29	7.71	19.85	22.86	-0.38	-1.26	-0.60	-1.39	0.08	0.95
	Period 3	79.28	16.38	18.47	8.41	20.29	24.47	0.96	0.22	1.09	0.68	0.88	0.89
	Period 4	67.33	16.69	18.97	8.76	20.89	24.80	0.22	0.11	0.23	0.2	0.17	0.22
Without Cooling	Period 1	80.25	16.23	18.55	8.26	20.30	24	0.01	0.06	0.01	0.06	0.02	0.04
	Period 2	99.67	15.20	15.48	6.69	16.86	23.38	0.16	0.34	0.19	0.39	0.04	0.27
	Period 3	69.63	16.49	19.37	8.61	21.19	23.96	-0.1	-0.11	-0.15	-0.16	0.09	0.13
	Period 4	62.63	14.17	13.84	5.52	14.90	21.73	0.53	2.18	0.81	2.33	0.05	1.52

Table 1: Means for firmness, color properties of fresh cantaloupe, which were stored after cooling at 10°C and 90±5% RH for 15 days.

and 14.17 for treatment with-cooling and without cooling, respectively. Meanwhile, mean values of luminosity (L*) of the skin fruit decreased over the postharvest storage time of the fruit. The decrease of color brightness in the skin cantaloupe was obviously associated with the ripeness stage of the fruit [19] and Figure 7 showed the hue angle (h˚) of the skin cantaloupe. The hue angle (h˚) values for fruit which prepared by cooling were recorded a largest values than sample without cooling during all storage period.

Yellowing of the peel and pulp based on increases in b* values proceeded at a relatively reduced rates in precooled fruits than in non-precooled ones (control). This was noted after 15 days of storage, with the 5-10°C precooling treatment causing consistently lower b* values and hence reduced rate of yellowing of both peel and pulp. Changes in L* values did not reflect the increasing degree of yellowing with ripening of the fruit. However, in terms of softening, precooling had no delaying effect.

Internal (flesh) color: In flesh fruit, intensity of yellow color is an important quality property. The color is primarily due to carotenoid pigments. Values of color parameters such L*, a* and b* varied under all treatments (Figure 8). The maximum values for L*, a*, b* and C* were 20.84, 20.15, 14.74 and 24.96 with treatment (cooling), and were 20.93, 19.37, 13.71 and 24.45 in treatment (without cooling). Data showed that values of a*, b* and C* were decreased with increasing in L* value. Variation in color parameters during storage period may be to the burst of ethylene gas that signals genes to transform chloroplasts, chlorophyll is gradually replaced by the carotenoids.

The results for chromaticity (C˚) as shown in Figure 8 were used to observe the orange color purity of the flesh cantaloupe. The chromaticity of the flesh cantaloupe prepared after two weeks of postharvest storage decreased by the end of the sample storage. After 15 days of storage, the chromaticity value of the flesh cantaloupe was dramatically decreased along with its brightness (L*). The decrease in the color purity may be caused by the occurrence of physical damage during fresh-cut processing. The lack of maintenance of the cutting utensils, for example using a blunt knife, can be the main cause of the development of surface darkness or a decrease in the yellow color of the cantaloupe [21]. From Table 2, mean values of firmness of flesh cantaloupe decreased from 11.98 to 4.32 N for stored precooled fruit and from 27.22 to 4.78 N for stored non-cooled fruit as the duration of postharvest storage increased. This indicates that the delay of fresh fruit processing prior to postharvest ripening resulted in texture quality degradation over the sample storage period.

Weight losses: Data showed that, lower physiological occurred in weight of fruits with each treatment. Meanwhile, Pre-cooled fruits exhibited relatively slower loss in weight on the corresponding dates as compared to the non-pre-cooled (Figure 9). This may be due to the fact that loss of moisture from pre-cooled commodities is slower if higher relative humidity is maintained in the storage atmosphere [22]. And generally, data showed that losses in weight of fruit increasing with the increasing of storage period then decreased at the end of storage period for different treatments. Data revealed that fruits which precooled generally had the minimum physiological weight loss. The lowest physiological weight loss value recorded (1.15%) with fruits

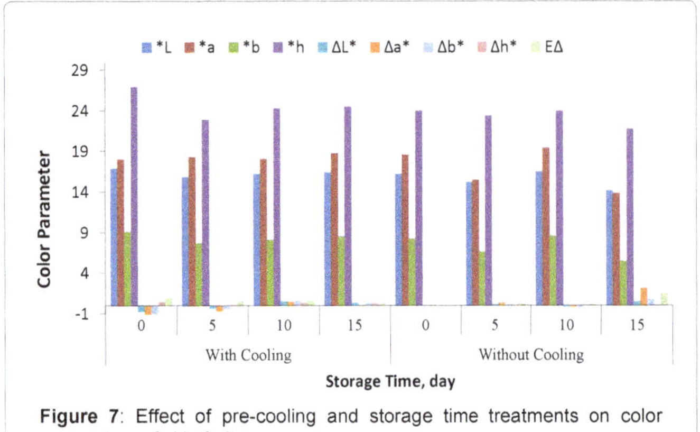

Figure 7: Effect of pre-cooling and storage time treatments on color parameters of skin fruit.

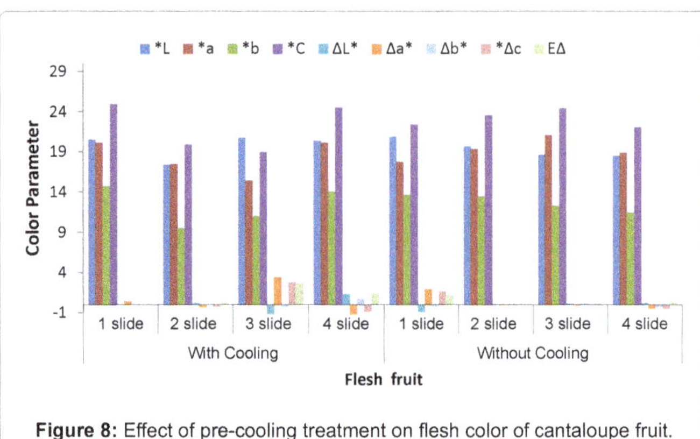

Figure 8: Effect of pre-cooling treatment on flesh color of cantaloupe fruit.

Treatment	Storage period	Firmness	L*	a*	b*	c*	h*	ΔL*	Δa*	Δb*	ΔC*	Δh*	ΔE
With Cooling	Period 1	4.79	20.55	20.15	14.74	24.96	36.18	-0.03	-0.01	-0.02	-0.01	0.01	0.02
	Period 2	11.98	17.44	17.47	9.54	19.90	28.64	0.18	-0.29	0.04	-0.23	0.18	0.22
	Period 3	6.78	20.84	15.43	11.05	18.98	35.60	-2.28	3.49	-0.24	2.81	2.08	2.69
	Period 4	4.32	20.41	20.15	14.07	24.57	34.92	1.32	-1.52	0.76	-0.81	1.55	1.41
Without Cooling	Period 1	13.9	20.93	17.76	13.71	22.44	37.67	-0.89	2.02	-0.17	1.73	1.06	1.16
	Period 2	27.22	19.71	19.37	13.51	23.61	34.90	0.03	-0.03	0.08	0.02	0.09	0.04
	Period 3	20.78	18.66	21.10	12.34	24.45	30.31	0.13	-0.05	0.14	0.02	0.14	0.11
	Period 4	4.78	18.55	18.89	11.49	22.11	31.31	0.21	-0.42	0.3	-0.51	0.04	0.31

Table 2: Means for firmness, color properties of flesh cantaloupe, which were stored after cooling at 10°C and 90 ± 5% RH for 15 days.

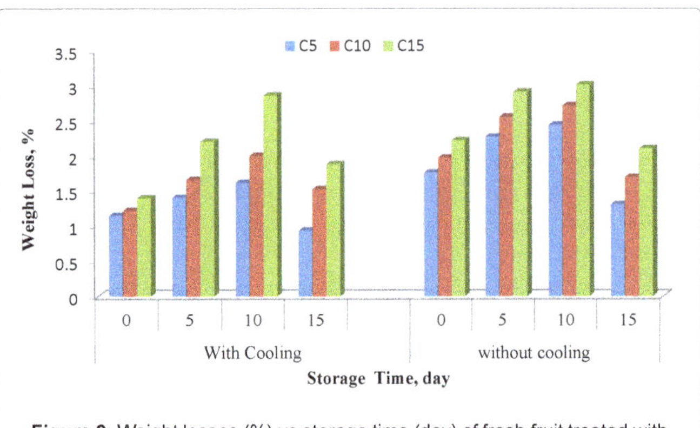

Figure 9: Weight losses (%) vs storage time (day) of fresh fruit treated with three levels of air precooling temperature (5°C, 10°C, 15°C).

which precooled at air precooling temperature (5°C), while the highest physiological weight loss value recorded (3.02%) with fruit which non precooled at air precooling temperature (15°C).

Conclusions

Forced air cooling could rapidly remove field heat from the fruit which is necessary to reduce refrigeration load during subsequent cold storage at 10°C. However, it seems to have no beneficial effect on fruit quality and shelf life. In certain cases, it may result to undesirable storage behavior such as the increased in weight loss when fruits were forced air precooled at higher temperatures of 10-15°C. Cantaloupe allows for postharvest storage at a low storage temperature of 15°C and 90 ± 5% RH. This is a potential alternative for fresh-cut processing extension, since it provides for reduced physicochemical and microbial changes during storage at cold environment. Firmness, luminosity (L*), were decreased by increasing the postharvest storage period of the fruit.

Acknowledgement

The research team would like to thank the Deanship of Scientific Research of King Faisal University, Hofuf, Saudi Arabia, for financial support.

References

1. Gil MI, Aguayo E, Kader AA (2006) Quality changes and nutrient retention in fresh-cut versus whole fruits during storage. Journal of Agricultural and Food Chemistry 54: 4284-4296.
2. Kader AA (2002) Postharvest technology of horticultural crops. Coop. Ext. Uni. of Ca. Division of Agriculture and Natural Resources. University of CA, Davis, CA.
3. Salunkhe DK, Desai BB (1984) Postharvest biotechnology of fruits. CRC Press, Inc. Boca Raton, Florida 1: 168.
4. Nunes MCN, Emond JP (2002) Storage temperature. In Bartz JA, Brecht JK Postharvest Physiology and Pathology of Vegetables. Marcel Dekker, Inc, New York.
5. Nunes MCN, Emond JP, Rauth M, Dea S, Chau KV, et al. (2009) Environmental conditions encountered during typical consumer retail display, affect fruit and vegetable quality and waste. Postharvest Biol. Technol 51: 232-241.
6. Talbot MT, Chau KV (2002) Precooling strawberries agricultural and biological engineering department, florida cooperative extension service. Gainesville: Institute of Food and Agricultural Sciences, University of Florida, USA.
7. Elansari AM (2009) Design Aspects in the Precooling Process of Fresh Produce. Global Science Books, Fresh Produce 3: 49-57.
8. Ferreira MD, Brecht JK, Sargent SA, Aracena JJ (1994) Physiological responses of strawberry to film wrapping and precooling methods. Proc. Fla. State Hort. Soc 107: 265-269.
9. Thompson JF, Mitchell FG, Rumsey TR, Kasmire RF, Crisosto CH, et al. (1998) Commercial cooling of fruits, vegetables, and flowers. Division of Agricultural and Natural Resources, University of California.
10. Émond JP, Mercier F, Sadfa SO, Bourré M, Gakwaya A, et al. (1996) Study of parameters affecting cooling rate and temperature distribution in forced-air precooling of strawberry. Trans. ASAE 39: 2185-2191.
11. Toivonen PM, Brummell DA (2008) Biochemical bases of appearance and texture changes in fresh-cut fruit and vegetables. Postharvest Biology and Technology 48: 1-14.

12. Clydesdale FM (1993) Color as a factor in food choice. Critical Reviews in Food Science and Nutrition 33: 83-101.

13. Guzmán M, Sánchez A, Valenzuela JL (2009) Postharvest Quality of Melon Fruits in Soil and Soilless Crops. Acta Hort 843: 211-218.

14. Lester G (1996) Calcium alters senescence rate of postharvest muskmelon fruit disks. Postharvest Biology and Technology 7: 91-96.

15. Luna-Guzmán I, Barrett DM (2000) Comparison of calcium chloride and calcium lactate effectiveness in maintaining shelf stability and quality of fresh-cut cantaloupes. Postharvest Biology and Technology 19: 61-72.

16. Silveira AC, Aguayo E, Chisari M, Artés F (2011) Calcium salts and heat treatment for quality retention of fresh-cut 'Galia' melon. Postharvest Biology and Technology 62: 77-84.

17. Yam KL, Papadakis S (2004) A simple digital imaging method for measuring and analyzing color of food surfaces. Journal of Food Engineering 61: 137-142.

18. Parveen S, Ali MA, Asghar M, Khan AR, Abdus Salam, et al. (2012) Physico-Chemical Changes In Muskmelon (*Cucumis Melo L.*) As Affected By Harvest Maturity Stage. J Agric Research.

19. Simandjuntak V, Barrett DM, Wrolstad RE (1996) Cultivar and maturity effects on muskmelon (*Cucumis melon*) colour, texture, and cell wall polysaccharide composition. Journal of the Science of Food and Agriculture 71: 282-290.

20. Soltani M, Alimardani R, Omid M (2011) Changes in physic-mechanical properties of banana fruit during ripening treatment. J Amer Sci 7: 14-19.

21. Machado FLC, Alves RE, Figueiredo RW (2008) Application of 1-methylcyclopropene, calcium chloride and calcium amino acid chelate on fresh-cut cantaloupe muskmelon. Pesq Agro-pec. bras. Brasília 43: 569-574.

22. Lurie SK, Ben AR (1990) Physiological changes in diphenyl treated 'Granny Smith' apples. Israel J Bot 38: 199-207.

Honeybee Production and Honey Quality Assessment in Guji Zone, Ethiopia

Birhanu Tesema Areda*

Faculty oAgricultural Sciences, Department oAnimal and Range Science, Blue Hora University, Ethiopia

Abstract

Assessment of Honeybee production practices and quality assessment were undertaken in Guji district of Ethiopia. Honey samples were collected from farmers' hives and local honey market for chemical analysis to determine its quality. Physicochemical analysis of honey was carried out at Haramaya University Animal Nutrition and Food Science technology. All physicochemical parameters analyzed lie within limits of local and international standard. Honey laboratory analysis was subjected to one way ANOVA of SAS. Above all, improving the low level of technological input and honey quality defects, address the skill gap on post-harvest handling of hive products, processing and packaging need a practical training to local beekeepers. Moreover, facilitating supply of quality apicultural equipment is crucial and further consistent practical training on bee and bee products management for community is recommended.

Keywords: Honey bee; Physio-chemical property; Honey quality

Introduction

Honey is a natural substance produced by bees and nutritious food of economic importance worldwide. It is a sweet viscous liquid that is composed of sugars, amino acids, proline, minerals, aromatic substances, pigment waxes and grains [1,2] and contains large amount of glucose but low in sucrose (<8%) [3]. Honey is easily digestible and a more palatable which supplies substantial energy with 75 to 85% fructose and glucose. The physicochemical composition, flavour and colour of honey vary due to climate, soil, flora, bee species and production methods. The precise composition variation depending on the plant species on which bee forages are the main constituents [4,5]. Storage conditions may also influence final composition, with the proportion of disaccharides increasing overtime [6]. Careless handling of honey can reduce quality like, high temperature, length of storage and moisture content which lead to fermentation, high levels of Hydroxy-methylfurfural (HMF), loss of enzymatic microbial growth [7]. Therefore, this study was designed to collect information on production system, productivity, bee flora and post managements of honey and determine its quality.

Materials and Methods

Sampling techniques and sample size

The study was conducted in beekeeping potential of Guji Zone. A total of 16 honey samples were purposively selected from four of beekeeper peasants of the zone and a sample of honey from market. A half (1/2) kg of honey samples were collected from two types of hives for laboratory analysis.

Collected data

The study was requiring wide range of information with reference to beekeeping. The chemical compositions of honey samples was determined accordingly [8,9] in the laboratory.

Moisture content: The moisture content of honey was determined by using the refractive index of the honey.

The table is derived from a formula developed by Wedmore E [10] and calculated by:

$$W = \frac{1.73190 - \log(R.I - 1)}{0.002243}$$

Where

W=Water content in g/100 g honey and R.I. is the refractive index

Mineral (ash) content: Ash content was determined after the sample burnt in an electric muffle furnace. Percent ash g/100 g honey was calculated by using:

$$\text{Ash }\% = \frac{M_1 - M_2}{M_0} \times 100$$

Where M_0=weight of honey, M_1=weight of dish + ash and, M_2=weight of dish.

pH and free acidity: The pH of sample was measured with pH meter and the solution was further titrated with 0.1M sodium hydroxide (NaOH) solution to pH 8.30.

Acidity =10V

Where

V = the volume of 0.1N NaOH in 10 g honey

Reducing sugar: This method is a modification of the Lane and [11] procedure, involving the reduction of Soxhlet's modification of Fehling's solution by titration at boiling point against a solution of reducing sugars in honey using methylene blue as an internal indicator and expression of result:

$$C = \frac{2 \times 1000}{W_2 Y_2}$$

Where

***Corresonding author:** Birhanu Tesema Areda, Faculty of Agricultural Sciences, Department of Animal and Range Science, Blue Hora University, Ethiopia
E-mail: berhan2m@gmail.com

C = g invert sugar per 100 g honey

W_2 = weight (g) of honey sample

Y_2 = volumes (ml) of diluted honey solution

Apparent sucrose: The procedure of determining reducing sugar was

ASC = (invert sugar content after inversion - invert sugar content before inversion) × 0.95

Where

ASC= Apparent sucrose content

The result will be expressed as g apparent sucrose per 100 g honey

Data Management and Statistical Analysis: Honey quality parameters were analyzed by using one way ANOVA and ± SD.

Results and Discussion

Feeding honey bees and flora condition

Honey bee colonies naturally withstand themselves and produce honey by foraging natural and cultivated crops and store honey for their own feeding during dearth period. But, beekeepers are harvesting honey which honey bees stored for them. The management for honey bees is very minimal in the study area. During the survey period it was observed that some farmers who have modern beehives (33.33%) were manage and 66.67% did not manage it properly. With regard to type of feed they provide, respondent feeding their bees (36.67%) use honey and pea flour (3.33%), pea flour and sugar syrup (6.67%), sugar syrup (10.0%) and pea flour (16.67.8%). 63.33% beekeepers were not give anything to honey bees as food.

Hive products harvesting and post handling

The frequency and amount of honey harvested varied depending on flowering condition of major bee forage, colony management practices and number of beehive [12]. In the study area, honey harvesting periods were from March to April and July to August where harvesting periods correlate with availability of moisture and peak flowering period. During honey harvested, beekeepers cut and pull the fixed combs one by one and then pollen, brood and honey combs were removed, and kept in a container and covered with a lid which affects quality of honey in relation to length of storage. According to Gichora [13], plastic container is the ideal one for the quality of honey. Accordingly plastic bucket and plastic sack were highly used and in some case they use nickels to store honey for both short and long period and which result rusting; deteriorates the honey and technically not appropriate for storage facilities.

Honey bee flora

Beekeeping is more dependable on ecological suitability of an area than any other livestock production [14] and, honey bee population and their productivities in general are mainly influenced by the nature of honey bee flora. Vegetation characteristics of the study areas are considered to be an important indicator for the potentialities of the area for beekeeping. Survey conducted showed that, the potential of cultivated and natural honey flora makes it very favorable for beekeeping. The respondents pointed that, even though there are different types of bee plants and flora seasons, there is a shortage of bee feed during the dry seasons where ground and surface water resources are limited. They also indicated that bee forages become declined as compared with the past period due to forest degradation, use of herbicides and expansion of cultivated lands in the area.

Honey quality laboratory result

Physicochemical properties of honey produced in the study area were analyzed compared to Quality and Standards Authority of Ethiopia (QSAE), Codex Alimentarius Commission (CAC) and European Union (EU) were described below.

Moisture: The mean moisture contents of honey samples collected from different locations and hive types are reported below which is depends on the environmental conditions such as temperature, relative humidity of the area and the manipulation of honey during harvesting period by beekeepers, and seasonal variation [15,16].

Ash: The minimum, maximum and mean ash contents of the honey samples analyzed in the present study was lower than the maximum limits (0.6%) set for ash content of the honey by EU, CA and QSAE and the average was within the national and international limits for ash content of honey.

Free acidity: The overall mean free acidity of honey samples analyzed was 24.08% which is within the acceptable limits (≤40 meq/kg) set by QSAE and CAC, whereas the limit for honey acidity according to EU (2002) honey standard is ≤50 meq/kg. None of the samples exceeded the limit set, which may be taken as indicative of freshness of all the honey samples of the study area. Variation in free acidity among different honeys can be attributed to floral origin or to variation in the harvest season [17].

pH: There was no significant difference ($p>0.05$) in pH between honey samples obtained from traditional (3.45) and modern hives (4.03) (Table 1), similarly, no significant difference in acidity observed was also other factor. Honey pH has great importance during storage of honey, as they influence the texture, stability and shelf life of honey [18]. pH of honey samples in the current study ranged from 4.13 to 5.02, with an average value of 4.45 (Table 3).

Reducing sugars: The overall mean reducing sugar content of the analyzed honey samples was 76% which within quality requirement limits (≥65%) (QSEA; CAC; EU). There were no significant differences ($P>0.05$) in reducing sugars content between honey samples obtained from the two hive types and locations (Tables 1 and 2). Similarly, the average reducing sugars content of honey obtained from market location (80.2%) was significantly higher ($p<0.05$) than the average moisture content of honey obtained from the two agro ecologies (collected directly from beekeepers).

Apparent sucrose: Apparent sucrose are set to be 5 g/ 100 g for the majority of honeys, which have higher limits (10 g/100 g), as well as

Variables	Location		
	Shakiso (n=8) Mean ± SD	Adola (n=8) Mean ± SD	Market (n=2) Mean ± SD
Moisture content (g/100 g)	15.12 ± 0.74	14.28 ± 0.89	13.20 ±0.76
Reducing sugars content (g/100 g)	60.8 ± 1.28*	60.45 ±1.12	67.72 ±1.28
Apparent sucrose (g/100 g)	3.0 ± 1.14	3.62 ±1.23	2.94 ±1.32
Ash content (g/100 g)	0.34 ± 0.48	0.36 ± 0.58	0.20 ± 0.34
Free acid (meqkg^{-1})	19.26 ± 0.00	16.38 ± 3.28	14.10 ± 2.00*
pH	3.22 ± 0.17	3.24 ± 0.21	3.29 ± 0.07

SD = standard deviation; * significantly different ($p<0.05$); ** significantly different ($p<0.01$). meq = milli equivalent; n=number of samples; means followed by different superscript letters in a row are significantly different.

Table 1: Comparison of physicochemical properties of honey samples collected from different locations.

Variables	Hive type	
	Traditiona (n=8) Mean ± SD	Movable frame (n=8) Mean ± SD
Moisture content (g/100 g)	13.67 ± 1.20	13.62 ± 1.40
Reducing sugars content (g/100 g)	60.84 ± 0.62	60.07 ± 0.78
Apparent sucrose (g/100 g)	3.22 ± 0.86	3.33 ± 1.39
Ash content (g/100 g)	0.22 ± 0.10	0.52 ± 0.75
Free acid (meqkg^{-1})	14.46 ± 3.40^{b}*	16.52 ± 3.76^{a}*
pH	3.40 ± 0.25	3.69 ± 0.19

SD=Standard deviation; meq=milli equivalent; n=Number of samples; Means followed by different superscript letters in a row are; *significantly different (p<0.05).

Table 2: Comparison of physicochemical properties of honey samples collected from different hive types.

No.	Variable	Unit	Current study	
			Range	Mean ± SD
1	Moisture	g/100 g	12.72-19.04	14.35 ± 1.50
2	Reducing sugars	g/100 g	59.52-61.60	68.4 ± 1.19
3	Apparent sucrose	g/100 g	1.61-4.81	2.22 ± 0.9
4	Ash	g/100 g	0. 26-1.42	0.34 ± 0.38
5	Free acid	meqkg^{-1}	12.33- 27.74	13.2 ± 5. 42
6	pH		3.82-4.02	4. 50 ± 0.33

CAC=Codex Alimentarius Commission; EU=European Union; meq=milliequivalent; n: Number of samples; SD=standard deviation; QSEA=Quality and Standards Authority of Ethiopia.

Table 3: Physicochemical properties of honey produced in the study area in relation to national and international standards.

lavender honeys (15 g/100 g) (EC Directive 2001/110). Higher sucrose contents could be the result of an early harvest of honey, i.e., the sucrose has not been converted to fructose and glucose [19,20]. The amount of sucrose in honey differs according to the degree of maturity and nectar compound of the honey (Tables 1-3).

Summary and Recommendation

Laboratory evaluation showed that the mean moisture, reducing sugars, sucrose, acidity, ash and pH contents of the honey samples collected from the study area revealed that, all the physicochemical parameters lie with-in limits of local and international standards set by Quality and Standards Authority of Ethiopia, Codex Alimentarius Commission and EU Council. There were significant differences for acidity and water insoluble solids (p<0.01) between hive types. But, there were no significant differences (p<0.05) between hive types and among locations for moisture, reducing sugar, sucrose, ash, and pH contents of honey samples tested. Therefore, to improve the low level of technological input utilization, it needs to be facilitate the supply improved bee-hives, honey processing materials and other beekeeping equipment, address the skill gap on bee colony management and post-harvest handling of hive products, further consistent practical training on bee and bee products management for community is recommended.

References

1. Bogdanov SV, Kilchenmann V, Fluri P (1998) American Bee Journal 139: 61-63.
2. Qiu PY, Ding HB, Tang YK, Xu RJ (1999) Determination of Chemical Composition of Commercial Honey by Near-Infrared Spectroscopy. Journal of Agriculture and Food chemistry 47: 2760- 2765.
3. USDA National honey report (2008).
4. Terrab AA, Gonzalez AG, Diez MJ, Francisco JH (2003) Characterisation of Moroccan unifloral honeys using multivariate analysis. Europe Food Research and Technology 218: 88-95.
5. Gulfra MF, Iftikhar MF, Raja S, Asif S, Mehmood S, et al. (2010) Quality Assessment And Antimicrobial Activity Of Various Honey Types Of Pakistan. African Journal of Biotechnology 9: 43-49.
6. White JW, Subers MH (1964) Studies on Honey inhibine: effects of heat. Journal of Applied Research 3: 45-50.
7. Moguel O, Carlos EG, Escobedo RM (2005) Physicochemicalquality of honey from honeybees. Apis mellifera produced in the State of Yucatanduring stages of the production process and blossoms 43: 323-334.
8. Bogdanov S (2002) Harmonized Methods of the International Honey Commission. Results for Amhara Region, Statistical Reports onLivestock and Farm Implants, Ababa, Ethiopia.
9. Bultosa G (2005) Food chemistry laboratory manual. Department of food science and post harvest technology, Haramaya University, Ethiopia.
10. Wedmore E (1955) The accurate determination of the water content of honeys. Taylor and Francis 24: 197-206.
11. Lane JH, Eynon L (1923) Determination of reducing sugar by means of Fehling's solution with methylene blue as internal indicator. Journal of Soc. Chem. India
12. Kajobe R, Godfrey AJ, Kugonza DR, Alioni V, Otim SA, et al. (2009) National beekeeping calendar, honeybee pest and disease control methods for improved production of honey and other hive products, Uganda.
13. Gichora M (2003) Towards Realization of Kenya's Full Beekeeping Potential. A Case Study of Baringo District. Cuvillier Verlag Gottingen, Germany.
14. Adgaba N (2002) Geographical races of the Honeybees of the Northern Regions of Ethiopia. The Tropical Agriculturalist. The Technical Center forAgriculture and Rural Cooperation (CTA), International Bee ResearchAssociation (IBRA) and Macmillan, Malaysia.
15. Acquarone C, Buera P, Elizalde B (2007) Pattern of pH and electrical conductivity upon honey dilution as a complementary tool for discriminating geographical origin of honeys. Food Chemistry 101: 695-703.
16. Cantarelli MA, Pellerano RG, Marchevsky EJ, Camina JM (2008) Quality of honey from Argentina: Study of chemical composition and trace elements. The Journal of the Argentine Chemical Society 96: 33-41.
17. Pe´rez-Arquillue C, Conchello P, Arin A, Juan T, Herrera A (1995) Physicochemical attributes and pollen spectrum of some unifloral Spanish honeys. Food Chemistry 54: 167-172.
18. Terrab A, Recamales AF, Hernanz D, Heredia FJ (2004) Characterisation of Spanish thyme honeys by their physicochemical characteristics and mineral contents. Food Chemistry 88: 537-542.
19. Azeredo LC, Azeredo MAA, Souza SR, Dutra VML (2003) Protein contents and physicochemical properties in honey samples of Apis mellifera of different floral origins. Food Chemistry 80: 249-254.
20. Codex Alimentarius Commission (2001) Revised Codex Standard for Honey. Joint FAO/WHO Food StandardsProgramme. FAO Headquarters Rome, Italy.

Influence of Storage Duration on Stability and Sensorial Quality of Dried Beef Product (*Kilishi*)

Iheagwara MC[1*] and Okonkwo TM[2]

[1]*Department of Food Science and Technology, Federal University of Technology, Owerri, Imo State, Nigeria*

[2]*Department of Food Science and Technology, University of Nigeria, Nsukka, Enugu State, Nigeria*

Abstract

This study was designed to investigate the influence of storage duration on the stability and sensorial quality of *kilishi* samples. The conventional traditional *kilishi* (TK) and sausage-type (SK) at varying percentage of ingredients were processed. Chemical and sensorial analyses were performed to investigate quality changes and to determine the shelf stability of the *kilishi* samples stored at 28 ± 2°C for 150 days. The proximate, peroxide value (PV), free fatty acid (FFA) and thiobarbituric acid (TBA) were found statistically significant ($P \leq 0.05$) in the *kilishi* samples throughout storage. The lowest PV (8.24 mEq/kg), FFA (3.12% oleic acid) and TBA (0.26 mgMDA/kg) were recorded in SK7 (115% ingredients), while the highest PV (35.11mEq/kg), FFA (11.18% oleic acid) and TBA (1.57 mgMDA/kg) occurred in SK2 (85% ingredients). Highest protein (55.84 ± 0.05%), fat (19.20 ± 0.09%) and ash (5.58 ± 0.08%) were obtained from SK7, SK2 and SK7 respectively and the organoleptic results showed that SK2 had the best acceptance and was significantly different ($P \leq 0.05$) compared to the other *kilishi* samples.

Keywords: Storage; Stability; Meat; Quality; *Kilishi*

Introduction

Meat is an animal tissue used as food [1]. It is composed of tissue or muscle fibre cells, fat and connective tissue, it can also be composed of pieces of bone [2,3]. Meat plays an important role in nutrition as a contributor of high quality protein. However, meat is a highly perishable food item due to abundance of a number of nutrients that favor the establishment, growth and multiplication of microorganisms [4]. In the tropics, meat spoils quickly a few hours after the outset of rigor-mortis and post-mortem handling, hence meat needs to be preserved. A major meat preservation technique whose use dates back to records from 12th Century is sun drying [5,6]. Today, a variety of sun-dried meat products exist. They include amongst others: the *Pemmican* that is prepared by exposing strips of lean meat to the sun, the *Charqui*, which is native to South American and *Biltong* found in South Africa [7]. In Nigeria, the dried meat product *kilishi* constitute one of daily delicacies and is equivalent to *Pemmican* and *Charqui*. *Kilishi* is a traditional sun dried Nigerian and Sahara African meat product processed using lean beef in combination with plant ingredients. It contains about 46% meat and 54% non-meat ingredients. A finished product contains about 50% protein, 75% moisture, 18% lipid and 9.8% fibre/ ash respectively [8,9]. It is a rich nourishing snack and a source of supplementary animal protein formulated using hurdle technology. Salting, dehydration or sun drying and packaging are hurdles applied in sequence to inhibit deteriorating microorganism [10]. Traditionally, *Kilishi* is prepared from boneless lean meat that is sliced in sheet of 2 mm thickness and partially dried in the sun followed by immersion in a slurry of ingredients before a second period of sun drying and brief roasting to stabilize the protein [11,12]. As a ready-to-eat convenience meat product, *Kilishi* possess an excellent shelf-life. According to Igene et al. [13] and Isah and Okubanjo [14], *Kilishi* has a shelf life of 12 months at room temperature. The ability of the product to keep for several months at room temperature is fast making the product a household name. This study therefore is aimed at evaluating the influence of storage duration on the stability and sensorial quality of *Kilishi*.

Materials and Methods

Raw materials procurement

The fresh beef muscles (*Longissimus dorsi*) used for the study was purchased from butchers at central Abattoir in Owerri, Nigeria. Spices such as ginger (*Zingiber officinale*), alligator pepper, (*Afromomum meleguata*), black pepper (*Piper guineense*), red pepper (*Capsicum frutescens*), sweet pepper (*Capsicum annum*), African nutmeg (*Monodora myristica*) as well as ingredients such as groundnut paste (*Arachis hypogea*), garlic (*Allium sativum*), onion (*Allium cepa*), sugar, salt and magi seasoning were bought from a grocery shop in Owerri, Nigeria.

Sample preparation and processing

Meat preparation: The semitendinosus muscle of beef (9 kg) used for the study was trimmed of all visible fat, bone and connective tissue and then weighed. About 1kg of the resultant lean beef (8.2 kg) was sliced into thin sheets of 0.17 cm to 0.20 cm thick and 60 cm to 80 cm long along the fibre direction and dried for the traditional *kilishi* (TK). While 1 kg each of the remaining resultant lean beef will be comminuted with different percentage of the infusing ingredients to form the sausage-type *kilishi* (SK).

Preparation of infusing ingredients: Infusion slurry was prepared following the procedures of Igene [8] using the ingredients as shown in Table 1. The fresh groundnut paste was prepared from grains of dry uncooked groundnut after extraction of oil by pressing. The various ingredients were ground and mixed thoroughly with water to form slurry.

Preparation of *kilishi*: The dried thin sheets of meat were soaked in the infusion slurry for about 30 min, after which it was taken out

***Corresponding author:** Iheagwara MC, Department of Food Science and Technology, Federal University of Technology, Owerri, P.M.B 1526 Owerri, Imo State, Nigeria, E-mail: marcquin.iheagwara@futo.edu.ng

Ingredients	TK	SK1	SK2	SK3	SK4	SK5	SK6	SK7
Ginger (*Zingiber officinale*)	3.30	3.30	2.81	2.97	3.14	3.47	3.63	3.80
Alligator pepper (*Afromomum meleguata*)	1.20	1.20	1.02	1.08	1.14	1.26	1.32	1.38
Black pepper (*Piper guineense*)	3.00	3.00	2.55	2.70	2.85	3.15	3.30	3.45
Red pepper (*Capsicum frutescens*)	2.00	2.00	1.70	1.80	1.90	2.10	2.20	2.30
Sweet pepper (*Capsicum annum*)	2.00	2.00	1.70	1.80	1.90	2.10	2.20	2.30
Onion (*Allium cepa*)	12.00	12.00	10.20	10.80	11.40	12.60	13.20	13.80
Garlic (*Allium sativum*)	0.50	0.50	0.43	0.45	0.48	0.53	0.55	0.58
African nutmeg (*Monodora myristica*)	1.00	1.00	0.85	0.90	0.95	1.05	1.10	1.15
Groundnut paste (*Arachis hypogea*)	31.50	31.50	26.78	28.35	29.93	33.08	34.65	36.23
Magi seasoning	1.50	1.50	1.28	1.35	1.43	1.56	1.65	1.73
Salt	3.00	3.00	2.55	2.70	2.85	3.15	3.30	3.45
Sugar	3.00	3.00	2.55	2.70	2.85	3.15	3.30	3.45
Water	36.00	36.00	30.60	32.40	34.20	37.80	39.60	41.40

TK - Traditional *kilishi* (100% ingredients)
SK1 - Sausage-type *kilishi* (100% ingredients)
SK2 - Sausage-type *kilishi* (85% Ingredients)
SK3 - Sausage-type *kilishi* (90% ingredients)
SK4 - Sausage-type *kilishi* (95% ingredients)
SK5 - Sausage-type *kilishi* (105% ingredients)
SK6 - Sausage-type *kilishi* (110% ingredients)
SK7 - Sausage-type *kilishi* (115% ingredients)

Table 1: Composition of infusion mixtures used in *kilishi* preparation (kg/100kg).

and spread out on flat steel trays on a raised platform till sun dried to generate the traditional *kilishi*. For the sausage-type *kilishi*, the meat was comminuted with various percentages of the infusing ingredients, spread into thin sheets of approximately 2mm thickness on steel trays, cut into long strips and dried. After drying, the traditional *kilishi* (TK) and respective sausage-type *kilishi* (SK) were roasted in an oven at a temperature of 100°C for 10-15min. Finally, the finished products were cooled at room temperature, packed and heat sealed in high density polyethylene (HDPE) bags and stored at ambient temperature (28 ± 2°C) for further analysis.

Storage stability and sampling

The traditional *kilishi* and sausage-type *kilishi* were stored for 150 days at ambient temperature and samples were drawn at specified days and subjected to proximate, chemical and organoleptic analysis.

Chemical analysis

The samples used for the analysis were assayed in duplicate. The moisture, protein, fat and ash contents of the *kilishi* samples were determined using the standard methods of AOAC [15] and Nielsen [16]. Peroxide value (PV), free fatty acid (FFA), and thiobarbituric acid (TBA) were determined according to the method described by Nielsen [16].

Organoleptic analysis

Organoleptic attributes of flavour, juiciness, tenderness, pungency and overall acceptability of the *kilishi* samples were evaluated by a 30-member in-house consumer panelist selected from among students and staff of Department of Food Science and Technology of the University. A 9-point hedonic scale was used with 9 for like extremely down to 1 for dislike extremely [17].

Statistical analysis

All the analysis was carried out in triplicates and data obtained were analyzed using analysis of variance (ANOVA) method. Where the variance ratio (F-values) proved significant, Fishers least significant difference (LSD) was used to separate the means.

Results and Discussion

Proximate composition

The proximate composition of the *kilishi* samples are presented in Table 2. There were significant differences ($P \leq 0.05$) in all the proximate parameters evaluated. The moisture content of the *kilishi* samples ranged from 10.02% - 12.02% with SK 6 having the highest moisture content of 12.02% and SK2 the lowest moisture content of 10.02%. The reduction in moisture content of SK2 is desirable as this can affect the quality of the sample positively in relation to other *kilishi* sample [12]. Generally, the moisture content of the *kilishi* samples indicates that the *kilishi* samples were sufficiently dried to minimize microbial growth though moisture values of 6.92%, 9.87% and 10.00% were recorded by Jones et al. [18], Apata et al. [12] and Olusola et al. [19] respectively. The protein content of the entire *kilishi* samples ranged from 51.62% in SK2 to 55.84% in SK7. The range of values obtained with regards to the crude protein content was similar to the values (53.41%-64.53%) reported by Isah and Okubanjo [14]. However, significant difference ($P \leq 0.05$) occurred in the protein content of the *kilishi* samples. SK7 had the highest protein content (55.84%) and SK2 the lowest protein content (51.62%). The high crude protein content obtained can be attributed to the various ingredients utilized in the *kilishi* preparation and is in agreement with report by Igene et al. [9]. The fat content of the *kilishi* samples differed significantly ($P \leq 0.05$) and ranged from 17.34% in TK to 19.20% in SK2. Generally, the fat content of the *kilishi* samples were high and this can be attributed to the groundnut cake powder which represent a considerable proportion of the product [13]. The ash content is an indicator of the mineral content of the meat. The ash content of the *kilishi* samples ranged from 4.54% in SK2 to 5.58% in SK7. The pattern of ash content observed in this study revealed that the higher the slurry infused into the meat during *kilishi* preparation, the higher the level of ash content in the product. This indicated that most of the ingredients in the slurry might have lost their mineral contents into the slurry hence, into the meat and product and this agrees with the report of Elizabeth [20] who observed that the ash content of any processed meat would be the content of the muscle tissue in addition to that of ingredients used.

Kilishi samples	Proximate Composition (%)			
	Protein	Fat	Ash	Moisture
TK	54.10 ± 0.08^{d}	18.38 ± 0.28^{cd}	4.82 ± 0.01^{b}	11.30 ± 0.01^{ab}
SK1	53.70 ± 0.01^{f}	18.57 ± 0.18^{bc}	5.01 ± 0.01^{bc}	11.78 ± 0.28^{ab}
SK2	51.62 ± 0.14^{h}	19.20 ± 0.23^{a}	4.54 ± 0.05^{c}	10.02 ± 0.14^{c}
SK3	52.30 ± 0.28^{g}	19.03 ± 0.42^{a}	4.68 ± 0.01^{bc}	12.01 ± 0.18^{a}
SK4	52.84 ± 0.24^{e}	18.02 ± 0.11^{d}	4.73 ± 0.11^{bc}	11.66 ± 0.01^{ab}
SK5	54.78 ± 0.08^{c}	18.94 ± 0.16^{ab}	5.22 ± 0.23^{ab}	11.00 ± 0.01^{b}
SK6	55.20 ± 0.14^{b}	17.83 ± 0.44^{d}	5.41 ± 0.16^{ac}	12.02 ± 0.11^{a}
SK7	55.84 ± 0.05^{a}	17.34 ± 0.09^{e}	5.58 ± 0.08^{a}	11.26 ± 0.04^{ab}

$^{a-h}$Means with different superscript along the column differ significantly at P ≤ 0.05.
TK - Traditional *kilishi* (100% ingredients)
SK1 - Sausage-type *kilishi* (100% ingredients)
SK2 - Sausage-type *kilishi* (85% Ingredients)
SK3 - Sausage-type *kilishi* (90% ingredients)
SK4 - Sausage-type *kilishi* (95% ingredients)
SK5 - Sausage-type *kilishi* (105% ingredients)
SK6 - Sausage-type *kilishi* (110% ingredients)
SK7 - Sausage-type *kilishi* (115% ingredients)

Table 2: Mean values of proximate composition of *kilishi* samples.

Free fatty acid (FFA)

The FFA content in a product is an indication of the quality of the product. Lipid hydrolysis development strongly depends on the hydrolytic enzyme content and it is also influenced by different external and internal factors. The FFA content of the *kilishi* samples are presented in Figure 1. The FFA values ranged from 3.12%-11.18% oleic acid and the FFA content significantly increased (P ≤ 0.05) in all the samples during the 150 days of storage. However, SK2 showed higher FFA values and it is significantly different (P ≤ 0.05) when compared with the other *kilishi* samples. Also, it was observed that the FFA content of the *kilishi* samples decreased progressively as the percentage of ingredients used in processing the *kilishi* increases. SK7 had the lowest FFA content and SK2 the highest FFA content in all the storage days. This result obtained suggests that the various ingredients (concentration) used in processing of the *kilishi* samples especially the spices inhibited FFA production and is in agreement with the findings of Ogbonnaya and Imidobah [6] and Mgbemere et al. [21].

Peroxide value

The changes of peroxide value as primary products of lipid oxidation of the *kilishi* samples are shown in Figure 2. The PV content significantly increased (P ≤ 0.05) in all the *kilishi* samples during the 150 days storage and there were significant differences (P ≤ 0.05) among the samples. The highest value (35/11 mEq/kg) of peroxide was recorded for SK2 while the lowest value (8.24 mEq/kg) was observed in SK7. Also, it was observed that the PV content decreased progressively as the percentage of ingredients inclusion used in processing the *kilishi* samples increases. Since peroxides are inversely related to development if rancidity, it is inferential that the sample with the highest inclusion of ingredients, SK7 (115% ingredients) was the most effective in slowing down primary oxidation when compared to the other *kilishi* samples. This result is in agreement with report by Mgbemere et al. [21]. It also agrees with the studies of Siripongvutikorn et al. [22] that spices activities as antioxidant are directly related to their concentration.

Thiobarbituric acid (TBA)

TBA is a widely used indicator for the assessment of degree of secondary lipid oxidation. It evaluates the second stage of autoxidation during which the peroxides are oxidized to aldehyde and ketones which impart the disagreeable rancid odours and flavour. The effect of storage duration on TBA value of the *kilishi* samples stored at ambient temperature (28 ± 2°C) is shown in Figure 3. The TBA values increased in all the samples over time particularly in SK2. The initial TBA values ranged from 0.26 mgMDA/kg in SK7 to 0.33 mgMDA/kg in SK2. After 150 days of storage at 28 ± 2°C, the TBA values ranged from 0.97 mgMDA/kg inSK7 to 1.57 mgMDA/kg in SK2, thus indicating that TBA of SK7 increased by 0.71 mgMDA/kg, while the TBA of SK2 increased by 1.24 mgMDA/kg after 150days. This result indicates that high concentration of spices is effective in retarding lipid oxidation and is in agreement with reports by Siripongvutikorn et al. [22]. However, the results obtained are within the acceptable limits of the maximum level of TBA value, indicating good quality of the *kilishi* samples during storage, which is 1-2 mgMDA/kg lipid.

Organoleptic analysis

The organoleptic evaluation of food products to any food processing technology is very important in determining the consumer acceptability. The results of sensorial analysis for the *kilishi* samples stored at ambient temperature for 150 days are presented in Tables 3-6. The results obtained shows that there were significant variations (P ≤ 0.05) in all the sensory parameters evaluated. With regards to flavour, the highest flavor of 8.43 ± 0.56 for day one, 7.60 ± 0.36 for day 50, 6.30 ± 0.47 for day 100 and 5.55 ± 0.31 were observed in SK2. Physiologically, the perception of flavour involves the detection of four basic sensations including saltiness, sweetness, sourness and bitterness by the nerve endings of the surface of the tongue [23]. This result reveals that the flavour of SK2 was higher probably because the fat content of the product was relatively high and the moisture content was also low hence the high flavour of the product compared to the other *kilishi* samples and this is in conformity with the findings of Olusola et al. [19] and Apata et al. [12]. It also supports the observation of Melton [24] that as the fat of meat increases so does the flavour.

In relation to pungency, there were significant variations (P ≤ 0.05) among the *kilishi* samples. According to Olusola et al. [19], the hotness of *kilishi* is an evaluation of the pungency of the product. The results of pungency as shown in Tables 3-6 reveals that SK7 had the highest pungent scores compared to the other *kilishi* samples in all the respective days of storage evaluated. This suggests that as the percentage of the spices used in the infusing mixtures increases, the pungent level also increases. This agrees with the report of Isah and Okubanjo [14].

12
FFA (% Oleic acid)
0
8
6
4
2
0
TK
SK1
SK2
SK3
SK4
SK5
SK6
SK7
1
50
100
150
Storage duration (days)

TK - Traditional *kilishi* (100% ingredients)
SK1 - Sausage-type *kilishi* (100% ingredients)
SK2 - Sausage-type *kilishi* (85% Ingredients)
SK3 - Sausage-type *kilishi* (90% ingredients)
SK4 - Sausage-type *kilishi* (95% ingredients)
SK5 - Sausage-type *kilishi* (105% ingredients)
SK6 - Sausage-type *kilishi* (110% ingredients)
SK7 - Sausage-type *kilishi* (115% ingredients)

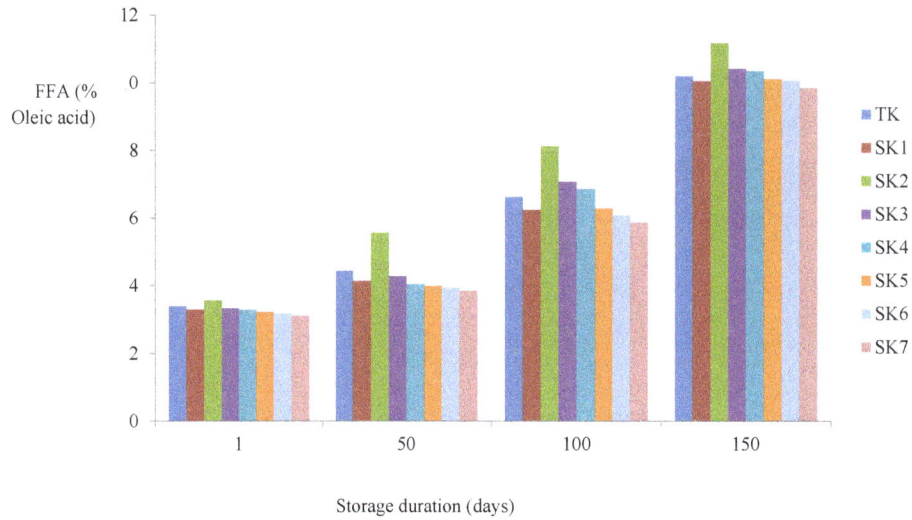

Figure 1: Effect of storage duration on free fatty acid (FFA) of *kilishi* samples.

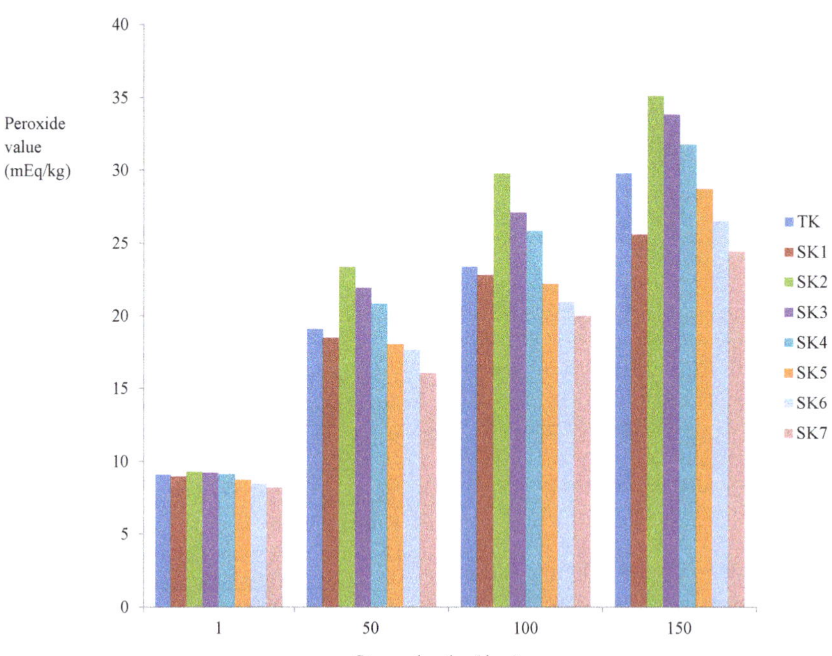

TK - Traditional *kilishi* (100% ingredients)
SK1 - Sausage-type *kilishi* (100% ingredients)
SK2 - Sausage-type *kilishi* (85% Ingredients)
SK3 - Sausage-type *kilishi* (90% ingredients)
SK4 - Sausage-type *kilishi* (95% ingredients)
SK5 - Sausage-type *kilishi* (105% ingredients)
SK6 - Sausage-type *kilishi* (110% ingredients)
SK7 - Sausage-type *kilishi* (115% ingredients)

Figure 2: Effect of storage duration on peroxide value (mEq/kg) of *kilishi* samples.

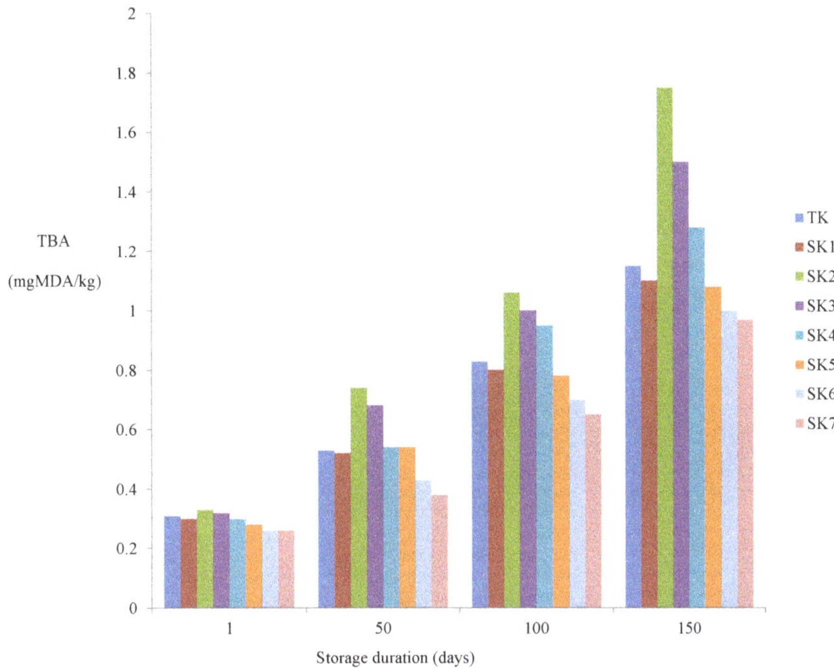

STK - Traditional *kilishi* (100% ingredients)
SK1 - Sausage-type *kilishi* (100% ingredients)
SK2 - Sausage-type *kilishi* (85% Ingredients)
SK3 - Sausage-type *kilishi* (90% ingredients)
SK4 - Sausage-type *kilishi* (95% ingredients)
SK5 - Sausage-type *kilishi* (105% ingredients)
SK6 - Sausage-type *kilishi* (110% ingredients)
SK7 - Sausage-type *kilishi* (115% ingredients)

Figure 3: Effect of storage duration on thiobarbituric acid (TBA) value of *kilishi* samples.

Kilishi samples	Sensory Parameter				
	Flavour	Pungency	Juiciness	Tenderness	Overall acceptability
TK	7.86 ± 0.52^{f}	7.18 ± 0.34^{e}	7.18 ± 0.45^{b}	7.15 ± 0.51^{f}	7.63 ± 0.28^{f}
SK1	8.06 ± 0.32^{d}	7.65 ± 0.42^{d}	7.13 ± 0.23^{c}	7.25 ± 0.37^{e}	7.75 ± 0.44^{d}
SK2	8.43 ± 0.56^{a}	7.64 ± 0.53^{d}	7.15 ± 0.33^{bc}	7.53 ± 0.49^{b}	8.00 ± 0.78^{a}
SK3	8.15 ± 0.46^{b}	7.78 ± 0'41^{c}	7.23 ± 0.48^{a}	7.62 ± 0.38^{a}	7.93 ± 0.37^{b}
SK4	8.10 ± 0.49^{c}	7.84 ± 0.42^{b}	7.12 ± 0.43^{c}	7.41 ± 0.52^{c}	7.88 ± 0.45^{c}
SK5	8.05 ± 0.34^{d}	7.85 ± 0.32^{b}	7.08 ± 0.23^{d}	7.26 ± 0.37^{e}	7.68 ± 0.49^{e}
SK6	8.00 ± 0.33^{e}	7.98 ± 0.41^{a}	7.00 ± 0.28^{e}	7.33 ± 0.32^{d}	7.50 ± 0.18^{g}
SK7	7.82 ± 0.48^{f}	8.00 ± 0.52^{a}	6.58 ± 0.49^{f}	7.04 ± 0.45^{g}	7.08 ± 0.33^{h}

$^{a-h}$Means with different superscript along the column differ significantly at P ≤ 0.05.
TK - Traditional *kilishi* (100% ingredients)
SK1 - Sausage-type *kilishi* (100% ingredients)
SK2 - Sausage-type *kilishi* (85% Ingredients)
SK3 - Sausage-type *kilishi* (90% ingredients)
SK4 - Sausage-type *kilishi* (95% ingredients)
SK5 - Sausage-type *kilishi* (105% ingredients)
SK6 - Sausage-type *kilishi*(110% ingredients)
SK7 - Sausage-type *kilishi* (115% ingredients)

Table 3: Mean sensory scores of *Kilishi* samples stored for one day.

According to Moloney [25], meat juiciness is an important component of meat tenderness and palatability and it has two major components; the first is the impression of wetness produced by the release of fluid from the meat during the first few chews, while the second is the more sustained juiciness that apparently results from the stimulating effect of fat on the production of saliva and coating of fat that builds up in the tongue, teeth and other parts of the mouth. The results of juiciness of the *kilishi* samples were significantly different (P ≤ 0.05). For day one, SK3 had the highest (7.23 ± 0.48) juiciness score while for day 50, SK4 was juicier (6.81 ± 0.22) than the other *kilishi* samples. For day 100 and 150 respectively, SK2 had the highest juiciness scores of 5.64 ± 0.28 and 5.02 ± 0.43 respectively compared to the other

Kilishi samples	Sensory parameter				
	Flavour	Pungency	Juiciness	Tenderness	Overall acceptability
TK	7.30 ± 0.44^{d}	6.84 ± 0.23^{h}	6.41 ± 0.32^{e}	6.62 ± 0.48^{e}	6.71 ± 0.33^{c}
SK1	7.53 ± 0.33^{b}	7.20 ± 0.52^{d}	6.52 ± 0.21^{cd}	6.65 ± 0.36^{de}	6.80 ± 0.44^{b}
SK2	7.60 ± 0.36^{a}	6.92 ± 0.33g	6.68 ± 0.42^{b}	6.76 ± 0.42c	7.10 ± 0.38^{a}
SK3	7.42 ± 0.18^{c}	7.04 ± 0.41^{f}	6.76 ± 0.33^{a}	6.84 ± 0.33^{b}	6.84 ± 0.49^{b}
SK4	7.40 ± 0.24^{c}	7.12 ± 0.28^{e}	6.81 ± 0.22^{a}	6.93 ± 0.47^{a}	6.53 ± 0.21^{d}
SK5	7.38 ± 0.48^{c}	7.43 ± 0.33^{c}	6.54 ± 0.38^{c}	6.68 ± 0.32^{d}	6.41 ± 0.32^{e}
SK6	7.27 ± 0.51^{de}	7.51 ± 0.42^{b}	6.48 ± 0.41^{d}	6.34 ± 0.24^{f}	6.23 ± 0.24^{f}
SK7	7.24 ± 0.38^{e}	7.74 ± 0.38^{a}	6.25 ± 0.35^{f}	6.20 ± 0.43^{g}	6.08 ± 0.32^{g}

$^{a-h}$Means with different superscript along the column differ significantly at $P \le 0.05$.
TK - Traditional *kilishi* (100% ingredients)
SK1 - Sausage-type *kilishi* (100% ingredients)
SK2 - Sausage-type *kilishi* (85% Ingredients)
SK3 - Sausage-type *kilishi* (90% ingredients)
SK4 - Sausage-type *kilishi* (95% ingredients)
SK5 - Sausage-type *kilishi* (105% ingredients)
SK6 - Sausage-type *kilishi* (110% ingredients)
SK7 - Sausage-type *kilishi* (115% ingredients)

Table 4: Mean sensory scores of *kilishi* samples stored for 50 days.

Kilishi samples	Sensory parameter				
	Flavour	Pungency	Juiciness	Tenderness	Overall acceptability
TK	6.11 ± 0.36^{c}	6.30 ± 0.48^{c}	5.18 ± 0.23^{c}	5.48 ± 0.32^{cd}	5.59 ± 0.21^{c}
SK1	6.18 ± 0.19^{b}	6.27 ± 0.35^{d}	5.20 ± 0.11^{c}	5.52 ± 0.22^{c}	5.68 ± 0.45^{b}
SK2	6.30 ± 0.47^{a}	6.18 ± 0.47f	5.64 ± 0.28^{a}	5.80 ± 0.43^{a}	5.88 ± 0.21^{b}
SK3	6.22 ± 0.51^{b}	6.20 ± 0.35^{ef}	5.43 ± 0.41^{b}	5.63 ± 0.35^{b}	5.70 ± 0.41^{a}
SK4	6.20 ± 0.22^{b}	6.23 ± 0.21^{e}	5.22 ± 0.32^{c}	5.51 ± 0.28^{c}	5.62 ± 0.28^{c}
SK5	6.18 ± 0.18^{b}	6.29 ± 0.31^{d}	5.11 ± 0.27^{d}	5.45 ± 0.34^{de}	5.53 ± 0.38^{d}
SK6	6.07 ± 0.34^{c}	6.35 ± 0.48^{b}	5.08 ± 0.12^{d}	5.40 ± 0.18^{e}	5.50 ± 0.42^{d}
SK7	5.96 ± 0.47^{d}	6.41 ± 0.32^{a}	5.01 ± 0.11^{e}	5.32 ± 0.21^{f}	5.48 ± 0.22^{d}

$^{a-h}$Means with different superscript along the column differ significantly at $P \le 0.05$
TK - Traditional *kilishi* (100% ingredients)
SK1 - Sausage-type *kilishi* (100% ingredients)
SK2 - Sausage-type *kilishi* (85% Ingredients)
SK3 - Sausage-type *kilishi* (90% ingredients)
SK4 - Sausage-type *kilishi* (95% ingredients)
SK5 - Sausage-type *kilishi* (105% ingredients)
SK6 - Sausage-type *kilishi* (110% ingredients)
SK7 - Sausage-type *kilishi* (115% ingredients)

Table 5: Mean sensory scores of *kilishi* samples stored for 100 days.

Kilishi samples	Sensory parameter				
	Flavour	Pungency	Juiciness	Tenderness	Overall acceptability
TK	5.32 ± 0.43^{cd}	5.43 ± 0.32^{b}	4.84 ± 0.21^{c}	4.70 ± 0.32^{c}	4.48 ± 0.24^{d}
SK1	5.40 ± 0.22^{b}	5.30 ± 0.25^{c}	4.91 ± 0.33^{b}	4.78 ± 0.26^{b}	4.56 ± 0.33^{c}
SK2	5.55 ± 0.31^{a}	5.20 ± 0.22^{e}	5.02 ± 0.43^{a}	4.98 ± 0.41^{a}	4.80 ± 0.22^{a}
SK3	5.35 ± 0.43^{c}	5.24 ± 0.31^{de}	4.94 ± 0.45^{b}	4.92 ± 0.34^{a}	4.68 ± 0.23^{b}
SK4	5.30 ± 0.21^{d}	5.29 ± 0.42^{cd}	4.80 ± 0.31^{c}	4.62 ± 0.25^{d}	4.61 ± 0.35^{c}
SK5	5.24 ± 0.22^{e}	5.32 ± 0.26^{c}	4.68 ± 0.42^{d}	4.50 ± 0.18^{e}	4.50 ± 0.22^{d}
SK6	5.20 ± 0.36^{e}	5.48 ± 0.33^{ab}	4.60 ± 0.22^{e}	4.42 ± 0.23^{f}	4.41 ± 0.26^{e}
SK7	5.08 ± 0.24^{f}	5.50 ± 0.24^{a}	4.53 ± 0.26^{f}	4.28 ± 0.33^{g}	4.35 ± 0.24^{f}

$^{a-h}$Means with different superscript along the column differ significantly at $P \le 0.05$.
TK - Traditional *kilishi* (100% ingredients)
SK1 - Sausage-type *kilishi* (100% ingredients)
SK2 - Sausage-type *kilishi* (85% Ingredients)
SK3 - Sausage-type *kilishi* (90% ingredients)
SK4 - Sausage-type *kilishi* (95% ingredients)
SK5 - Sausage-type *kilishi* (105% ingredients)
SK6 - Sausage-type *kilishi* (110% ingredients)
SK7 - Sausage-type *kilishi* (115% ingredients)

Table 6: Mean sensory scores of *kilishi* samples stored for 150 days.

kilishi samples. However, the juiciness of all the *kilishi* samples were inversely proportional to the storage duration though the juiciness as detected by the consumer are dependent on the intramuscular lipids and water content of the meat [19].

The tenderness of meat can be defined as the secondary manifestation of the structure of meat and the manner in which this structure reacts to the force applied during biting and the specific senses involved in eating [25]. It is how meat feels in the mouth during manipulation and mastication [19]. The result of tenderness of the *kilishi* samples as shown in Tables 3-6 shows that there were significant differences ($P \leq 0.05$) in the tenderness of the *kilishi* samples. For day one, SK3 was more tenderly (7.62 ± 0.38) than the other *kilishi* samples. For day 50, SK4 had the highest tenderness score of 6.93 ± 0.47compared to others. For days 100 and 150, SK2 had the highest tenderness of 5.80 ± 0.35 and 4.98 ± 0.34 respectively compared to the other *kilishi* samples. This variation in tenderness of the *kilishi* samples can be attributed to a large extent on the variations in the ingredients used since the same semitendinosus muscle and processing method were adopted throughout the experiment especially for the sausage-type *kilishi* samples. Also, this result confirms the report of Moloney [25] who said that meat juiciness is an important component of meat tenderness and palatability. Similarly, the tenderness of the *kilishi* samples was inversely proportional to the storage duration.

The mean panel ratings for the overall acceptability of the *kilishi* sample as shown in Tables 3-6 reveals that there were significant ($P \leq 0.05$) differences among the *kilishi* samples. The result obtained in relation to the overall acceptability indicates that the panelist preferred SK2 greatly compared to other *kilishi* samples in all the storage days. Among the *kilishi* samples, SK7 had the least rating in overall acceptability. This result obtained indicates that the level of inclusion of the spices used in the *kilishi* production affected the consumer preference as the panelist preferred SK2 (8.5% ingredients) to the hot pungent SK7 (115% ingredient). This result is in agreement with the findings of Isah and Okubanjo [14].

Conclusion

The present study has demonstrated that inclusion of infusion ingredients to a level of 115% possesses anti-oxidative property that retards oxidative rancidity thus extending the shelf stability of the *kilishi*. This is justified by the low peroxide, free fatty acid and thiobarbituric acid levels of *kilishi* sample processed using 115% ingredient inclusion compared to other percentage ingredients inclusion.

References

1. Forrest JC, Aberle ED, Gerrard DE, Mills WE, Hedrick HB, et al. (2001) The principles of meat science. (4thedn), Kendall/Hunt Publishing company, USA.
2. Ikeme AI (1990) Meat science and technology-A comprehensive approach. Africana FEP Publishers Ltd., Onitsha, Nigeria.
3. Miller RK (2002) Factors affecting the quality of raw meat processing: Improving quality. In: Kerry J, Lediord D (eds.) CRC Press LLC., USA.
4. Fonkem DN, Tanya VN, Ebangi AL (2010) Effect of season on the microbiological quality of Kilishi, a traditional Cameroonian dried beef product. Tropicultura. 28 (1): 10-15.
5. FAO (2007) Manual on meat cold store operations and management, Agriculture and consumer protection. FAO Corporate Document Repository, FAO Rome.
6. Ogbonnaya C, Imodiboh LI (2009) Influence of storage conditions on shelf-life of dried beef product (Kilishi). World Journal of Agricultural sciences, 5: 34-39.
7. Lawrie RA (1989). Meat science, Pergamon Press, London.
8. Igene JO (1988). Lipid, fatty acid composition and storage stability of Kilishi - a sundried meat product. Tropical Science. 28:153-161.
9. Igene JO, Abubakar U, Akanbi CT, Negbenebor CA (1993). Effect of sodium tripolyphosphate and moisture level on the drying characteristics and yield of Kilishi. Journal of Agriculture, Science and Technology. 3(2): 166-173.
10. Ogunsola OD, Omojola AB (2008) Qualitative evaluation of kilishi prepared from beef and pork. African Journal of Biotechnology. 7: 1753-1758.
11. Musonge P, Njolai EN (1994) Drying and infusion during the traditional processing of kilishi. Journal of Food Engineering 23: 159-168.
12. Apata ES, Osidibo OO, Apata OC, Okubanjo AO (2013). Effects of different solar drying methods on quality attributes of dried meat production (kilishi). Journal of Food Research 2: 80-86
13. Igene JO, Farouk MM, Akanbi CT (1990) Preliminary studies on the traditional processing of Kilishi. Journal of Science, Food and Agriculture. 50: 89-98.
14. Isah OA, Okunbanjo AO (2012). Effect of various additives on proximate composition and acceptability of Kilishi made from semitendinosus muscle of white Fulani cattle. Pacific Journal of Science and Technology, 13 (1): 506-511.
15. AOAC (2007). Official methods of analysis (20thedn), Association of Analytical Chemist International, Gathersburg, MD, USA.
16. Nielsen SS (2003). Food analysis (3rdedn) Kluwer Academic/Plemem publishers New York, USA.
17. Carbonell I, Izquierdo L, Costell E (2002). Sensory profiling of cooked gilthead sea bream (*Sparus aurata*): Sensory evaluation procedures and panel training. Food Science and Technology International. 8(3): 169-177
18. Jones MJ, Tanya VN, Mbofung CMF, Fonkem DN, Silverside DE (2001). A microbiological and nutritional evaluation of the West African dried meat product, Kilishi Journal of Food Technology in Africa. 6:126-129.
19. Olusola OO, Okubanjo AO, Omojola AB (2012) Nutritive and organoleptic characteristics of Kilishi as affected by meat type and ingredients formulation. Journal of Animal Production Advances. 2(5): 221-232.
20. Elizabeth B (1995). Ingredients in processed meat products paper presentation. Department of Animal Science and Industry Kansas state University, USA.
21. Mgbemere VN, Akpapunam MA, Igene JO (2011). Effect of groundnut flour substitution on yield, quality and storage stability of Kilishi - a Nigerian indigenous dried meat product. African Journal of Food, Agriculture, Nutrition and Development 11 (2): 4718-4738.
22. Siripongvutikorn S, Patawatchai C, Usawakesmanee W (2009). Effect of herb and spice pastes on the quality changes in minced salmon flesh waste during chilled storage. Asian Journal of Food and Agro - Industry. 2: 481-492.
23. Hedrick HB, Aberle ED, Forrest JC, Judge MD, Merkel RA (1994) Principles of Meat Science, (3rdedn) Kendall Hunt Publishing Co. Dubuque, Iowa.
24. Melton SL (1990) Effects of feed on flavor of red meat: a review. J Anim Sci 68: 4421-4435.
25. Moloney A (1999) The quality of meat from beef cattle - is it influenced by diet? In R & Hall, Technical Bulletin issue No.4.

5

Moisture Sorption Isotherm of Preconditioned Pressure Parboiled Brown Rice

Naveen Kumar M and Das SK*

Department ofAgricultural and Food Engineering, Indian Institute ofTechnology, Kharagpur, India

Abstract

The pressure parboiling of paddy was carried out at 294.204 kPa for 7 min and preconditioning of brown rice was carried out in fluidized bed dryer at 60-80°C. The moisture sorption isotherms of pressure parboiled preconditioned brown rice at different salt concentrations (0, 2, 3, 3.5 and 4%) were obtained at 20 ± 1°C, 25 ± 1°C, and 30 ± 1°C. The experimental data of sorption isotherm were fitted with some of sorption models (GAB, MGAB, MCPE, MOSE, MHEE, and MHAE models). According to the statistical results, the MGAB model gave the best fit to the experimental sorption data and MHAE model was the least adequate. Sorption isotherm data were used to determine the some thermodynamic functions. The net isosteric heat of sorption was determined from the best fitting equation using the Claussius-Calpeyron equation. The net isosteric heat of sorption decreased with increasing moisture content and increased with increasing salt concentration same trend was observed in entropy of sorption. The spreading pressure increased with increasing water activity and salt concentration and decreased with increasing temperature. The net integral enthalpy decreased with increasing moisture content and increased with increasing salt concentration and reverse trend was observed in integral entropy.

Keywords: Brown rice; Equilibrium moisture content; Pressure parboiling; Sorption isotherms; Spreading pressure

Notation

A,B,C	:	Model constants
Am	:	Area of water molecule, 1.06×10^{-19} m^2
a,b,c	:	Constants
aw	:	Water activity
°C	:	Degree Celsius
CG, k_G	:	GAB model constants
K	:	Boltzmann's constant, 1.38×10^{-23} J/K
K	:	Net isosteric heat constant
Mo	:	Monolayer moisture content, kg water/ kg dry solids
Me	:	Equilibrium moisture content, kg water/ kg dry solids
Mr	:	Constant
R	:	Universal gas constant, 8.314 kJ/mol K
R2	:	Coefficient of determination
T	:	Temperature, K
Q_{st}	:	Net isosteric heat of sorption, kJ/mol
Q_o	:	Constant, kJ/mol
Q_{eq}	:	Net integral enthalpy, kJ/mol
ΔS_{eq}	:	Net integral entropy, kJ/mol K
ΔS	:	Entropy of sorption, J/mol K
Ø	:	Spreading pressure (J/m^2)

Introduction

Rice ranks first in terms of global production (603 million tonnes) and used as a staple food for approximately 400 million people in the developing countries [1]. India exports 5% of the produced rice to the international market and compete with Thailand, Vietnam, and Pakistan. About 10% of the production of paddy is converted to three rice products, namely, puffed rice, popped rice and flaked rice [2-4]. Among the rice based breakfast cereals, puffed rice is largely demanded product for centuries in India because of its lightness and crispness. When grains such as rice, paddy, corn, gram etc. are heated, vapour pressure of water inside the grain increases. At a certain temperature and after certain duration of time the vapour pressure becomes high which causes expansion of the grain and the process is called puffing [5,6].

Preconditioning of rice is the most critical factor for obtaining highly expanded smooth-surface puffed rice [7]. The process comprises of uniform and slow heating of moisture-salt-conditioned parboiled rice with continuous turning until optimum moisture content of puffing (10% w.b.) is attained. Non uniform heating of grain severely impairs the quality of product with less expansion ratio in addition to rough and blistered surface. Further, puffing efficiency for rice grain depends on several factors, including the nature and concentration of salts diffused into the kernel.

Water activity of food material is determined as the ratio of vapour pressure of water in the food to vapour pressure of pure water at the same temperature. Many food deterioration reactions and the growth of important microorganisms depend on the water activity of the food and water activity is thus an important parameter to predict food stability. In order to determine the storage conditions, it is necessary to know the relationship between the equilibrium moisture content

***Corresonding author:** Das SK, Department of Agricultural and Food Engineering, Indian Institute of Technology, Kharagpur, India, E-mail: skd@agfe.iitkgp.ernet.in

(EMC) in the rice and equilibrium relative humidity of the aeration air at a given temperature. This relationship is described by the sorption isotherm equations [8]. The knowledge of the EMC of rice at several temperatures would allow specifying the storage conditions of rice.

Thermodynamic properties that describe the relationship between water and food are helpful in evaluating the energy requirements in concentration and drying processes and in predicting optimal storage conditions for maximum stability of dry foods. In addition, the evaluation of several thermodynamics properties (enthalpy, entropy, Gibbs free-energy, etc.) is important in the design and optimization of dryers. The isosteric heat of sorption gives a measure of the water-solid binding strength [9]. Knowledge of the differential heat of sorption is useful for designing drying equipment and the understanding of the state of water on food surface [10]. Net integral enthalpy and entropy are used to explain the modes of moisture sorption by foods.

Studies on the sorption behaviour of different varieties of rice were carried out by different authors, jasmine rice crackers, rice [10-13], rough rice [14-16], rice kernel components [17], but to the best of my knowledge, no study has been reported on the moisture sorption of preconditioned pressure parboiled brown rice at different salt concentrations. Moisture content of preconditioned rice is the most critical factor for achieving the best quality expanded product. After preconditioning immediately rice has to be puffed otherwise rice should store in moisture proof packaging material to maintain constant moisture content of preconditioned rice. During storage absorption of moisture may deteriorate the puffing quality. Data on MSI of such preconditioned rice will help to evaluate energy required for dehydration and design of packaging systems for storage. Considering these aspects and identifying the existing knowledge gap, the present study has been undertaken with the following objectives.

a) To determine the effect of salt concentration and temperature on the moisture sorption isotherm of preconditioned brown rice.

b) To evaluate several models and compare their goodness of fit.

c) To determine thermodynamic properties and effect of salt concentration on thermodynamic properties such as isosteric heat of sorption, differential enthalpy and entropy, spreading pressure and integral enthalpy and entropy.

Materials and Methods

Experimental procedure

IR 1010 variety of pressure parboiled brown rice were collected from local rice mill located at Balichak, West Bengal, India. The paddy was soaked for nearly 8 min in cold water and then steamed under pressure at 294.204 kPa for 7 min after that paddy was dried until the moisture content of paddy was reached to 12-14% (w.b.). After reaching the optimum moisture content the paddy was milled by using rubber roll sheller. 150 ml of water per kg of rice was mixed with salt (NaCl) at pre-determined rate so as to arrive its concentration in the final dried mass 0-4% (w/w). This was followed by tempering for about 6 to 8 hours to facilitate diffusion of both water and salt into the kernel. Preconditioning was carried out using hot air at specific temperature (60-80°C) using fluidized-bed dryer (Lab dryer, Basic technology Pvt. Ltd, India). This process was continued until the moisture content reaches around 10% (w.b.). To maintain different moisture contents of sample, preconditioned sample was kept in desiccators containing water to maintain moisture content more than 10% and silica gel to maintain moisture content lower than 10%. The moisture contents of these samples were checked frequently to obtain required moisture content. After attaining desired moisture contents water activity of pre-conditioned rice samples having different moisture contents were noted at different room temperatures (20 ± 1°C, 25 ± 1°C, and 30 ± 1°C) using water activity meter (Rotronic, HygroLab C1).

Data analysis

Isotherm models: Experimental moisture sorption data can be described by many sorption models but for this study six isotherm equations were chosen to fit experimental sorption data these are shown in Table 1. Non-linear regression analysis was carried to find out model constants. The extent of fitting of models was evaluated and compared based on three statistical criteria namely coefficient of determination (R^2), Root mean square error (RMSE), and reduced-χ^2.

Root mean square error

$$\mathrm{RMSE} = \sqrt{\frac{\sum_{i=1}^{n}\left(X_{obs,i} - X_{model,i}\right)^2}{n}} \quad (1)$$

$$\mathrm{Reduced}-\chi^2 = \frac{1}{d.f}\sum\frac{\left(X_{obs} - X_{model}\right)^2}{\sigma^2} \quad (2)$$

Where X_{obs} is observed values and X_{model} is predicted value.

In general, low values of the correlation coefficient, high values of reduced-χ^2, and RMSE, means that the model is not able to explain the variation in the experimental data. It is also evident that a single statistical parameter cannot be used to select the best model and the model must always be assessed based on multiple statistical criteria.

Net isosteric heat of sorption: The net isosteric heat of sorption (Q_{st}) (differential) is defined as the total heat of sorption in the food minus the heat of vaporisation of water at the system temperature [18]. This thermal property can be determined from calorimetric measurements or more easily from moisture sorption data. The usual procedure to evaluate isosteric heat of sorption from moisture isotherm is based on a Clasius-Clapeyron equation derived equation [19].

$$\ln a_w = -\left(\frac{Q_{st}}{R}\right)\left(\frac{1}{T}\right) + k \quad (3)$$

Model	Mathematical expression
MGAB	$M_e = \dfrac{A.B.a_w.\frac{C}{T}}{(1-B.a_w)[1-B.a_w+a_w.B.(\frac{C}{T})}$
MCPE	$M_e = -\dfrac{1}{C}\ln\left[-\left(\dfrac{T+B}{A}\right)\ln a_w\right]$
MOSE	$M_e = (A+B.T).\left(\dfrac{a_w}{1-a_w}\right)^{\frac{1}{C}}$
MHEE	$M_e = \left[-\dfrac{\ln(1-a_w)}{A.(T+B)}\right]^{\frac{1}{c}}$
MHAE	$M_e = \left[-\dfrac{\exp(A+B.T)}{\ln a_w}\right]^{\frac{1}{c}}$
GAB	$m = \dfrac{m_o C_G k_G a_w}{(1-k_G a_w)(1-k_G a_w + k_G a_w C_G)}$

Table 1: Mathematical models used to fit experimental data of preconditioned brown rice.

Where, R is universal gas constant (8.314×10^{-3} kJ $mol^{-1} K^{-1}$), T is temperature in K and k is integral constant. The net isosteric heat of sorption is obtained from the slope of the graph representing ln a_w versus (1/T) at a particular equilibrium moisture content. This is carried out for several equilibrium moisture content determined by the best fitting sorption model. The correlations between Q_{st} and M_e have been reported by various authors [14,20,21].

$$Q_{st} = Q_o \exp\left(\frac{-M_e}{M_r}\right) \tag{4}$$

$$Q_{st} = \left(\frac{a.\ M_e^{\ b}}{c + M_e^{\ d}}\right) \tag{5}$$

Differential entropy: The differential entropy (ΔS) of sorption of water at each equilibrium moisture content was obtained by fitting to Equation (5) for various equilibrium moisture contents calculated from the best-fitting equation [22,23].

$$-\ln a_w = \frac{Q_{st}}{RT} - \frac{\Delta S}{R} \tag{6}$$

By plotting ln a_w versus 1/T, for given equilibrium moisture content, ΔS was determined from the intercept (ΔS/R). Many authors discuss about the relationship between ΔS and equilibrium moisture content McMinn and Magee [9] gave a power law relationship.

Spreading pressure: It is also called as surface potential, it is the force applied in the surface plane perpendicular to each unit length of edge to keep the surface from spreading. It represents the surface excess free energy [24]. It acts as a second pressure [12] and results in increase in surface tension on bare sorption sites due to the sorbed molecules on them [25]. It was calculated using an analytical procedure

$$\varnothing = \frac{kT}{A_m}\int_0^{a_w} \frac{m}{a_w m_o} da_w \tag{7}$$

Where k is Boltzman constant (1.38×10^{-23}, J/K); A_m is the area that occupies one single water molecule at monolayer (1.06×10^{-19} m^2). Combine the GAB equation with the above equation mathematically expressed the Ø in the form of GAB equation [26].

$$\varnothing = \frac{kT}{A_m}\ln\left[\frac{1 - k_G a_w + k_G C_G a_w}{1 - k_G a_w}\right] \tag{8}$$

Net integral enthalpy and entropy: The net integral enthalpy (q_{eq}) represents the total energy available to do the work. It gives an indication of the binding strength of water molecules to the solid. The net integral enthalpy is calculated in a similar manner to the isosteric heat but at constant spreading pressure. A plot of ln a_w versus 1/T at a constant spreading pressure (Ø) gives the net integral enthalpy from the slope [19].

$$q_{eq} = -R\left[\frac{d(\ln a_w)}{d\frac{1}{T}}\right]_{\varnothing} \tag{9}$$

Net integral entropy (ΔS_{eq}) indicates the degree of disorder and randomness of motion of water molecules [27], it is calculated by using the following equation.

$$\Delta S_{eq} = \frac{-q_{eq}}{T} - R\ln(a_w)^* \tag{10}$$

Where $(a_w)^*$ is the geometric mean water activity obtained at a constant spreading pressure at different temperatures [9].

Results and Discussions

Sorption curves

The moisture sorption isotherms of preconditioned pressure parboiled rice at salt different concentrations at 20 ± 1°C, 25 ± 1°C and 30 ± 1°C are shown in Figure 1. The results reveal that the water activity

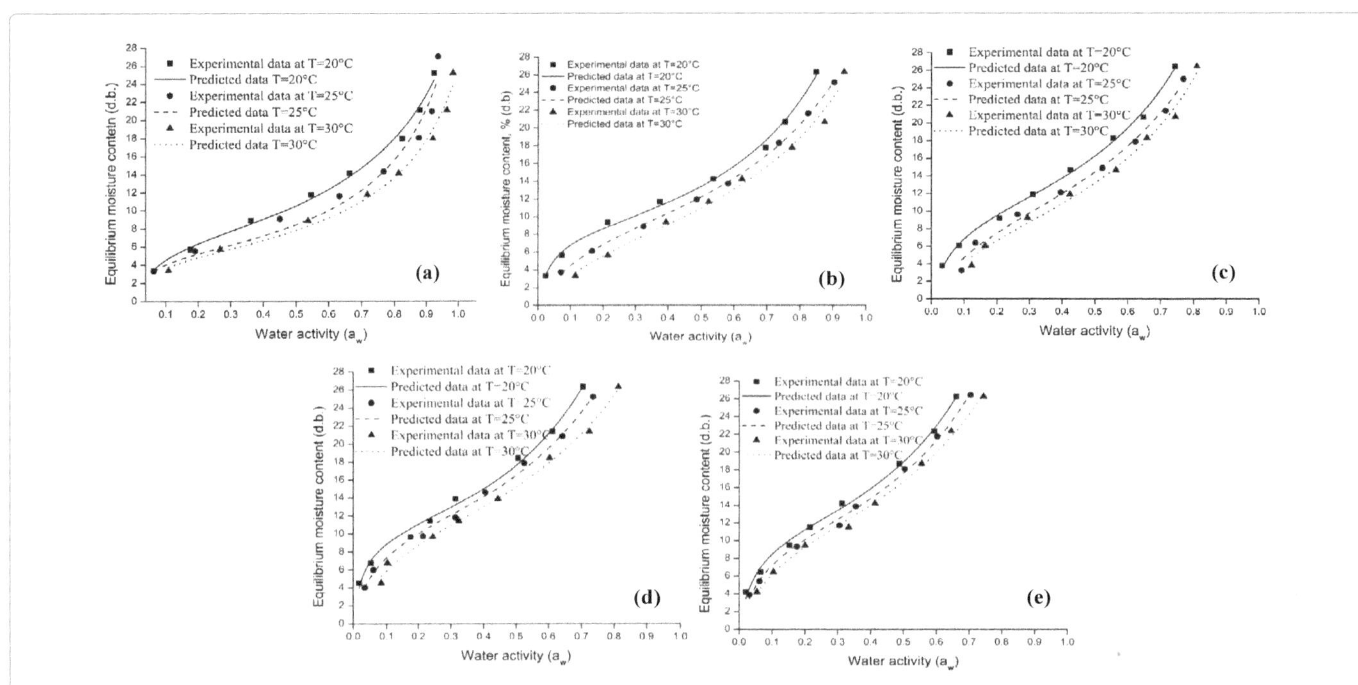

Figure 1: Effect of temperature on moisture sorption isotherm of pre conditioned pressure parboiled brown rice at (a) 0% salt, (b) 2% salt, (c) 3% salt, (d) 3.5% salt and (e) 4% salt.

increased with increasing EMC at constant temperature. These changes in EMC are due to the inability of the food to maintain vapour pressure at unity with decreasing moisture content. As moisture content decreased, moisture in the food tend to show a lower vapour pressure, acting as if in solution, changing with atmospheric humidity. These changes in vapour pressure in the food with atmospheric humidity result in the characteristic sigmoid shape of water sorption isotherms [10]. The isotherm presented a sigmoid shape (Type II according to BET classification) which is common for most of the hygroscopic foods. Sorptive behaviour depends on the temperature, it is decreasing with increasing temperature due to the activation of water molecules, at higher temperatures causes them to break away from the water binding sites, thus lowering the equilibrium moisture content [9] these results were compared with various researchers [14,15,28-30]. Effect of salt concentration on moisture sorption isotherm of preconditioned brown rice sample is shown in Figure 2. Where the water activity is seen to decreases with increasing salt concentration due to the ability of sodium and chloride ions to associate with water molecules [31,32]. These results are compared with the effect of salt and glucose concentration on water activity is explained by Martin Chaplin [33].

Fitting sorption models to experimental sorption data

The model coefficients and corresponding statistical results for the sorption models are listed in Tables 2-4. The goodness of fit was evaluated by using higher values of R^2 and lowest values of RMSE and reduced- χ^2. The average coefficients of determination (R^2) in all cases were greater than 0.940, the RMSE ≤ 2.152 and the Reduced -$\chi^2 \leq 4.629$. Among six models MGAB, MHEE and MOSE models gave better curve fitting compared to other models at all temperatures throughout the entire range of water activity. Least RMSE and reduced-χ^2 values were obtained in case of GAB model and maximum RMSE and reduced-χ^2 values were obtained in case of MHAE model over the temperature and salt concentration range. MGAB model gave the least value of RMSE and reduced-χ^2 and higher value of R^2 was considered the best model in case of all samples and MHAE gave the poor fitting results. The goodness of fit has been reported by several authors studying sorption behaviour of different food materials, GAB, BET and Halsey models [34] for blue berry, GAB model [21] for dent corn, GAB model [35] for capsicum, Modified Oswin [36] for millet, Strohman-Yoerger equation [13] for rice and Modified GAB [14] for rice.

Net isosteric heat of sorption and sorption entropy

Net isosteric heat of sorption and sorption entropy of preconditioned brown rice were determined by using Clausius-Clapeyron equation (Equation 3) to the experimental equilibrium isotherm data. The isosteric heat of sorption is strong moisture dependent. The net isosteric heat of sorption decreased with increasing moisture content is shown in Figure 3. It is due to the fact that sorption initially occurs on the most active primary sites giving rise to higher exothermic interaction energies than those released when these sites become occupied [21]. Similar trends have been reported for isosteric heat of sorption of jasmine rice crackers [11], rough rice [14], rice [10,30], potato [37]. It is depicted from Figure 3, Isosteric heat of sorption increased with increasing salt concentration. This confirms the fact that at higher salt concentrations, the strength of water binding increases. Most of the authors described the relation between Q_{st} and M_e the model constants and stastical data of models suggested by Tsami et al. [38] and Kechau and Maalej [20] were shown in Table 5, from these results, it can be revealed that Kechau and Maalej [20] model gave the best relationship between Q_{st} and M_e. Some more others express relations between Qst

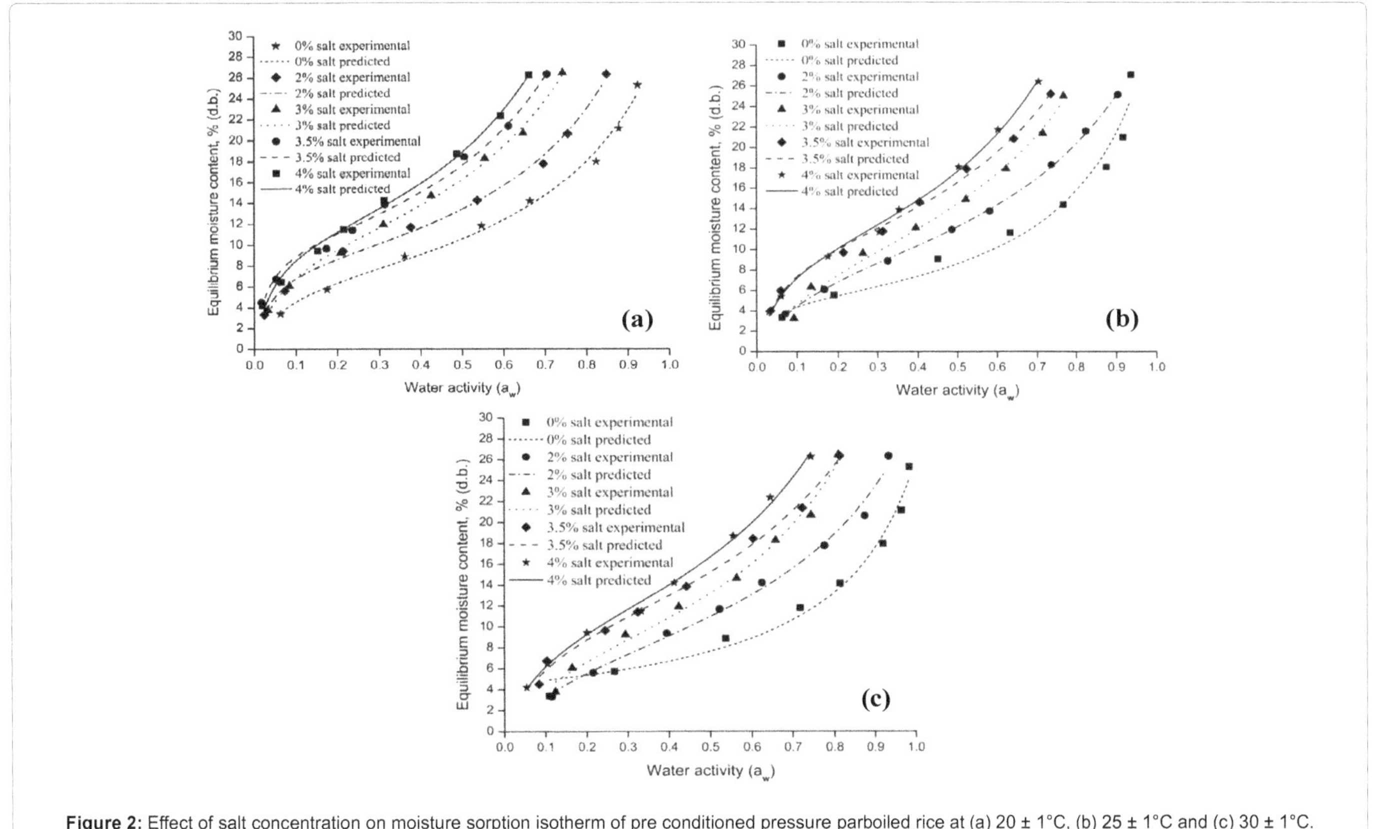

Figure 2: Effect of salt concentration on moisture sorption isotherm of pre conditioned pressure parboiled rice at (a) 20 ± 1°C, (b) 25 ± 1°C and (c) 30 ± 1°C.

Models		0% salt	2% salt	3% salt	3.5% salt	4% salt
MGAB	A	7.037	8.403	10.153	10.197	10.911
	B	0.778	0.802	0.844	0.880	0.901
	C	345.825	646.899	346.239	782.635	482.052
	Reduced -χ^2	0.490	0.294	0.216	0.457	0.325
	R^2	0.994	0.997	0.997	0.994	0.996
	RMSE	0.700	0.542	0.465	0.676	0.570
MCPE	A	454.796	501.715	431.728	529.803	436.290
	B	81.135	69.692	72.185	68.967	62.001
	C	0.167	0.144	0.111	0.115	0.103
	Reduced -χ^2	0.436	1.292	1.079	1.352	1.037
	R^2	0.995	0.985	0.987	0.983	0.988
	RMSE	0.660	1.137	1.039	1.163	1.018
MOSE	A	-4.178	12.300	13.774	15.221	15.254
	B	0.741	0.075	0.139	0.161	0.213
	C	2.868	2.759	2.332	2.618	2.414
	Reduced -χ^2	0.319	0.365	0.169	0.375	0.163
	R^2	0.996	0.996	0.998	0.995	0.998
	RMSE	0.565	0.604	0.411	0.613	0.404
MHEE	A($\times 10^{-5}$)	7.097	4.096	4.527	2.403	2.839
	B	142.230	71.128	81.860	44.102	51.351
	C	1.700	1.948	1.762	2.075	1.949
	Reduced -χ^2	0.719	1.890	0.815	1.340	0.717
	R^2	0.991	0.978	0.990	0.983	0.992
	RMSE	0.848	1.375	0.903	1.158	0.847
MHAE	A	6.945	6.647	6.138	6.323	6.092
	B	-0.099	-0.103	-0.113	-0.104	-0.110
	C	2.302	1.945	1.543	1.611	1.451
	Reduced -χ^2	1.959	1.535	1.378	0.859	1.041
	R^2	0.976	0.982	0.983	0.989	0.988
	RMSE	1.400	1.239	1.174	0.927	1.020
GAB	m_o	0.070	0.082	0.100	0.102	0.106
	c	17.291	17.311	24.100	39.130	39.132
	k	0.778	0.811	0.849	0.880	0.916
	Reduced -χ^2($\times 10^{-5}$)	4.904	16.580	5.716	4.574	9.239
	R^2	0.994	0.980	0.993	0.994	0.989
	RMSE	0.007	0.013	0.008	0.007	0.010

Table 2: Estimated parameters of preconditioned pressure parboiled rice at different salt concentration solutions for different isotherm models at 20°C.

and Me for different food materials, power function [9] for potato, exponential function for onions and exponential form [17] for rough rice.

The entropy of sorption of preconditioned pressure parboiled brown rice as a function of moisture content is shown in Figure 4, from this figure it can be observed that entropy of sorption decreased with increasing moisture content, similar trends have been reported for differential entropy of starch powders [39], pepper [40], crushed chillies [8] but for rice stored under controlled humidity chamber reported that sorption entropy decreased with increasing moisture content in desorption process but in adsorption process increased up to certain moisture content later it decreased with increasing moisture content [10]. Differential entropy increased with increasing salt concentration (Figure 4) from this we can reveal that number of sorption sites were more at higher salt concentration at a specific energy level. The ΔS versus M_e results are adequately represented by a exponential relation as represented by the equation

0% salt $\Delta S=0.307e^{-0.134Me}$ $R^2 = 0.994$

2% salt $\Delta S=0.919e^{-0.214Me}$ $R^2 = 0.978$

3% salt $\Delta S=0.392e^{-0.130Me}$ $R^2 = 0.993$

3.5% salt $\Delta S=0.504e^{-0.156Me}$ $R^2 = 0.957$

4% salt $\Delta S=0.345e^{-0.110Me}$ $R^2 = 0.984$

Many authors describe a relation between ΔS and M_e, power function [9] for potato, exponential form [26] for Mexican mennonite-style cheese and exponential [39] for the starch powders.

Net integral enthalpy and entropy

Net integral enthalpy given by Equation 9, the experimental sorption data were first represented in the form of spreading pressure isotherm. Spreading pressure values were calculated using Eq. 8 the constants C_G and K_G were determined from the GAB equation (Table 1). Spreading pressure increased with increasing water activity and salt

Models		0% salt	2% salt	3% salt	3.5% salt	4% salt
MGAB	A	5.100	9.005	10.285	10.510	10.724
	B	0.849	0.732	0.796	0.814	0.871
	C	700.000	267.807	208.718	506.351	413.963
	Reduced-χ^2	2.843	0.052	0.555	0.288	0.247
	R^2	0.969	0.999	0.993	0.996	0.997
	RMSE	1.686	0.228	0.745	0.537	0.497
MCPE	A	443.457	455.305	389.890	99.419	420.237
	B	83.842	78.834	82.662	-5.570	65.252
	C	0.175	0.148	0.110	0.116	0.103
	Reduced-χ^2	3.857	0.170	0.823	0.679	1.252
	R^2	0.957	0.998	0.989	0.991	0.986
	RMSE	1.964	0.413	0.907	0.824	1.119
MOSE	A	-7.394	11.221	12.368	14.113	13.978
	B	0.664	0.029	0.075	0.105	0.159
	C	2.669	2.821	2.145	2.485	2.237
	Reduced-χ^2	1.536	1.111	0.799	0.195	0.137
	R^2	0.983	0.986	0.990	0.997	0.998
	RMSE	1.239	1.054	0.894	0.442	0.370
MHEE	A($\times 10^{-5}$)	10.140	6.153	6.664	3.307	4.054
	B	213.857	119.535	133.194	68.685	83.617
	C	1.492	1.727	1.550	1.897	1.733
	Reduced-χ^2	4.115	0.138	0.886	0.612	0.706
	R^2	0.954	0.998	0.989	0.992	0.992
	RMSE	2.029	0.371	0.941	0.782	0.840
MHAE	A	6.916	7.136	6.235	6.502	6.244
	B	-0.100	-0.082	-0.107	-0.096	-0.102
	C	2.193	2.235	1.513	1.619	1.430
	Reduced-χ^2	1.819	3.730	2.116	1.672	1.413
	R^2	0.980	0.953	0.973	0.978	0.984
	RMSE	1.349	1.931	1.455	1.293	1.189
GAB	m_o	0.042	0.091	0.100	0.109	0.107
	c	7.827	9.349	10.712	16.559	20.254
	k	0.883	0.728	0.806	0.798	0.874
	Reduced -χ^2($\times 10^{-5}$)	94.920	1.062	8.350	3.684	4.194
	R^2	0.895	0.999	0.989	0.995	0.995
	RMSE	0.031	0.003	0.009	0.006	0.006

Table 3: Estimated parameters of preconditioned pressure parboiled rice at different salt concentration solutions for different isotherm models at 25°C.

Models		0% salt	2% salt	3% salt	3.5% salt	4% salt
MGAB	A	4.966	8.825	9.322	10.881	11.150
	B	0.807	0.723	0.817	0.744	0.812
	C	688.789	204.869	225.580	370.829	351.994
	Reduced -χ^2	0.975	0.769	0.734	0.374	0.319
	R^2	0.988	0.991	0.991	0.995	0.996
	RMSE	0.987	0.877	0.857	0.612	0.565
MCPE	A	495.111	434.463	377.374	76.310	410.615
	B	72.002	83.955	86.273	-13.109	67.366
	C	0.228	0.154	0.110	0.120	0.104
	Reduced -χ^2	0.076	0.401	1.169	0.465	0.847
	R^2	0.999	0.995	0.986	0.994	0.990
	RMSE	0.275	0.633	1.081	0.682	0.921
MOSE	A	-10.107	10.470	11.532	12.972	12.828
	B	0.630	0.009	0.051	0.066	0.128
	C	3.770	2.870	2.092	2.548	2.245
	Reduced -χ^2	1.530	1.971	0.818	0.581	0.342
	R^2	0.981	0.977	0.990	0.993	0.996
	RMSE	1.237	1.404	0.904	0.762	0.585
MHEE	A($\times 10^{-5}$)	8.955	7.544	8.033	3.996	4.527
	B	187.440	156.616	170.688	89.839	98.821
	C	1.675	1.620	1.437	1.806	1.675
	Reduced -χ^2	0.307	0.472	1.095	0.556	0.560
	R^2	0.996	0.994	0.987	0.993	0.993
	RMSE	0.554	0.687	1.046	0.746	0.749
MHAE	A	8.427	7.462	6.470	6.930	6.560
	B	-0.050	-0.074	-0.098	-0.082	-0.091
	C	3.372	2.387	1.550	1.826	1.518
	Reduced -χ^2	3.609	4.629	1.944	2.327	1.864
	R^2	0.955	0.945	0.976	0.971	0.978
	RMSE	1.900	2.152	1.394	1.526	1.365
GAB	m_o	0.049	0.087	0.093	0.109	0.111
	c	6.742	7.219	7.519	11.733	12.361
	k	0.810	0.727	0.817	0.744	0.812
	Reduced -χ^2($\times 10^{-5}$)	30.358	7.728	7.344	3.974	3.376
	R^2	0.962	0.991	0.991	0.995	0.996
	RMSE	0.017	0.009	0.009	0.006	0.006

Table 4: Estimated parameters of preconditioned pressure parboiled rice at different salt concentration solutions for different isotherm models at 30°C.

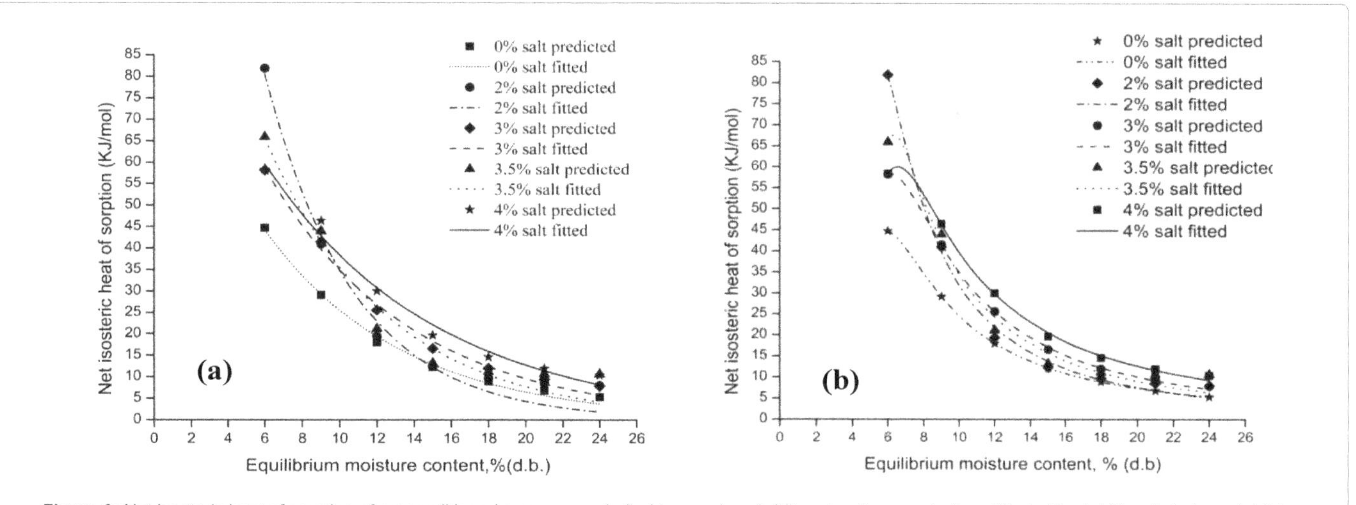

Figure 3: Net isosteric heat of sorption of preconditioned pressure parboiled brown rice at different salt concentrations fitted with (a) Tsami et al. model (b) Kechau and Maalej model.

Models		0% salt	2% salt	3% salt	3.5% salt	4% salt
Kechau and Maalej [24]	a	1470.814	3830.65	2625.3	2949.984	2044.665
	b	3.884	6.552	3.371	8.539	4.162
	c	9614.959	675433.616	7235.502	56946200	24450.6
	d	5.652	8.63	5.228	10.471	5.859
	Reduced-χ^2	0.003	6.907	0.309	14.766	0.833
	R^2	1	0.995	1	0.984	0.999
	RMSE	0.058	2.628	0.555	3.843	0.913
Tsami et al. [44]	Q_o	100.769	282.688	127.133	164.095	115.55
	M_r	7.282	4.775	7.734	6.542	9.072
	Reduced-χ^2	1.326	18.354	2.817	20.404	5.551
	R^2	0.995	0.979	0.993	0.963	0.987
	RMSE	1.151	4.284	1.678	4.517	2.356

Table 5: Fitted parameters of net isosteric heat of sorption versus equilibrium moisture content for different models.

aw	0% salt			2% salt			3% salt			3.5% salt			4% salt		
	20°C	25°C	30°C	20°C	25°C	30°C	20°C	25°C	30°C	20°C	25°C	30°C	20°C	25°C	30°C
0.100	0.034	0.022	0.018	0.035	0.021	0.018	0.045	0.026	0.020	0.060	0.035	0.026	0.061	0.042	0.029
0.200	0.055	0.038	0.033	0.056	0.037	0.032	0.068	0.043	0.036	0.085	0.055	0.044	0.087	0.065	0.048
0.300	0.070	0.052	0.046	0.072	0.050	0.044	0.085	0.058	0.049	0.103	0.071	0.058	0.105	0.082	0.063
0.400	0.083	0.065	0.057	0.085	0.061	0.054	0.099	0.070	0.061	0.118	0.084	0.071	0.121	0.096	0.077
0.500	0.095	0.077	0.068	0.097	0.072	0.065	0.112	0.082	0.072	0.132	0.096	0.082	0.135	0.109	0.089
0.600	0.106	0.089	0.079	0.109	0.082	0.074	0.124	0.093	0.083	0.145	0.108	0.093	0.148	0.122	0.100
0.700	0.117	0.102	0.090	0.121	0.092	0.084	0.137	0.105	0.095	0.159	0.120	0.103	0.163	0.136	0.112
0.800	0.129	0.116	0.102	0.133	0.102	0.095	0.151	0.117	0.107	0.173	0.132	0.115	0.179	0.150	0.125
0.900	0.142	0.134	0.117	0.148	0.114	0.106	0.167	0.131	0.122	0.191	0.146	0.127	0.199	0.168	0.140

Table 6: Effect of temperature and salt concentration on spreading pressure of pre conditioned brown rice at constant water activity.

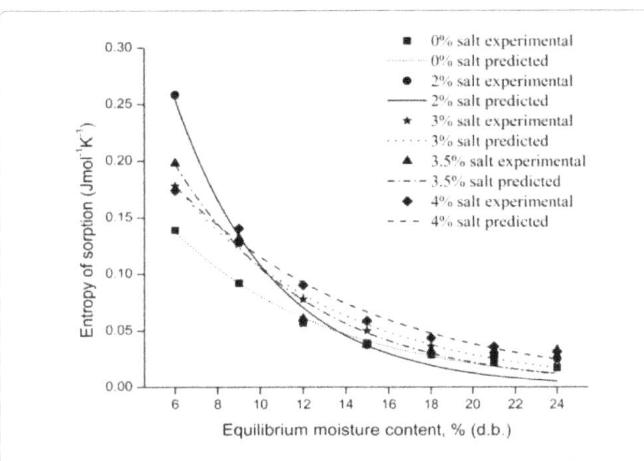

Figure 4: Entropy of sorption of preconditioned pressure parboiled brown rice at different salt concentrations fitted with exponential model.

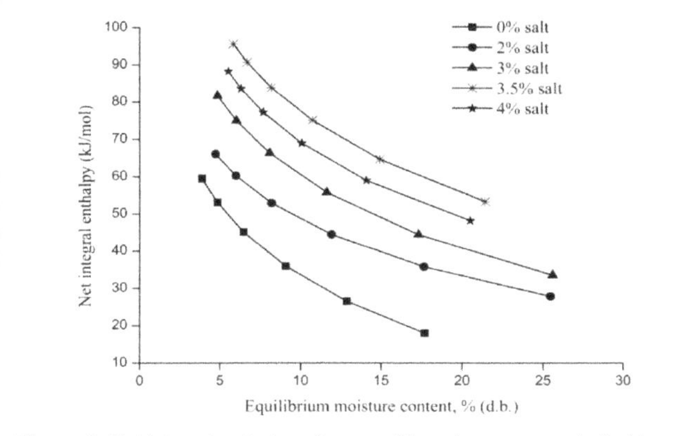

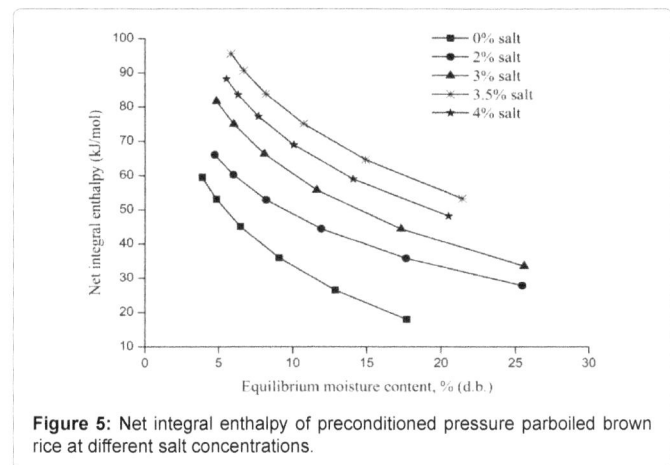

Figure 5: Net integral enthalpy of preconditioned pressure parboiled brown rice at different salt concentrations.

concentration and decreased with increasing temperature at a specific water activity (Table 6). The trends of spreading pressure with respect to temperature and water activity are comparable with [41] starchy materials [42], cereal grains [10], rice and [39] starch powders.

Effect of moisture content on net integral enthalpy of preconditioned pressure parboiled brown rice is shown in Figure 5. This graph clearly shows that net integral enthalpy decreased with increasing moisture content. At low moisture contents, water is adsorbed on the most accessible locations on the exterior surface of the solid. The net integral enthalpy then starts to decline as less favourable locations are covered and multiple layers of sorbed water form [39]. Similar trends have been reported for integral enthalpy of rice [30], maize, rough rice and wheat [42], rice [10] and for starch powders [39]. The Figure 5 shows the effect of salt concentration on spreading pressure from this graph we can reveal that integral enthalpy increased with increasing salt concentration up to 3.5% salt but in case of 4% salt treated sample integral enthalpy low as compared to 3.5% salt.

Net integral entropy of preconditioned brown rice was calculated by applying Equation (10). Net integral entropy increased with increasing moisture content shown in Figure 6. The net integral entropy negative in magnitude being low at low moisture content while increasing moisture content integral entropy moves towards 0 kJ/mol K. The initial highly negative value of integral entropy at low moisture content was due to the first few percent of water is very tightly

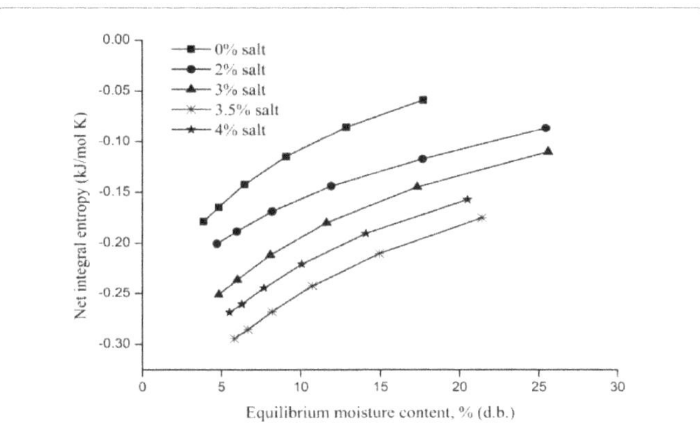

Figure 6: Net integral entropy of preconditioned pressure parboiled brown rice at different salt concentrations.

bound [39]. Similar trends were observed for integral entropy of starch powders [39], rice [10], potato [9] and cow pea [43]. Integral entropy decreased with increasing salt concentration (Figure 6) but in 4% salt treated sample net integral entropy high as compared to 3.5% salt treated sample.

Conclusions

On the basis of this work the following conclusion can be drawn. A non-linear regression analysis was used to evaluate the constants of the sorption models. The moisture sorption isotherm of preconditioned pressure parboiled brown rice at different salt concentrations adequately described by the MGAB equation. Net isosteric heat of sorption, calculated using the Clausis-Clapeyron equation. The model suggested by Kechau and Maalej gave the best relationship between net isosteric heat of sorption and equilibrium moisture content. Net isosteric heat of sorption decreased with increasing moisture content and increases with increasing salt concentration. Differential entropy of sorption can be characterised by an exponential model. Differential entropy decreased with increasing moisture content and increase with increasing salt concentration. Spreading pressure increased with increasing water activity and decrease with temperature. Net integral enthalpy decreased with increasing moisture content and increases with increasing salt concentration. Net integral entropy increased with increasing moisture content and decrease with increasing salt concentration.

References

1. FAO (1998) Rice in human nutrition.chap.6. Major processed rice products.
2. Chattopadhyay PK (2004) Post-harvest technology for rice in India: a changing scenario, Proceedings of the World Rice Research Conference held in Tokyo and Tsukuba, Japan, 4-7, November 2004. Los Banos, Philippines 294-296.
3. Narayanswamy CK (1956) The rice we eat. All India Khadi and Village Industries Board, Bombay, India.
4. Ghose RLM, Ghatge MB, Subrahmanyan, V (1960) Rice in India, revised edition. Indian Council of Agricultural Research, New Delhi, India.
5. Das H (2005) Food Processing Operation Analysis. Asian Books Private Limited, New Delhi, India.
6. Arya SS (1992) Convenience foods-emerging scenario. Indian Food Industry 11: 31-41.
7. Chinnaswamy R, Bhattacharya KR (1983) Studies on expanded rice, optimum processing conditions. Journal of Food Science Technology 48: 1604-1608.
8. Arslan N, Togrul H (2005) Moisture Sorption Isotherms for Crushed Chillies Biosystems Engineering 90: 47-61.
9. McMinn WAM, Magee TRA (2003) Thermodynamic properties of moisture sorption of potato. Journal of Food Engineering 60: 157-165.
10. Togrul H, Arslan N (2006) Moisture sorption behaviour and thermodynamic characteristics of rice stored in a chamber under controlled humidity. Biosystems Engineering 95: 181-195.
11. Siripatrawan U, Jantawat P (2006) Determination of Moisture Sorption Isotherms of Jasmine Rice Crackers Using BET and GAB Models. Food Science and Technology International 12: 459-465.
12. Rizvi SSH, Benado AL (1984) Thermodynamic properties of dehydrated foods. Food Technology 38: 83-92.
13. Da-Wen Sun (1999) Comparison and selection of EMC/ERH isotherm equations for rice. Journal of Stored Products Research 35: 249-264.
14. Iguaz A, Virseda P (2007) Moisture desorption isotherms of rough rice at high temperatures. Journal of Food Engineering 79: 794-802.
15. Basunia MA, Abe T (1999) Moisture adsorption isotherms of rough rice. Journal of Food Engineering 42: 235-242.
16. Ondier GO, Siebenmorgen TJ (2009) Equilibrium moisture contents of rough rice Dried using high-temperature, fluidized-bed conditions. BR Wells rice research studies; AAES research series.
17. Ondier GO, Siebenmorgen TJ, Mauromoustakos A (2011) Equilibrium moisture contents of rice kernel components. BR Wells rice research studies; AAES research series.
18. Tolaba MP, Peltzer M, Enriquez N, Pollio ML (2004) Grain sorption equilibria of quinoa grains. Journal of Food Engineering 61: 365-371.
19. Rizvi SSH (1986) Thermodynamic properties of food in dehydration. In: Rao MA and Rizvi SSH (Eds.). Engineering Properties of Foods. New York.
20. Kechau N, Maalej M (1999) Desorption isotherms of imported banana. Drying Technology 17: 1201-1213.
21. Samapundo S, Devlieghere F, Meuleneur B, Atukwase A, Lamboni Y, et al. (2007) Sorption isotherms and isosteric heat of sorption of whole yellow dent corn. Journal of Food Engineering 79: 168-175.
22. Fasina OO (2006) Thermodynamic properties of sweet potato. Journal of Food Eng 75: 149-155.
23. Liebanes MD, Aragon JM, Palancar MC, Arevalo G, Jimenez D, et al. (2006) Equilibrium moisture isotherms of two-phase solid olive oil byproducts: adsorption process thermodynamics. Colloid Surface 282: 298-306.
24. Noshad M, Shahidi F, Mohebbi M, Mortazavi S (2013) Desorption isotherms and thermodynamic properties of fresh and osmotic-ultrasonic dehydrated quince. Journal of Food Processing and Preservation 37: 381-390.
25. Fasina O, Ajibola OO, Tyler R (1999) Thermodynamics of moisture sorption of winged bean seed and Gari. Journal of Food Process Engineering 22: 405-418.
26. Sergio I, Monteagudo M, Fierro FS (2014) Moisture sorption isotherms and thermodynamic properties of mexican mennonite-style cheese. Journal of Food Science Technology 51: 2393-2403.
27. Mazza G, LeMaguer M (1978) Water sorption properties of yellow globe onion. Canadian Institute of Food Science and Technology 11: 189-193.
28. Aviara NA, Ajibolab OO, Aregbesolab OA, Adedejib MA (2006) Moisture sorption isotherms of sorghum malt at 40 and 50°C. Journal of Stored Products Research 42: 290-301.
29. Al-Muhtaseb AH, McMinn WAM, Magee TRA (2004) Water sorption isotherms of starch powders Part 1: Mathematical description of experimental data. Journal of Food Engineering 61: 297-307.
30. Benado AL, Rizvi SSH (1985) Thermodynamic Properties of Water on Rice as Calculated from Reversible and Irreversible Isotherms. Journal of Food Science 50: 101-105.
31. Fennema OR (1996) Food chemistry. 3rd ed. New York: Marcel Dekker.
32. Potter NN, Hotchkiss JH (1998) Food science. Food science texts series. 5th ed. New York: Chapman and Hall.
33. Martin Chaplin (2015) Water structure and science, chapter: water activity.
34. Vega-Galvez A, Lopez J, Miranda M, Scala KD, Yagnam F, et al. (2009) Mathematical modelling of moisture sorption isotherms and determination of

isosteric heat of blueberry variety O'Neil. International Journal of Food Science and Technology 44: 2033-2041.

35. Akin A, Ozbalta N, Gungor A (2009) Equilibrium moisture content and equations for fitting sorption isotherms of capsicum annuum. GIDA 34: 205-211.

36. Rajia AO, Ojediranb JO (2011) Moisture sorption isotherms of two varieties of millet. Food and Biproducts Processing 89: 178-184.

37. Wang N, Brennan JG (1991) Moisture Sorption Isotherm Characteristics of Potatoes at Four Temperatures. Journal of Food Engineering 14: 269-287.

38. Tsami E, Marinos-Kouris D, Maroulis ZB (1990) Water sorption isotherms of raisins, currants, figs, prunes and apricots. Journal of Food Science 55: 1594-1597.

39. Al-Muhtaseb AH, McMinn WAM, Magee TRA (2004b) Water sorption isotherms of starch powders. Part 2: Thermodynamic characteristics. Journal of Food Engineering 62: 135-142.

40. Kaymak-Ertekin F, Sultanoglu M (2001) Moisture sorption isotherm characteristics of peppers. Journal of Food Engineering 47: 225-231.

41. Tolaba MP, Suarez C, Viollaz P (1995) Spreading pressure water activity and moisture relationships in starchy materials. Drying Technology 13: 2097-2111.

42. Tolaba MP, Suarez C, Viollaz P (1997) Heats and entropies of sorption of cereal grains: A comparison between integral and differential quantities. Drying Technology 15: 137-150.

43. Ajibola OO, Aviara NA, Ajetumobi OE (2003) Sorption equilibrium and thermodynamic properties of cowpea (Vigna unguiculata). Journal of Food Engineering 58: 317-324.

6

Formation of Glycidol Fatty Acid Esters in Meat Samples Cooked by Various Methods

Ryo Inagaki, Chikako Hirai, Yuko Shimamura and Shuichi Masuda*

Laboratory of Food hygiene, Graduate School of Nutritional and Environmental Sciences, University of Shizuoka, Shizuoka, Japan

Abstract

Glycidyl fatty acid esters (GEs) are found in some refined edible oils. It is thought that GEs may be broken down by lipase and release glycidol which has been classified as a genotoxic and carcinogenic compound. GEs are formed during deodorization step in the oil refining process. The deodorizing temperature occurs at temperatures of about 200 to 250°C. The cooking temperature is also around 200°C or higher. The aim of this study was to evaluate the formation of GEs in edible meat patties cooked using two methods in order to clarify the intake source of GEs. Three ground meat (beef, pork and chicken) patties were heated by gas fired and char-grilling cooking methods. GEs were formed in meat samples cooked with both heating treatments. In particular, a high concentration of GEs was contained in meat samples heated at high temperature using a charcoal grill. The concentration of each GE compound formed by heating treatment contributed to the amount of each corresponding fatty acid in non-treated raw meat samples. From these results, it is suggested that we may normally ingest GE compounds through cooked meat on a daily basis.

Keywords: Glycidol fatty acid ester; Glycidol; Edible meat; Cooking; Risk assessment

Introduction

Glycidyl fatty acid esters (GEs), such as glycidyl palmitate (C16:0-GE), glycisyl stearate (C18:0-GE), glycidyl oleate (C18:1-GE), glycidyl linoleate (C18:2-GE), and glycidyl linolenate (C18:3-GE), were contained in Diacylglycerol (DAG) oil at high concentration [1]. GEs were also found to be present in small amounts in other common oils rich in triacylglycerol [2]. It was thought that GEs may be carcinogenic [3,4]. This is because it is understood that GEs are broken down by the action of lipase to produce equimolal glycidol (G), which has a reactive epoxy site in the structure [5,6]. G was confirmed as a rodent carcinogen in National Toxicology Program (NTP) study [7]. The International Agency for Research on Cancer (IARC) also classifies G in Group 2A (Probably carcinogenic to humans) [8]. If we ingest 10.0 g per day of DAG oil containing GEs (269 μg/g) [1], the margin of exposure (MOE) of GEs is calculated as 342 value based on comparison of human exposure (0.012 mg/kg bw per day). A benchmark dose lower confidence limit (BMDL10) of G is estimated as 4.06 mg/kg bw per day from the data of mesotheliomas induction of the tunica vaginalis in rats [3]. Therefore, the sale of DAG oil was halted. It is important to estimate the human exposure and toxicities of GEs and G.

Hemoglobin (Hb) adducts have been applied for estimating human exposure to various reactive chemicals as biomarkers [9]. N-(2,3-dihydroxy-propyl) valine (diHOPrVal), which is a Hb adduct in the red blood cells of humans with exposure to G, is an useful biomarker for G and GEs exposure [10,11]. Honda et al. [11] demonstrated that there was no significant difference in diHOPrVal levels in the blood of DAG oil consumers and non-consumers. There was a report that the values of diHOPrVal in the blood of German subjects without G exposure were higher than those of the Japanese DAG oil user [10,12]. These results suggest that we might be exposed to GEs through different food sources other than DAG oil in daily life. Some studies demonstrated that GEs are formed during deodorization step in the oil refining process. The deodorizing temperature occurs at temperatures of about 200 to 250°C. The cooking temperature is also around 200°C or higher. It is reported that some mutagens and carcinogens, such as heterocyclic amines and acrylamide, are formed from meat and fish or potatoes cooked at high temperature [13-16]. Because edible meat and fish contain fatty materials, GEs may be generated in meat and fish heated at high cooking temperature. In the present study, we estimated the formation of GEs in some kinds of meat patties heated by two cooking methods.

Materials and Methods

Reagents

Methanol, 2-propanol, tert-butyl methyl ether, boron trifluoridemethanol complex methanol solution were purchased from Wako PureChemical Industries (Osaka, Japan). n-Hexane, ethyl acetate, diethyl ether, and dichloromethane were bought from Kanto Chemical (Tokyo, Japan). Lauric acid was purchased from Tokyo Chemical Industries (Tokyo, Japan). All reagents were used based on analytical grade reagent.

Standard materials

The standard materials of glycidyl palmitate (C16:0-GE, purity 98.0%), glycidyl stearate (C18:0-GE, purity 98.0%), glycidyl oleate (C18:1-GE, purity 98.0%), glycidyl linoleate (C18:2-GE, purity 90.0%), and glycidyl linolenate (C18:3-GE, purity 85.0%) were purchased from Wako Pure Chemical Industries (Osaka, Japan). Individual GE solutions were prepared at 5.0 mg/ml using methanol/2-propanol(1:1 v/v). Each solution, mixed in equal proportion, was diluted with methanol/2-propanol (1:1 v/v). Standard mix solutions (0.005-1.0 ppm) were used for LC-MSanalysis. The standard materials of methyl palmitate (C16:0-MF, purity 98.0%), methyl stearate (C18:0-MF, purity 99.5%), and methyl oleate (C18:1-MF, purity 99.0%) were purchased from Wako Pure Chemical Industries (Osaka, Japan). Methyl linoleate (C18:2-MF,

***Corresponding author:** Shuichi Masuda, Laboratory of Food hygiene, Graduate School of Nutritional and Environmental Sciences, University of Shizuoka, Shizuoka 422-8526, Japan, E-mail: masudas@u-shizuoka-ken.ac.jp

purity 99.0%) and methyl linolenate (C18:3-MF, purity 98.0%), were bought from Tokyo Chemical Industries (Tokyo, Japan). Individual MF solutions were prepared at 5.0 mg/ml in equal quantity and diluted with dichloromethane. Finally, the standard mix solution was prepared by addition of methyl laurate (Tokyo Chemical Industries, Tokyo, Japan, C12:0-MF purity 99.5%) as an internal standard. Standard mix solutions (1.0-1000 ppm) were used for GC-FID analysis.

Heat treatment of edible meat samples by cooking methods

The pork, beef and chicken ground meats were purchased from a local supermarket in Shizuoka city and stored at -20°C before the experiment. At time of the experiment, ground meat samples were allowed to reach room temperature and thoroughly mixed by hand. One hundreds grams of mixed meat samples were used to form circular patties. The size of each patty was (1.0 cm thick and 10 cm indiameter). Gas fired frying pan cooking and charcoal barbecue cooking were used to cook meat patty samples. The pan cooking method was carried out with a commercial Teflon-coated flying pan (metal), which was preheated until the surface temperature was attained at 150 (low temperature condition) or 250°C (high temperature condition). Then, the meat patty samples were cooked for 2 min per side, for a total cooking time of 20 min (low temperature condition) or 10 min (high temperature condition), without adding oil. For the charcoal barbecue cooking, approximately 1.0 kg of charcoal was placed in the bottom of a barbecue oven. A firelighter was poured onto charcoal to start the fire. When all the flames had subsided, the charcoal was leveled by raking. The meat patty samples were then barbecued over the charcoal for 2 min per side, a total cooking time of 5 min. The distance between the samples and the charcoal was about 2 cm. Cooking temperature on surfaces of each heated meat patty sample was monitored with a thermometer (Hioki 3412-50 Temperature HiTester, HIOKI E.E. Corp, Nagano, Japan) for 1 min. The average temperature of charcoal open fire was about 400°C. The gas fired frying pan cooking experiments and charcoal barbecue cooking were respectively performed in quadruplicate and decaplicate.

Purification of meat samples for the instrumental analysis

Heated meat samples were crushed in a blender and freeze-dried with lyophilizer. Subsequently, 10 g of dry samples were oil-extracted by soxhlet extraction with diethyl ether. Extracted oil samples were evaporated with vacuum concentrator. Evaporated samples of 1.0 g were weighted accurately and dissolved in 5.0 ml tert-butyl methyl ether/ethyl acetate (4:1, v/v). Each solution was cleaned according to double solid-phase extraction (SPE) as described in the previous study [17]. The solution of 500 µl was loaded on the first reverse-phase (RP) SPE (Sep-Pak Vac RC C18 cartridge 500 mg, Waters) column preconditioned with 4 ml methanol just prior to use. Three elutions of 2.0 ml methanol each were then applied. The combined eluates (total volume: 6.0 ml) were dried using a nitrogen stream. The dried residues were dissolved in 2.0 ml n-hexane/ethyl acetate (95:5 v/v), and the solutions were loaded on the second normal-phase (NP) SPE (Sep-Pak Vac RC Silica cartridge 500 mg, Waters) preconditioned with 4.0 ml n-hexane/ethyl acetate (95:5 v/v) just prior to use. Three eluates of 2.0 ml n-hexane/ethyl acetate (95:5 v/v) each were then applied. The combined eluates (total volume: 6.0 ml) were dried using a nitrogen stream. The dried residues were then carefully dissolved in 1.0 ml methanol/2-propanol (1:1 v/v) and mixed by ultrasonic wave. The solutions were centrifuged (4000 rpm × 5 min) and the supernatants were used for the analysis of GEs using LC-MS.

LC-MS analysis

LC-MS was used with AOCS and JOCS conjunction testing methodology CD 28-10 to perform the measurement of GEs [18]. The mobile phases A (methanol/ distillated water 92:8) and B (2-propanol) were consecutively time-programmed as follows: isocratic elution of A 0% (B 100%) between 0 and 4 min, an isocratic elution of A 100% (B 0%) between 4 and 30 min, finally, an isocratic elution of A 0% (B 100%) between 30 and 60 min. For the selected ion monitoring measurement, each of the protonated molecular ions [M + H]+ were used: m/z 313.3 for C16:0-GE, m/z 341.4 for C18:0-GE, m/z 339.4 for C18:1-GE, m/z 337.4 for C18:2-GE and m/z 335.4 for C18:3-GE. Other parameters were as follows: instrument; API2000 LC-MS system (column; L-column ODS (4.6 mm diameter, 150 mm length, 5 µm packing materials, Chemical Evaluation and Research Institute, Tokyo, Japan), flow rate of mobile phases; 1.0 ml/min, injection volume; 2.0 µl, column temperature; 40°C, atmospheric pressure chemicalionization; positive ion mode, vaporizer temperature; 500°C, heater temperature of nitrogen gas; 350°C, flow of heated dry nitrogen gas; 5.0 L/min, nebulizer gas pressure; 0.241 MPa, corona current; 8.0 mA, and fragmenter voltage; 150 V.

Derivatization of samples and GC-FID analysis

Extracted oil samples by soxhlet extraction were derivatized by methylation for analysis fatty acids using gas chromatography. About 20 mg of extracted oil samples were dissolved in 0.5 M sodium hydrate methanol solution containing lauric acid as an internal standard. These samples were heated up at 100°C for 9.0 min using dried block bus. Continuously, samples of 2.0 ml of boron trifluoridemethanol complex methanol solution were added and heated up for 7.0 min. After heating, samples were refrigerated in ice and 2.0 ml of dichloromethane with vortex were added. After mixing for 3.0 min by ultrasonic agitation, 5.0 ml of saturated saline were added to the samples and they were centrifugalized for 10 min (4000 rpm). Lower layers were diluted tenfold with dichloromethane and used as GC-FID analysis test sample. The GC-FID system was a GC-18A-FID (Shimazu Tokyo, Japan) equipped with an auto sampler (AOC-20 Series (Shimazu Tokyo, Japan)) and a hydrogen generator (OPGU-2200S (Shimazu Tokyo, Japan)). Separations were conducted on a SLB-IL100 Capillary Column(60 m × 0.25 mm × 0.2 µm film thickness) from Supelco (Bellefonte, PA, USA). Helium (99.999%) was used as a carrier gas in constant flow mode of 0.4 ml/min. Injections (1.0 µl) were done at 240°C in the split mode (1.5:4.1). The oven temperature was held at 140°C for 5 min. Then it was increased at a rate of 2°C/min to 240°C (held for 10 min). Hydrogen gas was generated with a hydrogen generator for FID at a flow rate of 2.0 kgf/cm^2. The flow rate of air for FID was 4.0 kgf/cm^2. Primary gas (nitrogen) flow rate was 3.0 kgf/cm^2. This experiments was performed in triplicate.

Statistical analysis

Significant differences between the experimentalmeans were calculated by t-test and Tukey-Kramer method. A paired t-test and Tukey–Kramer method were used for the comparison of cooked meat samples in terms of GE and fatty acid contents. The evaluation of linearity was achieved by applying Microsoft Excel.

Results and Discussion

Figure 1 shows the typical LC-MS chromatogram of each GEs standard along with their characteristic ion. Each GE was clearly separated on LC-MS chromatogram. The recovery rates varied from 72 to 91%, which may be attributed to differing GEs. Figures 2-4 shows the

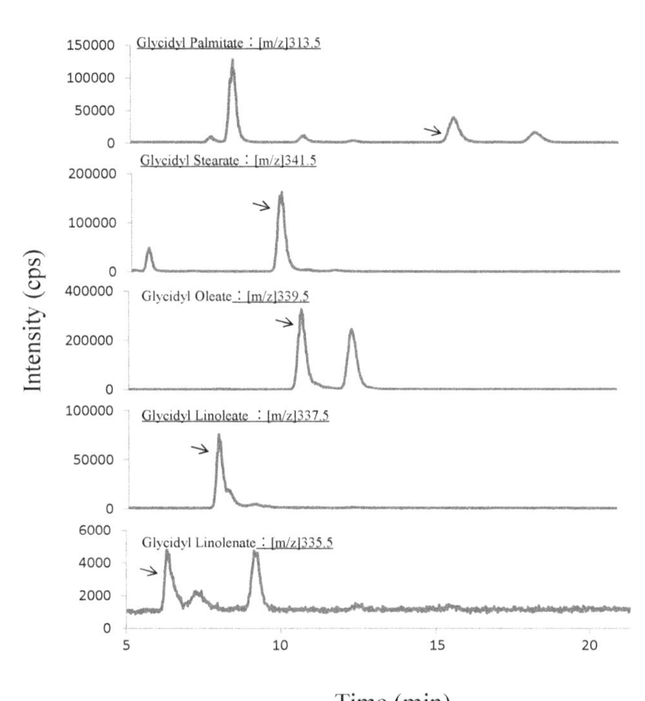

Figure 1: LC-MS chromatogram of each GEs standard (1 ppm) along with their characteristic ion.
For the selected ion monitoring measurement, each of the protonated molecular ions [M + H]+ were used: m/z 313.3 for C16:0-GE, m/z 341.4 for C18:0-GE, m/z 339.4 for C18:1-GE, m/z 337.4 for C18:2-GE and m/z 335.4 for C18:3-GE.

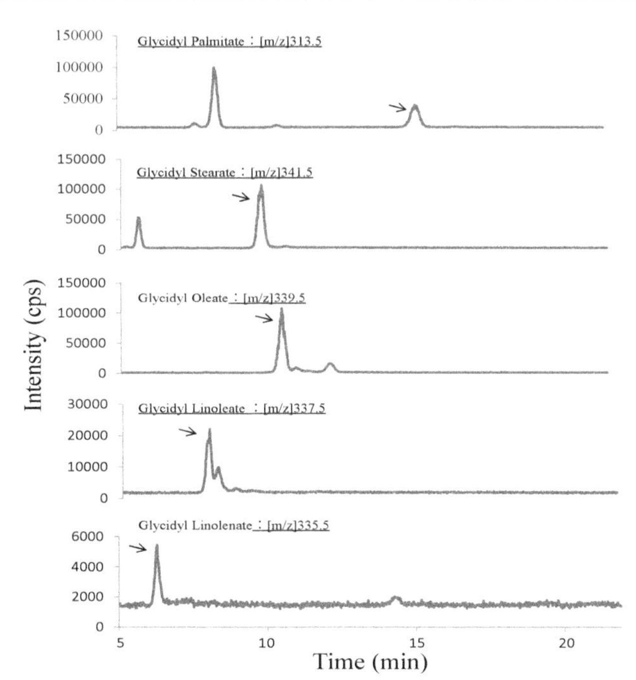

Figure 2: LC-MS chromatogram of each GEs compound in 0.05 g of extracted oil from pork meat patties cooked by charcoal grill. For the selected ion monitoring measurement, each of the protonated molecular ions $[M + H]^+$ were used: m/z 313.3 for C16:0-GE, m/z 341.4 for C18:0-GE, m/z 339.4 for C18:1-GE, m/z 337.4 for C18:2-GE and m/z 335.4 for C18:3-GE.

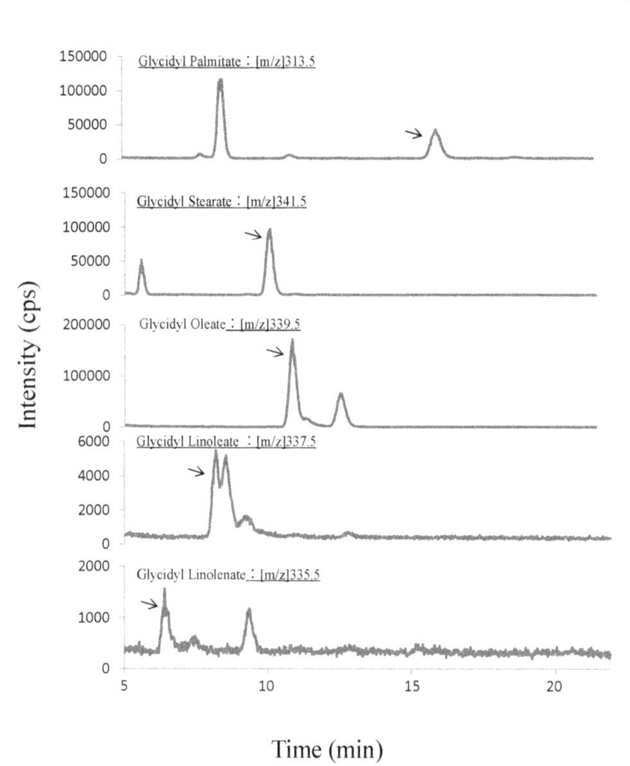

Figure 3: LC-MS chromatogram of each GEs compound in 0.05 g of extracted oil from beef meat patties cooked by charcoal grill. For the selected ion monitoring measurement, each of the protonated molecular ions $[M + H]^+$ were used: m/z 313.3 for C16:0-GE, m/z 341.4 for C18:0-GE, m/z 339.4 for C18:1-GE, m/z 337.4 for C18:2-GE and m/z 335.4 for C18:3-GE.

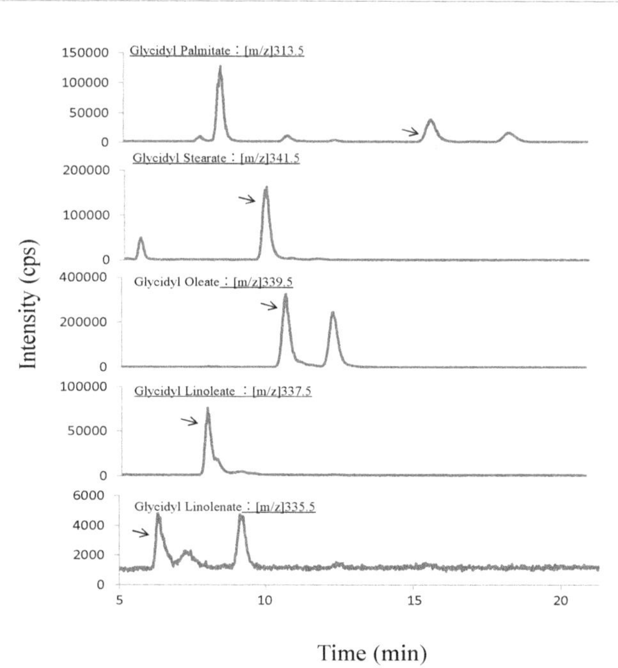

Figure 4: LC-MS chromatogram of each GEs compound in 0.05 g of extracted oil from chicken meat patties cooked by charcoal grill. For the selected ion monitoring measurement, each of the protonated molecular ions [M + H]+ were used: m/z 313.3 for C16:0-GE, m/z 341.4 for C18:0-GE, m/z 339.4 for C18:1-GE, m/z 337.4 for C18:2-GE and m/z 335.4 for C18:3-GE.

representative LC-MS chromatogram of each GEs compound in pork, beef and chicken meat patties cooked by charcoal grill by along with their characteristic ion. Peaks of each GE in heated meat samples was confirmed and clearly separated on LC-MS chromatogram.

Table 1 shows the concentration of total GEs in pork and beef meat samples cooked by gas fired frying pan. No GEs were detected in the pork and beef meat samples at low cooking temperature (150°C). At high temperature (250°C), GEs were determinedin pork meat samples cooked for 5 min (34.4 ± 1.5 ng) and 10 min (166.1 ± 6.8 ng/g) and beef meat samples cooked for 10 min (65.7 ± 6.2 ng/g), respectively. These results showed that the amount of GEs formed in meat samples might rise with increasing cooking temperature and time. However, GEs were not detected in chicken cooked at high temperature (250°C). The contents of fat and water in raw pork, beef and chicken meat samples were 0.15, 0.14 and 0.08 g fat/g, and 0.64, 0.65 and 0.70 g water/g respectively. The fat content is lower and water content is higher in raw chicken meat samples than those in pork and beef meat samples. Chicken meat samples could possibly not be sufficiently heated inside under gas fired frying pan cooking condition. Therefore, GEs could not be formed in chicken meat samples.

Table 1 shows the concentration and content percentage of each glycidyl fatty acid in cooked meat samples. Glycidyl oleate was present in pork meat samples cooked for 5 min at the highest content (13.4 ± 0.4 ng/g) and content percentage (39.1%), followed by palmitate (11.0 ± 0.7 ng/g (32.0%)), stearate (7.7 ± 0.2 ng/g (22.3%)), linoleate (1.8 ± 0.0 ng/g (5.2%)) and linolenate (0.5 ± 0.2 ng/g (1.5%)). In the case of cooking for 10 min, the concertation and content percentage of glycidyl oleate was also highest (57.3 ± 2.4 ng/g (34.5%)), followed by palmitate (53.0 ± 1.9 ng/g (32.1%)), stearate (41.8 ± 1.7 ng/g (25.1%)), linoleate (12.3 ± 0.7 ng/g (7.4%)) and linolenate (1.3 ± 0.5 ng/g (0.8%)). In beef meat samples cooked for 10 min, glycidyl oleate was contained at the highest content (28.7 ± 3.6 ng/g) and content percentage (43.6%), followed by palmitate (26.7 ± 3.9 ng/g (30.7%)), stearate (20.2 ± 2.1 ng/g (24.2%)), linolenate (0.6 ± 0.2 ng/g (0.9%)), linoleate (0.3 ± 0.1 ng/g (0.5%)). From these results, the concentration of each GE compound varied widely in cooked pork and beef meat samples.

Oleic acid, palmitic acid, stearic acid, linoleic acid and linolenic acid are contained in raw pork, beef and chicken at high concentrations. GEs might be generally formed from corresponding fatty acids in raw meat samples. Therefore, the concentrations of five main fatty acids in raw meat samples were detected using GC-FID. Table 2 shows the concentration and content percentage of each fatty acid in unheated raw meat samples. Oleate acid was present in raw pork and beef meat samples at the highest content (50.4 ± 1.0 mg/g (46.6%), 48.4 ± 1.5 mg/g (52.9%)), followed by palmitate acid (27.6 ± 0.5 mg/g (25.5%), 27.3 ± 1.0 mg/g (29.8%)), stearate acid (16.4 ± 0.3 mg/g (15.2%), 13.8 ± 0.4 mg/g (15.1%)), linoleate (12.3 ± 0.2 mg/g (11.4%), (1.6 ± 0.2 mg/g (1.7%)) and linolenate (1.4 ± 0.2 mg/g (1.3%), 0.4 ± 0.1 mg/g (0.5%)), respectively. In raw chicken meat samples, oleate acid was also detected at the highest concentration (25.4 ± 2.2 mg/g (52.1%)), followed by palmitate acid (12.2 ± 1.1 mg/g (25.0%)), linoleate (7.2 ± 0.6 mg/g (14.9%)), stearate acid (3.4 ± 0.3 mg/g (6.9%)) and linolenate (0.5 ± 0.1 mg/g (1.0%)). As compared with the content percentage of each GE in cooked meat samples and fatty acid in raw meat samples, GEs might be generally formed from corresponding fatty acids in raw meat samples.

In the next experiments, the formation of GEs in meat samples cooked by charcoal grill was examined. The temperature under the charcoal grill condition was higher than that under the pan frying condition. The range of cooking temperature using charcoal grill was from 350°C to 600°C. As it was difficult to control the temperature of charcoal grill fire, experiments were done under the same condition ten times. Table 3 shows the amount of total GEs and each GE in three ground meat samples cooked by charcoal grill. The average contents of total GEs in cooked pork meat samples was 1083.5 ± 602.9 ng/g meat sample. Regarding the average contents of each GE compound: glycidyl oleate was present in cooked pork meat samples at the highest content (481.4 ± 279.2 ng/g) and content percentage (44.7%), followed by palmitate (339.8 ± 213.8 ng/g (31.4%)), stearate (163.5 ± 82.7 ng/g (15.0%)), linoleate (80.0 ± 39.4 ng/g (7.4%)) and linolenate (18.8 ± 11.3 ng/g (1.8%)). The average contents of total GEs in beef meat samples cooked by charcoal grill were 669.5 ± 526.0 ng/ g meat sample. Glycidyl oleate was detected in cooked beef meat samples at the highest content (311.1 ± 262.7 ng/g) and content percentage (46.4%) of GEs compounds, followed by palmitate (219.6 ± 177.1 ng/g (32.8%)), stearate (109.2 ± 84.9 ng/g (16.3%)), linoleate (19.3 ± 11.1 ng/g (2.8%)) and linolenate (10.3 ± 9.0 ng/g (1.5%)). The average contents of total GEs in cooked chicken meat samples were 1106.6 ± 475.3 ng/ g meat sample. Glycidyl

Samples	Cooking treatment		Concentration of Glycidol Fatty Acid Esther (GEs) (ng/g meat samples) (Each GE content persentage in total GEs (%))					
	Temp (°C)	Time (min)	Total GEs	Palmitate	Stearate	Oleate	Linoleate	Linolenate
Pork	Low (150)	5	N.D.	N.D.	N.D.	N.D.	N.D.	N.D.
		10	N.D.	N.D.	N.D.	N.D.	N.D.	N.D.
		15	N.D.	N.D.	N.D.	N.D.	N.D.	N.D.
		20	N.D.	N.D.	N.D.	N.D.	N.D.	N.D.
	High (250)	5	34.4 ± 1.5^{b}	11.0 ± 0.7^{b} (32.0%)	7.7 ± 0.2^{b} (22.3%)	13.4 ± 0.4^{b} (39.1%)	1.8 ± 0.0^{b} (5.2%)	0.5 ± 0.2^{a} (1.5%)
		10	166.1 ± 6.8^{a}	53.0 ± 1.9^{a} (32.1%)	41.8 ± 1.7^{a} (25.1%)	57.3 ± 2.4^{a} (34.5%)	12.3 ± 0.7^{a} (7.4%)	1.3 ± 0.5^{a} (0.8%)
Beef	Low (150)	5	N.D.	N.D.	N.D.	N.D.	N.D.	N.D.
		10	N.D.	N.D.	N.D.	N.D.	N.D.	N.D.
		15	N.D.	N.D.	N.D.	N.D.	N.D.	N.D.
		20	N.D.	N.D.	N.D.	N.D.	N.D.	N.D.
	High (250)	5	N.D.	N.D.	N.D.	N.D.	N.D.	N.D.
		10	65.7 ± 6.2^{ab}	26.7 ± 3.9^{ab} (30.7%)	20.2 ± 2.1^{ab} (24.2%)	28.7 ± 3.6^{ab} (43.6%)	0.3 ± 0.1^{ab} (0.5%)	0.6 ± 0.2^{a} (0.9%)

N.D. : Not Detected; Each value is expressed as mean ± standard deviation (n=4); Values are shown as mean ± standard deviation. Symbols bearing different letters in the same column are significantly different ($p<0.05$).

Table 1: Concentration and content percentage of glycidol fatty acids in meat samples cooked by gas fired frying pan.

Meat samples	Concentration of fatty acid (mg/g meat samples) (Each fatty acid content percentage in total fatty acids (%))					
	Total Fatty acids	Palmitate	Stearate	Oleate	Linoleate	Linolenate
Pork	108.1 ± 2.2^{a}	27.6 ± 0.5^{a} (25.5%)	16.4 ± 0.3^{a} (15.2%)	50.4 ± 1.0^{a} (46.6%)	12.3 ± 0.2^{a} (11.4%)	1.4 ± 0.2^{a} (1.3%)
Beef	91.5 ± 3.2^{a}	27.3 ± 1.0^{a} (29.8%)	13.8 ± 0.4^{b} (15.1%)	48.4 ± 1.5^{a} (52.9%)	1.6 ± 0.2^{c} (1.7%)	0.4 ± 0.1^{b} (0.5%)
Chicken	48.7 ± 0.5^{b}	12.2 ± 1.1^{b} (25.0%)	3.4 ± 0.3^{c} (6.9%)	25.4 ± 2.2^{b} (52.1%)	7.2 ± 0.6^{b} (14.9%)	0.5 ± 0.1^{b} (1.0%)

Each value is expressed as mean ± standard deviation (n=3).
Values are shown as mean ± standard deviation. Symbols bearing different letters in the same column are significantly different ($p<0.05$).

Table 2: Concentration and content percentage of fatty acids in raw meat samples.

Meat samples	Concentration of Glycidol Fatty Acid Esther (GEs) (ng/g meat samples) (Each GE content persentage in total GEs (%))					
	Total GEs	Palmitate	Stearate	Oleate	Linoleate	Linolenate
Pork	1083.5 ± 602.9^{b*}	339.8 ± 213.8^{a} (31.4%)	163.5 ± 82.7^{a} (15.0%)	481.4 ± 279.2^{a} (44.4%)	80.0 ± 39.4^{a} (7.4%)	18.8 ± 11.3^{a} (1.8%)
Beef	669.5 ± 526.0^{c*}	219.6 ± 177.1^{b} (32.8%)	109.2 ± 84.9^{a} (16.3%)	311.1 ± 262.7^{b} (46.4%)	19.3 ± 11.1^{b} (2.8%)	10.3 ± 9.0^{a} (1.5%)
Chicken	1106.6 ± 475.3^{a}	332.8 ± 160.8^{c} (30.1%)	98.1 ± 54.9^{b} (8.9%)	518.4 ± 230.6^{c} (46.8%)	139.4 ± 57.7^{c} (12.6%)	17.8 ± 6.6^{a} (1.6%)

Each value is expressed as mean ± standard deviation (n=10).
Values are shown as mean ± standard deviation. Symbols bearing different letters in the same column are significantly different ($p<0.05$). (*vs gas fired frying pan High 10 min $p<0.05$)

Table 3: Concentration and content percentage of glycidol fatty acids in meat samples cooked by charcoal grill.

oleate was detected in cooked beef meat samples at the highest content (518.4 ± 230.6 ng/g meat samples) and content percentage (46.8%), followed by palmitate (332.8 ± 160.8 ng/g (30.1%)), linoleate (139.4 ± 57.7 ng/g (12.6%)), stearate (98.1 ± 54.9 ng/g (8.9%)) and linolenate (17.8 ± 6.6 ng/g (1.6%)). From these results of cooked pork, beef and chicken meat samples, the levels of GEs in meat samples cooked by charcoal grill were very higher than those in meat samples cooked by gas fired frying pan. This was because the cooking temperature in a charcoal grill was much higher than that of a gas fired frying pan. The temperature range was from 350°C to 600°C. The formation of GEs in cooked meat samples might be increased with the temperature rise, especially under very high temperature conditions such as charcoal grill cooking. The level of each GE in meat samples cooked by charcoal grill might also be proportional to the contents of each fatty acid in raw meat samples as with gas fired frying pan. Therefore, GEs might be formed from each corresponding fatty acid directly under high temperature conditions. Cooking meat by charcoal fire was improper for eating in terms of visual presentation, burn deposits, hardness and the formation of GEs. However, meat cooking by frying pan is conducive to a good diet when done on a routine basis and decreases the risk of exposure of mutagens. It was reported that rice oil contains approximately 2 μg/g of GEs, and 1 g of DAG oil has approximately 200 μg/g of GEs. The concentration of GEs in 100 g meat patties cooked by charcoal grill in the present study is equivalent to the amount of GEs in 1 g of DAG. Though the sale of DAG oil has been discontinued, the present results show that GEs are contained in cooked meat even without DAG oil or other refined oils.

From these results, it was demonstrated that GEs were formed in meat samples cooked at high temperature. There are many studies about the examinations of mutagens and carcinogens formed from heat cooking from the 1970s. Derived mutagens and carcinogens were identified and characterized as heterocyclic amines (HCAs) polycyclic aromatic hydrocarbon (PAH) from cooked meat and fish. It was reported that char-grilled beef cooked in similar heating condition to ours contained HCAs and PAH at high concentration [19,20]. Some scientists demonstrated that lengthening cooking time and raising temperatures increase the formation of HCAs in meats. In the present study, the formation of GEs was increased with the cooking time elongation and temperature rise. Therefore, excessive heating and feeding treatment might generate not only HCAs and PAH but also GEs as novel mutagens in cooked meat. 3-Monochloropropane-1,2-diol (3-MCPD) esthers, which is a GE-related compound and has been classified as possiblly carcinogenic to humans by IARC, is also formed from a model oil under deodorization conditions at high temperature [21]. Before now, ecological studies have shown that high exposure to cooked meat containing mutagens, such as HCAs and PAH, increases the risk of human cancers. Ngoan reported a high intake frequency of cooked food is well correlated with sundry cancer [22]. From our results, GEs in cooked meats might also be involved in an increased risk of human cancer.

G is liberated from GEs by action of lipase within an organism. Released G is metabolized by conjugation with glutathione and is excreted as conjugates in urine [5,23]. Formed G also combines with hemoglobin to form a hemoglobin adduct as a biomarker of G invivo. Eckert et al. [24] also reported that 2,3-dihydroxypropyl mercapturic acid (DHPMA) is a useful biomarker as a biological exposure index of G. However, it is reported that 3-MCPD esthers is metabolized to G by the pathway of dehalogenation *in vivo* [3]. We have also to examine the formation of 3-MCPD esther in cooked meats in order to research the precursor of G in cooked meat. and to research G *in vivo*. In the present study, it is thought that we might be exposed by GEs through cooked meats. Exposure to GEs from regular diets, such as cooked meat dishes using these biomarkers of G and GEs in humans, should be examined to estimate the risk of mutagens containing GEs in diets holistically.

Conclusions

Overall, the present study shows that meat cooked by heat treatment contains GEs. Especially, highly concentrated GEs were generated in meat by cooking at high temperature using an open fire charcoal grill. Therefore, we might be exposed to GEs through regular diet other than refined edible oils on a daily basis. As a result, assessment of the risk of GEs for humans must involve in research into the concentration of GEs in various foodstuffs and the level of comprehensive GE intake in regular diets.

References

1. Masukawa Y, Shiro H, Nakamura S, Kondo N, Jin N, et al. (2010) A new analytical method for the quantification of glycidol fatty acid esters in edible oils. J Oleo Sci 59: 81-88.

2. Michael RB, Mark WC, Richard C, Hiroki S, Yoshinori M, et al. (2013) Collaborative Study for the Analysis of Glycidyl Fatty Acid Esters in Edible Oils using LC-MS. Journal of the American Oil Chemists' Society 90: 493-500.

3. Risiken E (2009) Initial evaluation of the assessment of levels of glycidol fatty acid esters detected in refined vegetable fats.

4. Bakhiya N, Abraham K, Gürtler R, Appel KE, Lampen A, et al. (2011) Toxicological assessment of 3-chloropropane-1,2-diol and glycidol fatty acid esters in food. Mol Nutr Food Res 55: 509-521.

5. Frank N, Dubois M, Scholz G, Seefelder W, Chuat JY, et al. (2013) Application of gastrointestinal modelling to the study of the digestion and transformation

of dietary glycidyl esters. Food additives & contaminants. Part A, Chemistry, analysis, control, exposure & risk assessment 30: 69-79.

6. Appel KE, Abraham K, Berger-Preiss E, Hansen T, Apel E, et al. (2013) Relative oral bioavailability of glycidol from glycidyl fatty acid esters in rats. Arch Toxicol 87: 1649-1659.
7. National Toxicology Program (1990) Toxicology and carcinogenesis studies of glycidol in F344/N rats and B6C3F1 mice. NTP Technical Report 374: 1-229.
8. Glycidol (2000) IARC Monogr Eval Carcinog Risks Hum 77: 469-486.
9. Ogawa M, Oyama T, Isse T, Yamaguchi T, Murakami T, et al. (2006) Hemoglobin adducts as a marker of exposure to chemical substances, especially PRTR class I designated chemical substances. Journal of Occupational Health 48: 314-328.
10. Hiroshi H, Kenkichi F, Tohru Y, Naohiro I, Naohiro N, et al. (2012) Glycidol exposure evaluation of humans who have ingested diacylglycerol oil containing glycidol fatty acid esters using hemoglobin adducts. Food and Chemical Toxicology 50: 4163-4168.
11. Honda H, Törnqvist M, Nishiyama N, Kasamatsu T (2014) Characterization of glycidol-hemoglobin adducts as biomarkers of exposure and in vivo dose. Toxicol Appl Pharmacol 275: 213-220.
12. Hindsq Landin H, Grummt T, Laurent C, Tates A, et al. (1997) Monitoring of occupational exposure to epichlorohydrin by genetic effects and hemoglobin adducts. Mutat Res 381: 217-226.
13. Rose M, Holland J, Dowding A, Petch SR, White S, et al. (2015) Investigation into the formation of PAHs in foods prepared in the home to determine the effects of frying, grilling, barbecuing, toasting and roasting. Food Chem Toxicol 78: 1-9.
14. Sugimura T, Wakabayashi K, Nakagama H, Nagao M (2004) Heterocyclic amines: Mutagens/carcinogens produced during cooking of meat and fish. Cancer Sci 95: 290-299.
15. Costa M, Viegas O, Melo A, Petisca C, Pinho O, et al. (2009) Heterocyclic aromatic amine formation in barbecued sardines (Sardina pilchardus) and Atlantic salmon (Salmo salar). J Agric Food Chem 57: 3173-3179.
16. Stadler RH, Blank I, Varga N, Robert F, Hau J, et al. (2002) Acrylamide from Maillard reaction products. Nature 419: 449-450.
17. Hiroki S, Naoki K, Nobuyuki K, Yoshinori M (2011) Direct method for quantification of glycidol fatty acid esters in edible oils. European Journal of Lipid Science and Technology 113: 356-360.
18. Joint AOCS/JOCS Official Method (2012) Glycidyl fatty acid esters in edible oils.
19. Viegas O, Yebra-Pimentel I, Martínez-Carballo E, Simal-Gandara J, Ferreira IM, et al. (2014) Effect of beer marinades on formation of polycyclic aromatic hydrocarbons in charcoal-grilled pork. J Agric Food Chem 62: 2638-2643.
20. Keskekolu H, Uren A (2014) Inhibitory effects of pomegranate seed extract on the formation of heterocyclic aromatic amines in beef and chicken meatballs after cooking by four different methods. Meat Sci 96: 1446-1451.
21. Freudenstein A, Weking J, Matthäus B (2013) Influence of precursors on the formation of 3-MCPD and glycidyl esters in a model oil under simulated deodorization conditions. European Journal of Lipid Science and Technology 115: 286-294.
22. Ngoan le T, Thu NT, Lua NT, Hang LT, Bich NN, et al. (2009) Cooking temperature, heat-generated carcinogens, and the risk of stomach and colorectal cancers. Asian Pac J Cancer Prev 10: 83-86.
23. Elisabeth E, Hans D, Thomas G (2010) Determination of six hydroxyalkyl mercapturic acids in human urine using hydrophilic interaction liquid chromatography with tandem mass spectrometry (HILIC-ESI-MS/MS). Journal of Chromatography B 878: 2506-2514.
24. Elisabeth E, Klaus S, Barbara S, Kerstin HK, Hans D, et al. (2011) Mercapturic acids as metabolites of alkylating substances in urine samples of German inhabitants. International Journal of Hygiene and Environmental Health 214: 196-204.

Nitroso-Hemoglobin Preparation and Meat Product Colorant Development

Hammad HHM[1]*, Meihu Ma[1], Guofeng Jin[1]* and Lichao He[2]

[1]National R&D Center for Egg Processing, College of Food Science and Technology, Huazhong Agricultural University, Wuhan, Hubei, PR China
[2]College of Food and Biotechnology, Wuhan Institute of Design and Science, Wuhan, Hubei, PR China

Abstract

Research focused on preparing new type curing pigment, which called Nitroso-hemoglobin prepared by use of by-products (beef blood), and its application prospect substitutes of nitrite, because of its a poisonous reagent, and can produce carcinogenic nitrosamine after reacting with amine and lead to carcinogenicity, nitrosamines toxicity in meet's product, and this project defined the development of the substitute of nitrite in the meat product from color to make meat color fixative with good stability and coloring stability, antioxidant, antimicrobials, increase nutrition value, and introduced as additives. Additionally, nitroso-hemoglobin pigment synthesis. Moreover, economic aids and environmental protection and its implementation possibility well developed.

Keywords: Nitrite; Hemoglobin; Meat by-products; Beef blood; Color development

Introduction

Hemoglobin is globular proteins (red blood cells) that ferry oxygen molecules and carbon dioxide molecules; all hemoglobin protein structure consists of four polypeptide subunits, which are held composed by hydrogen bonds, van der Waals forces, hydrophobic fundamental interactions, and ionic bonds, as well as four *heme* pigments, one in each of the subunits, heme groups contain positively-charged (Fe^{2+}) molecules reversibly bind to oxygen molecules and transport them to several areas within the body and release their oxygen loads; the complete hemoglobin undergoes conformational changes, which modified their affinity for oxygen [1-3]. Color is the most intimate characteristic indicator used by customers to judge meat freshness the color of meat may indicate microbial spoilage, but it is not the very good meter of nutritionary quality [4,5]. Color is the mainly consequential factor of meat quality products that significant influence's customer purchasing conclusion and materially affects their perception of the freshness of the product [6-8]. The effect color is very prominent clear to understanding troubles when they occur fresh and cured meat color both be contingent on myoglobin, but are substantially very distinctive from each other in terms of how they are successfully created and their comprehensive stability [9]. The first strong impression customers have of any meat commodity is its color and hence color is of furthermost importance and responsive; the color of meat could vary from the deep purplish-red of freshly cut beef to the light gray of faded cured beef [9]. Fortunately, everyone business practices with meat commodities must have a functioning comprehension about the color. Moreover, blood is a rich source of iron and proteins of high nutrition and functional quality [10]. Meats have great potential for delivering important nutrients such as antioxidants, minerals, dietary fiber fatty acids, and bioactive peptides into the diet [11,12]. However, to manufacture prosperous and quality products with these constituents, advanced technology's necessity be fully developed to significant. Increase their constancy and decrease their flavor impact on muscle foods. In addition, many regulatory hurdles must be overcome for the commercial production of meats with supplemented nutrients. These contain redefinition of high standard of identities and policies that allow front of the package nutritional claims [13]. Without these regulative modifications manufacture of healthier meat quality products won't come to be a fact since these products would not have a competitive benefit over unfortified meats [12].

Blood is incredibly a rich source of elevated nutritional values and functional high quality due to comprise iron and proteins. Consequently, blood used as food supplements and obtained products. It is occasionally mentioned by to as fluid protein [14,15]. *Myoglobin* is the main protein known as the control of meat pigment, although another hem protein such as cytochrome C, and *hemoglobin* may also play important part in beef, lamb, poultry, and pork pigment [4]. Economic benefits and environmental protection can be both achieved by processing in meat color development, and this review paper discussed the progress in substitute of nitrite in meat produce from colorant, antioxidant, antioxidant, and antimicrobial [16]. Consumer's judge meat due to its color that is the most common quality indicator to use as freshness, fresh and cured meat depended on myoglobin, but are significantly different from each other in terms of by what method they are formed and their overall stability [17]. Our objectives were to review publish *nitroso-hemoglobin* preparation and meat product color development substitutes of nitrite in meet's product from color fixative, antioxidant, antimicrobials, increase nutrition value, and introduced as additives. Research conducted for raw meat pigment in under the last few years ago. The objective was there exactly indicated what is new but rather to talk over the present meat pigment study. As such, our valuation of new studied. Literature proposes that the subsequent was the key of interest to researchers. Nitrite is greatly used as the chromogenic reagent in meat product at present, because it can perform color in the meat product enchanting and prevent *Clostridium botulinum* in meat products. Nitrite is a toxic reagent and can yield carcinogenic nitrosamine after responding with amine. So, people are always looking for the substitutes of nitrite. This project looking to the development of the substitute of nitrite in the meat product from colorant, antioxidant and increase nutrition value. And suggestions in prospective furtherance meat product [18].

***Corresponding author:** Guofeng Jin, National R&D Center for Egg Processing, College of Food Science and Technology Huazhong Agricultural University, 1-Shizishan Street, Wuhan, Hubei 430070, PR China
E-mail: jgf@mail.hzau.edu.cn

Title	Results/Conclusion
Preparation of (Hb) imprinted polymer by Hb catalyzed ATRP and its application in biosensor.	Displayed a broader linear range and a lower detection limit for hemoglobin (Hb) determination when it was compared to those Hb sensors based on molecularly imprinted polymer (MIP) and in improve the polymer growth, mixed Self-assembled monolayer (SAM), Hb, and MIP was surface of electrode (MIP/Au).
Development and Application of the Substitute of Nitrite in the Meat Product	Preparing new type curing pigment used as substitute of nitrite in the meat and meat product from colorant, antioxidant and antimicrobial which called Glycosylation nitroso - hemoglobin and its application and natural, pigment was treated with polysaccharide to prepare glycosylated nitroso - hemoglobin, stabilities and enhanced meat color.
New type curing pigment glycosylation nitroso research progress of hemoglobin.	Preparing glycosylated nitroso-hemoglobin by the use of by-products (blood) in livestock and poultry processing and its application prospect were elaborated and used as curing pigment.
Influence of nitrosohaemoglobin with monascus on the color and	The using of nitroso-hemoglobin with monascuscolors had the ability to improve the quality of red sausage.
The research progress of nitrosohemoglobin pigment synthesis and its application	Nitroso-hemoglobin is used in meat processing to substitute of nitrite residue, improve the nutritional value of the meat product, synthesis and application of nitroso-hemoglobin and glycosylated nitroso- hemoglobin in meat products.

(Hb) = hemoglobin
(eATRP) = electrode by electrochemically mediated atom transfer radical Polymerization.
(MIP) = Molecularly imprinted polymer
(SAM) = Self-assembled monolayer.
(MAP) =Modified atmosphere packaging

Table 1: Summary of research evaluating (Nitroso-hemoglobin preparation) and used in meat colorant development.

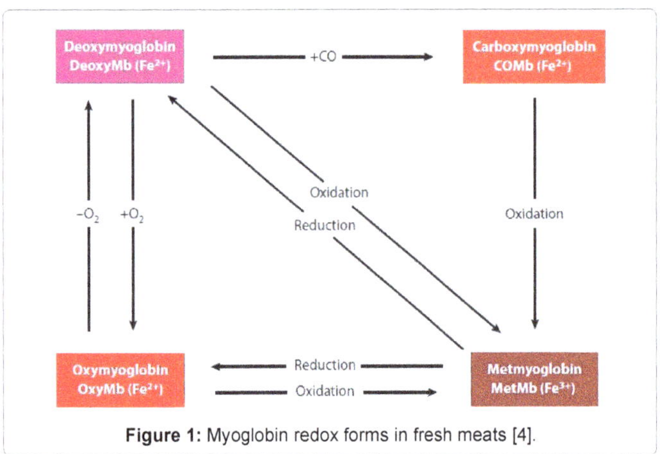

Figure 1: Myoglobin redox forms in fresh meats [4].

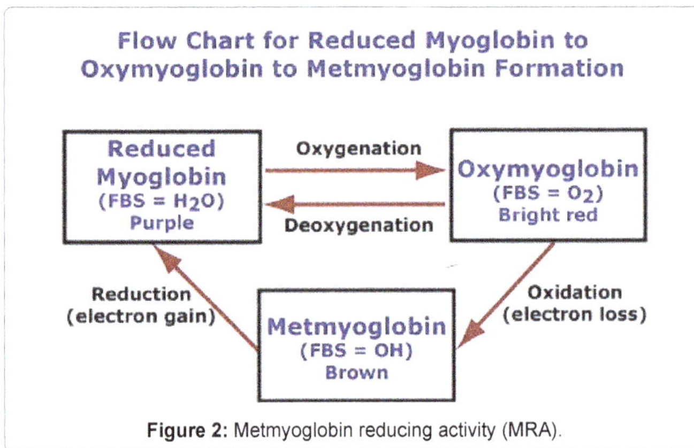

Figure 2: Metmyoglobin reducing activity (MRA).

Meat color and meat product colorants status and development trend

Nitroso-hemoglobin prepared by use of by-products (beef blood), (*Hb*) preparation condition as modified starch and maltodextrin were taken as raw materials, color fixative was microencapsulated and spray dried reacting with nitrogen monoxidum on definite condition and its application prospect substitutes of nitrite in meet's product from color fixative, antioxidant, antimicrobials, increase nutrition value, and introduced as additives. Additionally, to make meat color fixative with good stability and coloring ability *nitroso-hemoglobin* pigment synthesis. Moreover, economic aids and environmental protection.

Meat color

Color of meat known as myoglobin proteins, and it's recognized in exudation by gel-electrophoresis, fully accounting in part for the modification on the pigment steadiness of meat after chilling conditions and melting [19,20]. The denaturation guidance's to an improved sensitivity of myoglobin to auto-oxidation and following diminution of optimum color production. This model possesses existed confirmed by various writers by associating the grade of bloom and the capability of the meat to resisting oxidization to metmyoglobin through cooled storehouse post freeze/thaw [21-25]. Moreover, previously it was reported that the denaturation of the globin moiety of the myoglobin molecule possesses the function at some stage through chilling, chilled storing and defrosting by Calvelo [26] (Table 1). Preparation of Molecularly imprinted polymer (MIP) was schemedin synthesis of molecularly imprinted polymer to the Au electrode surface through an electrode by *Hb* with (*eATRP*) both as catalytic agent and template [27]. Then the elect trade was injected to protein solution containing *AM, Hb, and MBA* as functioning electrode. When a potential was applied tore-duce *Hb–Fe (III)* to *Hb–Fe (II)* catalyst for *(eATRP)*, polymerization would be triggered. After the removal of *Hb, MIP* was efficaciously ready on the surface of the electrode (*MIP/Au*) [27]. The functional monomer and cross-linker were acrylamide (AM) and N. N-methylene bis-acrylamide (MBA), respectively. In order to improve the polymer growth, mixed SAM of initiator and HTP was self- assembled on the surface of Au electrode (Br/Au) [27]. Reaction reversible dependent of enzyme's and the compounds (Figure 1). Alteration *myoglobin oxymyoglobin* vice usually, quiet, Likewise, reaction produces brown *metmyoglobin* quiet but reverse this more in meat, is dynamic in company oxygen three myoglobin, and is interconverted; three are evenness one (Figure 1). Moreover, meat chemistry pigments showed that the converted of myoglobin from a dark purple pigment to a bright-red pigment merely from *oxymyoglobin* or to a brown pigment by losing electrons due to present of oxygen depending on the store situations the meat pigment's *oxymyoglobin*, metmyoglobin, and *myoglobin*, can be transformed from one to the other [4]. *Oxymyoglobin* and *myoglobin* they able to lose an electron (*oxidation*) changes the pigment

Title	Results/Conclusion
The Oxford Companion to Food. 2nd ed. UK: Oxford University Press, p. 81-82.	Preparing a traditional dishes, it is made with meat, blood, and spices all encased in the animal's entrails or intestinal in Kenya.
The Oxford companion to food, OUP, p.104.	Make black pudding generally made from pork blood mix with oatmeal, occasionally flavored with pennyroyal, and barley instead of onions to absorption blood. And it is ready to eat but is habitually grilled, and fried or boiled in its surface.
Mutura - The traditional Kenyan sausage. On1 May 2014.	Preparing traditional dish, it is made from meat blood, and spices all coated in the animal's intestines or stomach (central of Kenya).
The Oxford Companion to Food. 2nd ed. UK: Oxford University Press, 2006, p. 81-82.	Used coagulated blood from (pig, duck, chicken, sheep, goat, and cow) as fried food or steamed as snack food rectangular pieces and cooked is most often used (In the Northeast China and Taiwan).
Stick Out Your Tongue Chatto and Windus London.	Used blood (enzyme's Hat takes part in blood clotting) as established required agents and additional to meat cuts or minced meat to creation portions of looked-for mass and formation.
Screening method for the addition of bovine blood-based binding agents to food using liquid chromatography tripled quadrupole mass spectrometry. Rapid Communications in Mass Spectrometry, 21(18): 2919–2925.	Used blood as based binding agents and supplementary (increase nutrition value) to minced meat or meat cuts to form portions of required mass and procedure in the binding process, and thrombin cleaves fibrinogen

Table 2: Summary of research evaluating to use's blood as food or additives.

called denatured *metmyoglobin* is shaped, which normally cannot be transformed to another pigment. To preserving this color needs that the meat surface must be free from any pollution reason a chemical reaction subsequent in the *metmyoglobin* pigment Oxygen must be obtainable for an enough attentiveness in order to the association with the myoglobin to procedure *oxymyoglobin* [4,28]. This reaction is reversible and depends on the current of oxygen, active enzymes and decreasing compounds in the muscle (Figure 2). The procedure or operating system of cooking consequences in change in the protein composition of soluble myoglobin (denaturation), and temperature persuaded myoglobin denaturation is held directly accountable for the lethargic umber pigment of cooked meats [29]. Nitrite is commonly used in meat products, but it is a poisonous reagent, and can produce carcinogenic *nitrosine* after reacting with amine. So people are looking for new type of pigment beef blood instead of nitrite. Also, a valuable source of protein is missing if animal blood is cast off as leftover and this is led to serious environmental contamination troubles, there are several nations need that the animal blood be willing to in a biologically kindly method, which is a wealth concentrated procedure [14].

Meat product colorants status and development

Nitroso-hemoglobin prepared by use of by-products (beef blood), (*Hb*) preparation condition as modified starch and malt dextrin were taken as raw materials, color fixative was microencapsulated and spray dried reacting with nitrogen monoxidum on definite condition and its application prospect substitutes of nitrite in meet's product from color fixative, antioxidant, antimicrobials, increase nutrition value, and introduced as additives. Additionally, to make meat color fixative with good stability and coloring ability *nitroso-hemoglobin* pigment synthesis. Moreover, economic aids and environmental protection. A promising approach to improving meat color care would be to produce food supply as a preventive color care strategy. The food supply could be improved by producing functional foods that have nutritional profiles that are healthier than conventional products. However, production of functional foods is not always easily accomplished since they must also taste good, be convenient and reasonably priced so that consumers will regularly purchase and use the products [12].

Hemoglobin profiles (source, structure and properties)

Structure and function: Sadava [3] reported that hemoglobin tetramers are contained of the four subunits, two α-globin chains and two *β-globin* chains all of which take the form of alpha helices. Found in each chain is a non-protein heme group, which is an assembly of cyclic ring structures surrounding an iron ion that is tied by nitrogen atoms and the heme group are characteristically hidden within the several subunits, is covalently bound to yet dissimilar atom longs for a *histidine* and is a chain, with hydrophobic stabilize heme with in subunit. Molecules opposite the *histidine*. Near opposite is a different chain, serves significant even it not bound the group (*Perultz, Histidine* itself charged; close to negatively iron averts iron from too, which inhibit binding oxygen *Heme* has greater affinity carbon than oxygen, of chain *histidine* allow groups to bind to waste to prevent the proteins providing with oxygen for activities 2006). Of *hemoglobins* of transport, *oxygen-bind* Fe^{2+}, not (Boyer, Also, size, and of the distal chain of molecules will to heme (Boyer, directly cutting, color quite-would a purplish-redtois (surface the blooms) this is *oxymyoglobin*, beef bright red associate freshness.

Hemoglobin utilization: To make attractive color using by consumers to judge meat freshness. Substitute of nitrite in the meat and meat products from colorant, antioxidant, and increase nutrition value. Enhance color changes, first converting to *nitroso-myoglobin* (bright red), then, on heating, change to *nitroso-hemochrome* (a pink pigment). Progress of *nitroso-hemoglobin* pigment synthesis and its application in meat product. Color is particularly an important quality guideline for both fresh meat and nitrite cured meat commodities and depends upon the oxidation-reduction status and ligand bound to heme iron in Mb. The fundamental knowledge of Mb chemistry is therefore, crucial, and the "inorganic chemistry" of meat had to include a quantitative accurate description of Mb complicated formation with small ligands, such as O_2, NO, H_2O and CO, and the kinetics of transformations of these complexes under transforming conditions of temperature, oxygen pressure, pH, ionic strength and light exposure. The practical aspect of colour solidity of meat and meat products in the meat industry and retail trade has started abundant examinations over the last 50 years and significantly an increased understanding of the complicated chemistry [30]. The detection of the physiological momentousness of NO and the quite possible function of hypervalent Mb and Hb throughout the oxidative stress possess added new perspectives to the dynamic description of electron transfer and ligand exchange reactions of these heme pigments [31]. Additional significant functions of Mb than oxygen conveyance and storage seem to be important, and these *in vivo* important functions contain activity as a pseudo-enzyme in quite specific muscle tissue similar to NO dioxygenase clearly recognized from microorganisms. In effect, Mb characterizes as a cellular defender against nitrosamine stress during excess production of NO [30]. Future investigation is supposed to focus on the role of Mb as a mediator of responses between small molecules most important as bio-regulators and contain researches of reaction dynamics hopefully happening within the protein structure using time-resolved spectroscopy [30].

Title	Results/ Conclusion
Comparison of color and thiobarbituric acid values of cooked hamburger patties after storage of fresh beef chubs in modified atmospheres.	High oxygen modified atmosphere packaging increases, while low oxygen and 0.4% CO avoid early change to brown color in ground beef, use this technique in meat manufactures to avoid meat color.
Internal premature browning in cooked steaks from enhanced beef round muscles packaged in high-oxygen and ultra-low oxygen modified atmospheres.	The improved of beef round steak packaging in high oxygen modified atmosphere packaging (MAP) are brown when heated to 71.1°C for the high oxygen MAP encourages oxy-hemoglobin creation in the product.
Effect of muscle source on premature browning in ground beef.	Muscle basis position, and inherent biochemical that can be affected of shape ground of beef cooked color.
Effect of erythorbate, storage and high-oxygen packaging on premature browning in ground beef.	Addition of erythorbate to ground beef rises decreasing movement previous to cooking and reductions the incidence of premature browning color.
Effects of succinate and pH on cooked beef color.	The effect of succinate is partially due to increase of pH, and it can protect myoglobin against thermal denaturation, and ground beef enhanced cooked redness due to pH.
Chitosan inhibits premature browning in ground beef.	Myoglobin's thermal constancy is dependent on several issues such as color concentration, myoglobin redox state, and muscle source, pH, packing kind, storing condition, as well as adding ingredients.
Factors that influence cooked meat color. Journal of Food Science, 71, R31-R40.	The cooked of meat color due to the myoglobin denaturation, and thermal stability is dependent on myoglobin redox formal, muscle source, pigment concentration, pH, packaging type, storage situation, and ingredients added.

Table 3: Summary of research assessing current affecting in meat product's color.

Progress and functional of Nitroso-hemoglobin preparation

Summary of investigation appraising to use's blood as food or additives was shown in Table 2. Numerous nations eat blood as diet, normally in mixture with meat used as supplementary (increase nutrition value) to form of blood sausage, as a thickener for sauces, a preserved salted form for times of diet lack, or used blood as the soup [31-35]. There is the traditional dish blood food among the people of dominant Kenya Mutura, it is made from mixing meat, blood, and spices all inside the animal's intestines or stomach. In United Kingdom produced black pudding usually made from ham blood and a comparatively in height prepared of oat flour in the previous it was sporadically flavored through pennyroyal, and different from continental European forms in its comparatively limited range of elements and dependence on oat flour and barley substitute of onions to absorption the blood, and it can be consumed uncooked, but is often grilled, fried or boiled in its skin [36]. In Asia, there were many people produce diet from solid animal blood and greatest of these kinds' diet of don't have been casing and might be careful a form of sliced sausage [37,38]. Although in Chinese cultures, as well as Taiwan whole coagulated blood is fried or steamed as a food or cooked in a hot jar and rectangular piece and cooked is most often made through pigs or duck's blood, chicken's, and cow's blood as well as used in the Northeast China, blood sausage is a traditional food which is cooked with goat or sheep blood in resource-poor, congealed yak's blood used as the traditional food in Tibet area [38]. Although used blood as additives and enzymes take part in blood coagulation these be able to be used as blood as binding agents and supplementary to meat slashes or minced meat to procedure portions of wanted mass and form in the binding method, thrombin cleaver's fibrinogen to fibrin peptides and used to detect bovine and porcine fibrin peptides concentrations down to 5% [39]. As the fibrin peptides, species exact the approaches are capable of discerning amid blood-based binding agents of bovine and porcine origin. The nutrition blood of livestock and poultry is very rich, most of the hemoglobin; deposit lies in the red blood cells, accounting for 80% of the total protein. However, hemoglobin color and stability are very poor; the *hemoglobin* (*Hb*) protein is made up of four molecules and molecules composed of four peptides globulin, and muscle of *myoglobin* (*Mb*) is composed of a molecule hemoglobin and a molecule of a single polypeptide chain globulin [40,41]. Animals are used as food either directly as meat, which comes from muscle systems or from tissues or indirectly includes milk produced via mammary glands, in numerous cultures is drunk or treated into dairy products such as yoghurts, butter, cheese, etc., [36]. Additionally, some cultures are eating blood forming of blood sausage, as a thickener for sauces, or in a cured, salted form times of food lack in the sometimes, also use blood in stews such as jugged hare birds and other animals lay eggs, which are often taken as food, and bees produce honey, a reduced nectar from flowers, which is use a widespread sweetener in many cultures [36,42]. The increase used of modified atmosphere packaging has carried about some modifications in heated meat pigment (Table 3) owing the consequence of packaging atmosphere on *myoglobin* chemical reaction (*redox*), known as principal factor of heated pigment. Seyfert et al. [43] they reported that packaging of the beef steaks in high oxygen MAP greater than before and heightened prevalence color thoroughly heats excessively low to kill pathogen bacteria. Exactly, injection-enhanced beef in MAP was prematurely brown col while heated to 71.1 Co, a high temperature that ought to have some pink pigment and be inoffensive to consumption and strong degrees' oxygen in this product exploited oxymyoglobin creation on the surface and in certain carrying cases, deep inside the product [43]. Whereas *MAP* was an advantage for fresh meat pigment, it disposed culinary to produce products to early browning color at temperatures less than that compulsory for protection of diet, since *oxymyoglobin* and *metmyoglobin* were more temperature labile than *deoxyhemoglobin* and *carboxymyoglobin* [42]. Additional reasons affecting cooked pigment contain, source of muscle, anatomical situation, and inherent biochemical shape, fat contented, store at a low temperature, endpoint temperature of frozen storage, and post-cooking heat increase [44]. *Myoglobin* denaturation resulting creation of each *ferro-/ferrihemochrome*. Though, *myoglobin's* thermal solidity is dependent on different issues for example *myoglobin redox state*, muscle sources, pH, storage condition, coloring concentration, added ingredients, and the type of packaging [29]. It expressly granted the function of myoglobin's oxidation-reduction state-run into heated pigment [42]. Obviously established that (CO) in improved atmospheres would carefully avoid early browning pigment of surface beef pies. Moreover, Suman et al. [45] recommended that erythorbate possibly will be commonly used to reduce premature browning. Metmyoglobin reducing activity, pH can keep myoglobin in contradiction of thermal denaturation the adding of succinate to grind beef enhance cooked red color the consequence of succinate is partly suitable

Title	Results/Conclusion
Changes in Color and Myoglobin of Minced Beef Meat Due to High Pressure Processing. Lebensm.-Wiss. u.-Technol., 28, (1995), 528-538.	The packet of red meat under vacuum power present of an oxygen scavenger in partially adequately meet color, and color values increased significantly.
Effects of carbon monoxide-modified atmosphere packaging and irradiation on E. coli K12 survival and raw beef quality. Meat Science 83 (2009) 358-365.	Aerobic packed beef was optically greener and reduced redder than carbon monoxide (CO) in heavily atmosphere packaging CO-MAP, and sufficient preserved and provide color. However, storage period was the main factor lead to reducing fresh beef and substantially increasing or considerably acid green and rancid odor.
Evaluation of carbon monoxide treatment in Modified atmosphere packaging or vacuum packaging to increase color stability of fresh beef Meat Science 59 (2001). 317-324.	Packaging or Vacuum packaging method is excessively important to development of raw beef color needed to maintain redness and extended color stability and decrease of microbial counts.
Carbon monoxide in modified atmosphere packaging affects color, shelf life, and microorganisms of beef steaks and ground beef. Journal of Food Science, 69(1), C45-C52	It determined that the use of 0.4% carbon monoxide (CO) through an ultra-low oxygen pack had to enhance the beef color immovability deprived of hiding decomposition and biochemical profile of muscle may result from reaction to (CO).
Evaluation of carbon monoxide treatment in modified atmosphere packaging or vacuum packaging to increase color stability of fresh beef. Meat Science, 59(3), 317-324.	The usage of 0.5% carbon monoxide, and 0.5% (CO-MAP) system with pretreating steaks can increase color stability through following vacuum packaging and significantly economic aid.

Table 4: Summary of research evaluating changes in Color and myoglobin effect atmosphere packaging.

Title	Results/Conclusion
Formation of red myoglobin derivatives and inhibition of spoilage bacteria in raw meat batters by lactic acid bacteria and *Staphylococcus xylosusLWT* - Food Science and Technology 68 (2016) 251-257.	Myoglobin derivatives by bacteria such us (*Staphylococcus xylosus* and *Lactobacillus fermentum*) using as substitution of nitrite and significantly progressing meat pigment and repressive micro- organism increase (spoilage bacteria) in curing meat.
Production of cured meat color in nitrite-free Harbin red sausage by *Lactobacillus fermentum* fermentation. Meat Science 77 (2007) 593-598.	The Lactobacillus fermented was capable of producing NO-Mb in this sausage product (replaced of nitrite to produce cured pink pigment in a Chinese-style sausage), without the addition of extraneous sodium nitrite.
Elucidation of the chemical structure of preformed cooked cured-meat pigment by electron paramagnetic resonance spectroscopy. Journal of Agriculture and Food Chemistry, 44 (1996) 416-421.	Used of nitrosyl-myoglobin, with proper antioxidant, and antimicrobial agent to yield good substitute pink, cured pigment is settled to microbial transformation of myoglobin.
Color structure in fermented sausages by meat-associated staphylococci with various nitrite- and nitrate-reductase activities. Meat Science 78 (2008) 492-501.	*Staphylococcus* strains in nitrite-cured sausages were of partial significance highly regarding color improvement, and color formation through initial fermentation stages.
Influence of high pressure on the color and microbial quality of beef meat. Lebensm.-Wiss. u.-Technol. 36 (2003) 625-631.	The valuation immediate effect of varying pressure values (50-600 MPa) and holding times (20-300S) on color improvement and microbiological quality of meat product (bovine muscle).
Synthesis process and application of nitroso hemoglobin in sausage decreased residues.	The use of (HbNo) as color additive and replace NaNo2 or NaNo3 in meat products like sausages were emphasized to obtain stable color.

Table 5: Summaries evaluating research of antimicrobial significantly affecting meat product color.

to significantly increased PH [46]. Moreover John et al. [42] and Suman et al. [47] report that the vacuum packaging beef devoid of oxygen (vacuum or CO-MAP) be able to efficiently the reduction of the incidence of early browning. The changes in Color effect atmosphere packaging (Table 4) minced beef meat was packed under vacuum power, air or oxygen; pigment great rates enhanced in the range the meat virtual becoming pink, whereas values significantly decreased the meat becoming grey-brown. Concomitantly, total extractible myoglobin reduced, at the same time as the quantity of metmyoglobin substantially enhanced at the write-off oxymyoglobin pressurization did not meaningfully raise the extractability of heme iron by an acid solution and meat packaging under vacuum in the company of an oxygen scavenger somewhat supported meat color [48]. Furthermore, color changes resultant from pressure handling are caused by two or three diverse mechanisms. In the attendance of nitrite, the extensive nitroso-myoglobin is defended against oxidization into the ferric form. The whitening consequence, however, is not hampered [49]. Vacuum packaging (VP) is excessively important new approaches to increase raw beef red pigment steadiness for carbon monoxide (CO)-treated beef steaks was wanted to take care of ruddiness succeeding re-packaging in (VP). Extended color relative stability lows of microbial counts in (VP) was successfully achieved by pretreatment with 5% CO [50]. Combined effects of irradiation and (CO) in modified atmosphere packaging (CO-MAP) on total plate counts, *Escherichia coli* K12, color, and odor of fresh beef through the cooled store. Aerobically or in CO-MAP, no substantial difference existed for visual green color scores due to gas atmosphere aerobically packaged beef was greener and less red than CO-MAP packaged beef and reduced microbic loads to protected levels through 28 days of storage. Red color of CO-MAP packaged samples significantly decreased slightly after irradiation [49]. The application of 0.4% (CO) through the store period in MAP enhanced the beef pigment without effectively masking spoiling upon removal of end product from (COP), meat pigment grouping of COMb and OMb) deterioration through the demonstration as a method not different as of produce the product exposed only to air [51]. Furthermore, upon take product from (COP), meat pigment the combination of COMb and OMb declined through presentation in a manner not distinctive from product quite increasingly exposed only to air [50]. They are Theyrted that the usage of packing steaks and surface beef in 0.5% (CO) leads to better-quality color constancy. While usage of a (COP) system that also contained oxygen prohibiting, and low levels of (CO) tested [50,51]. The evaluating research of antimicrobial meaningfully affecting meat color (Table 5). Effectively a probable efficient method for nitrite substitution by considerably improving meat pigment and preventing microorganism spoiling in meat miraculously curing by using the change of *metmyoglobin (MbFeIII)* to red myoglobin derivatives by micro-organisms such as *Staphylococcus xylosus* and *Lactobacillus fermentum* and their positive influence on the major impediment of spoilage organisms in crude meat batters as essentially defensive cultures fully investigate dependent and independent culture (PCR-denaturing gradient gel electrophoresis) implement approaches were functioning in disclose the microbic generation [52]. Moreover, *L. fermentum* creative ferment donated to the structure of *NO-Mb* in cured meat deprived of the adding nitrite was financed by the lack of

Title	Results/Conclusion
The potential of increased meat intake to improve iron nutrition in rural Kenyan school children. International Journal for Vitamin and Nutrition Research, 77(3), 193-198.	Risk associated with treated meat's consumption accompanied by other significant risk matters connected with colorectal cancer, micro-nutrients such as zinc, iron and vitamins.
Meat, fish and fat intake in relation to subsite-specific risk of colorectal cancer: The Fukuoka Colorectal Cancer Study. Cancer Science, 98(4), 590-597.	The measurement of colorectal cancer occurrence was not found to be directly connected to meet or treated meat product's intake in a Japanese limited population.
Food, nutrition, physical activity, and the prevention of cancer: A global perspective. Washington, DC: American Institute for Cancer Research.	The majors increase of considerable alarms approximately the cancer risks correlated with consumption of red and processed meats.
Meat and meat mutagens and risk of prostate cancer in the agricultural health study. Cancer Epidemiology, Biomarkers and Prevention, 17(1), 80-87.	The consumption of well or very well done total meat was accompanying accompanied by advanced disease were of borderline significance for increased prostate cancer risk.
"Healthier meat products as functional foods." Meat science 86(1): 49-55.	Healthiness care possibly will be proactively considerably improved by manufacturing a in good health food provide as a preventative healthiness care strategy and food security classification through both positive and negative nutritionarycharacteristics.

Table 6: Summaries of research evaluating processed meat and colorectal cancer.

any quantifiable rosy pigment if the sausage was handled with neither supplemented nitrite nor *L. fermentum*, and the suggested that the *L. fermentum* implementation level was particularly important; a 108 CFU/g vaccination was strictly necessary to produce the pink color high intensity comparable to that produced by 60 mg/kg of nitrite [53]. Using of *nitrosylmyoglobin*, with proper antioxidant, and antimicrobial agent to yield good substitute pink, cured pigment is settled to microbial transformation of myoglobin [54]. Nitrosylated hemin has been technologically advanced for nitrite-free cured meats. When *pre-synthesized nitrosyl myoglobin*, with accepted antioxidant, and antimicrobic agents, were supplemented to fresh meat, the eventually consequent meat productions reportedly resembled nitrite-cured meats [54]. *Staphylococcus* strains, such as (*S. carnosus, S. simulans* and *S. saprophyticus*), chosen due to their highy variable nitrite and/or nitrate-*reductase* activities, were used to formally initiate pigment formation during sausage ferment the rate of nitrosyl myoglobin formation in sausages through additional of nitrate depended on the limited specific *Staphylococcus*, high strain nitrate-reductase activity showed a faster rate of pigment development of product stability for the sliced, packed up sausage was fully assessed as surface pigment and oxidization by auto fluorescence and hexane content [55]. Presentation of record high intensity of pressure for a short period dominances to a substantial reduction of total flora quickly growing in meat product a delay of one week; this delay pre-eminences to let the ripening of the meat longer, and perhaps will be improving the meat tenderness. Though, this intensity of pressure is not perfectly adequate to change the microbiology of beef meat, therefore, a select basic essentials continually be possessed among a color development and a microbiological enhancement hinging on the final meat product formed exploitation development though, this advantageous good result of the act is attended by a splotch of meat [56]. The influence of pressure is larger substantial than that of the retention period on the pigment factors and metmyoglobin content. The pressures (130 MPa) reason an enhancement in meat pigment quality with and significantly increase in ruddiness whatever is preserved through several days of the storage period. Indeed, 130 MPa resolve be preferred to sell raw meat and meat products [56]. The using of nitroso-hemoglobin with monascus colors lade to reduce the total plate counts, ability to improve the presence and quality of red sausage, and nitroso-hemoglobin used as instead of sodium nitrite in the application of sausage, furthermore, use of nitroso-hemoglobin to replace the amount of residual nitrite meat products, significantly improve the security of the sausage [57].

Grillenberger et al. [58] mentioned that dissimilarity the stages of intake related to and raised hazard for increasing cancer in the company of the degrees essential to source micro-nutrients such as (zinc, iron and vitamins). Exhibition that the new development in meat product's consumption in Japan and Korea as considered the colorectal cancer rate was not found to be associated to meet product or processed meat intake in a Japanese limited population-based case-control study 782 cases and 793 controls [59]. Furthermore, Koutros et al. [60], the publication in 2007 of the elevated abundant alarms about the cancer risks correlated with red and processed meats, in determining that they are an utterly convincing cause of colorectal cancer (CRC) people ought to be the consuming further-up limit of 500 g of cooked red meat per week, and narrowly averting processed meats (Table 6). The majority of the present indication of hazard from such organic compounds depends upon close association of cancer risk with particular in good health cooked meats, likely to have high carcinogenic heterocyclic amines HCA degrees prospectively examined the association between meat types; meat cookery approaches meat doneness, and meat mutagens and the risk for prostate cancer in 197,017 person-years of follow-up in the Agricultural Health care [61]. Constitution may well be proactively better-quality by industrial manufacturing a healthier food supply as a protective health-care policy and food classification with both positive and negative nutritionary characteristics [12] reported that every year, in the united states medicinal payments for great chronic diseases, including cardiovascular disease, cancer, osteoporosis, diabetes and obesity, spend to exceed 400 billion dollars, lots of these disorders are commonly known to be directly linked to the human diet that is meant that many challenges in health care could be proactively enhanced by producing a healthier food supply as a preventive health-care policy.

Conclusion and Future Prospects

Meat systems have a great potential for delivering important nutrients into the diet. Significant fresh meat color research has been shown in the last few years; Nitroso-hemoglobin prepared to use by-product's beef (blood) extracting hemoglobin from blood processing, and its application prospect were the expounded substitutes of nitrite in meat product from color fixative, color ability and stability, antioxidant, antimicrobials and increase nutrition value and introduced as additives. Prospective research study might make use of health systems and implementations to perform significantly advances in our knowledge of meat color constancy compared to our current, more applied product research.

Acknowledgement

I wish to express my special appreciations and gratitude to my supervisor Prof. Meihu Ma, and Prof. Guofeng Jin for their helpful supervision proper guidance, patience, kindness, attitudes, advices and encouragement to carry out this work. I am very much indebted to china government scholarship. Special thanks are due to all my friends for their assistance and encouragement.

References

1. Benfatto DM (2006) A combined computational and experimental approach in the structural investigations of metalloproteins.
2. Corbett SA, Foty RA (2008) Cell structure, function, and genetics, in surgery. Springer pp: 37-73.
3. Sadava DE (2008) Understanding genetics: DNA, genes, and their real-world applications. Teaching company.
4. Mancini R, Hunt M (2005) Current research in meat color. Meat sci 71: 100-121.
5. Herrero AM (2008) Raman spectroscopy a promising technique for quality assessment of meat and fish: A review. Food Chem 107: 1642-1651.
6. Carpenter CE, Cornforth DP, Whittier D (2001) Consumer preferences for beef color and packaging did not affect eating satisfaction. Meat Sci 57: 359-363.
7. Krystallis A, Chryssohoidis G (2005) Consumers' willingness to pay for organic food: Factors that affect it and variation per organic product type. British Food J 107: 320-343.
8. Bredahl L (2004) Cue utilisation and quality perception with regard to branded beef. Food Qual Pref 15: 65-75.
9. Boles JA, Pegg R (2010) Meat color, Montana State University and Saskatchewan Food Product Innovation, Program University of Saskatchewan.
10. Reilly C (2008) Metal contamination of food: Its significance for food quality and human health. John Wiley & Sons, USA.
11. Mc Clements DJ, Decker EA, Park Y, Weiss J (2009) Structural design principles for delivery of bioactive components in nutraceuticals and functional foods. Critical Rev Food Sci Nutri 49: 577-606.
12. Decker EA, Park Y (2010) Healthier meat products as functional foods. Meat Sci 86: 49-55.
13. Swapna CH, Amit KR, Vinod KM, Bhaskar N (2012) Characteristics and consumer acceptance of healthier meat and meat product formulations - A review. J Food Sci Technol 49: 653-664.
14. Ofori JA, Hsieh HA (2012) The use of blood and derived products as food additives. INTECH Open Access Publisher.
15. Koemseang S (2015) The Improving stability of hemoglobin-based natural food colorant by means encapsulation with calcium alginate.
16. Fu-Man Y, Hong-Zhan Y, Wu-Bian J, Yu-ming C, Jin-Feng L (2011) A review on applications of intensively processed pig blood products in meat products. Meat Res 7: 9-15.
17. Mohan A (2009) Myoglobin redox form stabilization: Role of metabolic intermediates and NIR detection. Kansas State University.
18. Zhang H, Kong B, Jiang Y (2012) Development and application of the substitute of nitrite in the meat product. Packag Food Machine 3: 16.
19. Paredi G, Sentandreu MA, Mozzarelli A, Fadda S, Hollung K, et al. (2013) Muscle and meat: New horizons and applications for proteomics on a farm to fork perspective. J Proteomics 88: 58-82.
20. Han J (2008) The effect of pre-rigor infusion of lamb with kiwifruit juice on meat quality. Lincoln University.
21. Ben-Abdallah M, Marchello JA, Ahmad HA (1999) Effect of freezing and microbial growth on myoglobin derivatives of beef. J Agricultur Food Chem 47: 4093-4099.
22. Lanari M, Zaritzky N (1991) Effect of packaging and frozen storage temperature on beef pigments. Int J Food Sci Technol 26: 629-640.
23. Leygonie C, Britz TJ, Hoffman TC (2011) Oxidative stability of previously frozen ostrich Muscularis iliofibularis packaged under different modified atmospheric conditions. Int J Food Sci Technol 46: 1171-1178.
24. Marriottz NJ, Garcia RA, Kurlan ME, Lee DR (1980) Appearance and microbial quality of thawed retail cuts of beef, pork and lamb. J Food Protect 43: 185-189.
25. Otremba M, Dikeman M, Boyle E (1999) Refrigerated shelf life of vacuum-packaged, previously frozen ostrich meat. Meat Sci 52: 279-283.
26. Calvelo A (1981) Recent studies on meat freezing. Development Meat Sci.
27. Sun Y, Du H, Lan Y, Wang W, Liang Y, et al. (2016) Preparation of haemoglobin (Hb) imprinted polymer by Hb catalyzed eATRP and its application in biosensor. Biosensor Bioelectronic 77: 894-900.
28. Drăghici O, Avram I, Hîrîciu D, Nan M, Toader A, et al. (2013) Study on the beef pigments. Acta Universitatis Cibiniensis, Series E: Food Technol 17: 47-52.
29. Whyte R (2006) Does it look cooked? A review of factors that influence cooked meat color. J Food Sci 71: 31-40.
30. Møller JK, Skibsted LH (2006) Myoglobins: The link between discoloration and lipid oxidation in muscle and meat. Quimica Nova.
31. Shimizu T, Huang D, Yan F, Stranava M, Bartosova M, et al. (2015) Gaseous O_2, NO, and CO in signal transduction: Structure and function relationships of heme-based gas sensors and heme-redox sensors. Chemical reviews 115: 6491-6533.
32. Lei C (2013) Research progress of new type pigment glycosylated nitroso-hemoglobin. J Anhui Agricultur Sci 7: 120.
33. Han K, Bao-hua K, Qian L (2012) Influence of nitrosohaemoglobin with monascus on the colour and quality of red sausages. Sci Technol Food Industry 8: 21.
34. Fei L, Xin S, Rui-Yao H, Xin-Xin Z (2014) The research progress of nitroso-hemoglobin pigment synthesis and its application. China Food Additive 5: 027.
35. Jawaid M (2013) Zabiha or halal meat?
36. Davidson A, Jaine T (2006) The oxford companion to food, segunda. Oxford University Press.
37. Aoma I (2014) Development of frankfurter sausage using spider plant as an alternative filler and extender. University of Nairobi.
38. Ma J (2007) Stick out your tongue. Random House Publishing.
39. Helen HG, Reece P, Mark DS, Julie AC, Audsley N, et al. (2008) Method to screen for the addition of porcine blood-based binding products to foods using liquid chromatography/triple quadrupole mass spectrometry. Rapid Communication Mass Spectrometry 22: 2006-2008.
40. Andago AA (2015) Improving the iron status of children in Kisumu county Kenya using porridge flour enriched with bovine blood. University of Nairobi.
41. Pazos M, Medina I (2013) Oxidants occurring in food systems. Food oxidants and antioxidants: Chemical, biological and functional properties.
42. John L, Cornforth D, Carpenter CE, Sorheim O, Pettee BC, et al. (2004) Comparison of color and thiobarbituric acid values of cooked hamburger patties after storage of fresh beef chubs in modified atmospheres. J Food Sci 69: 608-614.
43. Seyfert M, Hunt MC, Mancini RA, Kropf DH, Stroda SL, et al. (2004) Internal premature browning in cooked steaks from enhanced beef round muscles packaged in high-oxygen and ultra-low oxygen modified atmospheres. J Food Sci 69: 142-146.
44. Suman SP, Faustman C, Lee S, Tang J, Sepe HA, et al. (2004) Effect of muscle source on premature browning in ground beef. Meat Sci 68: 457-461.
45. Suman SP, Faustman C, Lee S, Tang J, Sepe HA, et al. (2005) Effect of erythorbate, storage and high-oxygen packaging on premature browning in ground beef. Meat Sci 69: 363-369.
46. Ramanathan R, Mancini RA, Dady GA, Van Buiten CB (2013) Effects of succinate and pH on cooked beef color. Meat Sci 93: 888-892.
47. Suman SP, Mancini RA, Joseph P, Ramanathan R, Konda MK, et al. (2011) Chitosan inhibits premature browning in ground beef. Meat Sci 88: 512-516.
48. Carlez A, Veciana-Nogues T, Cheftel JC (1995) Changes in colour and myoglobin of minced beef meat due to high pressure processing. LWT-Food Sci Technol 28: 528-538.
49. Ramamoorthi L, Toshkov S, Brewer M (2009) Effects of carbon monoxide-modified atmosphere packaging and irradiation on *E. coli* K12 survival and raw beef quality. Meat Sci 83: 358-365.
50. Jayasingh P, Cornforth DP, Carpenter CE, Whittier D (2001) Evaluation of carbon monoxide treatment in modified atmosphere packaging or vacuum packaging to increase color stability of fresh beef. Meat Sci 59: 317-324.
51. Hunt MC, Mancini RA, Hachmeister KA, Kropf DH, Merriman M, et al. (2004) Carbon monoxide in modified atmosphere packaging affects color, shelf life, and microorganisms of beef steaks and ground beef. J Food Sci 69: 45-52.
52. Li P, Luob H, Baohua KC, Qian LC, Chen C, et al. (2016) Formation of red myoglobin derivatives and inhibition of spoilage bacteria in raw meat batters

by lactic acid bacteria and *Staphylococcus xylosus*. LWT-Food Sci Technol 68: 251-257.

53. Zhang X, Kong B, Xiong YL (2007) Production of cured meat color in nitrite-free Harbin red sausage by *Lactobacillus fermentum* fermentation. Meat Sci 77: 593-598.

54. Ronald BP, Shahidi F, Niall JG, Sheila ID (1996) Elucidation of the chemical structure of preformed cooked cured-meat pigment by electron paramagnetic resonance spectroscopy. J Agri Food Chem 44: 416-421.

55. Gøtterup J, Olsen K, Knöchel S, Tjener K, Stahnke LH, et al. (2007) Relationship between nitrate/nitrite reductase activities in meat associated *staphylococci* and nitrosylmyoglobin formation in a cured meat model system. Int J Food Microbiol 120: 303-310.

56. Jung S, Ghoul M, Lamballerie-Anton M (2003) Influence of high pressure on the color and microbial quality of beef meat. LWT-Food Sci Technol 36: 625-631.

57. Meihu M (2001) Studies on synthesis of HbNO and its application to reducing residual NO_2 in sausages. Food Sci 8: 17.

58. Grillenberger M, Murphy SP, Neumann CG, Bwibo NO, Verhoef H, et al. (2007) The potential of increased meat intake to improve iron nutrition in rural Kenyan school children. Int J Vitamin Nutri Res 77: 193-198.

59. Kimura Y, Kono S, Toyomura K, Nagano J, Mizoue T, et al. (2007) Meat, fish and fat intake in relation to subsite-specific risk of colorectal cancer: The Fukuoka Colorectal Cancer Study. Cancer Sci 98: 590-597.

60. Fund WCR (2007) Research, food, nutrition, physical activity, and the prevention of cancer: A global perspective. Amer Inst for Cancer Res.

61. Koutros S, Amanda JC, Dale PS, Jane AH, Xiaomei M, et al. (2007) Meat and meat mutagens and risk of prostate cancer in the Agricultural Health Study cohort. Cancer Res 67: 858-859.

8

Insecticidal Effects of Natural Preservatives on Insect Pests of Smoked African Mud Catfish, *Clarias gariepinus* (Burchell, 1822)

Ayeloja AA[1]* and George FOA[2]

[1]*Fisheries Technology Department, Federal College of Animal Health and Production Technology, Moor Plantation, Ibadan, Nigeria*
[2]*Department of Aquaculture and Fisheries Management, Federal University of Agriculture, Abeokuta (FUNAAB), Abeokuta, Nigeria*

Abstract

The insecticidal affects some of natural preservatives: ginger, garlic, pepper, garlic-ginger and homogenate of garlic-ginger-pepper on insect pests of smoked cat fish (*Dermestes maculatus* and *Necrobial rufipes*) were investigated. Twenty each of *Necrobial rufipes, Dermestes maculatus* and larvae of these insects were introduced into each smoked fish product with the various natural insecticides, screened with mosquito net after which the mortality were observed and counted at every 3 days interval for a period of 7 weeks (49 days). The mortality of different species of insect pests and larvae were counted, sensory attributes of the products were evaluated using hedonic scale and ranking method. The entomological data as well as sensory data obtained by ranking the products were converted to percentage while the data obtained using the hedonic scale was subjected to kruscal wallis (*H*) analysis. The study revealed ginger had the highest insecticidal effect against *Necrobia rufipes*. However, garlic, pepper and garlic-ginger-pepper are very effective against *Dermestes maculatus*. The Kruskal wallis analysis of the hedonic scale indicated that there was no significant difference (p>0.05) in the taste panelist perception of the odour, flavour and texture of the various smoked catfish products during the period of study. The study established these natural preservatives are specie selective which should be considered when they are intended to be used as insecticide as garlic-ginger-pepper homogenate had more potency against *Dermestes maculatus* (1.7%) while ginger had more potency against *Necrobia rufipes* (0.2% mortality). However, the odour, flavour and of unpreserved smoked catfish was ranked as the best by the taste panelist during the period of this study.

Keywords: Insecticidal effect; Natural preservatives; Insect pests; Smoked catfish

Introduction

Fish is highly perishable especially in the tropics where high temperature and humidity accelerate spoilage of fish immediately after catch as a result of which efforts are primarily directed towards the preservation of fish for human consumption [1,2]. However, poor handling, inadequate processing facilities, lack of ice or storage facilities, remoteness of the fishing villages to urban market centers, poor transportation system and poor distribution channels have drastically reduced fish utilization in the tropics [3]. A number of simple high temperature preservation techniques suitable for small scale preservation in the tropics such as sun drying, frying and smoking have been reviewed [4,5]. However, smoking is the most common method of fish preservation employed in the tropics; it increases fish shelf life, gives the product a desirable taste and odour, it also provides antibacterial and oxidative effects, lowers pH, impacts coloration as well as accelerating the drying process and acting as antagonist to spoilage agents [6]. In spite of the desirable effects of smoke on fish quality, high incidence of insect pest infestation has been reported to cause substantial losses in the nutritive value of fish during storage [5]. Insect pests such as *Dermestes maculatus* and *Necrobia rufipes* are insect pests that destroy smoked fish during storage just as microbes, enzymes and fat oxidation accelerates rates of spoilage [7]. Akpotu and Adebote [8] also stated that dermestis beetle, *Dermestes maculatus*, is a very important pest of smoke-dried fish as it destroys the flesh of stored fish. However, efforts to reduce losses through insect infestation by the use of synthetic insecticides and pesticides have not been fully adopted due to the hazardous nature of these chemicals to health and toxicity at high doses to users [9]. In order to eliminate much of these problems, many researchers are now working on plant based insecticides which are biodegradable, environment friendly, cheap, available and affordable to fish processors thereby justifying the use of plant based insecticide in this study.

Materials and Methods

Two hundred and fifty (250) live catfish (*Clarias gariepinus*) were collected from a concrete pond of PRO square Fish Farm Odo-Onaelewe, Orita Challenge, Ibadan, Oyo state Nigeria with an average weight of the fish was 150 ± 10g, and transported by road to the fish processing unit of the Federal College of Animal Health and Production Technology (FCAH & PT) Moor Plantation Ibadan where they were slaughtered immediately and prepared in the sequence described by Ayeloja [2] which is presented in Figure 1. The fish were smoked using NIOMR (Nigeria Institute for Oceanography and Marine Research) smoking kiln installed in the fish processing unit of FCAH and PT Moor Plantation Ibadan Oyo State Nigeria at temperature of 90°C ± 10°C for 72 hours using charcoal as source of fuel. After smoking, the fish allowed to cool 12 hours in the smoking kiln after which they were stored at ambient temperatures (27°C ± 3°C) in the fish processing laboratory of FCAH and PT. Three replicate of each smoked catfish fish products were kept inside transparent polyethelne terephtalates (plastics) covered with mosquitoes net with rubber bound to prevent insects from escaping. Thereafter, twenty *Necrobial rufipes, Dermestes maculatus* and insect larvae each were introduced into each fish product with the various natural insecticides, screened with mosquito net after which the mortality were observed and counted at every 3 days

***Corresponding author:** Ayeloja AA, Fisheries Technology Department, Federal College of Animal Health and Production Technology, Moor Plantation, PMB 5029 Ibadan, Nigeria, E-mail: ayeloja2@gmail.com

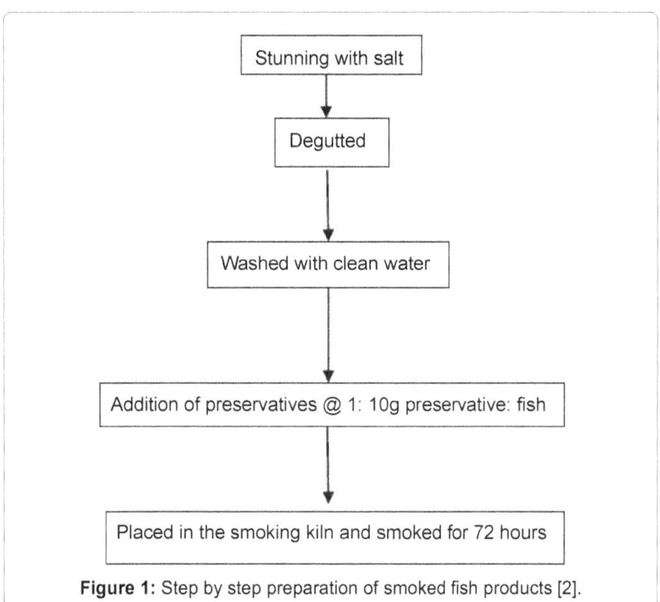

Figure 1: Step by step preparation of smoked fish products [2].

interval for a period of 7 weeks (49 days). The insects and larvae present were identified with the aid of hand lens while forceps were used to extract the insect pests.

Counting and identification of insect pest and larvae

The identification of the insect pests and insect larvae was based on the morphological features of the insects; hand lens was also employed for the proper view in order to clearly identify their features. The mortality of different species of insect pests and larvae were counted and recorded.

Collection of fish samples for organoleptic assessment

The fish samples were collected at every week (7 days) for organoleptic assessment for a period of 7 weeks. The sensory quality attributes that were evaluated was based on 5-point hedonic scale modified. Odour, flavour and texture were examined, the following grades were allotted depending on their qualities: $8 \leq 10$ = Excellent, $6 \leq 8$ = Very good, $4 \leq 6$ = good, $2 \leq 4$ = bad and ≤ 2 = worst. Ranks between 1st to 6th were also allotted to the various products for odour, flavor, texture and colour acceptance by semi trained taste panelists from Federal College of Animal Health and Production Technology Ibadan. The hedonic scale used for the study is presented on Table 1.

Statistical analysis

The entomological data as well as sensory data obtained by ranking the products were converted to percentage while the subjective sensory evaluation data obtained by using the hedonic scale was subjected to kruscal wallis (*H*) analysis.

Results and Discussion

The result of the percentage mortality of the various insect pests in smoked catfish products (Table 1) indicated that garlic, pepper and garlic-ginger-pepper homogenate are very effective insecticide against *Dermestes maculatus* as 25% percent mortality were recorded for the insect (*Dermestes maculatus)* when preserved with these preservatives during 49days of storage while ginger had the lowest potency (5% mortality) against *Dermestes maculatus.* However, ginger had the highest potency against *Necrobia rufipes* where 50% mortality was recorded during the period of the experiment followed by garlic and pepper where 25% mortality were recorded, while garlic-ginger, garlic-ginger-pepper and control had the least percent mortality of *Necrobia rufipes.* All the preservatives added were able to kill insect larvae with the highest larvae mortality observed in pepper preserved smoked catfish (28.5%) followed by ginger preserved smoked catfish (25.9%), followed by garlic preserved smoked catfish (20.9%). The potency of the preservatives against each insect pest during ambient storage as presented in Table 2 indicated that the highest mortality of *Dermestes maculatus* was recorded in garlic-ginger-pepper homogenate (1.7%) while the lowest mortality of *Dermestes maculatus* was recorded in ginger (0.09% mortality). This indicates that garlic-ginger-pepper homogenate is the most effective preservative against *Dermestes maculatus* while ginger is the least effective. However, ginger and garlic are the most effective preservative against *Necrobia rufipes* (0.2% mortality) during the period of storage while garlic-ginger homogenate and garlic-ginger-pepper homogenate are the least effective against *Necrobia rufipes.* All the preservatives were however effective against insect larvae.

The result in Tables 3-5 which shows the panelist perception of the odour, flavor and texture of the differently preserved smoke catfish products during the period of the experiment indicated that the taste panelists observed no significant difference ($p>0.05$) in the odour, flavor and texture of the various smoked catfish products. This implies

Preservatives	*Dermestes maculatus*	*Necrobia rufipes*	Insect larvae
Ginger	5%	50%	25.90%
Garlic	25%	25%	20.90%
Pepper	25%	25%	28.50%
Garlic-Ginger	10%	0%	4.20%
Garlic-Ginger-Pepper	25%	0%	7.30%
Control	10%	0%	13.10%

Percentage presented along the column.

Table 1: The percentage mortality of the various insect pests in smoked catfish products.

Preservatives	*Dermestes maculatus %*	*Necrobia rufipes %*	Larvae %
Ginger	0.09%	0.20%	99.70%
Garlic	0.60%	0.20%	99.30%
Pepper	0.50%	0.10%	99.50%
Garlic-Ginger	1.30%	0%	98.80%
Garlic-Ginger-Pepper	1.70%	0%	98.30%
Control	0.40%	0%	99.60%

Percentage presented across the row.

Table 2: The potency of the preservatives against each insect pest during ambient storage.

Week	Ginger	Garlic	Pepper	Garlic-Ginger	Garlic-Ginger-Pepper	Control	*H*-value	P-value
1	35.79	32.92	35.58	41.29	41.04	32.38	0.814	2.25
2	28.33	36.58	29.21	38.13	42.29	41.46	0.284	6.233
3	36.63	34.88	31.79	31.54	34.21	43.96	0.772	2.527
4	42.92	34.63	25	34.29	41.21	40.96	0.266	6.438
5	41.42	23.13	45.96	31.71	34.29	42.5	0.069	10.226
6	41.88	35.42	31.13	31.46	29.04	42.08	0.515	4.24
7	32.17	31	50.17	32.21	39.04	34.42	0.167	7.814

Table 3: Panellist perception of the odour of differently spiced catfish preserved with different preservatives for seven weeks.

Week	Ginger	Garlic	Pepper	Garlic-Ginger	Garlic-Ginger-Pepper	Control	*H*-value	P-value
1	43.29	28.63	9.33	44.79	36.04	36.92	0.648	6.655
2	34.83	31.46	29.63	44.17	36.58	42.38	0.432	4.866
3	31.33	37.46	46.29	25.71	32.42	45.71	0.074	10.047
4	36.5	39	25.96	33.88	35.33	48.33	0.169	7.77
5	23.92	37.71	47.54	32.13	31.67	46.04	0.035	11.963
6	34.08	36 .58	36.54	25.92	36.42	49.46	0.146	8.197
7	41.67	31.33	35.42	39.33	27.37	43.88	0.318	5.877

Table 4: Panellist perception of the texture of differently spiced catfish preserved with different preservations for seven weeks.

Week	Ginger	Garlic	Pepper	Garlic-Ginger	Garlic-Ginger-Pepper	Control	*H*-value	P-value
1	37.63	31.21	43.79	43.29	31	32.08	0.401	5.127
2	38.58	35.46	29.83	30.71	41.17	43.25	0.502	4.335
3	44.38	23.96	42.96	43.63	26.29	37.79	0.03	12.408
4	40.46	30.33	39.46	33.83	35.25	49.67	0.06	10.614
5	35.58	40.42	41.54	29.96	30.83	40.67	0.578	3.8
6	40.96	23.67	34.58	57.58	30.96	41.25	0.062	10.507
7	32.79	33.88	31.38	40.67	30.88	43.42	0.604	3.63

Table 5: Panelists perception on the flavour of differently preserved smoked catfish product.

Preservatives	1st (%)	2nd (%)	3rd (%)	4th (%)	5th (%)	6th (%)
P1	8.33	16.67	25	25	-	25
P2	16.67	8.33		8.33	33.33	33.33
P3	16.67	41.67	25	8.33	8.33	-
P4		25	8.33	33.33	16.67	8.33
P5	8.33	8.33	41.67	16.67	25	-
P6	50				16.67	33.33

Table 6: Panellist perception of odour for week 1.

Preservatives	1st (%)	2nd (%)	3rd (%)	4th (%)	5th (%)	6th (%)
P1	33.33			25	25	16.67
P2	8.33	16.67	16.67	8.33	41.67	8.33
P3	25	16.67	8.33	16.67	8.33	25
P4	-	50	16.67	8.33	8.33	16.67
P5	16.67	8.33	33.33	16.67	16.67	8.33
P6	16.67	8.33	25	25	-	25

Table 7: Panellist perception of odour for week 2.

Preservatives	1st (%)	2nd (%)	3rd (%)	4th (%)	5th (%)	6th (%)
P1	16.67	25	16.67	16.67	16.67	8.33
P2	-	-	33.33	25	25	16.67
P3	25	25	16.67	33.33	-	-
P4	-	33.33	16.67	8.33	25	16.67
P5	25	8.33	16.67	8.33	25	16.67
P6	33.33	8.33	-	8.33	8.33	41.67

Table 8: Shows the panellist perception of odour for week 3.

that the type of preservative used have no significant effect ($p>0.05$) on the odour, flavor and texture of smoked catfish.

Tables 6-12 show the panelists ranking of the odour of the various smoked fish products during the period of the experiment. At the onset of the experiment, most of the panelists (33%) ranked the odour of smoked C. *gariepinus* preserved with ginger as the best, Fasakin and Aberejo [7] gave similar report in their study of the anti-oxidative and antifungal effects of fresh ginger (*Zingiber officnale*) treatment on the shelf life of hot-smoked catfish (*Clarias garieplinus*, Burchell, 1822). However, taste panel preferred the odour of unpreserved smoked C. *gariepinus* more than other products during 2nd, 3rd and the 6th weeks of storage (50.00%, 33.33% and 41.67% respectively). The result of the study is similar to the report of Ayeloja [1] where it was reported that consumers had more preference for un-spiced smoked catfish, an attitude that was attributed to non-familiarity of the spiced products by the consumers.

Tables 13-19 show the panelists ranking of the texture of the various smoked fish products during the period of the experiment. At the onset of the experiment, most of the panelists (50%) ranked the texture of smoked C. *gariepinus* preserved with ginger as the best, Fasakin and

Preservative	1st (%)	2nd (%)	3rd (%)	4th (%)	5th (%)	6th (%)
P1	8.33	16.67	25	8.33	33.33	8.33
P2	-	8.33	8.33	33.33	16.67	33.33
P3	41.67	25	25	-	-	8.33
P4	16.67	25	-	25	33.33	-
P5	-	8.33	25	16.67	8.33	41.67
P6	33.33	16.67	16.67	16.67	8.33	8.33

Table 9: Shows the panellist perception of odour for week 4.

Preservatives	1st (%)	2nd (%)	3rd (%)	4th (%)	5th (%)	6th (%)
P1	16.67	25	16.67	16.67	16.67	8.33
P2	-	-	33.33	25	25	16.67
P3	25	25	16.67	33.33	-	-
P4	-	33.33	16.67	8.33	25	16.67
P5	25	8.33	16.67	8.33	25	16.67
P6	33.33	8.33	-	8.33	8.33	41.67

Table 10: Shows the panellist perception of odour for week 5.

Preservatives	1st (%)	2nd (%)	3rd (%)	4th (%)	5th (%)	6th (%)
P1	-	33.33	8.33	-	50	8.33
P2	16.67	25	33.33	-	-	16.67
P3	33.33	8.33	41.67	16.67	-	-
P4	-	16.67	-	41.67	8.33	33.33
P5	8.33	-	8.33	33.33	25	25
P6	41.67	16.67		8.33	16.67	16.67

Table 11: Shows the panellist perception of odour for week 6.

Preservatives	1st (%)	2nd (%)	3rd (%)	4th (%)	5th (%)	6th (%)
P1	25	16.67	33.33	-	16.67	8.33
P2	-	8.33	25	33.33	25	8.33
P3	50	-	16.67	8.33	8.33	25
P4	-	8.33	16.67	50	16.67	8.33
P5	8.33	25	8.33	8.33	8.33	33.33
P6	16.67	41.67	-	-	25	16.67

Table 12: Shows the panellist perception of odour for week 7.

Preservatives	1st (%)	2nd (%)	3rd (%)	4th (%)	5th (%)	6th (%)
P1	16.67	-	25	-	25	33.33
P2	-	33.33	8.33	8.33	16.67	33.33
P3	25	-	8.33	33.33	33.33	-
P4	8.33	25	41.67	16.33	-	8.33
P5	-	25	8.33	41.67	16.67	8.33
P6	50	25	-	-	8.33	16.67

Table 13: Shows the panellist perception of texture Week 1.

Preservatives	1st (%)	2nd (%)	3rd (%)	4th (%)	5th (%)	6th (%)
P1	50	25	8.33	8.33	8.33	-
P2	-	8.33	33.33	16.67	33.33	8.33
P3	-	25	25	16.67	16.67	16.67
P4	33.33	8.33	16.67	8.33	8.33	25
P5	8.33	16.67	8.33	33.33	8.33	25
P6	8.33	16.67	8.33	16.67	25	25

Table 14: Shows the panellist perception of texture for week 2.

Preservatives	1st (%)	2nd (%)	3rd (%)	4th (%)	5th (%)	6th (%)
P1	25	8.33	16.67	16.67	8.33	25
P2	8.33	16.67	16.67	16.67	25	16.67
P3	16.67	33.33	8.33	16.67	16.67	8.33
P4	-	16.67	33.33	33.33	8.33	8.33
P5	-	8.33	8.33	16.67	33.33	33.33
P6	41.67	16.67	16.67	-	8.33	8.33

Table 15: Shows the panellist perception of texture for week 3.

Preservatives	1st (%)	2nd (%)	3rd (%)	4th (%)	5th (%)	6th (%)
P1	16.67	8.33	16.67	16.67	33.33	8.33
P2	16.67	33.33	16.67	16.67	-	16.67
P3	8.33	8.33	25	33.33	16.67	8.33
P4	16.67	8.33	8.33	25	16.67	33.33
P5	16.67	25	16.67	-	25	16.67
P6	33.33	16.67	16.67	8.33	8.33	16.67

Table 16: Shows the panellist perception for texture week 4.

Preservatives	1st (%)	2nd (%)	3rd (%)	4th (%)	5th (%)	6th (%)
P1	-	16.67	16.67	8.33	16.67	41.67
P2	33.33	8.33	33.33	16.67	8.33	-
P3	16.67	33.33	16.67	33.33	-	-
P4	-	8.33	25	8.33	16.67	41.67
P5	8.33	8.33	-	25	33.33	8.33
P6	33.33	16.67	8.33	8.33	25	8.33

Table 17: Shows the panellist perception for texture week 5.

Preservatives	1st (%)	2nd (%)	3rd (%)	4th (%)	5th (%)	6th (%)
P1	33.33	16.67	33.33	8.33	-	8.33
P2	16.67	25	25	16.67	8.33	8.33
P3	8.33	33.33	8.33	41.67	8.33	-
P4	-	-	16.67	33.33	41.67	8.33
P5	8.33	-	8.33	-	33.33	50
P6	33.33	25	8.33	-	8.33	25

Table 18: Shows the panellist perception for texture week 6.

Preservatives	1st (%)	2nd (%)	3rd (%)	4th (%)	5th (%)	6th (%)
P1	16.67	8.33	25	8.33	25	33.33
P2	8.33	25	8.33	16.67	16.67	25
P3	41.67	16.67	16.67	25	-	-
P4	8.33	16.67	25	8.33	16.67	25
P5	8.33	8.33	8.33	33.33	25	16.67
P6	16.67	25	33.33	8.33	16.67	-

Table 19: Shows the panellist perception for texture week 7.

Aberejo [7] gave similar report in their study of the anti-oxidative and antifungal effects of fresh ginger (*Zingiber officnale*) treatment on the shelf life of hot-smoked catfish. Similar result as observed in the 1st week was also observed at the 6th week of the storage. However, taste panel preferred the texture of unpreserved smoked C. *gariepinus* more than other products during the period of this study as it was ranked to be the best by most of the taste panel during the 2nd, 3rd, 4th, 5th and the 6th weeks of storage (50.00%, 41.67%, 33.33%, 33.33% and 33.33% respectively). The result of the study is similar to the report of Ayeloja [1] where it was reported that consumers had more preference for un-spiced smoked catfish, an attitude that was attributed to non-familiarity of the spiced products by the consumers.

Tables 20-26 show the panelists ranking of the flavour of the various smoked fish products during the period of the experiment. The result of this study indicated that pepper preserved smoked *C. gariepinus* was best preferred by the taste panel during the 4th and 7th week of storage (50.00% and 41.67% respectively). However, similar trend as observed for the odour and texture of these products were observed by the taste

Preservatives	1st (%)	2nd (%)	3rd (%)	4th (%)	5th (%)	6th (%)
P1	33.33	16.67	25	25	-	-
P2	16.67	8.33	8.33	16.67	41.67	8.33
P3	16.67	25	25	8.33	8.33	16.67
P4	-	16.67	16.67	33.33	16.67	16.67
P5	8.33	16.67	8.33	16.67	25	25
P6	33.33	16.67	16.67	8.33	8.33	16.67

Table 20: Shows the panellist perception for flavour week 1.

Preservatives	1st (%)	2nd (%)	3rd (%)	4th (%)	5th (%)	6th (%)
P1	25	25	41.67	-	8.33	-
P2	-	16.67	25	16.67	25	16.67
P3	25	8.33	8.33	33.33	8.33	16.67
P4	8.33	16.67	-	16.67	25	33.33
P5	8.33	16.67	-	33.33	16.67	25
P6	33.33	16.67	25	-	16.67	8.33

Table 21: Shows the panellist perception for flavour week 2.

Preservatives	1st (%)	2nd (%)	3rd (%)	4th (%)	5th (%)	6th (%)
P1	33.33	25	16.67	8.33	8.33	8.33
P2	8.33	8.33	41.67	8.33	16.67	16.67
P3	8.33	16.67	25	33.33	8.33	8.33
P4	-	8.33	16.67	41.67	8.33	25
P5	8.33	8.33	-	33.33	33.33	16.67
P6	41.67	33.33	-	-	-	25

Table 22: Shows the panellist perception for flavour week 3.

Preservatives	1st (%)	2nd (%)	3rd (%)	4th (%)	5th (%)	6th (%)
P1	16.67	25	25	-	8.33	25
P2	-	8.33	25	41.67	25	-
P3	16.67	16.67	8.33	25	16.67	16.67
P4	16.67	16.67	16.67	16.67	33.33	-
P5	-	8.33	16.67	16.67	8.33	50
P6	50	25	8.33		8.33	8.33

Table 23: Shows the panellist perception for flavour week 4.

Preservatives	1st (%)	2nd (%)	3rd (%)	4th (%)	5th (%)	6th (%)
P1	-	25	33.33	16.67	16.67	8.33
P2	8.33	33.33	25	8.33	-	25
P3	41.67	8.33	8.33	16.67	16.67	8.33
P4	-	8.33	8.33	25	50	8.33
P5	16.67	-	25	8.33	8.33	41.67
P6	33.33	16.67	-	33.33	8.33	8.33

Table 24: Shows the panellist perception for flavour week 5.

Preservatives	1st (%)	2nd (%)	3rd (%)	4th (%)	5th (%)	6th (%)
P1	16.67	25	25	-	8.33	25
P2	-	8.33	25	41.67	25	-
P3	16.67	16.67	8.33	25	16.67	16.67
P4	16.67	16.67	16.67	16.67	33.33	-
P5	-	8.33	16.67	16.67	8.33	50
P6	50	25	8.33	-	8.33	8.33

Table 25: Shows the panellist perception for flavour week 6.

Preservatives	1st (%)	2nd (%)	3rd (%)	4th (%)	5th (%)	6th (%)
P1	33.33	25	16.67	8.33	8.33	8.33
P2	8.33	8.33	41.67	8.33	16.67	16.67
P3	8.33	16.67	25	33.33	8.33	8.33
P4	-	8.33	16.67	41.67	8.33	25
P5	8.33	8.33	-	33.33	33.33	16.67
P6	41.67	33.33	-	-	-	25

Table 26: Shows the panellist perception for flavour week 7.

panel for the flavour as the result indicated that the taste panel preferred the flavour of unpreserved smoked *C. gariepinus* more than other products during 1st, 2nd, 3rd, 5th and the 6th weeks of storage (41.67%, 41.67%, 33.33%, 50.00%, and 33.33% respectively). This is similar to the report of Ayeloja [1] where it was reported that consumers had more preference for un-spiced smoked catfish, an attitude that was attributed to non-familiarity of the spiced products by the consumers.

Conclusion

This study revealed that all the preservatives have high potency against insect larvae, while ginger had the highest insecticidal effect against *Necrobia rufipes.* However, garlic, pepper and garlic-ginger-pepper are very active against *Dermestes maculatus.* The Kruskal wallis analysis of the hedonic scale indicated that there was no significant difference ($p>0.05$) in the taste panelist perception of the odour, flavour and texture of the various smoked catfish products during the period of study. The preservatives are specie selective; this should be considered when they are intended to be used as insecticide as garlic-ginger-pepper homogenate had more potency against *Dermestes maculatus* (1.7%) while ginger had more potency against *Necrobia rufipes* (0.2% mortality). However, the odour, flavour and of unpreserved smoked catfish was ranked as the best by the taste panelist during the period of this study.

References

1. Ayeloja AA, George FOA, Awobifa OM, Sodeeq AE, Jimoh WA, et al. (2015) Effect of insect infestation on the economic value of smoked fish sold in selected markets within Oyo State, South West Nigeria. Delta State University, Asaba Campus, Delta State, Nigeria.
2. Ayeloja AA, George FOA, Akinyemi AA, Atanda OO (2015) Proximate and mineral composition of spiced, smoked catfish *Clarias gariepinus* (Burchell, 1822). J Agri Sci Environ 15: 68-74.
3. Fasakin EA (2003) Use of some plant oil extracts as surface protectant against storage insect pest, *Dermestes maculatus Degeer*, on smoked fish. UNAAB Journal 3 :1-6.
4. Ames GR (1992) Traditional and modern post-harvest technologies for increased food and supply from inland fisheries in Africa. Proceedings of the symposium of post-harvest fish technology, Cairo, Egypt. FAA-CIFA Technical Paper 19: 11-17.
5. Ikenweiwe NB, Bolaji BO, Bola GA (2010) Fabrication and performance assessment of a locally developed fish smoking kiln. Ozean J Appl Sci 3: 1-12.
6. Akintola AJ, Oyegoke OO, Lawal OS (2013) Medical implications of bio-deteriorating agents in stored fish samples in Nigeria. Africa J Microbiol Res 7: 3429-3434.
7. Fasakin EA, Aberejo O (2000) Effect of some pulverized plant materials on the developmental stages of fish beetle, *Dermestes maculatus Degeer* in smoked catfish (*Clarias gariepinus*) during storage. Biores Technol 85: 173-177.
8. Akpotu JO, Adebote DA (2013) Repellency effect of five plants extracts against the larvae of *Dermestes maculatus* larvae on smoke-dried *Clarias gariepinus* fish. Res J Chem Environ Sci 1: 1-4.
9. Balogun AM (1992) Fish handling and quality control: Aquaculture development in Africa, training and references manual for aquaculture extensionists. Commonwealth Secretariat, London.

Nutritional Evaluation of the Mineral Composition of Chocolate Bars: Total Contents vs. Bioaccessible Fractions

Peixoto RRA[1]*, Villa JEL[1], Silva FF[2] and Cadore S[1]

[1]*Institute of Chemistry, University of Campinas, Campinas, SP, Brazil*

[2]*Agilent Technologies Brasil, Av. Dr. Marcos Penteado Ulhoa, Barueri, SP, Brazil*

Abstract

The total contents and the bioaccessibility of essential (Co, Cu, Fe, Mn, Zn, Cr and Se) and non-essential elements (Al, Ba, Sr, As, Ni and V) were estimated in dark, milk and white chocolates. The analytical determinations were made by Inductively Coupled Plasma Mass Spectrometry (ICP-MS) after microwave-assisted acid mineralization. The bioaccessibility was determined for each type of chocolate by applying an *in vitro* digestion method and the results were in the range of 45-63% for Co, 26-89% for Cu, 4-67% for Fe, 29-89% for Mn, 8-89% for Zn, 28-36% for Cr, 9-89% for Se, 3-5% for Al, 13-55% for Ba, 25-86% for Sr, 42-62% for As and 50-63% for Ni. Although the highest total elemental concentrations were found in dark chocolate, white chocolate exhibited bioaccessible fractions significantly higher (55-89%) when compared to milk (5-63%) and dark (3-50%) chocolate. The results also pointed out that the consumption of milk and dark chocolate contributes to a high daily intake of Cr for children.

Keywords: Bioaccessibility; Chocolate; ICP-MS; Trace elements

Abbreviations: ANVISA: Agência Nacional de Vigilância Sanitária; AOAC: Association of Official Agricultural Chemists; ICCO: International Cocoa Organization

Introduction

Chocolate bars are the most consumed cocoa-derived product worldwide. The regular and moderate consumption of chocolate bars and other cocoa derivatives can benefit health due to their antioxidant activity, high carbohydrate content and presence of some essential metallic elements such as Ca, Cu, Fe, K, Mg, Mn and Zn [1-3]. Despite their positive health effects, the pleasant palatability of chocolate products is probably the main reason for their popularity [4,5]. As a result, between the years 2013 and 2014, the cocoa beans world production was estimated at 4370 thousand tons [6], highlighting its wide consumption and the global importance of the cocoa and chocolate sector. Cocoa, the seeds of the *Theobroma cacao* tree, is the primary basis of chocolate [1,7]. For the production of dark and milk chocolate, cocoa mass or liquor is used, and the tablets are normally classified according to their cocoa content. According to Brazilian legislation, to be considered as a chocolate a food product must contain at least 25% of cocoa liquor or 20% of cocoa butter [8]. Dark chocolates present the highest cocoa contents, varying between 50 and 70%, although some products with 80% or higher cocoa contents can also be found in the market, whereas in milk chocolate bars the cocoa content is normally between 25 and 50%. For the production of white chocolate, cocoa butter is the main ingredient and usually no information about the cocoa content is provided. The chemical composition of chocolate products has already been investigated in some papers [9-11]. Most of studies concerning the presence of elements in cocoa and derivatives are devoted to the determination of their total concentrations, an approach usually adopted to evaluate the presence of essential and non-essential elements in foodstuffs. In a recent paper, Villa et al. [11] determined the total contents of Cd and Pb in samples of chocolate bars, showing that the concentrations of these elements increase linearly with the cocoa content and, as expected, dark chocolates presented the highest concentrations of Cd and Pb, suggesting that the principal source of Cd and Pb in chocolate is the cocoa used in their production, which is also the main source of essential elements in chocolate products [3]. In spite of the important data raised in studies of total elemental concentrations, a more accurate evaluation of the benefits and risks associated with chocolate consumption can be done by evaluating the bioaccessible and bioavailable fractions. The oral bioaccessibility of a food component can be described as the fraction that is released from the food matrix and is soluble in the gastrointestinal tract, being available for intestinal absorption [12,13]. The estimate of the bioaccessible fraction can be used as indicative of maximum bioavailability, the fraction of the ingested compound that could be absorbed by the human organism. Taking this into account, Mounicou et al. [14] studied the bioaccessibility of Cd and Pb from cocoa. According to their results, Cd presented bioaccessibility in the range of 10-50%, and bioaccessible fractions below 10% were obtained for Pb in the main raw material used for the production of chocolate. Bioaccessibility studies have also been conducted for chocolate drink powder [15], in which essential elements presented higher bioaccessibility than potentially toxic elements at trace levels. For chocolate bars, despite their significant consumption, there is still no data available about the bioaccessibility of the elements. Thus, in order to provide a better evaluation of the presence of elements in chocolate bars, the aim of this work was to estimate the bioaccessible fractions of Al, Ba, Cu, Fe, Mn, Sr, Zn, As, Co, Cr, Ni, Se and V in dark, milk and white chocolates.

Materials and Methods

Reagents, samples and certified materials

Deionized water, obtained using a Milli-Q system (Millipore,

***Corresponding author:** Peixoto RRA, Institute of Chemistry, University of Campinas, P.O. Box 6154, 13083-970, Campinas, SP, Brazil
E-mail: rafaella_peixoto@hotmail.com

Bedford, MA), with a conductivity of 18.2 mΩ cm, was used throughout. Nitric acid (65%, w/w) and hydrogen peroxide (30%, w/w) were purchased from Merck (Darmstadt, Germany). Standard monoelemental solutions (1000 mg/L) of Al, Ba, Cu, Fe, Mn, Sr, Zn, As, Co, Cr, Ni, Se and V from Sigma Aldrich (TraceCERT®, Fluka Analytica, St. Louis, USA) were used in the preparation of calibration standards. A 100 mg/L solution of Ge (Specsol®, São Paulo, Brazil) was used as internal standard. Pepsin (enzyme activity 944 U/mg of protein), porcine pancreatin (activity equivalent to 4XUS Pharmacopeia specifications/mg pancreatin) and porcine bile extract (glycine and taurine conjugates of hyodeoxycholic and other bile salts), all acquired from Sigma Aldrich and were used in the digestion simulations. Three chocolate bars from the same brand, labeled by the manufacturer as white, milk and dark (55% of solid cocoa content) were analyzed in this study. They were purchased in a local market of Campinas, State of São Paulo - Brazil in April 2014. A certified reference material of Baking Chocolate (NIST 2384) from the National Institute of Standards and Technology - NIST (Gaithersburg, USA) was used to verify the accuracy of the analytical method. All the glassware and plastic materials were decontaminated with a 10% (v/v) HNO_3 solution for 24 h and rinsed with deionized water prior to use.

Instrumentation

An inductively coupled plasma mass spectrometer (ICP-MS) instrument, model 7700x (Agilent Technologies, Hachioji, Japan), equipped with octapole collision/reaction system third generation (ORS^3) using He as collision gas, was used for the determination of all elements. The operational conditions are shown in Table 1. The analyte solutions used for external calibration were prepared by doing appropriate dilution of standard stock solutions and the internal standard $^{72}Ge^+$ (12.5 µg/L) was added *on line* to compensate the possibility of instrument drift and matrix effects. Liquid argon and helium gas (99.996%), purchased from White Martins (São Paulo, Brazil), were used for all measurements. Sample treatment was performed using a microwave digestion system (ETHOS1, Milestone, Sorisole, Italy) equipped with closed polytetrafluoroethylene vessels and sensors for temperature and pressure control. A pH-meter (model Q400RS), a water bath with linear shaking and temperature control (Dubnoff, model Q226M) and a centrifuge (model Q222TM), all from Quimis® (Diadema-SP, Brazil), were used in the *in vitro* digestion method.

Procedure for *in vitro* digestion method

The *in vitro* digestion method employed was based on the method used by Laparra et al. [16] and adapted for chocolate samples. Briefly, 4.0 g of the chocolate sample were weighed and 50 g of water were added. Next, the pH was adjusted to 2.0 with 6 mol/L HCl and freshly prepared pepsin solution (0.1 g/mL in 0.1 mol/L HCl) was added to provide a proportion of 0.01 g of pepsin/50 g of solution. The sample was incubated in a water bath with shaking, at 37°C, for 2 h. Prior to the intestinal digestion step, the pH of the gastric digests was set to pH 6.5 by adding 1 mol/L NH_4OH. Then, a pancreatin-bile extract mixture was added to provide a proportion of 2.5×10^{-3} g pancreatin and 1.5×10^{-2} g bile/50 g of solution, then incubation at 37°C continued for an additional 2 h. The pH of digests was then adjusted to 7.2 and centrifuged at 4000 rpm for 30 min. In order to achieve a better separation of the soluble and insoluble fraction, the samples were filtered through Whatman filter paper (No. 42) and the soluble fractions were submitted to analysis. Blank digestion assays were run in parallel with the samples. The bioaccessibility of elements is defined as the proportion of the total element content that is available for absorption, and it was calculated as:

$$\text{Bioaccessibility (\%)} = ([\text{Element}]_{\text{soluble fraction}}/[\text{Element}]_{\text{total}}) \times 100 \qquad (1)$$

Microwave-assisted acid digestion

Chocolate samples (500 mg) or the soluble content obtained in the *in vitro* digestion (5.0 mL) were put into the microwave vessels and 2.0 mL of deionized water, 2.0 mL of H_2O_2 (30%, w/w) and 3.0 mL of 14 mol/L HNO_3 were added. The vessels were then placed inside the microwave oven, and the heating program was performed over six steps: (1) heat from room temperature to 80°C in 3 min, (2) hold for 3 min at 80°C, (3) heat from 80°C to 140°C in 3 min, (4) hold for 4 min at 140°C, (5) heat from 140°C to 200°C in 7 min, and (6) hold for 15 min at 200°C. The resulting solutions were left to cool, transferred to polyethylene flasks and diluted to 25.0 mL with deionized water for the determination of the total concentration and to 13.0 mL to determine the bioaccessible fraction of the elements.

Method validation and statistical analysis

The method validation was performed by evaluating the accuracy,

Parameters		
Radio frequency power	1550 W	
Sample depth	10.0 mm	
Plasma flow rate	15 L/min	
Nebulizer gas flow rate	1.09 L/min	
Nebulizer pump	0.10 rps	
Spray chamber	Scott (double pass) at 2°C	
Interface	Platinum cones	
Sampling cone	1 mm	
Skimmer	0.4 mm	
Resolution	1.5 u	
Selected isotopes for analytes	$^{27}Al^+$, $^{51}V^+$, $^{52}Cr^+$, $^{55}Mn^+$, $^{56}Fe^+$, $^{59}Co^+$, $^{60}Ni^+$, $^{63}Cu^+$, $^{68}Zn^+$, $^{75}As^+$, $^{80}Se^+$, $^{88}Sr^+$ and $^{137}Ba^+$	
On line internal standard	^{72}Ge (250 µg/L diluted *on line* 20 x)	
ORS^3*	Collision	
Gas flow	He Mode	HEHe Mode
	4.3 mL/min	10 mL/min
Energy discrimination	3.0 V	7.0 V

* ORS^3: Octapole collision/reaction system third generation

Table 1: Operating conditions used in the ICP-MS.

precision, LOD, LOQ and the correlation coefficients. A certified reference material (Baking Chocolate, NIST 2384) was used to verify the accuracy for the determination of Fe, Cu, Mn and Zn. For Al, Ba, Sr, As, Co, Cr, Ni, Se and V, analyte addition and recovery experiments were performed. Analytical blanks were prepared following the same procedure used for sample preparation and the LOD and LOQ were calculated as 3 and 10 times the standard deviations obtained in the analysis of the analytical blanks. The analyses for the determination of the total concentrations were performed in triplicate and the *in vitro* digestion assays in quintuplicate. All the results are expressed as the mean of independent experiments ± standard deviation. The differences obtained for the bioaccessible fractions were analysed by using one-way ANOVA with Tukey HSD post-hoc tests (Statistica 7) and differences were considered as significant for $p < 0.05$.

Results and Discussion

Analytical determinations

The determination of elements in the fraction of food soluble in the gastrointestinal tract (bioaccessible fraction) almost always requires the use of a sensitive technique for detection, considering that the concentrations to be determined in these types of assays are often lower than the total concentrations found in the food matrix. Besides the low concentrations, a lot of attention has to be paid to possible interferences caused by the organic compounds of the food matrix, by the enzymes added during the digestion simulation, or by the salts of easily ionized elements that are commonly added in high concentrations during the digestion described for some protocols [12,17]. In the protocol used in this work, there was no addition of salts of easily ionized elements and the main concern remains the elimination of the organic matter of the food matrix and bioactive compounds. So, all the samples were treated by acid mineralization assisted by microwave radiation prior to the analysis made by ICP-MS. The operating conditions to be used in the ICP-MS also has to be carefully evaluated, especially for some elements such as Al, As and Co, which are mono-isobaric and whose analytical determination might be severely hampered by mass interferences, especially from polyatomic and molecular species [18]. Thus, the use of a strategy to minimize possible interferences is frequently mandatory for obtaining reliable results. In this case, the use of an octapole collision/reaction system third generation (ORS^3), using He, as collision gas, and kinetic energy discrimination, was the strategy adopted to reduce spectral interferences. The ORS^3 was used in two different tunes mainly defined by He flow rate: 4.3 (He) and 10 mL/min (High Energy Helium mode, HEHe). The use of He gas in the highest flow rate (HEHe) was necessary for the determination of most elements (As, Cr, Cu, Fe, Mn, Ni, Se and Zn), whereas He at a flow rate of 4.3 mL/min was used for the determination of Al, Ba, Co, Sr and V. The choice of the best He flow rate was made based on the recovery and the repeatability obtained in the *in vitro* digestion assay, considering that the digested portion of chocolate is the most complex matrix to be analysed in this study. In the selected conditions, method validation was performed and the analytical features are presented in Table 2. The results obtained for the CRM were in agreement with the certified levels ($p > 0.05$) for Cu, Fe, Mn and Zn and recoveries in the range 92-122% were obtained for all elements in the addition and recovery experiments, which is in agreement with AOAC recommendations [19]. Analytical curves presented correlation coefficients of $r > 0.999$ or higher and the precision of the measurements were better than 8%. Therefore, considering the low levels to be determined and the complexity of the matrix to be analysed in digestion studies, the analytical method was considered suitable to be applied to the determination of the bioaccessible fractions of metallic elements in chocolate.

Total elemental concentrations in chocolates

The results obtained for total concentrations of Co, Cu, Fe, Mn, Zn, Cr, Se, Al, Ba, Sr, As, Ni and V in the three types of chocolate are shown in Table 3. Concentration ranges of Al, As, Cr, Cu, Fe, Mn, Ni, Co and Zn were consistent with measurements reported in previous papers [20-22] and the contents of Ba, Se, Sr and V were determined by the first time in chocolate bars. As can be seen from the results, there were significant differences in total elemental concentrations between white, milk and dark chocolate, with the latter presenting the highest concentrations of all elements studied, which is probably due to the capability of the cocoa plant to absorb minerals from the soil, accumulating them in the cacao seeds, which in turn are the main ingredient of dark chocolates. The most abundant essential and non-essential elements determined in chocolate bars were Fe and Ni, respectively. Concentrations of As in the three types of chocolate were below the maximum level allowed in Brazil (0.2 mg/kg for cocoa products with less than 40% of cocoa and 0.4 mg/kg for products with a cocoa content higher than 40%) [23]. With regard to other elements studied, there are no maximum admissible levels specifically established for chocolate bars.

Element	Addition experiments		CRM baking chocolate (mg/kg)		LOD	LOQ
	Spiked (µg/L)	Recovery (%)	Certified	Found	(µg/kg)	(µg/kg)
Al	500	92	n. r.*	-	300	1000
Ba	500	106	n. r.*	-	50	170
Cu	500	100	23.2 ± 1.2	22.8 ± 0.2	20	70
Fe	500	114	132 ± 11	131 ± 1	160	530
Mn	500	107	20.3 ± 1.3	19.9 ± 0.2	11	37
Sr	20	96	n. r.*	-	4	15
Zn	500	103	36.6 ± 1.7	36.1 ± 0.3	250	850
As	20	116	n. r.*	-	0.2	0.6
Co	20	99	n. r.*	-	1.5	5
Cr	500	103	n. r.*	-	27	90
Ni	20	102	n. r.*	-	0.9	3.2
Se	20	122	n. r.*	-	0.5	1.9
V	20	106	n. r.*	-	1.3	4.3

* n. r.: not reported

Table 2: Analytical features obtained in the method validation.

As determined in this study, differences in the chocolate matrix can lead to significantly different elemental concentrations. However, in studies in which only the total concentrations are considered, the effects of the matrix on the nutritional value of food cannot be assessed, nor can risk assessment analysis be made. In this case, the application of a simple *in vitro* digestion method could provide a more reliable approach for the evaluation of the presence of essential and potentially toxic elements in food.

Evaluation of bioaccessible fractions

The bioaccessible fractions (%) determined for essential and non-essential elements in white, milk and dark chocolate [Calculated by Eq. (1)] are presented in Table 4 and Figure 1. The bioaccessibility of some elements (Al, Co, Cr, Ni in white chocolate and V in milk and dark chocolate) are not represented since the concentrations found in the bioaccessible fractions were below the LOD of the analytical method. It was possible to estimate the bioaccessibility of V but only in white chocolate, for which a bioaccessibility of 89% was obtained. The results presented show that the solubility of the elements in the simulated gastric and intestinal juices varies depending on the type of chocolate being analyzed, except for As, for which the bioaccessible fractions obtained were similar in the three types of chocolate ($p < 0.05$). Although the highest total concentrations of all elements were found in dark chocolate, this type of chocolate presented the lowest bioaccessible fractions for the majority of elements, including essential elements (Cu, Fe, Mn, Se and Zn) and some contaminants (As, Ba, Ni and Sr), for which bioaccessible fractions lower than 50% were obtained. An opposite behavior was observed for white chocolate, which presented the lowest total concentrations and the highest bioaccessible fractions. The bioaccessibility of metallic elements in white chocolate was shown to be significantly higher, considering that the bioaccessible fractions of 8 of 12 elements studied exceeded values of 50%. Metallic elements in milk chocolate presented an intermediate bioaccessibility between dark and white chocolates. In this way, in general, bioaccessible fractions for most elements were obtained in the ascending order: dark < milk < white chocolate. The bioaccessibility of metallic elements from foodstuffs depends on several factors such as the type of food, mainly due to the presence of different food components that would interact in different ways with the elements; the gastrointestinal conditions and physical and chemical characteristics of the element under study. The chemical form of the element and its solubility in the aqueous environment of the human gastrointestinal tract, particularly, play important roles in bioaccessibility. In the case of chocolate, the main source of metallic elements is cocoa, and a possible explanation for the differences observed in the bioaccessible fractions, is the form in which cocoa is added to the food matrix. Cocoa butter and cocoa mass or liquor could present elements in different chemical forms, considering that the production process may change the chemical form of some metallic elements. The results suggest that the metallic elements in white chocolate are more easily released from the food matrix and present higher solubility in the human gastrointestinal tract when compared with milk and dark chocolate. Considering the influence of the element bioaccessibility in the nutritional value of food, the contribution of chocolate consumption to the Recommended Daily Intake (RDI) of essential elements (Cr, Cu, Fe, Mn, Se and Zn) were calculated for the three types of chocolate, using the total contents and their bioaccessible fractions, for children of 7-10 years [24], which are important consumers of this type of food. The results obtained are shown in Table 5. Cobalt is not represented since there is no RDI for this element in the Brazilian legislation. In general, a reduction in the contribution of chocolate consumption was observed when the bioaccessible fractions were taken into account. Despite the high total concentrations found in dark chocolate, the consumption of milk chocolate contributes most to the daily intake of Cr, Fe, and Zn, whereas for Se a higher contribution was observed when white chocolate is consumed, considering the bioaccessibility of the elements. In this way, milk and white chocolate contributes more to the ingestion of Cr, Fe, Se and Zn, despite their lower total concentrations. Additionally, the results also

Element	White	Milk	Dark
Al	<LOQ	25.35 ± 1.40	26.7 ± 2.86
Ba	0.46 ± 0.06	1.76 ± 0.13	5.63 ± 0.34
Cu	0.17 ± 0.01	3.30 ± 0.02	12.18 ± 0.20
Fe	1.60 ± 0.23	56.44 ± 0.63	77.06 ± 1.81
Mn	0.23 ± 0.01	4.03 ± 0.04	13.36 ± 0.18
Sr	1.10 ± 0.07	2.54 ± 0.07	6.04 ± 0.34
Zn	3.89 ± 0.19	10.65 ± 0.04	23.06 ± 0.34
As*	6.52 ± 0.29	12.37 ± 0.56	14.06 ± 1.11
Co*	<LOQ	87.50 ± 1.72	358.68 ± 16.37
Cr*	<LOQ	456.73 ± 12.45	515.59 ± 23.65
Ni*	7.6 ± 1.1	195.97 ± 5.36	815.43 ± 15.37
Se*	17.62 ± 2.31	40.46 ± 1.67	70.77 ± 1.48
V*	9.70 ± 0.42	50.46 ± 2.77	76.37 ± 6.44

* Total content in µg/kg

Table 3: Total content (mg/kg ± s, n=3) of essential and non-essential elements in white, milk and dark chocolate.

Element	White	Milk	Dark
Al	n.d.*	5.3 ± 0.2	2.7 ± 1.1
Ba	54.8 ± 4.2	34.7 ± 2.5	13.3 ± 1.2
Cu	88.8 ± 9.1	35.7 ± 2.4	25.5 ± 0.7
Fe	66.7 ± 8.8	8.3 ± 1.3	3.5 ± 0.4
Mn	88.9 ± 6.8	57.6 ± 4.1	28.6 ± 1.9
Sr	86.0 ± 8.7	56.9 ± 3.8	25.0 ± 1.5
Zn	88.7 ± 7.4	37.0 ± 2.7	7.9 ± 1.1
As	61.6 ± 15.1	61.5 ± 27.4	41.9 ± 8.1
Co	n.d.*	62.9 ± 4.0	45.3 ± 3.4
Cr	n.d.*	35.9 ± 7.5	27.7 ± 5.2
Ni	n.d.*	62.8 ± 7.7	50.1 ± 1.8
Se	89.1 ± 8.1	21.8 ± 2.8	8.7 ± 0.7
V	89.3 ± 5.0	n.d.*	n.d.*

* n.d.: not determined

Table 4: Bioaccessible fractions (%) of essential and non-essential elements in three types of chocolate bars (n=5).

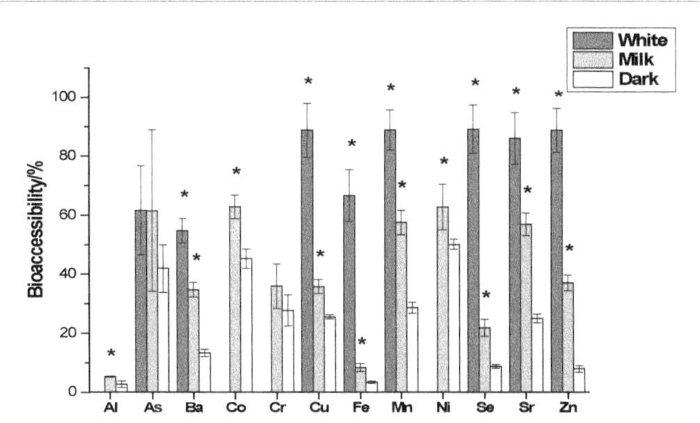

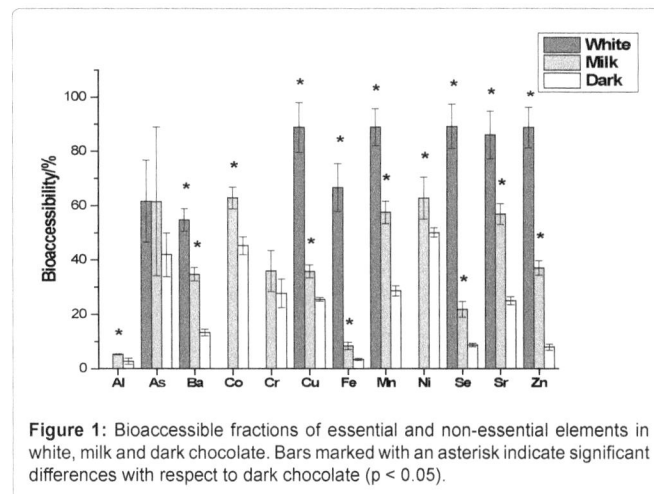

Figure 1: Bioaccessible fractions of essential and non-essential elements in white, milk and dark chocolate. Bars marked with an asterisk indicate significant differences with respect to dark chocolate ($p < 0.05$).

Element	Type of chocolate	Total content	Bioaccessible fraction
Cr	White	n.d.	n.d
	Milk	122	73
	Dark	138	64
Cu	White	1.5	1.4
	Milk	30	11
	Dark	111	29
Fe	White	0.7	0.5
	Milk	25	2.0
	Dark	34	1.4
Mn	White	0.6	0.55
	Milk	11	6.1
	Dark	36	10
Se	White	3.4	3.0
	Milk	7.7	1.7
	Dark	14	1.2
Zn	White	2.8	2.5
	Milk	7.6	2.8
	Dark	17	1.3

n.d.: not determined, RDI for children 7-10 years: Cr: 15 µg; Cu: 440 µg; Fe: 9 mg; Mn: 1.5 mg; Se: 21µg e Zn: 5.6 mg [8].

Table 5: Contribution (%) of chocolate consumption (40 g/day) to the RDI* of essential elements for children (7-10 years).

pointed out that the consumption of milk and dark chocolate would also contribute to a high daily intake of Cr for children, even when the bioaccessible fractions are considered, representing 73 and 64% of the RDI, respectively. Although Cr is considered an essential element, it is important to mention that there are some recent studies that refute the essentiality of Cr (III), mainly due to the lack of experimental evidence of biomolecules that would interact with Cr (III) at a cellular level [25,26]. However, Cr (III), the predominant specie of Cr in foods of animal and plant origin, is not toxic to humans when present at trace levels [27].

Conclusions

According to the results, dark chocolate presented higher total concentrations of essential and non-essential elements than milk and white chocolate, probably due to its high cocoa content. As already demonstrated in another work, the concentrations of some elements in chocolate bars are related to their cocoa content [11]. However, the evaluation of the bioaccessibility showed that the highest bioaccessible fractions for all elements studied were found in white chocolate, indicating that the fractions of the elements released into the gastrointestinal tract have a high dependence on the type of chocolate being considered. The results obtained show that information about bioaccessibility of elements should be considered in conjunction with their total concentrations in studies of risk assessment or nutritional evaluation of foodstuffs.

Acknowledgements

The authors acknowledge Dr. Carol H. Collins for assistance with English in this manuscript; the funding agencies Fundação de Amparo à Pesquisa do Estado de São Paulo (Processes nº 2013/26855-2, 2011/21023-3 and 2011/07759-7), Conselho Nacional de Desenvolvimento Científico e Tecnológico (Cnpq) and INCTAA (Instituto de Nacional de Ciências e Tecnologias Analíticas Avançadas); and ITAL (Instituto de Tecnologia de Alimentos).

References

1. Grivetti LI, Shapiro H (2009) Chocolate: History, Culture and Heritage, Wiley, Hoboken.
2. Crozier SJ, Preston AG, Hurst JW, Payne MJ, Mann J, et al. (2001) Cacao seeds are a "super fruit": A comparative analysis of various fruit powders and products. Chemistry Central J 5: 1-6.
3. Sager M (2012) Chocolate and cocoa products as a source of essential elements in nutrition. J Nutritional & Food Sci 2: 1-10.
4. Bruinsma K, Taren DL (1999) Chocolate: food or drug. J the Am Dietetic Association 99: 1249-1256.
5. Parker G, Parker I, Brotchie H (2006) Mood state effects of chocolate. J Affective Disorders 92: 149-159.
6. International Cocoa Organization, ICCO Quarterly Bulletin of Cocoa Statistics, Vol. XLI, nº3, Cocoa year 2014/15.
7. Watson RR, Preedy VR, Zibadi S (2013) Chocolate in Health and Nutrition, Humana Press, London.
8. ANVISA (2005) Diretoria colegiada-September 2005. Imprensa National, Brazil.
9. Dahiya S, Karpe R, Hedge AG, Sharma RM (2005) Lead, cadmium and nickel in chocolates and candies from suburban areas of Mumbai, India. J Food Composition and Analysis 18: 517-522.
10. Rehman S, Husnain SM (2012) Assessment of trace metal contents in chocolate samples by atomic absorption spectrometry. J Trace Element Analysis 1: 1-11.
11. Villa JEL, Peixoto RRA, Cadore S (2014) Cadmium and lead in chocolates commercialized in Brazil. J Agri and Food Chemistry 62: 8759-8763.
12. Oomen AG, Hack A, Minekus M, Zeijdner E, Cornelis C, et al. (2002) Comparison of five in vitro digestion models to study the bioaccessibility of soil contaminants. Environ Sci and Technol 36: 3326-3334.
13. Moreda-Piñero J, Moreda-Piñero A, Romarís-Hortas V, Moscoso-Pérez C, López-Mahía P, et al. (2011) In-vivo and in-vitro testing to assess the bioaccessibility and the bioavailability of arsenic, selenium and mercury species in food samples. Trends in Analytical Chemistry 30: 324-345.
14. Mounicou S, Szpunar J, Andrey JD, Blake C, Lobinski R, et al. (2003) Concentrations and bioavailability of cadmium and lead in cocoa powder and related products. Food Additives and Contaminants 20: 343-352.
15. Peixoto RRA, Mazon EAM, Cadore S (2013) Estimation of the bioaccessibility of metallic elements in chocolate drink powder using an in vitro digestion method and spectrometric techniques. J Brazilian Chemical Soc 24: 884-890.
16. Laparra JM, Vélez D, Montoro R, Barberá R, Farré R (2003) Estimation of arsenic bioaccessibility in edible seaweed by an in vitro digestion method. J Agri and Food Chemistry 51: 6080-6085.
17. Versantvoort CHM, Oomen AG, Kamp EV, Rompelberg CJM, Sips AJAM (2205) Applicability of an in vitro digestion model in assessing the bioaccessibility of mycotoxins from food. Food and Chemical Toxicol 43: 31-40.
18. Thomas, R (2008) Practical guide to ICP-MS - A tutorial for beginners. (2nd ed.), CRC Press, Boca Raton.
19. AOAC (2002) AOAC guidelines for single laboratory validation of chemical methods for dietary supplements and botanicals.
20. Ieggli CVS, Bohrer D, do Nascimento PC, de Carvalho LC (2011) Determination of sodium, potassium, calcium, magnesium, zinc and iron in emulsified chocolate samples by flame atomic absorption spectrometry. Food Chemistry 124: 1189-1193.
21. Jalbani N, Kazi TG, Afridi HI, Arain MB (2009) Determination of toxic metals in different brands of chocolates and candies, marketed in Pakistan. Pakistan J Analytical & Environ Chemistry 10: 48-52.
22. Yanus RL, Sela H, Borojovich EJC, Zakon Y, Saphier M, et al. (2013) Trace elements in cocoa solids and chocolate: an ICPMS study. Talanta 119: 1-4.
23. ANVISA (2013) Diretoria colegiada-August 2013. Imprensa National, Brazil.
24. ANVISA (2005) Diretoria colegiada-September 2005. Imprensa National, Brazil.
25. Di Bona KR, Love S, Rhodes NR, McAdory D, Sinha SH, et al. (2011) Chromium is not an essential trace element for mammals: effects of a low-chromium diet. J Biological Inorganic Chemistry 16: 381-390.
26. Vincent JB (2013) Chromium: is it essential, pharmacologically relevant, or toxic? Metal Ions in Life Sci 13: 171-198.
27. Novotnik B, Zuliani T, Ščančar J, Milačič R (2013) Chromate in food samples: an artefact of wrongly applied analytical methodology? J Analytical Atomic Spectrometry 28: 558-566.

10

Impact of Packaging and Storage on Quality of Watermelon Honey

Abker AM[1]*, Madwi HA[2], Dawood SY[3] and Elkhedir AE[1]

[1]*Industrial Research and Consultancy Center (IRCC), Khartoum, Sudan*

[2]*Department of Food Science and Technology, Faculty of Agriculture, University of Khartoum, Sudan*

[3]*Agriculture Ministry, Khartoum, Sudan*

Abstract

The objectives of this work were directed towards the study of the suitability of honey produced from watermelon and the impact of packaging materials and storage temperature on quality of watermelon honey. The processing operations included: boiling, filtration and concentration. The concentration was carried out using an open jacketed pan at (100°C) temperature. Effects of storage temperatures (35°C and 4°C) and duration (1, 2 and 3 months) were studied for watermelon honey packaged in glass bottles and plastic jercans. The analysis involved physical properties, chemical composition (moisture, pH, TSS, ash, total sugars, reducing sugars and non-reducing sugars). It was found that there was a big difference between the percentage of honey and melon crude product; also there was a significant increase in the percentage of chemical parameters except pH and moisture, where there was a decrease. There were no significant differences (at $P \leq 0.05$) between samples of honey melon, which were packaged in glass bottles or in plastic jercans. However, there were significant differences (at $P \leq 0.05$) between products stored at 35°C and that at 4°C. Samples that were stored at a temperature of 35°C showed, a low shelf stability which was reflected in the low proportion of TSS, Total sugars, non-reducing sugars.

Keywords: Honey; Quality; Watermelon

Introduction

Watermelon (*Citrullus lanatus*) is one of the important horticultural crops belonging to the family Cucurbitaceae. It is a native of dry parts of Africa, and it is a popular thirst-quencher during hot summer weather. Watermelon juice is proven to be a very concentrated source of carotenoid, namely lycopene [1]. When ripe, the sweet juicy pulp is eaten fresh; seeds are roasted as a snack or ground into an ingredient in oils or sauces. Juice from the red part of a watermelon contains 8 to 10 percent solids, of which 20 to 50 percent is sucrose. An edible syrup or fermented beverage can be made from the juice. Honey consists essentially of different sugars. Predominantly glucose and fructose . The color of honey varies from nearly colorless to dark brown. The consistency can be fluid, viscous or partly to entirely crystallized. The flavor and aroma vary, but usually derived from the plant origin [2]. Honey has antibacterial and antifungal properties and will not rot or ferment when stored under normal conditions. However honey can crystallize with time but crystallized honey is not damaged or defected in any way for human use.

Humans use the honey for many different purposes. Honey can be substitute for sugar in many foods. Long ago priests use honey and cake sweetened. Watermelon honey was not produced commercially in the Sudan; it is limited at home level preparation within production areas. Utilizing watermelon during production season for the production of honey will help in solving marketing problem of excess watermelon during the season, reduce watermelon losses, generate income, sustain food security, improve nutritional status and preserve watermelon by transforming it to honey [3].

The Objectives of the present study are to investigate the suitability of watermelon honey for storage and to determine the most appropriate packaging material (glass bottle or plastic jercan).

Materials and Methods

Materials

Twenty Fresh watermelons pieces were purchased from Khartoum central market with an average weight of 3.5 kg and had no bruises on the skin. Watermelon rinds were washed prior to water extraction. The watermelon was cut into quarters and the flesh was scooped out and cut into small cubes.

Physical properties of watermelon

The percentage of skin, flesh, water extract and honey was calculated as described by Saeed and Kattab [4].

Preparation of water melon for processing

Watermelon fruits were first weighed washed, then cleaned fruits were peeled, cut and sliced using sharp, clean, stainless steel knives, then boiled for thirty minutes, and then flirted to separate the water from insoluble solids. The raw watermelon that already boiled was based through a muslin bag (juice cloth) to increase filtration rate and hence shorten the filtration cycle, the bag was squeezed by hand.

Processing of watermelon (Honey)

The processing was carried out in the Food Research Center (FRC). Watermelon juice was concentrated at atmospheric pressure in an open jacketed pan (Model OSK1602), its capacity was about 60 liters, and its evaporation rate was 0.143 Kg/minute. The pan was heated by home liquid gas, boiling was done slowly to evaporate the water without burning the product especially during the later stage when thick treacle was formed. The processing time was recorded when the concentration of honey reached 73 Brix.

***Corresponding author:** Abker AM, Industrial Research and Consultancy Center (IRCC), Shambat- Altyar Alkadro, Khartoum, Sudan
E-mail: abdeensiddig@yahoo.com

Chemical evaluation methods

Total soluble solids (TSS): Total soluble solids (TSS) were determined at room temperature using digital refractometer with degree °Brix scale 0-100 according to AOAC [5].

pH-value: The pH was determined using a pH-meter (model HI 8521 microprocessor bench pH/MV/C° meter). Two standard buffer solution of pH 4.00 and 7.00 were used for calibration of the pH meter at room temperature. The pH meter was allowed to stabilize for one minute and then the pH of the samples was directly measured.

Reducing and total sugars: The reducing and total sugars were determined according to the method described by Schneider [6].

Sucrose (%): The sucrose (%) was determined by the following equation:

Sucrose (%) = Total sugars – Reducing sugars × 0.95

Moisture content: The moisture content of the sample was determined according to the modified method of AOAC [5].

Ash content: The ash content of the sample was determined according to the AOAC [5].

Statistical analysis: The data were subjected to Statistical Analysis System (SAS) Software and Randomized Complete Design (RCD) was used with factorial design and then means separated according to Gomez and Gomez [7].

Results and Discussion

Physical properties of watermelon

Table 1 shows the percentage of skins, flesh and seeds, juice extract and honey in relation to the whole fruit weight. These were skins (33.5%), flesh and seeds (45.2%), water extract (16%) and honey (5.3%).

Skin (%)	Flesh and seeds (%)	Flesh and seeds (%)	Honey (%)
33. 5	45.2	45.2	5.3

Table 1: Physical properties of watermelon.

The chemical characteristics of watermelon before and after processing

Table 2 shows comparison between the chemical composition of watermelons before and after processing, as shown in the Table, there were decrease in moisture content and pH, and increase in total soluble solids, ash, total sugars, reducing sugars and non-reducing sugars.

Effect of storage conditions on the properties of watermelon honey

The watermelon honeys have been packaged in two different packaging materials i.e., plastic jercan and glass bottles and then were stored at two different temperature i.e., 35°C (room temperature) and 4°C (domestic refrigerator), for three months.

Total soluble solids of watermelon honey (TSS)

Table 3 shows total soluble solids, there is a significant differences (at $P \leq 0.05$) for decrease in T.S.S. during storage periods temperature for the two packaging materials. Glass bottle reported 70.60% at the end of storage while plastic jercans 71.30%. This can be attributed to the concentration of some soluble solids during storage [8].

Total sugars of watermelon honey

Table 4 shows the total sugars (%) during storage periods at different storage temperatures. There are significant differences ($P \leq 0.05$). The total sugars of watermelon honey at refrigerator temperature and room temperature, packaged in glass bottle after one month decreased but increased at 2 months and 3 months, storage period to higher levels 53.06% (room temperature) and 54.90% (refrigerator) respectively.

Whereas watermelon honey stored at room temperature and refrigerator, packaged in plastic jercan were decreased after one month

TSS (%)	Total Sugar (%)	Reducing Sugars (%)	Non Reducing sugar (%)	pH	Moisture (%)	Ash (%)
7.2	5.1	4.51	0.55	5.5	91.86	0.13
73.4	48.73	35.88	12.53	4.94	27.2	1.68

Table 2: Chemical composition of watermelon before and after processing.

Type of container	Temperature							
	Room				Refrigerator			
	Storage period (month)							
	0	1	2	3	0	1	2	3
Glass bottle	73.40^{a} (± 0.04)	70.90^{e} (± 0.02)	70.40^{h} (± 0.02)	70.60^{g} (± 0.02)	73.40^{a} (± 0.04)	71.00^{d} (± 0.03)	70.40^{h} (± 0.02)	70.20^{i} (± 0.02)
Plastic jercan	73.40^{a} (± 0.04)	70.70^{f} (± 0.02)	69.40^{k} (± 0.01)	71.20^{c} (± 0.03)	73.40^{a} (± 0.04)	69.70^{j} (± 0.01)	71.00^{d} (± 0.03)	71.30^{b} (± 0.03)
Lsd 0.05	0.0005259*							
SE ±	0.0001826							
Mean ± SD value(s) bearing different superscript(s) within columns and rows (interactions) are significantly different ($P \leq 0.05$)								

Table 3: Effect of storage temperature and packaging materials on total soluble solids of watermelon honey over period of three months.

Type of container	Temperature							
	Room				Refrigerator			
	Storage period (month)							
	0	1	2	3	0	1	2	3
Glass bottle	48.73^{e} (± 1.13)	47.33^{f} (± 1.18)	53.40^{b} (± 1.58)	53.06^{b} (± 1.51)	48.73^{e} (± 1.13)	42.72^{i} (± 1.09)	51.14^{d} (± 1.53)	50.98^{d} (± 1.49)
Plastic jercan	48.73^{e} (± 1.13)	45.4^{h} (± 1.21)	54.80^{a} (± 1.61)	54.90^{a} (± 1.62)	48.73^{e} (± 1.13)	46.36^{g} (± 1.18)	52.06^{c} (± 1.54)	52.09^{c} (± 1.50)
Lsd 0.05	0.4141*							
SE ±	0.1438							
Mean ± SD value(s) bearing different superscript(s) within columns and rows (interactions) are significantly different ($P \leq 0.05$)								

Table 4: Effect of storage temperature and packaging materials on total sugars of watermelon honey over period of three months.

but increased after 2 months and 3 months to higher levels 50.98% (room temperature) and 52.09% (refrigerator) respectively. This decrease and increase of total sugars may be due to inversion of sucrose to glucose and fructose sugars during storage time [7]. The production of invert sugars may cause non enzymic browning.

Reducing sugars of watermelon honey

Table 5 shows high significant differences (at $P \le 0.05$) in content of reducing sugars during storage time. The glass bottles stored at room temperature reported (35.88%, 31.72%, 40.78% and 44.36%) and that stored in refrigerator recorded (35.88%, 29.60%, 45.10% and 45.10%) they decreased after one month and increased after 2 and 3months.

Reducing sugars in plastic jercan stored at room temperature and refrigerator decreased after one month (32.36% and 31.63%). Then increased after 2 and 3 months gradually (42.46%, 43.34%, 39.86% and 39.70%). This decrease and increase may be due to non-enzymatic browning reactions Hulme [9] reported that many interactions are possible in fruit juice concentrates, notably between sugars, organic acids, nitrogenous components and phenol compounds, the change being greatest at higher temperature of storage.

Non reducing sugars of watermelon honey

Table 6 Shows, that there are significant differences (at $P \le 0.05$) in non-reducing sugars content values. They were found to range between (14.83%, 11.99%, 8.34%, 12.46%, 5.74% and 5.59% in glass bottles and 12.39%, 11.72%, 10.98%, 13.68%, 11.61% and 12.07% in plastic jercans) at 35°C and 4°C, respectively. The results obtained shows gradually decrease during storage periods; this could be due to the inversion of sucrose with notably slightly increased in plastic jercan stored in refrigerator after 3months of storage. This could be due to container.

pH of watermelon honey

Table 7 shows pH values, there is a significant difference at ($P \le 0.05$) during storage periods; the high level of pH was at glass bottles stored at refrigerator.

Moisture content of watermelon honey

Table 8 Shows the moisture content (%) during storage periods at different storage temperatures. There are significant differences (at $P \le 0.05$) decrease in moisture content of watermelon honey. This could be due to evaporation and concentration of water during storage.

Ash content of watermelon honey

Table 9 shows ash content; there is a significant difference (at $P \le 0.05$) in ash content during storage periods at both storage temperatures and package containers of watermelon honey.

Conclusion

From the interpretation of the results obtained in this study it can be concluded that Watermelon showed suitability in processing to give watermelon honey and there was increase in total sugars most of time

Type of container	Temperature							
	Room				Refrigerator			
	Storage period (month)							
	0	1	2	3	0	1	2	3
Glass bottle	35.8^{g} (± 1.17)	31.7^{i} (± 1.11)	40.78^{e} (± 1.23)	44.36^{b} (± 1.35)	35.88^{g} (± 1.17)	29.60^{j}(± 1.07)	45.10^{a} (± 1.37)	45.10^{a} (± 1.37)
Plastic jercan	35.8^{g} (± 1.17)	32.36^{h} (± 1.13)	42.46^{d} (± 1.31)	43.34^{c} (± 1.32)	35.88^{g} (± 1.17)	31.63^{i} (± 1.12)	39.86^{f} (± 1.22)	39.70^{f} (± 1.20)
Lsd 0.05	0.2467**							
SE ±	0.08563							
Mean ± SD value(s) bearing different superscript(s) within columns and rows (interactions) are significantly different (P ≤ 0.05)								

Table 5: Effect of storage temperature and packaging materials on reducing sugars of watermelon honey over period of three months.

Type of container	Temperature							
	Room				Refrigerator			
	Storage period (month)							
	0	1	2	3	0	1	2	3
Glass bottle	12.53^{c} (± 0.97)	14.83^{a} (± 1.10)	11.99^{d} (± 1.09)	8.34^{g} (± 0.65)	12.53^{c} (± 0.97)	12.46^{c} (± 0.97)	5.74^{h} (± 0.33)	5.59^{h} (± 0.31)
Plastic jercan	12.53^{c} (± 0.97)	12.39^{c} (± 1.08)	11.72^{e} (± 1.07)	10.98^{f} (± 0.91)	12.53^{c} (± 0.97)	13.68^{b} (± 1.02)	11.61^{e} (± 1.06)	12.07^{d} (± 0.94)
Lsd 0.05	0.2292*							
SE ±	0.07958							
Mean ± SD value(s) bearing different superscript(s) within columns and rows (interactions) are significantly different (P ≤ 0.05)								

Table 6: Effect of storage temperature and packaging materials on Non-reducing sugars of watermelon honey over period of three months.

Type of container	Temperature							
	Room				Refrigerator			
	Storage period (month)							
	0	1	2	3	0	1	2	3
Glass bottle	4.94^{a} (± 0.36)	4.72^{d} (± 0.31)	4.51^{j} (± 0.25)	4.58^{h} (± 0.27)	4.94^{a} (± 0.36)	4.71^{e} (± 0.29)	4.55^{i} (± 0.24)	4.62^{f} (± 0.28)
Plastic jercan	4.94^{a} (± 0.36)	4.76^{b} (± 0.31)	4.61^{g} (± 0.28)	4.75^{c} (± 0.30)	4.94^{a} (± 0.36)	4.76^{b} (± 0.31)	4.61^{g} (± 0.28)	4.75^{c} (± 0.32)
Lsd 0.05	0.0005259*							
SE ±	0.0001826							
Mean ± SD value(s) bearing different superscript(s) within columns and rows (interactions) are significantly different (P ≤ 0.05)								

Table 7: Effect of storage temperature and packaging materials on pH-value of watermelon honey over period of three months.

Type of container	Temperature							
	Room				Refrigerator			
	Storage period (month)							
	0	1	2	3	0	1	2	3
Glass bottle	27.20^{a} (± 1.25)	25.07^{b} (± 1.23)	23.72^{c} (± 1.19)	24.87^{b} (± 1.22)	27.20^{a} (± 1.25)	24.51^{bc} (± 1.21)	24.38^{b}c (± 1.19)	24.92^{b} (± 1.23)
Plastic jercan	27.20^{a} (± 1.25)	24.4b^{c} (± 1.21)	24.42^{bc} (± 1.21)	24.76^{b} (± 1.22)	27.20^{a} (± 1.25)	24.45^{bc} (± 1.20)	24.22^{bc} (± 1.17)	24.46^{bc} (± 1.20)
Lsd 0.05	0.8464*							
SE ±	0.2938							
Mean ± SD value(s) bearing different superscript(s) within columns and rows (interactions) are significantly different (P ≤ 0.05)								

Table 8: Effect of storage temperature and packaging materials on Moisture content of watermelon honey over period of three months.

Type of container	Temperature							
	Room				Refrigerator			
	Storage period (month)							
	0	1	2	3	0	1	2	3
Glass bottle	1.580^{bcde} (± 0.45)	1.810^{ab} (± 0.49)	1.730^{abcd} (± 0.46)	1.610^{abcde} (± 0.45)	1.580^{bcde} (± 0.45)	1.423^{e} (± 0.41)	1.493^{cde} (± 0.45)	1.777^{ab} (± 0.42)
Plastic jercan	1.580^{bcde} (± 0.45)	1.743^{abc} (± 0.47)	1.470^{de} (± 0.44)	1.447e (± 0.44)	1.580^{bcde} (± 0.45)	1.857^{a} (± 0.50)	1.447^{e} (± 0.43)	1.467^{de} (± 0.43)
Lsd 0.05	0.2352*							
SE ±	0.08165							
Mean ± SD value(s) bearing different superscript(s) within columns and rows (interactions) are significantly different (P ≤ 0.05)								

Table 9: Effect of storage temperature and packaging materials on Ash content of watermelon honey over period of three months.

during storage time. The best storage condition should be Packaging in tightly sterilized glass bottles or plastic jerkins.

Acknowledgment

The authors express their sincere gratitude and thanks to the staff and technicians of The Food Research Centre-Shambat for their valuable and unlimited assistance which they presented during implementation of the research.

References

1. Edwards AJ, Vinyard BT, Wiley ER, Brown ED, Collins JK, et al. (2003) Consumption of watermelon juice increases plasma concentrations of lycopene and beta-carotene in humans. J Nutr 133: 1043-1050.
2. Codex Alimentarius Commission (1994) Codex Standard for Honey. Ref 11: 21-24.
3. Baker AM (2012) Personal Communication.
4. Saeed A, Kattab A (1974) Suitability of Mango Cultivar for nectar manufacture. Journal of food since and technology. 6: 24-31.
5. AOAC (1984) Official methods of analysis. Association of Official Analytical Chemist. Washington DC, USA.
6. Schneider JM, Curet LB, Olson RW, Schneider JM, Zachman RD (1979) Effect of diabetes mellitus on amniotic fluid lecithin/sphingomyelin ratio and respiratory distress syndrome. Am J Obstet Gynecol 135: 10-13.
7. Gomez A, Gomez AA (1984) Statistical procedure for agriculture research (2ndedn). John Wiley and Sons.
8. Herschdoereer SM (1972) Food Science and Technology: Quality Control in the Food Industry. Academic press, London.
9. Hulme AC (1971) Biochemistry of fruits and their products. Academic press, London.

Enhancement of Non-Thermal Treatment on Inactivation of Glucoamylase and Acid Protease Using CO_2 Microbubbles

Pokhum C[1]*, Chawengkijwanich C[1] and Kobayashi F[2]

[1]*National Nanotechnology Center, National Science and Technology Development Agency, Pathumthani, Thailand*
[2]*Nippon Veterinary and Life Science University, Musashino, Tokyo, Japan*

Abstract

Thermal treatment is usually used for food pasteurization and enzyme inactivation. However, it has an adverse effect on the quality of thermal-sensitive food such as fruit juice, Japanese sake, milk, yogurt and jam. In this study, we presented an alternative method for a non-thermal treatment with at 45 and 50°C for glucoamylase and protease inactivation using pressurized carbon dioxide (CO_2). Twelve liters of enzyme solution (0.004% glucoamylase or 0.015% protease) was fed into a low pressure (2 MPa) CO_2 mixing vessel. CO_2 microbubbles (MB-CO_2) were generated by introducing the mixture through a swiveling microbubble generator. The mixture containing MB-CO_2 was flowed to incubate in a heating coil at various conditions (temperature at 45 or 50°C and pressure 2, 4, or 6 MPa). After incubation, the mixture was sampled at 10, 20 and 30 min from the sampling valve. The relative residual activities of glucoamylase and acid protease were measured by a spectrophotometer at the absorbance of 400 nm (Abs_{400}) and 660 nm (Abs_{660}), respectively. Relative residual activity of glucoamylase with MB-CO_2 treatment at 50°C and 4 MPa was 15.01% whereas 74.83% of glucoamylase activity was found from treatment without MB-CO_2 at same temperature. For acid protease, relative residual enzyme activity with MB-CO_2 treatment at 45°C and 4 MPa was 2.29% whereas that without MB-CO_2 treatment at 45°C was 81.25%. These results suggested that glucoamylase and acid protease could be inactivated effectively at 45 and 50°C present of MB-CO_2.

Keywords : Enzyme inactivation; Carbon dioxide; Microbubbles; Glucoamylase; Protease

Introduction

Enzyme inactivation in various foods and beverages are desired in the food industry. Its activity produced undesirable chemical changes in food attributes; e.g. color, texture, and flavor during storage and distribution [1]. For example, glucoamylase degrades the quality of sake by producing an excessive amount of glucose [2] which causes a bitter taste. Acid protease degrades casein components in milk [3]. Polyphenol oxidase causes enzymatic browning in fruit juice [4] and also degrades the sensory quality of wine during the aging by polymerization polyphenol compounds [5]. Pectin esterase (PE) causes undesired cloudy instability in orange juice [4]. So far, thermal treatment is used widely for enzyme inactivation in the food industry [6,7]. However, thermal treatment (≥65°C) can alter the nutrition and qualities of thermal-sensitive food [7,8] such as fruit juice, Japanease sake, milk, yogurt and jam. Poor solubility of calcium was observed in the thermal treatment of milk [9]. Tanimoto et al. [2] reported that unpasteurized sake treated thermally at 65°C lost its fresh flavor. The negative effect of thermal treatment on the aroma and flavor of beer was also reported [10]. The sensory quality and nutritional content of coconut juice processed by heat treatments also changed [11].

To eliminate the disadvantages of thermal treatment, developments of innovative non-thermal technologies for enzyme inactivation have been encouraged [8]. In the last decade, high pressure carbon dioxide (HP-CO_2) has emerged as a non-thermal treatment which is able to inactivate enzyme activity [12]. It has been applied in both solid and liquid food matrices [11]. The use of CO_2 treatment has become an attractive technique in food processing for enzyme inactivation because it is nontoxic, non-flammable, inexpensive and no residue [11,12]. Furthermore, CO_2 is guaranteed as a substance that can be used safely on food products [11]. Peroxidase (PO) and polyperoxidase (PPO) in red beet and carrot juice were inactivated using HP-CO_2 treatment at 10-30 MPa [13,14]. Yoshimura et al. [15] reported the inactivation kinetics of acid protease and α-amylase with supercritical CO_2 technique (SC-CO_2) at 30 MPa. The inactivation of α-glucosidase, glucoamylase, α-amylase and carboxypeptidase in fresh sake using HP-CO_2 treatment at 2 MPa was investigated [16]. Gui et al. [17] showed that PPO in cloudy apple juice was inactivated by SC-CO_2 at 30 MPa. The inactivation of acid protease and glucoamylase using SC-CO_2 bubbles at 30 MPa were reported [18-20]. However, SC-CO_2 require a high pressure conditions about 10-30 MPa. Furthermore, both HP-CO_2 and SC-CO_2 induced loss of flavor due to the fact that they extract some constituents including phospholipids and hydrophobic compounds [21,22]. Gasperi et al. [23] reported that the HP-CO_2 treatment induced a reduction in the concentration of many volatile compounds (ester and aldehydes) responsible for the observed change in odor and flavor of treated apple juice. In the last decade, microbubbles (MB-CO_2), very fine bubbles with a diameter of 10-50 µm, had also been investigated for its potential as an alternative process for enzyme inactivation [24]. MB-CO_2 was studied at the present for different characteristics from ordinary bubbles because of their reduced diameter size [24]. Some advantages of microbubbles are their highly specific area (surface area per volume) and high stagnation in the liquid phase, which increases the gas dissolution [24,25]. The key differences between ordinary bubbles, micro bubbles and nano bubbles were explained by Takahashi [26]. The ordinary bubbles go up rapidly and burst at the surface whereas nano bubbles (diameter<200 µm), remain for months and do not burst out at once [26]. In the case of microbubbles, they tend to gradually

***Corresonding author:** Pokhum C, National Nanotechnology Center, National Science and Technology Development agency, Pathumthani, Thailand, E-mail: chonlada@nanotec.or.th

decrease in size and subsequently collapse due to the long stagnation and dissolution of interior gases into surrounding water. Furthermore, hydroxyl radicals occur when microbubbles collapse due to the high density of ions in gas-liquid interface just before the collapse [27]. The type of gas used for the generation of microbubbles can also affect the quantity of free radicals occurring [24].

Enzyme inactivation using low pressurized MB-CO_2, which required lower pressure conditions SC-CO_2, have been studied by several researchers [5,28-31]. Kobayashi et al. [28] showed that esterase, esterase lipase, leucine arylamidase, valine arylamidase and acid phosphatase from *Saccharomyces pastorianus* were completely inactivated by MB-CO_2 at 2 MPa and 50˚C. The quality of Japanese sake, which α-amylase, glucoamylase and acid carboxypeptides enzyme were inactivated by MB-CO_2, retained good taste and flavor [29]. Kobayashi et al. [30] reported that score of sweetness and total aroma of MB-CO_2 treated Japanese sake were higher than unpasteurized Japanese sake. Kobayashi et al. [5] reported the inactivation kinetics of PPO using MB-CO_2 and a decrease in decimal reduction time (D value) and activation energy in MB-CO_2 treatment. In this study, we analyzed the capability of glucoamylase and acid protease inactivation at 45-50˚C by using a MB-CO_2 continuous system.

Materials and Methods

Enzyme solution

Glucoamylase and acid protease extracted from *Aspergillus niger* was purchased from HBI enzyme INC., Shiso, Japan. Glucoamylase and acid protease enzyme powder was dissolved in a concentration of 0.004% and 0.015% w/v, respectively. The concentration of sodium acetate buffer (pH 4) was 10 mM.

Enzyme inactivation with MB-CO_2 treatment

The MB-CO_2 continuous system used in this study is shown in Figure 1. This system has been previously reported [5,30,31]. Twelve liters of glucoamylase (0.004% w/v) or acid protease (0.015% w/v) enzyme solution were added into the mixing vessel (volume 15 L with 300 mm dia. 220 mm height) then, the mixing vessel was tightly closed. The temperature of the mixing vessel was controlled at 10°C. Valve A and B were opened, gaseous CO_2 was fed to replace air in the headspace of the mixing vessel until pressure increased to 2 MPa. Then, valve B was closed and the regulator valve (valve C) was opened, gaseous CO_2 was fed into the mixing vessel at a rate of 2 L/min controlling by valve C. MB-CO_2 were generated by introducing the mixture of enzyme solution and gaseous CO_2 to a swiveling microbubbles generator (56 mm diameter x 86 mm long, outlet and inlet diameter are 5 mm and 13 mm, respectively; BT-50; Bubble tank Co., Japan) by circulating pump at rate 15 L/min. MB-CO_2 was generated for 10 min, and then valves A and C were closed and the circulating pump was stopped. Next, valve D was opened, the enzyme solution saturated with CO_2 was continuously fed to a heating coil (volume 170 ml with 5 mm dia x 850 mm length) using a metering pump at flow rate of about 13-17 L/min. In this study, the enzyme inactivation was performed under the following conditions: the temperatures of the heating coil were 45 or 50°C, and the pressure of the heating coils was 2, 4 or 6 MPa. The treated enzymes were collected from the sampling valve (valve E) at 0, 10, 20 and 30 min and then immediately taken to measure the level of enzyme activity (%).

Enzyme inactivation with thermal treatment

The efficacy of conventional thermal treatment was compared with MB-CO_2. Five milliliters of glucoamylase (0.004% w/v) or acid protease (0.015% w/v) enzyme solution were placed in a test tube and incubated in a water bath. The temperatures of the water bath were varied between 45 and 70°C. The incubation times were 10, 20 and 30 min.

Measurement of glucoamylase activity

The activity of the glucoamylase was measured using a glucoamylase assay kit (Kikkoman Co., Chiba, Japan). The enzyme activity was determined at pH 4. The assay medium contained 250 μl of β-glucosidase and 250 μl of 4-nitrophenyl β-maltoside as substrate. The substrate was incubated at 37°C for 5 min and then 100 μl of enzyme sample was added to the substrate and incubated for 10 min. Next, 1 ml of sodium carbonate ($NaCO_3$) was added to stop the reaction. The activity of the glucoamylase enzyme was determined with a spectrophotometer at the absorbance of 400 nm (Abs_{400}). The blank sample was prepared by adding the stopping reagent to the mixture before the sample of enzyme. The relative activities of glucoamylase were calculated using the following equation (1):

$$\text{Relative residual activity} = \frac{Abs_{400} \text{ of } t_t \times 100}{Abs_{400} \text{ of } t_0} \quad (1)$$

Where t_t is incubation time (min) and t_0 is initial time (0 min)

Measurement of acid protease activity

The acid protease activity was determined at pH 4. One ml of casein solution (2% w/v) of casein and sodium acetate buffer (pH 4) were mixed and incubated at 37°C for 5 min. Then, 500 μl of protease enzyme was added and incubated for 10 min. The reaction was stopped by 3 ml of 0.4 M trichloroacetic acid to stop the reaction. The mixture was incubated at 37°C for 30 min. After that, the mixture was filtrated. One ml of the filtrated mixture was mixed with 1 ml of 20% (v/v) phenol reagent and 5 ml of 0.55 M Na_2CO_3. The mixture was incubated at 37°C for 30 min. Acid protease activity was determined at the absorbance of 660 nm (A_{660}) using spectrophotometer. The blank sample was prepared by adding the stopping reagent to the mixture before the sample of enzyme. The relative activity of protease was calculated as followed equation (2):

$$\text{Relative residual activity (\%)}: \frac{Abs_{660} \text{ of } t_t \times 100}{Abs_{660} \text{ of } t_0} \quad (2)$$

Where t_t is incubation time (min) and t_0 is initial time (0 min)

Kinetic analysis

The kinetic data on the inactivation of glucoamylase and acid protease was calculated using a conventional 1st-order kinetic model [16] was shown in equation (3):

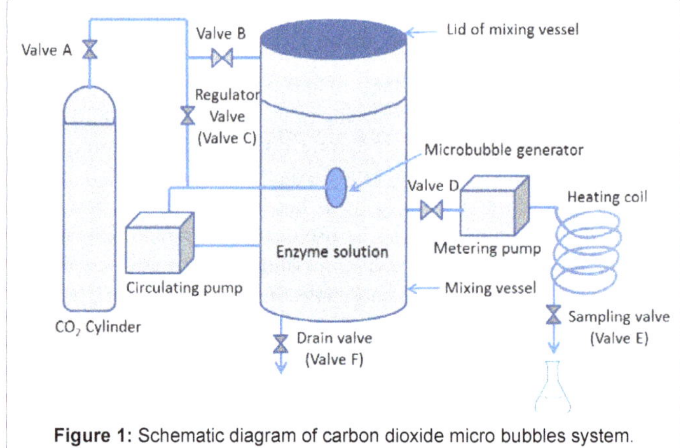

Figure 1: Schematic diagram of carbon dioxide micro bubbles system.

$$\ln(A) = -kt \quad (3)$$

Where A is the residual activity at any residence time (t, min) and k is the reaction rate constant (min^{-1}) at a given condition. The value of k was obtained from the slope of the regression of ln (A) versus time.

The decimal reduction time (D value), defined as the treatment time needed for 90% inactivation of initial activity at a given condition [16] was shown in equation (4):

$$D = 2.303/k \quad (4)$$

The Z value defined as the temperature increment needed for a 90% reduction in the D value at a given condition [16] was shown in equation (5):

$$Z = (T_2-T_1)/ (\log D_1-\log D_2) \quad (5)$$

Where T_1 and T_2 are temperature at a given condition.

Statistical analysis

The data were measured in triplicate. Data were analyzed using the software package SPSS 17.0 for Windows (SPSS Inc., Chicago, IL). Significant differences between mean values were determined using Turkey's honestly significant different (HSD) multiple range test (P=0.05).

Results and Discussion

Table 1 shows the relative activities of glucoamylase at 45 and 50°C with MB-CO_2 treatments in comparison with thermal treatments of 45-70°C. It shows that the inactivation of glucoamylase could not be achieved by temperature lower than 65°C without MB-CO_2. In contrast, the inactivation of glucoamylase was successfully achieved at 45 and 50°C using MB-CO_2. The residual activity of glucoamylase treated by MB-CO_2 at 50°C was lower than that by MB-CO_2 at 45°C. The thermal treatment inactivated glucoamylase by denaturation of enzyme structure. The absorption and co-existence of pressurized CO_2 into the C-terminal affect the decomposition of the α-helix enzyme structure [19,32,33]. The result showed that MB-CO_2 treatment increased the ability of non-thermal treatment on glucoamylase inactivation. The efficacy of MB-CO_2 on enzyme inactivation was increased by increasing the temperature [34] which was possibly caused by increased diffusivity of MB-CO_2 molecules to enzyme structure at higher temperatures. In addition, the efficacy of MB-CO_2 at different pressures in the heating coil on glucoamylase inactivation showed in Figure 2. The residual activity of glucoamylase at 2 MPa was much higher than that at 4 MPa, while glucoamylase activity was slightly decreased when pressure was increased from 4 MPa to 6 MPa (from 39.74% to 36.59%). The correlation coefficient between residual activity of glucoamylase and pressure was - 0.9147. This indicates that the residual activity of glucoamylase was associated with pressure, residual activity of glucoamylase decreases as the pressure increases. In order to save energy, thus, the condition operation at 4 MPa of the heating coil was preferable. The pressure in the heating coil would be responsible for the stability of dissolved MB-CO_2 in enzyme solution [29]. The present data suggests that the pressure of 4 MPa in the heating coil was enough to maintain the stability of dissolved MB-CO_2 when the pressure of mixing vessel was operated at 2 MPa. This result was similar to that reported by Kobayashi et al. [5] who studied on PPO inactivation using MB-CO_2 at 1-4 MPa.

Table 2 shows the rate constants (k), decimal reduction time (D) and Z value. The Z value means as the temperature increase needed for a 90% reduction of D value. The K value of treatments without MB-CO_2 at 45 and 50°C were 0.0093 and 0.0371 min^{-1}, respectively whereas those treated with MB-CO_2 were much higher (0.1493-0.6150 min^{-1}). The lowest D value (3.75 min) was obtained from condition MB-CO_2 at 50°C and 4 MPa whereas the D value of treatment at same temperature (50°C) without MB-CO_2 was up to 177.04 min. The recommended MB-CO_2 condition for glucoamylase inactivation in this study was temperature at 50°C and 4 MPa pressure. Furthermore, Z values of glucoamylase inactivation were 16.05°C and 44.81°C for treatments with and without MB-CO_2 respectively. Lower Z value indicated the lower temperature required. The enzyme inactivation at lower temperature could maintain the sensory quality than at higher temperature.

Table 3 shows the relative activities of acid protease at 45 and 50°C with MB-CO_2 at 4 MPa in comparison with thermal treatments of 45-65°C. The results show that the inactivation of protease by thermal treatments could not be achieved at temperature lower than 65°C. In contrast, the inactivation of acid protease was achieved at 45 and 50°C using MB-CO_2. The residual enzyme activity treated without MB-CO_2 at 50°C was 51.81%, whereas that treated with MB-CO_2 treatment at 50°C was greatly decreased to 1.12%. The residual activity treated using MB-CO_2 at 50°C was similar to that treated using MB-CO_2 at 45°C. Thus, the optimum operating temperature for acid protease inactivation by MB-CO_2 at 45°C was recommended.

Table 4 shows the k, D and Z values of acid protease inactivation. It showed that k value of treatments without MB-CO_2 at 45 and 50°C were 0.0660 min^{-1} and 0.2016 min^{-1}, respectively whereas those treated with MB-CO_2 at 45 and 50°C were higher (1.3866-1.489 min^{-1}). The lowest D value was obtained from MB-CO_2 at 50°C (1.01 min) while a much higher D value (11.46-35.80 min) were obtained from treatment without MB-CO_2 at the same temperature. Z values of acid protease inactivation were 111.24°C and 10.35°C for with and without MB-CO_2, respectively. The high Z value supported that there was no difference

Treatment No.	MB-CO_2	Temperature (°C)	Relative residual activity at 30 min (%)
1	No	45	74.83 ± 1.73[a]
2	No	50	75.03 ± 0.66[a]
3	No	60	65.50 ± 1.95[b]
4	No	65	2.04 ± 0.38[e]
5	No	70	0.37 ± 0.01[f]
6	Yes	45	39.74 ± 0.43[c]
7	Yes	50	15.01 ± 0.96[d]

[a-f] Data in the column with different superscript letters are significantly different (P<0.05).

Table 1: Inactivation of glucoamylase by thermal and MB-CO_2 treatments at 4 MPa.

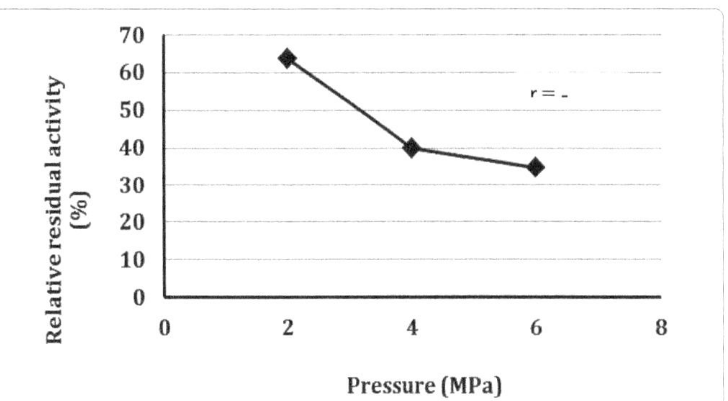

Figure 2: Effect of pressure in a heating coil on relative residual activity of glucoamylase by MB-CO_2 at 45°C.

Treatment	$MB-CO_2$	Temperture (°C)	Pressure (MPa)	k (min^{-1})	D (min)	Z (°C)
1	No	45	No	0.0093 ± 0.00	248.02 ± 21.21	44.81 ± 14.76
2	No	50	No	0.0371 ± 0.05	177.04 ± 132.04	
6	Yes	45	4	0.2996 ± 0.00	7.69 ± 0.12	16.05 ± 0.92
7	Yes	50	4	0.6150 ± 0.02	3.75 ± 0.11	
8	Yes	45	2	0.1493 ± 0.01	15.48 ± 1.10	
9	Yes	45	6	0.3123 ± 0.01	7.38 ± 0.16	

Table 2: Rate constant (k), decimal reduction time (D) and Z values of glucoamylase inactivation at 45 and 50°C with and without $MB-CO_2$.

Treatment No.	$MB-CO_2$	Temperature (°C)	Relative residual activity at 30 min (%)
1	No	45	81.25 ± 3.24[a]
2	No	50	51.81 ± 1.63[ab]
3	No	60	35.80 ± 3.03[bc]
4	No	65	22.01 ± 0.69[bc]
6	Yes	45	2.29 ± 0.73[c]
7	Yes	50	1.12 ± 0.45[c]

[a-c] Data in the column with different superscript letters are significantly different (P<0.05).

Table 3: Inactivation of acid protease by thermal and $MB-CO_2$ treatments at 4 MPa.

Treatment.	$MB-CO_2$	Temperature (°C)	Pressure (MPa)	k (min^{-1})	D (min)	Z (°C)
1	No	45	No	0.0660 ± 0.01	35.80 ± 6.80	10.35 ± 1.55
2	No	50	No	0.2016 ± 0.01	11.46 ± 0.75	
6	Yes	45	4	0.3866 ± 0.11	1.67 ± 0.13	111.24 ± 4.36
7	Yes	50	4	1.4289 ± 0.07	1.01 ± 0.08	

Table 4: Rate constant (k) decimal reduction time (D) and Z values of protease inactivation at 45°C and 50°C with and without $MB-CO_2$.

in acid protease inactivation between $MB-CO_2$ at 45°C and 50°C. Thus, condition operating for acid protease inactivation at 45°C and 4 MPa of $MB-CO_2$ treatments was recommended.

Conclusion

In the present study, the efficacy of non-thermal treatment on glucoamylase and acid protease inactivation were increased by addition of $MB-CO_2$. The inactivation of glucoamylase and acid protease without $MB-CO_2$ could not be achieved at temperature lower than 65°C. The pressure at 4 MPa in the heating coil was suitable to maintain the stability of dissolved $MB-CO_2$. The relative residual activity of glucoamylase treated by $MB-CO_2$ at 50°C was significantly lower than that by $MB-CO_2$ at 45°C. For acid protease, the relative residual activity treated by $MB-CO_2$ at 45 and 50°C was not significantly different. The recommended $MB-CO_2$ condition for glucoamylase and acid protease inactivation in this study were at 50°C and 45°C with 4 MPa, respectively. Accordingly, the present study demonstrated that $MB-CO_2$ could enhance the efficacy of temperature at 45-50°C on enzyme inactivation.

Acknowledgements

The authors gratefully acknowledge the financial support granted to this work from Meiji University, Japan. We are also grateful to Professor Yasuyoshi Hayata at school of agriculture, Meiji University, Japan.

References

1. Hu W, Zhou L, Xu Z, Zhang Y, Liao X (2013) Enzyme inactivation in food processing using high pressure carbon dioxide technology. Critical reviews in Food Science and Nutrition 53: 145-161.
2. Tanimoto S, Matsumoto H, Fujii K, Ohdoi R, Sakamoto K, et al. (2004) Thermal inactivation behavior of enzyme in fresh sake during pasteurization. Journal of the Brewing Society of Japan 99: 208-204.
3. Kaminogawa S, Yamauchi K, Miyazawa S, Koga Y (1980) Degradation of casein components by acid protease of bovine milk. Journal of Dairy Science 63: 701-704.
4. Bayindirli A, Alpas H, Bozoglu F, Hizal M (2006) Efficiency of high pressure treatment on inactivation of pathogenic microorganisms and enzyme in apple, orange, apricot and sour cherry juices. Food Control 17: 52-58.
5. Kobayashi F, Ikeura H, Odake S, Hayata Y (2013) Inactivation kinetics of polyphenol oxidase using a two-stage method with low pressurized carbon dioxide microbubbles. Journal of Food Engineering 114: 215-220.
6. Anthon GE, Barrett DM (2002) Kinetic parameters for the thermal inactivation of quality related enzymes in carrots and potato. Journal of Agricultural and Food Chemistry 50: 4119-4125.
7. Liu X, Gao Y, Xu H, Hao Q, Liu G, et al. (2010) Inactivation of peroxidase and polyphenol oxidase in red beet (*Beta vulgaris* L.) extract with continuous high pressure carbon dioxide. Food Chemistry 119: 108-113.
8. Savadkoohi S, Bannikova A, Hao TT, Kasapis S (2014) Inactivation of bacterial protease and food borne pathogens in condensed globular proteins following application of high pressure. Food Hydrocolloids 42: 244-250.
9. Buzrul S (2012) High hydrostatic pressure treatment of beer and wine: A Review. Innovative Food Science and Emerging Technology 13: 1-12.
10. Cappelletti M, Ferrentino G, Endrizzi I, Aprea E, Betta E, et al. (2015) High pressure carbon dioxide pasteurization of coconut water: A sport drink with high nutritional and sensory quality. Journal of Food Engineering 145: 73-81.
11. Seiquer L, Delgado AC, Haro A, Navarro MP (2010) Assessing the effects of severe heat treatment of milk on calcium bioavailability: In vitro and in vivo studies. Journal of Dairy Science 92: 5635-5643.
12. Tanimoto S, Matsumoto H, Fujii K, Ohdoi R, Sakamoto K, et al. (2005) Inactivation of enzymes in fresh sake using a continuous flow system for high pressure carbonation. Bioscience, Biotechnology, and Biochemistry 69: 2094-2100.
13. Zhou L, Wang Y, Hu X, Wu J, Liao X (2009) Effect of high pressure carbon dioxide on the quality of carrot juice. Innovative Food Science and Emerging technology 10: 321-327.
14. Liu X, Gao Y, Peng X, Yang B, Xu H, et al. (2008) Inactivation of peroxidase and polyphenol oxidase in red beet (*Beta vulgarisL.*) extract with high pressure carbon dioxide. Innovative Food Science and Emerging Technology 9: 24-31.
15. Yoshimura T, Shimoda M, Ishikawa H, Miyake M, Hayakawa I, et al. (2001) Inactivation kinetics of enzyme by using continuous treatment with microbubbles of supercritical carbon dioxide. Journal of Food Science 66: 694-697.

16. Tanimoto S, Matsumoto H, Fujii K, Ohdoi R, Sakamoto K, et al. (2005) Inactivation of enzyme in fresh sake using a continuous flow system for high-pressure carbonation. Bioscience, Biotechnology and Biochemistry 69: 2094-2100.

17. Gui F, Wu J, Chen F, Liao X, Hu X, et al. (2007) Inactivation of polyphenol oxidase in cloudy apple juice exposed to supercritical carbon dioxide. Food Chemistry 100: 1678-1685.

18. Ishikawa H, Shimoda M, Kawano T, Sakamoto K, Osajima Y (1995) Inactivation of enzyme in namesake using micro-bubbles supercritical carbon dioxide. Bioscience, Biotechnology and Biochemistry 59: 1027-1031.

19. Ishikawa H, Shimoda M, Yonekura A, Osajima Y (1996) Inactivation of enzyme and decomposition of α-helix structure by supercritical carbon dioxide microbubble method. Journal of Agricultural and Food Chemistry 44: 2646-2649.

20. Ishikawa H, Yoshimura T, Kajihara M, shimoda M, Matsumoto K, et al. (2000) pH-Dependent inactivation of enzymes by microbubbling of supercritical carbon dioxide. Food Science and Technology Research 6: 212-215.

21. Tanimoto S, Matsumoto H, Fujii K, Ohdoi R, Sakamoto K, et al. (2008) Enzyme inactivation and quality preservation of sake by high pressure carbonation. Bioscience, Biotechnology and Biochemistry 72: 22-28.

22. Chen J, Zhang J, Feng Z, Song L, Wu J, et al. (2009) Influence of thermal and dense-phase carbon dioxide pasteurization on physicochemical properties and flavor compounds in Hami melon juice. Journal of Agricultural and Food Chemistry 57: 5805-5808.

23. Gasperi F, Aprea E, Biasioli F, Carlin S, Endrizz I, et al. (2009) Effect of supercritical CO_2 and N_2O pasteurization on the quality of fresh apple juice. Food Chemistry 115: 129-136.

24. Agarwal A, Jern WN, Liu Y (2011) Principle and applications of microbubble and nanobubble technology for water treatment. Chemosphere 84: 1175-1180.

25. Ushikubo FY, Furukawa T, Nakagawa R, Enari M, Makino Y, et al. (2010) Evidence of the existence and the stability of nano-bubbles in water. Colloids and Surfaces A: Physicochemical and Engineering Aspects 361: 31-37.

26. Takahashi M (2009) Base and technological application of microbubble and nanobubble. Materials Integration 22: 2-19

27. Takahashi M, Chiba K, Li P (2007) Free radical generation from collapsing microbubbles in the absence of a dynamic stimulus. Journal of Physical Chemistry 111: 1343-1347.

28. Kobayashi F, Sugiura M, Ikeura H, Sato M, Odake S, et al. (2014) Comparison of a two-stage system with low pressure carbon dioxide microbubbles and heat treatment on the inactivation of *Saccharomyces pastorianus* cells. Food Control 46: 35-40.

29. Kobayashi F, Sugawara D, Takatomi T, Ikeura H, Odake S, et al. (2012) Inactivation of *Lactobacillus fructivorans* in physiological saline and unpasteurised sake using CO_2 microbubbles at ambient temperature and low pressure. International Journal of Food Science and Technology 47: 1151-1157.

30. Kobayashi F, Ikeura H, Odake S, Hayata Y (2013) Inactivation of enzyme and *Lactobacillus fructivorans* in unpasteurised sake by a two-stage method with low-pressure CO_2 microbubbles and quality of the treated sake. Innovative Food Science and Emerging Technology 18: 108-114.

31. Kobayashi F, Ikeura H, Odake S, Hayata Y (2014) Inactivation of *Saccharomyces cerevisiae* by equipment pressurizing at ambient temperature after generating CO_2 microbubbles at lower temperature and pressure. Food Science and Technology 56: 543-547.

32. Hiran P, Kerdchoechuen O, Laohakunjit N, Sukrong S (2013) The inhibition of α-glycosidase for Type 2 Diabets from Maize (*Zea mays*). Journal of Agricultural Science 44: 449-452.

33. Seiquer C, Delgado AC, Haro A, Navarro MP (2010) Assessing the effects of severe heat treatment of milk on calcium bioavailability: In vitro and in vivo studies. Journal of Dairy Science 92: 2010-3469.

34. Tanaka Y, Ashikari T, Nakamura N, Kiuchi N, Shibona Y, et al. (1986) Comparison of amino acid sequence of three glucoamylase and their structure function relationships. Agricultural and Biological Chemistry 50: 965-969.

12

Optimisation of Deep-Fat Frying of Plantain Chips (*Ipekere*) using Response Surface Methodology

Adeyanju JA[1]*, Olajide JO[1] and Adedeji AA[2]

[1]*Department of Food Science and Engineering, Ladoke Akintola University of Technology, Ogbomoso, Oyo State, Nigeria*
[2]*Department of Biosystems and Agricultural Engineering, University of Kentucky, Lexington, KY, USA*

Abstract

Deep-fat frying of plantain chips (ipekere) was investigated with the aim of predicting optimum operating conditions for plantain chips to minimize oil content in order to produce healthy products. The effect of frying temperature and time on moisture content, oil content, breaking force and colour difference of plantain chips was evaluated. Response surface methodology was used to analyze the results of the central composite design of the frying processes for the responses as a result of variation in the levels of frying temperature (150°C-190°C) and frying time (2-4 min). Response surface regression analysis shows that responses were significantly ($p<0.05$) correlated with frying temperature and time. Regression model was developed for the investigation of the effect of frying temperature and time on the responses. The polynomial regression models were validated with statistical tool whose values of coefficients of determination (R^2) were 0.9949, 0.9817, 0.9709 and 0.9966 for moisture content, oil content, breaking force and colour intensity, respectively. The optimum values of moisture content, oil content, breaking force and colour difference were 3.73%, 1.18%, 17.66 N and 65.53 respectively, at frying temperature of 183°C and frying time of 3 minutes. Therefore, frying conditions had a significant effect on the quality attributes of chips produced from plantain.

Keywords: Plantain chips; Deep-fat frying; Regression models; Texture and colour

Introduction

Plantain (*Musa* AAB) is one of the most important food crops in the world. It is a major source of carbohydrate, antioxidants and minerals like potassium and calcium and caters for the calorific needs of many people in developing countries [1,2]. Nigeria is one of the largest plantain producing countries in the world with an estimated production at 2,722,000 metric tons in 2009 and an average consumption level of 190 kg/person/year [3,4]. In spite of its prominence, the country does not feature among plantain exporting nations because it produces more for local consumption than for export. Plantain consumption has risen greatly in Nigeria in recent years because of the rapidly increasing urbanization and the great demand for easy and convenient foods by the non-farming urban populations. Besides being the staple for many people in more humid regions, it is a delicacy and flavoured snack for people even in other regions. Plantain chips called Ipekere in south western part of Nigeria. It can be produced from green and slightly ripen green plantain with yellow patch, and fried to almost bone dry with golden yellow colouration.

Deep-Fat Frying (DFF) is a multifunctional unit operation of food transformation that can be described as cooking of food by immersion in edible oil or fat at a temperature higher than the boiling point of water [5]. DFF can be considered as a high temperature and a short time process which involves both mass transfer, mainly represented by water loss and oil uptake, and heat transfer [6]. It is one of the major value addition processes for plantain which results in products with a unique flavour-texture combination [7,8]. The primary reason for the popularity of DFF foods may be the characteristics like soft, juicy interior as well as thick and crispy outer crust [9]. Texture, colour and oil content are the main quality parameters of fried products [10].

Response surface methodology (RSM) is a useful technique for optimisation studies. This is a collection of mathematical and statistical techniques that is useful for modeling and analysis in applications where a response is influenced by several factors [11]. RSM is important in designing, formulating, developing, and analyzing new scientific studies and products. The most common applications of RSM are in industrial, biological, clinical, social, food, physical and engineering sciences. Optimisation is therefore required to ensure rapid processing while maintaining optimum product quality especially in term of the quality characteristics. The quality attributes for frying of food materials may include moisture content, oil content, breaking force and colour parameters while the process parameters to be optimised include frying temperature and time.

Most published studies on plantain over the years have dealt primarily with final product quality and physico-chemical changes in the oil medium during frying [12-15] and none of these reports have particularly optimised frying conditions to obtain fried plantain chips of acceptable quality attributes. However, there is little or no published work on statistical approach in RSM during DFF of plantain chips experiment for obtaining optimum conditions for quality characteristics of plantain. Therefore, the objective of this work was to optimize the DFF conditions with respect to quality attributes like moisture and oil contents, breaking force and colour parameters.

Materials and Methods

Matured plantain (*Musa paradisiacal* AAB) fruits harvested at green stage were procured from a local farm in Ogbomoso, South West Nigeria. Plantains were identified in the Department of Crop and Environmental Protection, LAUTECH, Oyo State, Nigeria. Plantains were washed with clean water, peeled manually and were cut into 2 mm thick slices using a stainless steel knife and a slicer. Refined vegetable oil

***Corresponding author:** Adeyanju JA, Department of Food Science and Engineering, Ladoke Akintola University of Technology, Ogbomoso, Oyo State, Nigeria
E-mail: biodunjames24@yahoo.com

(Devon King's®) obtained from Ace Supermarket, Ogbomoso was used as the frying medium in the deep-fat frying process.

Frying Operation

Frying was carried out in a deep-fat fryer (model MC-DF 1031, Cool Touch deep fryer, General Electric, Hong Kong, China) adapted with a PID temperature controller to maintain the set frying temperature within ±1°C. The fryer was filled with 2.5 L of oil and equipped with a 2 kW electric heater. Plantain slices to oil ratio was kept at 1:10 was used. The oil was preheated prior to frying and discarded after 2 h. Before each frying test, the oil level was checked and replenished as required. Samples were fried at temperatures 150°C, 160°C, 170°C, 180°C and 190°C, for 120, 150, 180, 210 and 240 seconds. After frying, excess oil was removed by shaking the baskets manually and the chips were placed on a rack to cool. Samples were stored in sealed, low density polyethylene bags and kept at room temperature until analyses were performed.

Experimental Design

Central composite of RSM for a two-variable experimental design was employed [11]. The independent factors considered were frying temperature (X_1: 150°C, 160°C, 170°C, 180°C and 190°C) and time (X_2: 2, 2.5, 3.0, 3.5 and 4.0 min) while the dependent factors were moisture content, oil content, breaking force and colour difference (ΔE) (Table 1).

Quality characteristics determination

Moisture content: Five gram was weighed into a pre-weighed moisture dish. The dish plus sample taken was transferred into the oven pre-set at 105°C to dry to a constant weight for 24 hours overnight. At the end of the 24 hours, the dish plus sample was removed from the oven and transferred to desiccator, cooled for 30 minutes and weighed [16].

Oil content: Oil content was determined according to method described by AOAC [16] using soxhlet extraction. Fried samples were ground using a grinder. Five gram of sample was weighed into thimbles for fat extraction in a solvent extractor using petroleum ether. Fat content was determined as the ratio of the mass of extracted fat and dry matter of the sample.

Breaking force: The texture (breaking force) of the chips was determined using a universal testing machine (model M500, Testometric AX, Rochdale, Lancashire, England) equipped with a 50 kN load cell. Fried plantain chips of uniform sizes were selected and then placed on a metal support with jaws at a distance of about 25 mm. They were pressed in the middle with a cylindrical flat end plunger (70 mm diameter) at a speed of 2.5 mm/min. The measurement was recorded by a computer connected directly to the equipment. The breaking force (N) interpreted as crispness was obtained as the peak force from the force-deformation curve [17].

Colour: Colour parameters lightness (L^*), redness (a^*) and yellowness (b^*) were measured using a colorimeter (Colour Tec-PCM, Hunterdon, NJ) as described by Krokida et al. [18]. The instrument was standardized and the samples were placed in the sample holder. Samples were scanned at different locations to determine (L^*, a^* and b^*) parameters. Colour difference (ΔE) was calculated according to the equation:

Colour difference

$$(\text{Hunter } \Delta E) = [(L-L_{ref})^2+(a-a_{ref})^2+(b-b_{ref})^2]^{1/2} \quad (1)$$

Where L_{ref}, a_{ref} and b_{ref} were the L, a and b values of fresh plantain slices which were used as references.

Statistical analysis

All data were analyzed using the Design-Expert Version 6.0.8 (State-ease software). Regression analysis and analysis of variance (ANOVA) were conducted by fitting the equation to the experimental data to determine the regression coefficients and statistical significance of model terms. The significance of the model terms was assessed by F-ratio at a probability $p<0.05$. Model adequacies were determined using model analysis, lack of fit test and coefficient of determination (R^2).

Optimisation procedure

Numerical optimisation was performed using Design Expert software V 6.0.8. Multiple responses were optimised simultaneously through the use of a desirability function that combines all the responses into one measurement. The method finds operating conditions (combination of independent variables) that maximizes the desirability function. The constraints were set to get the value of a variable for an optimum response (a minimum and maximum level must be provided for each variable included). The optimisation of the

Run	Coded		Actual		Responses			
	x_1	x_2	X_1	X_2	MC (%)	OC (%)	BF (N)	ΔE
1	-1.0	-1.0	190	4	3.67	1.22	21.41	34.56
2	1.0	-1.0	170	4	3.69	1.20	15.97	55.89
3	-1.0	1.0	180	3	3.77	1.20	15.75	51.39
4	1.0	1.0	170	3.5	3.5	1.19	17.78	52.64
5	0.0	1.414	190	2	3.76	1.22	24.90	56.80
6	0.0	-1.414	150	4	4.37	1.14	24.11	66.99
7	1.414	0.0	160	3.5	4.32	1.18	18.59	60.24
8	-1.414	0.0	170	3	3.69	1.20	15.97	55.89
9	0.0	0.0	180	2.5	3.91	1.21	18.88	54.55
10	0.0	0.0	190	3	3.86	1.20	15.67	47.73
11	0.0	0.0	170	3	4.06	1.19	22.40	49.19
12	0.0	0.0	170	2	4.19	1.19	17.78	52.64
13	0.0	0.0	170	3	4.02	1.22	23.01	56.42
Where: X1 = Frying temperature (oC), X2 = Frying time (min), MC = Moisture content, OC = Oil content, BF = Breaking force and ΔE = Colour difference								

Table 1: Experimental design arrangement and responses.

DFF process was aimed at finding the levels of frying temperature and time, which could minimize the moisture content, oil content, breaking force and moderate colour.

Results and Discussion

Estimation of the model parameters for plantain chips (*Ipekere*)

The experimental data and their corresponding responses are shown in Table 1. Multiple regression and correlation analysis are used as tools for assessment of the effects of two or more independent variables on the dependent variables. The coefficients of determination (R^2) for moisture content, oil content, breaking force and colour are 0.9647, 0.9844, 0.9303 and 0.9773, respectively. These values are quite high for response surfaces and indicated that the fitted quadratic models accounted for more than 95% of the variance in the experimental data, which were found to be highly significant. The only regression coefficients significant at 95% levels according to p-values were selected for developing the models (Table 2).

Adequacy test of the models for plantain chips

The fitted models were tested for adequacy and consistency by analysis of variance (ANOVA). The results from the statistical analysis revealed that the F-value for moisture content (38.21), oil content (88.19), breaking force (18.69) and colour difference (129.08) were significant at the 95% confidence level. Analysis of variance for response surface quadratic models of plantain chips is shown in Table 2. The empirical models obtained for the quality characteristics of plantain chips are stated as equations 1 to 4. Quadratic model was found to satisfactorily explain the relationship for moisture content, oil content and breaking force while 2FI model was used to represent the relationship between colours of *ipekere* as affected by frying conditions. The model satisfied lack of fit test. Significant model terms at $p<0.05$ were frying temperature (X_1), frying time (X_2), second order of frying temperature and time (X_1^2, X_2^2) and interaction of frying temperature and time (X_1X_2). Therefore, the model is appropriate for predicting moisture content, oil content, breaking force and colour as influenced by frying temperature and time.

Effect of Variables on the Responses

Moisture content

Equation 2 indicates that at linear level, temperature and time had negative effect on the moisture content. At quadratic level, all the variables had positive effect on the moisture content, while interaction between temperature and time had negative effect. Effect of moisture content with frying temperature and time are shown in Figure 1. Significant interaction suggests that the level of one of the interaction variables can be increased while that of other decreased for constant value of the response. Shyu and Hwang [19] reported that the unbound water in the fried food can be rapidly removed when the oil temperature reaches the boiling point of water. The effects of frying temperature and time were significant ($p<0.05$) on the moisture content. Response plot (Figure 1) revealed that the rate of moisture loss in plantain chips increased by increasing the frying temperature and time. The moisture changes during frying showed the typical progressive decrease with increasing frying time. The result obtained was similar to the findings of [20,21] on French fries and chicken meat respectively.

$$MC = 33.51 - 0.31X_1 - 2.11X_2 + 8.91E-004X_1^2 + 0.25X_2^2 - 2.73E-003X_1X_2 \quad (2)$$

$$OC = 1.01 + 5.87E-003X_1 - 0.24X_2 - 2.32E-005X_1^2 + 9.48E-003X_2^2 + 9.98E-004X_1X_2 \quad (3)$$

$$BF = 322.02 - 2.82X_1 - 34.94X_2 + 7.79E-003X_1^2 + 4.87X_2^2 - 0.02X_1X_2 \quad (4)$$

$$\Delta E = -70.18 + 0.78X_1 + 64.40X_2 - 0.39X_1X_2 \quad (5)$$

Responses	Sources of Variation	DF	Sum of Squares	Mean		
				Squares	F-value	R^2
MC	Regression	5	0.650	0.130	38.211	0.9647
	Residual	7	0.023	0.003		
	Total	12	0.674			
OC	Regression	5	0.005	0.001	88.194	0.9844
	Residual	7	9.01E-05	1.29E-05		
	Total	12	0.005			
BF	Regression	5	125.966	25.193	18.695	0.9303
	Residual	7	9.433	1.347		
	Total	12	135.399			
ΔE	Regression	5	660.2538	220.084	129.08	0.9773
	Residual	7	675.599			
	Total	12	0.650	0.130	38.211	

Where: MC = Moisture content, OC = Oil content, BF = Breaking force and ΔE = Colour difference

Table 2: Analysis of variance for the responses.

Oil content

As shown in Equation 3, at linear level, temperature and time had negative effect on the oil content. At quadratic level, temperature had negative effect and time had positive effect, while the interaction between temperature and time had positive effect on the oil content. Significant interaction suggests that the level of one of the interaction variables can be increased while that of other decreased for constant value of the response. The relationship of oil content with frying temperature and time are shown in Figure 2. The increased oil absorption during deep frying of plantain may be accredited to changes in porosity and molecular size redistribution. The oil content in fried products needs to be optimised because it affects the texture and appearance of the chips [22]. The final oil content of the chips is affected by the plantain's physicochemical characteristics and the processing conditions. Moreover, oil content in a product can depend on the frying temperature, time as well as moisture content [23-25]. The result obtained was similar to the findings of [26] who reported increase in fat content at longer frying time.

Breaking force

Breaking force was used to represent the crispness of fried plantain chips. Frying temperature and time had negative linear effect on the crispness of the chips. At quadratic level, temperature had negative effect and time had a significant positive effect, while interaction between temperature and time had negative effect on the crispness. The interaction effect of temperature and time on crispness is shown in Figure 3. As shown in Equation 4, at the two independent variables, second order derivatives and interactions between the variables are significant. Results showed that the effects of temperature were significant ($p<0.05$) on the crispness of the plantain chips. As temperature of frying increases, the rate of moisture removal becomes faster, thus favoring the frying process. The breaking force of the chips decreases as frying temperature and time increases. Crispness of chips is a vital measure that determines the consumers' acceptance [27]. It is the most important textural attribute which denotes freshness and high quality. The peak compression force increased with increasing frying

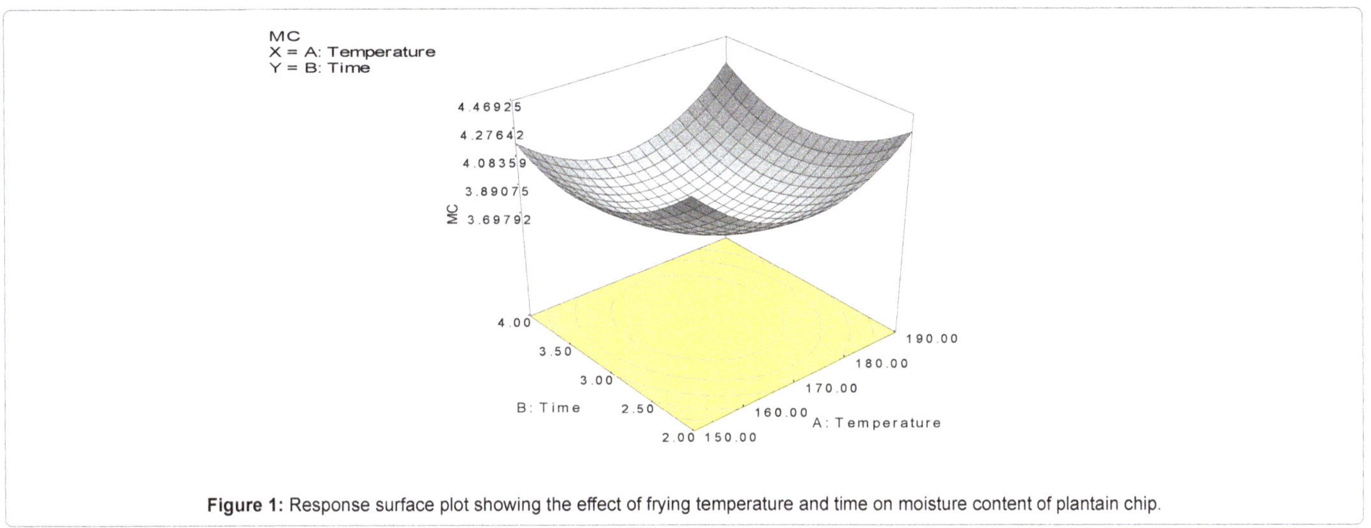

Figure 1: Response surface plot showing the effect of frying temperature and time on moisture content of plantain chip.

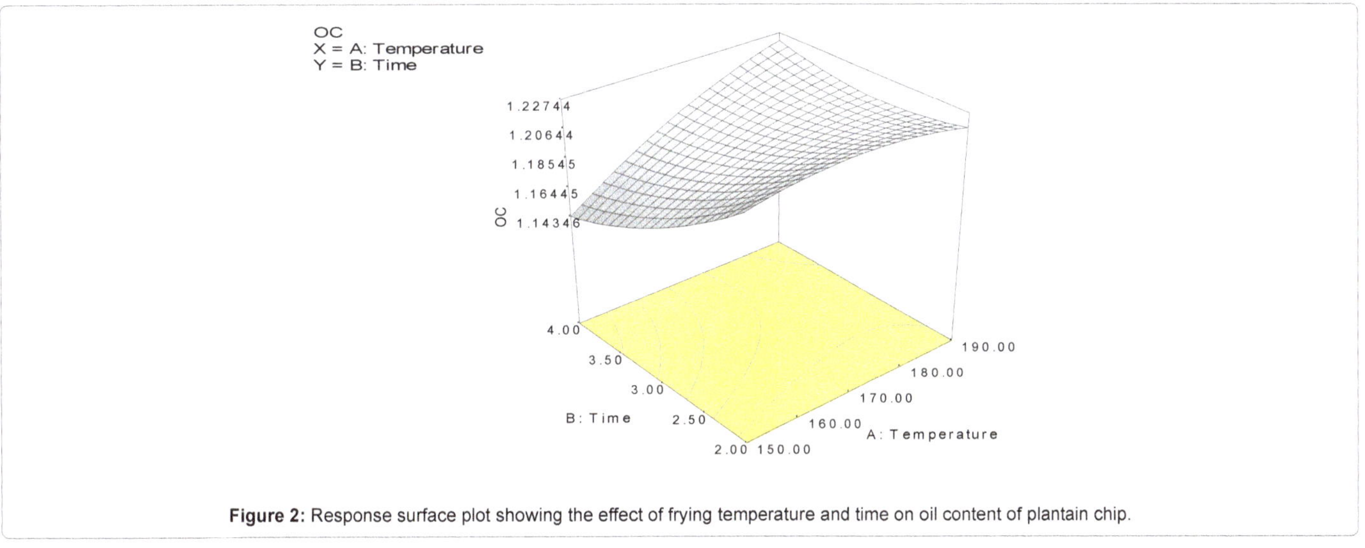

Figure 2: Response surface plot showing the effect of frying temperature and time on oil content of plantain chip.

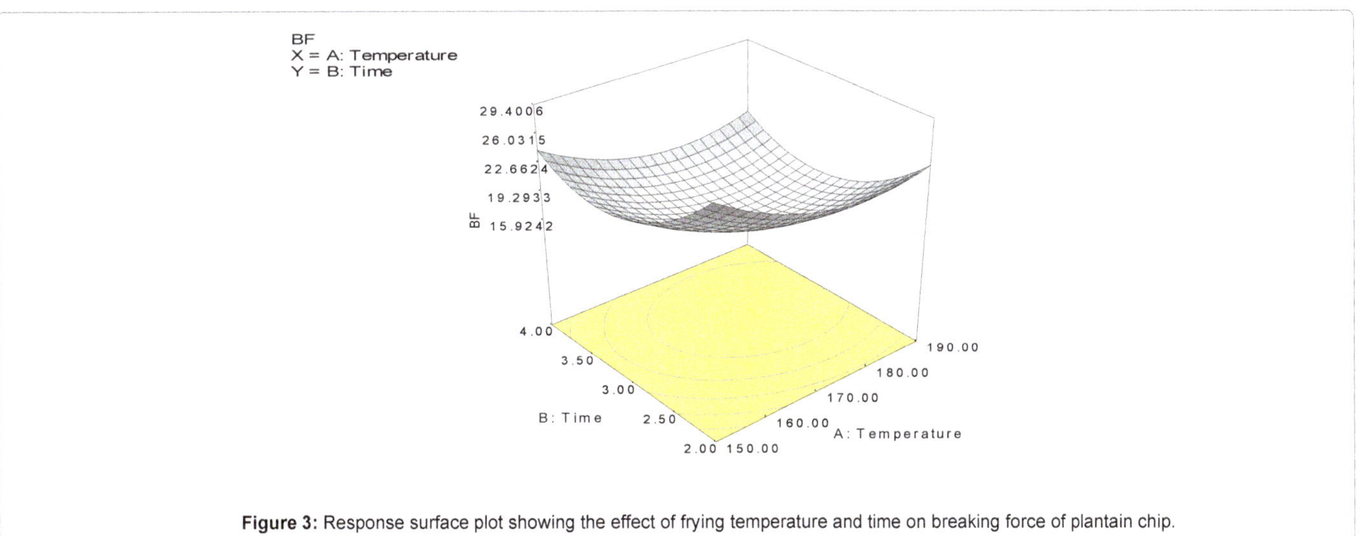

Figure 3: Response surface plot showing the effect of frying temperature and time on breaking force of plantain chip.

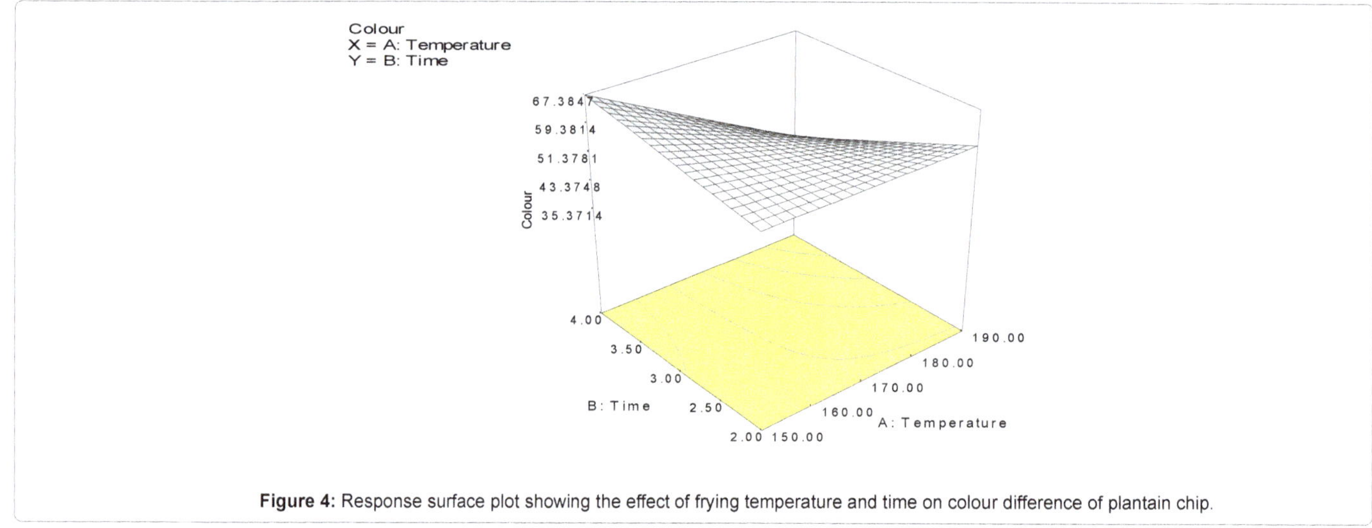

Figure 4: Response surface plot showing the effect of frying temperature and time on colour difference of plantain chip.

conditions or when the initial plantain moisture content was lower. The result obtained was in agreement with the findings that the hardness of fried chickpea flour-based snack increased at lower moisture content of the pre-fried product [28].

Colour difference

The effect of interaction of temperature and time on colour intensity is shown in Figure 4. As reflected in Equation 5, at linear level, temperature had negative effect; time had a significant positive effect on the colour difference. At quadratic level, temperature had positive effect and time had negative effect, while interaction between temperature and time had negative effect on the colour. The colour intensity of plantain chips decreased progressively as the frying temperature and time increased (Figure 4). This indicates that higher ΔE values will lead to lower colour quality on the surface of plantain chips. It was observed that an increase of temperature led to a significant formation of brown products. The modification in fried plantain colour is mainly due to the effect of temperature on heat-sensitive compounds such as carbohydrates, proteins and vitamins, which cause colour degradation during drying process. The formation of brown compounds in plantain may be related to both enzymatic and essentially non-enzymatic (maillard reaction) reactions [29]. This result is similar to the report of [30] that colour changes also indicated more mallard reaction with frying time which utilized the abundant reducing sugars in plantain.

Optimisation of the frying process

RSM was used for the optimisation of DFF of plantain chips and for understanding the factors affecting the frying process. Based on response surface regression analysis of each model, the models (MC, OC, BF and ΔE) were useful for indicating the direction in which to change variables in order to minimize moisture content, fat content and texture (breaking force) and moderate colour. The desirability lies between 0 and 1; and it represents the closeness of a response to its ideal value. Desirability of the solution was 0.76. Three possible optimum solutions were found with desirability ranging from 0.75 to 0.79. The best of the three conditions was frying at 183.61°C for 3.19 minutes which gave 3.73% MC, 1.18% OC, 17.66 N crispness and 65.53 colour.

Conclusion

The RSM was effectively used to investigate the effects of frying temperature and time on the quality characteristics of plantain chips. The optimum values for moisture content, oil content, breaking force and colour difference from the surface plot was 3.73%, 1.18%, 17.66 N and 65.53, respectively. Therefore, temperature and time were determined to have significant effect on the quality attributes of plantain chips. Thus, the optimized frying conditions were able to produce high quality plantain chips.

References

1. Kanazawa K, Sakakibara H (2000) High content of dopamine, a strong antioxidant, in Cavendish banana. J Agric Food Chem 48: 844-848.
2. Mohapatra D, Mishra S, Sutar N (2010) Plantain and its byproduct utilization: An overview. J Sci Ind Res 69: 323-329.
3. FAO (2011) Production, commodity by country; FAOSTAT Data. Food and Agriculture Organisation of the United Nations, Rome.
4. FAO (2013) Food and Agriculture Organization of the United Nations. Crop yield.
5. Farkas BE, Hubbard LJ (2000) Analysis of heat transfer during immersion frying. Dry Technol 18: 1269-1285.
6. Vitrac O, Dufour D, Trystram G, Raoult-Wack A (2002) Characterization of heat and mass transfer during deep-fat frying and its effect on cassava chip quality. J Food Eng 53: 161-176.
7. Mellema M (2003) Mechanism and reduction of fat uptake in deep-fat fried foods. Tr Food Sci Technol 14: 364-373.
8. Pedreschi F (2012) Frying of potatoes: Physical, chemical and microstructural changes. Dry Technol 30: 707-725.
9. Garcia MA, Ferrero C, Bertola N, Martino M, Zaritzky N (2002) Edible coatings from cellulose derivatives to reduce oil uptake in fried products. Inn Food Sci Emerg Technol 3: 391-397.
10. Hindra F, Baik OD (2006) Kinetics of quality changes during food frying. Crit Rev Food Sci Nutr 46: 239-258.
11. Montgomery DC (2005) Design and analysis of experiments: Response surface method and designs. John Wiley & Sons, Inc, New Jersey, USA.
12. Onyejegbu CA, Olorunda AO (1995) Effects of raw materials, processing conditions and packaging on the quality of plantain chips. J Sci Food Agri 68: 279-283.
13. Akubor PI, Adejo EE (2000) Physicochemical, microbiological and sensory changes in stored plantain chips. Plant Foods Hum Nutr 55: 139-146.
14. Agunbiade SO, Olanlokun JO, Olaofe OA (2006) Quality of chips produced from rehydrated dehydrated plantain and banana. Pak J Nutri 5: 471-473.
15. Avallone S, Rojas-Gonzalez JA, Trystram G, Bohuon P (2009) Thermal sensitivity of some plantain micronutrients during deep-fat frying. J Food Sci 74: C339-347.

16. AOAC (2000) Official methods of analysis, Association of Official Analytical Chemists. Virginia, USA.

17. DaSilva PF, Moreira RG (2008) Vacuum frying of high-quality fruit and vegetable-based snacks. LWT-Food Sci Technol 41: 1758-1767.

18. Krokida MK, Oreopoulou V, Maroulis ZB, Marinos-Kouris D (2001) Colour changes during deep-fat frying. J Food Eng 48: 219-225.

19. Shyu S, Hwang L (2001) Effects of processing conditions on the quality of vacuum fried apple chips. Food Res Int 34: 133-142.

20. Krokida MK, Oreopoulou V, Maroulis ZB, Marinos-Kouris D (2001) Effect of pre-drying on quality of french fries. J Food Eng 49: 347-354.

21. Kassama L, Ngadi M (2005) Pore development and moisture transfer in chicken meat during deep-fat frying. J Dry Technol 23: 907-923.

22. Dobarganes MC, Velasco J, Dieffenbache A (2000) Determination of polar compounds, polymerized and oxidized triacylglycerols, and diacylglycerols in oils and fats. Pure Appl Chem 72: 1563-1575.

23. Moreira RG, Sun X, Chen Y (1997) Factors affecting oil uptake in tortilla chips in deep-fat frying. J Food Eng 31: 485-98.

24. Sharma SK, Mulvaney SJ, Rizvi SSH (2000) Frying of foods. In: Food Process Engineering Theory and Laboratory Experiments. Wiley, New York, NY.

25. Yagua CV, Moreira RG (2011) Physical and thermal properties of potato chips during vacuum frying. J Food Eng 104: 272-283.

26. Nantawan T, Phaisan W, Anuvat J, Chatlada K (2007) Optimisation of vacuum frying condition for shallot. Kas J Natur Sci 41: 338-342.

27. Setiady D, Tang J, Younce F, Swanson BA, Rasco BA, et al. (2009) Porosity, colour, texture using microwave vacuum, heated air, and freeze drying. ASABE Appl Engi Agri 25: 719-724.

28. Debnath S, Bhat KK, Rastogi NK (2003) Effect of pre-drying on kinetics of moisture loss and oil uptake during deep fat frying of chickpea flour-based snack food. LWT - Food Sci Technol 36: 91-98.

29. Miranda M, Maureira H, Rodriguez K, Vega-Galvez A (2009) Influence of temperature on the drying kinetics, physicochemical properties, and antioxidant capacity of Aloe Vera (Aloe Barbadensis Miller) gel. J Food Eng 91: 297-304.

30. Onwuka GI, Onwuka ND (2005) The effects of ripening on the functional properties of plantain and plantain based cake. Int J Food Prop 8: 1021-1026.

13

Impact of Wheat-Barley Blending on Rheological, Textural and Sensory Attributes of Leavened Bread

Fiza Nazir[1] and Nayik GA[2*]

[1]*Division of Post-Harvest Technology, SKUAST-Kashmir, India*

[2]*Department of Food Engineering and Technology, SLIET, Punjab, India*

Abstract

The aim of the study was to develop a healthy alternative to wheat-leavened bread by using wheat-barley blended flour. The leavened breads were prepared by blending wheat and barley flour at different levels and varying the MSG concentration. A significant increase was observed in the extensibility of the dough with increase with increasing MSG concentration before fermentation while after fermentation it showed a significant decreasing effect. The highest value of color, taste and appearance was observed for T_2M_1. Results for the flavor of the breads revealed that the highest flavor score was observed for T_2M_2. The results showed that the overall acceptability score decreased with increasing barley flour and MSG level.

Keywords: Wheat flour; Barley flour; Leavened bread; MSG; Sensory analysis

Introduction

Barley (*Hordeum vulgare L.*) is an ancient and important cereal grain crop ranking fifth among all crops in dry matter production in the world. It is one of the most widely cultivated cereal crops that can provide valuable nutrients required by humans and domestic animals. The high adaptability of barley to various climates and growing conditions has led to its increased worldwide production. Barley is arguably the most widely adapted cereal grain species with production at higher latitudes and altitudes and farther in deserts than any other cereal crop. It is in extreme climates that barley remains a principal food source today, e.g. Himalayan nations, Ethiopia, and Morocco [1]. Whole barley grain consists of about 65% to 68% starch, 10% to 17% protein, 4% to 9% β-glucan, 2% to 3% free lipids and 1.5% to 2.5% minerals [2-4]. Human consumption of barley and barley-containing food products has been insignificant as compared to other cereal grains, the development of new processes and food products has been neglected and there has been little effort to define quality requirements for food uses. The significance of β-glucan and tocols for human nutrition is well known, but little is known about the functional properties of β-glucan for making food products. Some of the traits preferred for specific food applications are known through investigations on incorporating barley into wheat-based food products. Barley flour, prepared from pearled grain through hammer milling or roller milling, has been incorporated into wheat based products, including bread, cakes, cookies, noodles and extruded snack foods [5]. Wheat bread with barley flour added at 15% to 20% was acceptable in overall flavour, appearance and texture, but an increased proportion of barley flour caused a decrease in loaf volume, dull brown colour and hard crumb texture [6,7]

Numbers of technologies have been developed for the development of different kinds of bread like to design and optimize processing, storage, and transport conditions and methodologies. Many cereals and their products required for bread making are efficient in one way but at the same time are lacking in other properties as well. The demand for the specialty breads and the role of wheat bran, barley and commercial celluloses and other cereals and their products in bread making is increasing day by day. In order to overcome these demands, various methods like blending of different cereals are used for the development of breads with specialized properties. With this background, the present work for the development of leavened bread by using wheat-barley blended flour was undertaken to study the rheological attributes of wheat-barley blended dough and to prepare breads from composite flours and evaluate the suitable level of barley flour supplementation.

Materials and Methods

The work was carried out in the department of food technology, Islamic University of Science and Technology Awantipora during the year 2010-2011. The wild barley was purchased from Ladakh and then milled in a mixer to obtain whole flour. The flour was stored in plastic air tight containers at refrigerated temperatures until used. Refined wheat flour, shortening, compressed yeast etc were purchased from local market of Srinagar. Composite flours were prepared as mentioned in Table 1. The formulations of the barley enriched breads were according to the Table 2. Breads were prepared from blended flour of wheat and barley according to the procedure shown in Figure 1. MSG was used at 0%, 0.3% and 0.5% level. The ingredients were weighed

Treatment		Concentration of MSG	Wheat flour (%)	Barley flour (%)
T_1	T_1M_1	0.0%	100%	0%
	T_1M_2	0.3%	100%	0%
	T_1M_3	0.5%	100%	0%
T_2	T_2M_1	0.0%	90%	10%
	T_2M_2	0.3%	90%	10%
	T_2M_3	0.5%	90%	10%
T_3	T_3M_1	0.0%	80%	20%
	T_3M_2	0.3%	80%	20%
	T_3M_3	0.5%	80%	20%
T_4	T_4M_1	0.0%	70%	30%
	T_4M_2	0.3%	70%	30%
	T_4M_3	0.5%	70%	30%

Table 1: Wheat-barley composite flours used for bread formulations.

***Corresponding author:** Nayik GA, Department of Food Engineering and Technology, SLIET, Punjab-148106, India, E-mail: gulzarnaik@gmail.com

Wheat flour/composite flour	150 g
Yeast	3.0 g
Sugar	6.0g
Salt	1.5g
Shortening	3.0g

Table 2: Wheat-barley composite bread formulations.

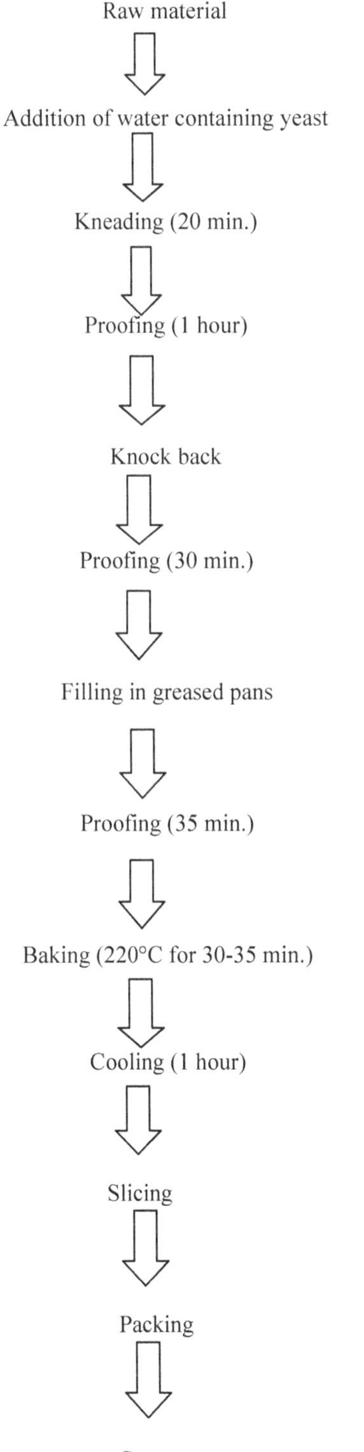

Figure 1: Flow chart for bread making.

accurately and the yeast was activated in hot water. All the ingredients were mixed in a vessel and yeast was added while taking into account the amount of water. The dough was then placed in an incubator at 37°C for fermentation. Dough was taken out after 1 hour and then knocked back to remove the excess gases. The dough was again placed in incubator for fermentation and removed after 30 min, knocked back, rolled and moulded into pans and then allowed to ferment for another 35 min. The pans were then placed in baking oven at 220°C for 30-35min. The breads were taken out, cooled and then sliced. The breads were stored at room temperature.

Rheological and textural analysis

A Texture Analyzer was used to measure the extensibility and firmness (TA. HD. Plus, Stable Micro Systems, Godalming, Surrey, UK). For extensibility, the dough strip was placed onto the grooved region of the sample plate and the plate was inserted into the rig while holding down the spring loaded clamp lever. The handle was released slowly. The analysis was done to a distance of 75 mm, at a pre-test speed of 2.0 mm/s, test speed of 3.3 mm/s, post-test speed of 10.0 mm/s using a 5 Kg load cell. For firmness, the sample was removed from its place of storage and was placed centrally over the supports just prior to testing. The texture profile analysis was done to a distance of 30 mm at pre-test speed of 1.0 mm/s, test speed of 1.7 mm/s and a post- test speed of 10.0 mm/s using a 5 Kg load cell.

Determination of organoleptic characteristics

A panel of 10 judges evaluated the organoleptic characteristics of prepared breads. They assessed crust colour, appearance, flavour, texture, taste and overall acceptability, using a 9-point hedonic rating scale (9-Like extremely, 8-Like very much, 7-Like moderately, 6-Like slightly, 5-Neither like nor dislike, 4-Dislike slightly, 3-Dislike moderately, 2-Dislike very much, 1-dislike extremely). Tap water was provided to the panelists to rinse between evaluations.

Statistical analysis

The data was statistically analysed on a computer using design factorial in Completely Randomized Design (CRD) as suggested by Snedecor and Cochran [8].

Results and Discussion

Table 3 indicated that the mean values for extensibility statistically showed a significant increasing trend from 24.76 to 27.98 with addition of MSG. Further the highest mean extensibility was observed in T_1 (33.51) and lowest mean extensibility was observed in T_4 (18.46). These results are in accordance with the findings of Sternberg et al. [9] who observed increase in extensibility of dough due to the reducing action of MSG. MSG unexpectedly caused weakening of gluten structure for a relatively short period (before fermentation). The extensibility is primarily due to elasticity of glutenin proteins present in wheat, which may dilute with the addition of non-wheat flour resulting in decreased extensibility [10]. The mean values for extensibility showed a significant decreasing trend from 22.39 to 19.18 with addition of MSG (Table 4). The highest mean extensibility was observed in T_1 treatment (28.19) and lowest mean extensibility was observed in T_4 treatment (14.42). Similar results were found by Sternberg et al. [9] who reported that MSG caused strengthening of gluten network after fermentation due to the oxidative action. The decrease in extensibility with addition of barley flour is due to glutenin dilution.

Table 5 indicated that the mean values for firmness elucidated statistically showed a significant increasing trend from 26.60 to 44.73

Treatment	Extensibility			Treatment mean
	M_1	M_2	M_3	
T_1	32.10	33.01	35.42	33.51
T_2	26.93	28.18	30.40	28.50
T_3	22.82	24.07	26.00	24.30
T_4	17.17	18.10	20.12	18.46
MSG mean	24.76	25.88	27.98	-

CD at 5%; Treatment mean (T) = 0.158; MSG mean (M) = 0.0132; T × M= 0.385; T_1 = Control bread (100% W.F); W.F = Wheat flour B.F = Barley flour; T_2 = 10% B.F: 90% W.F; T_3 = 20% B.F: 80% W.F; T_4 = 30% B.F: 70% W.F; M_1 = 0% MSG; M_2 = 0.3% MSG; M_3 = 0.5% MSG

Table 3: Extensibility (mm) of wheat-barley dough at different concentrations before fermentation.

Treatment	Extensibility			Treatment mean
	M_1	M_2	M_3	
T_1	30.12	28.34	26.10	28.19
T_2	25.10	23.15	21.32	23.19
T_3	19.40	17.42	16.10	17.64
T_4	14.97	15.12	13.17	14.42
MSG mean	22.39	21.01	19.18	-

CD at 5%; Treatment mean (T) = 0.15; MSG mean (M) = 0.048; T × M = 0.152 T_1 = Control bread (100% W.F); W.F = Wheat flour; B.F = Barley flour; T_2 = 10% B.F: 90% W.F; T_3 = 20% B.F: 80% W.F; T_4 = 30% B.F: 70% W.F; M_1 = 0% MSG; M_2 = 0.3% MSG; M_3 = 0.5%MSG

Table 4: Extensibility (mm) of wheat-barley dough at different concentrations after fermentation.

Treatments	Firmness		
	Day 1	Day 2	Day 3
T_1M_1	7.68	18.11	28.38
T_1M_2	11.00	23.92	36.83
T_1M_3	17.10	27.55	42.17
T_2M_1	10.04	15.20	27.52
T_2M_2	29.54	36.18	43.50
T_2M_3	33.10	41.73	47.02
T_3M_1	13.52	33.78	42.52
T_3M_2	30.28	43.55	46.60
T_3M_3	38.69	48.27	55.31
T_4M_1	28.05	34.51	42.62
T_4M_2	43.62	48.19	53.28
T_4M_3	56.56	62.35	71.03

Table 5: Raw data for change in firmness (N) of bread at different time intervals and with addition of MSG.

Treatment	Firmness			Treatment mean
	Day1	Day2	Day3	
T_1	11.93	23.19	35.78	23.63
T_2	24.23	31.04	39.36	31.54
T_3	27.49	41.84	48.14	39.16
T_4	42.74	48.35	55.64	48.91
Storage mean	26.60	36.10	44.73	-

CD at 5%; Treatment mean (T) = 3.50; Storage mean (S) = 2.93; T × S = 7.38

Table 6: Change in firmness (N) of bread during storage (Temperature 13°C to 18°C and relative humidity 64% to 63%).

during three days of storage period. Further, the highest mean firmness was observed in T_4 treatment (48.91) and lowest mean firmness was observed in T_1 treatment (23.63). The results are in accordance with the findings of Goesaert et al. [11] who reported increase in firmness of bread during storage due to the transformation of gelatinised starch (amylopectin) network into an extensive, partially crystalline, permanent amylopectin network, with amylopectin crystallites acting as junction zones. The increase in firmness with addition of barley flour is due to gluten dilution. These results are in alignment with the findings of Gill et al. [12] who further reported that β-glucan in barley flour, when added to wheat flour during bread making, could tightly bind to appreciable amounts of water in the dough, suppressing the availability of water for the development of the gluten network. An underdeveloped gluten network can lead to increased bread firmness.

Table 6 indicated that the mean values for firmness elucidated statistically showed a significant increasing trend from 25.15 to 45.07 during storage with the addition of MSG. Further, the highest mean firmness was observed in T_4 treatment (48.91) and lowest mean firmness was observed in T_1 treatment (23.63). These results are in accordance with the findings of Gill et al. [12] who observed increase in firmness of bread with addition of barley flour due to gluten dilution.

Organoleptic characteristics

Crust colour: The highest value of colour was observed for T_2M_1 (7.5) and the lowest value of colour was observed for T_4M_2 (5.4) (Table 7). As the level of barley was increased in blends, the crust colour of the breads changed from creamy white to dull brown. Similar results were reported by Gupta et al. [13] and Dingra et al. [6].

Appearance: The appearance score for the control breads decreased significantly upon increasing the blending level to 20 and 30% with barley. Among the blended breads, the highest appearance score was observed for T_2M_1 (7.1), whereas the lowest score was observed for T_4M_3 (5.0) (Table 7). These results are in quite similar with the findings of Gupta et al. [13].

Flavor: Results for the flavor of the breads revealed that the highest flavor score was observed for T_2M_2 (7.3), whereas the lowest score was observed for T_4M_1 (5.5). The flavor of barley-blended breads might be affected by the fibrous flavour of barley flour (Table 7). A similar decline in acceptability of breads due to strong flavor was observed by noticed by Dingra et al. [13].

Crust texture: The crust texture was related to the external appearance of the breads, i.e., smoothness or roughness of the crust. Crust texture score also decreased with increase in the substitution of barley and MSG in wheat flour as compared with the control sample. Among the blended bread, the highest score was observed for T_2M_1

Treatments	Colour	Texture	Taste	Flavor	Appearance	Overall acceptability
T_1M_1	7.0	7.1	7.0	7.2	6.8	7.0
T_1M_2	7.0	7.1	7.0	7.0	6.7	7.1
T_1M_3	7.0	7.0	7.0	7.1	6.7	7.1
T_2M_1	7.5	7.2	7.6	7.2	7.1	7.4
T_2M_2	7.4	7.1	7.4	7.3	7.0	7.3
T_2M_3	7.4	7.0	7.3	7.2	7.0	7.2
T_3M_1	6.5	6.4	6.0	6.0	6.2	6.3
T_3M_2	6.4	6.2	6.1	6.2	6.0	6.2
T_3M_3	6.3	6.1	6.2	6.3	6.1	6.2
T_4M_1	5.8	5.3	5.1	5.5	5.3	5.4
T_4M_2	5.4	5.2	5.0	5.7	5.1	5.3
T_4M_3	5.5	5.0	5.1	5.6	5.0	5.1

Table 7: Effect of blending on sensory quality of breads.

(7.2), whereas the lowest score was observed for T_4M_3 (5.0) (Table 7). Carson et al. [14] observed similar deterioration in the texture of wheat bread on supplementation.

Taste: Taste evaluation suggested that control and various supplemented breads had most satisfactory taste scores up to the 10% level. Results indicated that the highest taste score was observed for T_2M_1 (7.6), whereas the lowest score was observed for T_4M_2 (5.0) (Table 7). The decrease in taste score was because of the different flavor of flour blends [13].

Overall acceptability: The results showed that the overall acceptability score of all the supplemented breads at the 10% level was at par with the control (Table 7). Breads made from wheat and barley flour up to the 10% level were found acceptable, but at more than 10% substitution, the overall acceptability score was significantly reduced as compared with the control. In overall profile, the flavor of the breads was malty and fibrous at the 30% level of substitution. Based on the above results, breads containing 10% barley flour were found to be most acceptable by the sensory panelists.

Conclusion

The studies revealed that before fermentation the extensibility of the dough showed significant increase with addition MSG and significant decrease with addition of barley flour. After fermentation the extensibility of the dough showed a significant decreasing trend with addition of barley flour and MSG. Firmness of bread showed a significant increasing trend with addition of barley, MSG and with storage. It was concluded from the present studies that breads made from 10% barley flour blend proved superior with respect to sensory characteristics and were better than other blended breads as far as firmness and extensibility was concerned. Further investigations are required to evaluate the compositional and the antioxidant properties of these breads.

Acknowledgement

The first author is thankful to Department of Food Technology, IUST Awantipora, Pulwama J&K, India for providing necessary facilities for lab work.

References

1. Baik B, Ullrich C (2008) Barley for food: Characteristics, improvement and renewed interest. J Cereal Sci 48: 233-242
2. Czuchajowska Z, Klamczynski A, Paszczynska B, Baik BK (1998) Structure and functionality of barley starches. J Cereal Chem 75: 747-754.
3. Izydorczyk MS, Storsley J, Labossiere D, MacGregor AW, Rossnagel BG, et al. (2000) Variation in total and soluble β-glucan content in hulless barley: Effects of thermal, physical, and enzymatic treatments. J Agri Food Chem 48: 982-989.
4. Quinde Z, Ullrich SE, Baik BK (2004) Genotypic variation in colour and discolouration potential of barley-based food products. J Cereal Chem 81: 752-758.
5. Newman RK, Newman CW (1991) Barley as a food grain. J Cereal Food World 36: 800-805.
6. Dhingra S, Jood S (2004) Effect of flour blending on functional, baking and organoleptic characteristics of bread. Int J Food Sci Technol 39: 213-222.
7. Malik H, Nayik GA, Dar BN (2015) Optimisation of process for development of nutritionally enriched multigrain bread. J Food Process Technol 7: 544.
8. Snedecor GW, Cochran WG (1967) Statistical methods. The Lowa State University Press, USA.
9. Sternberg G, George P (1976) Preparation of yeast leavened dough products. J Food Addit 44: 100-110
10. Khatkar BS (2004) Dynamic rheological properties and breadmaking qualities of wheat gluten: Effects of urea and dithiothreitol. J Sci Food Agri 85: 337-341.
11. Goesaert H, Slade L, Levine H, Delcour AJ (2009) Amylase and bread firming-an integrated view. J Cereal Sci 50: 345-352.
12. Gill S, Vasanthan T, Ooraikul B, Rossnagel B (2002) Wheat bread quality as influenced by the substitution of waxy and regular barley flours in their native and extruded forms. J Cereal Sci 36: 219-237.
13. Gupta M, Bawa AS, Semwal AD (2011) Effect of barley flour blending on functional, baking and organoleptic characteristics of high-fiber rusks. J Food Process Preserv 35: 46-63.
14. Carson L, Setser C, Sun XS (2000) Sensory characteristics Food Sci Technol 35: 465-471.

Functional Properties of Restructured Surimi Gel Product Prepared from Low Valued Short Nose White Tripod Fish (*Triacanthus brevirosterus*)

Hema K[1]*, Shakila RJ[2], Shanmugam SA[3] and Jawahar P[4]

[1]*Department of Fish Processing Technology, Fisheries College and Research Institute, Thoothukudi, Tamil Nadu, India*
[2]*Department of Harvest and Post Harvest Management, Fisheries College and Research Institute, Thoothukudi, Tamil Nadu, India*
[3]*Tamil Nadu Fisheries University (TNFU), Nagapattinam, Tamil Nadu, India*
[4]*Department of Fisheries Biology and Resource Management, Fisheries College and Research Institute, Thoothukudi, Tamil Nadu, India*

Abstract

Low valued short nose white tripod fish (*Triacanthus brevirosterus*) was used for the preparation of minced meat, surimi and restructured surimi gel products. Eight different restructured surimi gel products (RS-1 to RS-8) were prepared using additives such as corn, egg white and casein in different proportions along with the control. Functional properties examined indicated that the control (RS-1) without additives had higher gel strength of 9.05 kgF than other products. RS-4 with egg white had more whiteness (74.75%) and got 'AA' class in folding test. Microstructure of surimi with egg white (RS-4) had less surface cracks and cavities contributing for good functional properties. Thus, RS-4 prepared with egg white could be best suited for surimi based products.

Keywords: Tripod; Restructured surimi gel; surimi; Minced meat; SEM

Introduction

Surimi is traditionally a Japanese product prepared from washed fish mince, in which myofibrillar proteins gets concentrated contributing for gel formation [1]. It is generally frozen at -20°C, with the addition of cryoprotectants, such as sucrose, sorbitol and polyphosphates to retain the functional properties [2]. It possesses some important functional properties such as gel forming ability, water holding capacity, foaming ability, emulsifying property and protein solubility [3].

Surimi based products mainly include restructured analogue or imitation products. Restructured products are prepared by the use of salt to solubilize and extract myofibrillar protein, to form sticky exudate that is responsible for the binding in these kinds of products to form desired shapes [4], but it however leads to salty product. It becomes therefore essential to form low salted restructured surimi gel products using some of the food additives, to improve the functional as well as mechanical properties of the resultant product. Egg white, casein, whey protein concentrate, beef plasma, thrombin and microbial transglutaminase are some of the additives that have been widely used to prepare restructured surimi gel products [5-7]. The textural characteristic of the restructured surimi gel products are based on the interaction between the myofibrillar proteins and different additives [8], Examination of the microstructures of these products becomes essential to understand the extent of cross linking and their possible effects on the functional properties. Additives at times cause discoloration in surimi gel products and thus, colour is another important property to be determined for these restructured surimi gel products.

Short nose white tripod fish (*Triacanthus brevirosterus*) belonging to the family Triacanthidae is one of the popular fish species available in surplus during certain season in South East coast of India and is found best suited for the surimi preparation due to their large availability, white flesh, low cost and less domestic consumption. It is an underutilized fish having silvery colour, dusky body on upper half, with or without darker blotches. So far, no work has been attempted in India on the preparation of restructures surimi gel products using surimi made out of trash fish and their functional as well as physical properties. Hence, in the present study, restructured surimi gel products are prepared from the white tripod fish and their functional properties in relation to microstructural changes were determined to examine their suitability in such products.

Materials and Methods

Materials

Short nose white tripod fish (*Triacanthus brevirosterus*) belonging to the family, Triacanthidae caught by the trawl net was procured fresh from the Fishing Harbour, Thoothukudi and brought to the laboratory in chilled condition with the ice to fish ratio (1:1) in insulated containers. The average length and weight of the fish were 28 cm and 330 g, respectively. The 3 ply laminated pouches of dimension 200 × 200 mm consisting of polyethylene, nylon and co-polyethylene were used for heat setting of the surimi gel product (Sealed Air India Ltd, Bangalore, India). Food additives (viz. Egg white, casein, corn starch and cryoprotectants (viz. Sucrose, sorbitol and polyphosphate used were food grade obtained from local merchants.

Preparation of restructured surimi gel product

Fish were washed with potable water to clean the dust, dirt, sand and other extraneous matter and dressed manually to remove the head, entrails and fins. The dressed fish were again washed thoroughly in chilled potable water. The temperature during all the processing steps was maintained between 5°C and 10°C by using sufficient flake ice made using flake ice maker (ZBE 150 Nr 940062, Orlando, Germany). The dressed fish were then fed into a mechanical deboner/mincer

***Corresponding author:** Hema K, Department of Fish Processing Technology, Fisheries College and Research Institute, Thoothukudi, Tamil Nadu, India
E-mail: vathi.hema79@gmail.com

(Baader/601, Berlin, Germany) to obtain minced fish meat. The minced meat was washed with cold water (5°C) at a mince/water ratio of 1:3 (w/v), stirred gently for 4 min and then filtered with a nylon screen having a pore size of 0.2 mm. The washing process was repeated thrice. In the third washing step, cold 0.5% NaCl solution (5°C) was also used. To the washed minced fish, 4% sucrose, 0.25% NaCl, 4% sorbitol and 2.5% NaCl were added and mixed well to obtain surimi.

The moisture content of the surimi was adjusted to around 80% to prepare the restructured surimi gel product. Different food additives were added to the surimi at appropriate concentrations as shown in Table 1 and mixed thoroughly for 30 min using a blender (National Super Mixer Grinder, Matsushita Appliances Co, Japan, India). The mixture was then placed in 3 ply laminated pouches and flattened on the top with a wooden roller. The pouches were then evacuated in a vacuum sealer (Sevana's Quick Seal Vac, Kochi, India) and heat set at 40°C for 30 min in an incubator (Secor, NewDelhi) to form the gel. Then, the pouches were heated in steam at 90°C in an electric steam cooker (Salzer, Chennai, India), for 45 min, to form the restructured surimi gel products.

Functional properties

Gel strength: Gel strength of the restructured surimi gel products was determined using the Universal testing machine (Texture analyzer, Lloyd instruments, UK) following the procedure of Benjakul et al. [9]. Cylinder shaped sample with a length of 2.5 cm was prepared and gel strength was measured using the cylindrical plunger having a 5 mm diameter operated at a depression speed of 60 mm/min and expressed as kgF.

Folding test: Binding structure of restructured surimi gel product was determined by a folding test as described by National Fisheries Institute [10]. The test was conducted by folding a 3 mm slice of gel slowly in half, and then in half again to examine the structural failure of the surimi gel. The number of folding required to crack the surimi specimen was then scored from 1 to 5 and assigned to 5 classes: AA, A, B, C or D. Class AA (5) categorizes good quality and D (1) for poor quality of surimi gels related to the cracking of the gel.

Color: The color of the restructured surimi gel product was determined using the Hunter Lab MainiScan® XE Plus Spectrocolorimeter. The L* (lightness), a*(redness/greenness) and b* yellowness/blueness) were measured and whiteness was calculated as described by Park et al. [11] as follows:

$$\text{Whiteness} = 100 - [(100 - L^*)^2 + a^2 + b^2]^{1/2}$$

Scanning electron microscopy (SEM)

Surimi gels with a thickness of 2-3 mm were first fixed with 2.5% (v/v) glutaraldehyde in 0.2 M phosphate buffer (pH-7.2). They were then rinsed for 1 h in distilled water before being dehydrated in ethanol with serial concentrations of 50%, 70%, 80%, 90% and 100% (v/v). The dried samples were then mounted on a bronze stub and sputter-coated with gold (Sputter coater SPI-Module, West Chester, PA, USA). The specimens were observed with a scanning electron microscope (Field Emission Scanning Electron Microscope S – 3400 N (Hitachi, Japan) at an acceleration voltage of 1000 kV to determine the microstructures of the different restructured surimi gel products.

Types	Composition
RS 1	Surimi
RS 2	Surimi + Corn (4%)
RS 3	Surimi + Casein (2%)
RS 4	Surimi + Egg white (1%)
RS 5	Surimi + Corn (4%) + Casein (2%)
RS 6	Surimi + Corn (4%) + Egg white (1%)
RS 7	Surimi + Casein (2%) + Egg white (1%)
RS 8	Surimi + Corn (4%) + Casein (2%) + Egg white (1%)

Table 1: Additives and their proportions for the preparation of restructured surimi product.

Statistical analysis

Statistical analysis was performed for the various functional and physical properties of the different restructured surimi gel products to examine their significance based on SPSS software (SPSS 10.0, Chicago, IL, USA). All the analysis was carried out in triplicates and the average mean ± standard deviations were calculated. The whole experiments were repeated twice to obtain concurrent results.

Results and Discussion

Gel strength

Hardness of the surimi gel products is generally evaluated using compression and puncture tests. Compression force (kgF) is also known as gel strength. The average gel strength of control surimi gel RS-1 was 9.05 kgF ($p<0.05$). Surimi gels prepared with egg white (RS-4) had slightly lower gel strength of 6 kgF than control but higher than other gel products. The egg white had more gel strength due to the coagulating capacity of ovalbumin [12]. Kuhn et al. [13] had also indicated that the gel strength of king weakfish significantly increased with the presence of protein additives such as bovine serum albumin (BSA) and egg white. Gel strengths of the Alaska Pollack and Pacific Whiting surimi gels prepared with potato starch and egg white were also lower than those gels prepared without any additives or with egg white alone [14], in accordance with the present study.

Water holding capacity

Water holding capacity of food refers to its ability to hold its own and added water during the application of forces, pressing, centrifugation and heating. It is a physical property and as the ability of a food structure to prevent water from being released from the three dimensional structure of the protein [15]. Water holding capacity (WHC) of the restructured surimi gel products differed with the addition of the different proteins (Figures 1 and 2). WHC of the white tripod surimi gel RS-1 was 99%. Alvarez and Tejada [16] have observed

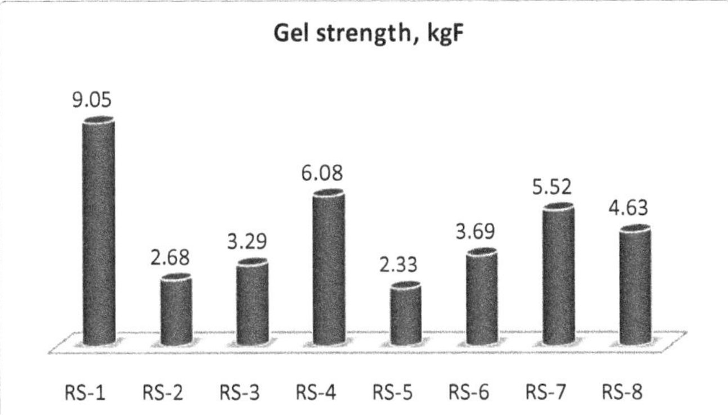

Figure 1: Gel strengths of the different restructured surimi gel products from short nosedwhite tripod fish (*Triacanthus brevirosterus*). (RS-1 Surimi, RS-2 Surimi with corn, RS-3 Surimi with casein, RS-4 Surimi with egg white, RS-5 Surimi with corn andcasein, RS-6 Surimi with corn and egg white, RS-7 Surimi with casein and egg white, RS-8 Surimi with corn, Casein and egg white).

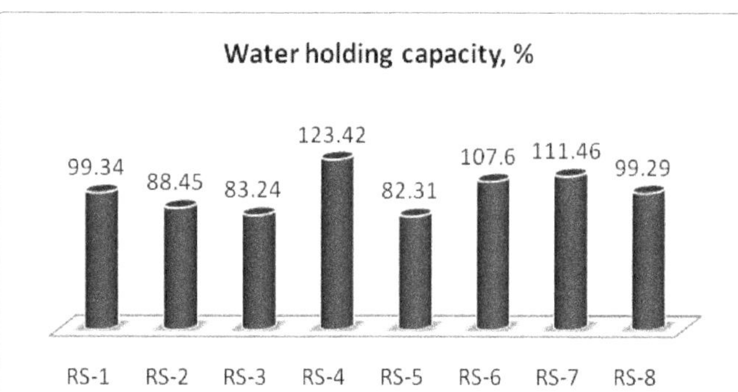

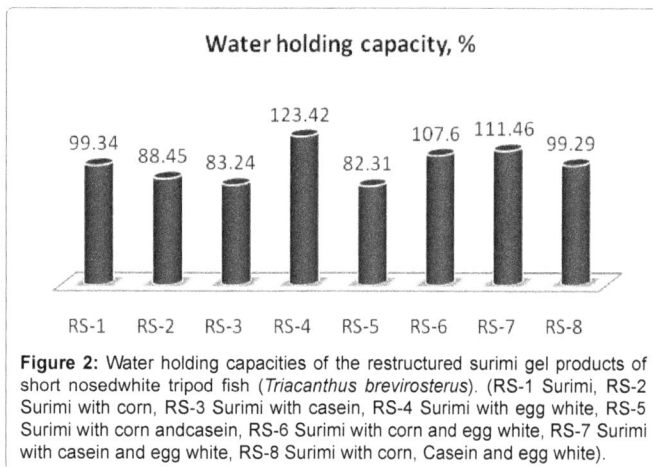

Figure 2: Water holding capacities of the restructured surimi gel products of short nosedwhite tripod fish (*Triacanthus brevirosterus*). (RS-1 Surimi, RS-2 Surimi with corn, RS-3 Surimi with casein, RS-4 Surimi with egg white, RS-5 Surimi with corn andcasein, RS-6 Surimi with corn and egg white, RS-7 Surimi with casein and egg white, RS-8 Surimi with corn, Casein and egg white).

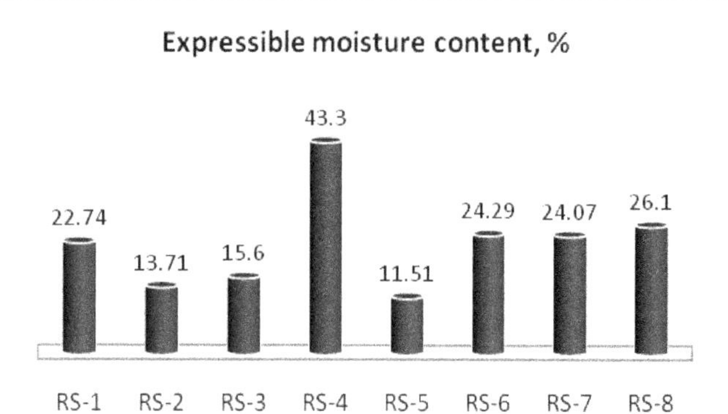

Figure 3: Expressible moisture contents of the restructured surimi gel products of short nosed white tripod fish (*Triacanthus brevirosterus*). (RS-1 Surimi, RS-2 Surimi with corn, RS-3 Surimi with casein, RS-4 Surimi with egg white, RS-5 Surimi with corn andcasein, RS-6 Surimi with corn and egg white, RS-7 Surimi with casein and egg white, RS-8 Surimi with corn, Casein and egg white).

that suwari gels have lower WHC than kamaboko gels. As the surimi gel products are processed at high temperature 90°C it is expected to contain more WHC, as heating induces the aggregation of protein after appropriate swelling and helps to imbibe the water in the gel matrix. But, the WHC's of the surimi gels prepared with corn (RS-2) and were lower than the control (RS-1). On the other hand, addition of egg white (RS-4) had increased WHC to 123% indicating that clefts and crevices in the coagulated egg protein having non-polar groups had helped in the retention of water as hydrodynamic water in the surimi gel matrix. Surimi gel products prepared with combination of different additives had intermediate WHC's.

Tabilo-Munizaga and Barbosa Canovas [14] have reported that Alaska polack surimi gels without additives Pacific Whiting surimi gel with egg white had the highest WHC stating that different fish surimi gels expressed different WHC based on the nature of fish proteins. The presence of sarcoplasmic proteins may also interfere with myosin cross linking during gel matrix formation because they do not form gels and have poor WHC [17]. It has also been reported that protein additives in surimi gels induce protein protein interaction which displace the water resulting in decreased WHC. Thus, it was clear, that WHC of surimi gels depends on gel setting temperature presence of sarcoplasmic proteins as well as additives.

Expressible moisture content (EMC)

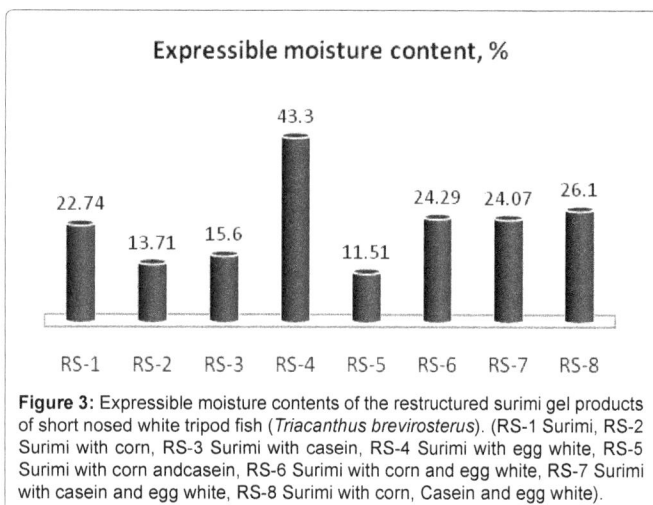

Average expressible moisture contents of the control gel; RS-1 was 23% (Figure 3). With the addition of corn, RS-2 and casein, RS-3 there was a significant reduction in the EMC's ($p < 0.05$). Balange and Benjakul [18,19] reported lower EMC in mackerel surimi gels added with oxidized tannic acid due to increased breaking force and deformation of surimi gels. Similarly, in this study, surimi gels prepared with corn and casein (RS-2, RS-3 and RS-5) had much lower EMCs, as they failed to pass folding test due to lower gel strengths and lower WHCs. Surimi gels prepared with egg white, RS-4 had very high EMC of 43% ($p < 0.05$) than other gel products. Egg white forms protein stabilized foam at neutral PH that are more stable due to lack of repulsive interactions. This promotes fish protein - egg protein interaction and formation of viscofilm at the interface, which further adsorbs more protein. Subsequent heating coagulate the egg protein leaving aside several clefts and crevices in gel matrix, in which, bulk water condenses as hydrodynamic water that had increased WHC. This bulk water can easily be removed on compression and expressed as high EMC in these gels. It has been stated that gels prepared by direct heating showed higher EMC than kamaboko gels indicating that protein network of suwari gels has lower WHCs [20].

Folding test

Folding test is a simple and fast method to measure the binding property of the surimi gel. All the restructured surimi gel products of white tripod fish did not break when dipped with finger (Table 1). Generally high quality surimi does not show any fracture because of the presence of cryoprotectants that improve the stability of myofibrillar protein and gel forming capacity. In folding test, the control surimi gel (RS-1) showed breakage and was graded as 'B' class. Such breaks were also exhibited by the surimi gels, RS-2 and RS-7, as they were made of corn starch. Starch is known to inhibit the gelation of fish proteins by competing for the available water [21]. Slight breakages were noticed in the surimi gels, RS-3 and RS-5, which contained casein. There were no breaks in surimi gels prepared with egg white (RS-4) providing them good gelling ability. Earlier, the folding test performed for surimi gel products of Alaska Pollack and Pacific whiting with egg white and potato starch scored the maximum of 5.0 with grade 'AA' indicated good gelling ability [22].

Color

Color is an important factor determining the quality and acceptability of surimi gels. The lightness, L* was high in surimi gel, RS-4 with 74.55 ($p<0.05$), followed by RS -3 prepared with casein (Table 2). The demand for surimi gels with high lightness (L*), low yellowness, (b*) and high whiteness (W), is generally high. Addition

Samples	Folding test		
	Finger dip	Folding	Class/Grade
RS-1	No	Yes	Class B/3
RS-2	No	Yes	Class B/3
RS-3	No	Slightly break	Class A/4
RS-4	No	No	Class AA/5
RS-5	No	Slightly break	Class A/4
RS-6	No	No	Class AA/5
RS-7	No	Yes	Class A/4
RS-8	No	No	Class AA/5

RS-1 Surimi, RS-2 Surimi with corn, RS-3 Surimi with casein, RS-4 Surimi with egg white, RS-5 Surimi with corn and casein, RS-6 Surimi with corn and egg white, RS-7 Surimi with casein and egg white, RS-8 Surimi with corn, Casein and egg white.

Table 2: Folding test grades and scores of the different restructured surimi gel products.

of egg white and casein had improved (L*) values, than corn starch. All protein additives have an impact on the colour of surimi gels with a slight reduction in (L*) values and a large increase of b* values [11]. Starch generally increases lightness as these granules absorb water and become fully swollen; and light can pass through them generating transparent gels [23], but this was not evidenced in case of corn starch due to improper gelatinization of starch in the protein gel. With the addition of corn, flaxseed, algae, menhaden, krill, and blend oils, the L*values reduced while with flaxseed and menhaden oils, the L* values increased due to minimal pigmentation in the oils [24]. Tabilo–Munizaga and Barbosa Canovas [14] reported that the L* values of heat induced Pacific Whiting surimi gels was 81.01, while it was 82.22 with potato starch, 81.36 with egg white and 80.86 with both egg white and potato starch. Corn starch added to white tripod surimi gels also did not increase the L*values, similar to potato starch. Tammatinna et al. [25] indicated that the L* values of shrimp surimi gels set at 24°C for 2 h were slightly higher than set at 40°C for 30 min. The processing time and temperature also affect the b* values of the Alaska Pollack surimi gels [26], similar to L* values.

The other two tri stimulus colour values, a* and b* are coordinates in which a* has positive or negative values for reddish or greenish hues, respectively and b* has positive or negative values for yellowish or bluish hues, respectively. White tripod surimi gels had greenish hue with negative a* values (Table 2), among which with RS-1 and RS-3 had higher negative values. Alaska Pollack surimi gels with no oil, corn, flaxseed, algae and menhaden oils also had slightly negative a* values [24]. However Park [27] indicated that there was no significant difference in the a* values at different heating and setting conditions of Alaska Pollack and Pacific Whiting surimi gels.

White tripod surimi gels had higher positive b* values indicating yellowness hues. The yellowness, b* was high in white tripod surimi gels with egg white than corn starch and casein (Table 3). But, in Pacific Whiting and Alaska Pollack surimi gels prepared with egg white, b* values were higher than potato starch [14].

The most important quality parameter in surimi seafood is the whiteness W, which was calculated as recommended by Park [24] due to its effectiveness to differentiate the behaviour of different additives. Whiteness of the surimi gels prepared with egg white (RS-4) was higher (74.75) than all other products ($p < 0.05$). Klesk et al. [28] had indicated that surimi gel prepared without additives exhibited higher whiteness values than those with additives. In Alaska Pollack and Pacific Whiting surimi gels with potato starch and egg white, whiteness was lower by 7% and 3% with respectively compared to control [14]. Addition of beef plasma protein (BPP) to the lizard fish surimi gel had resulted in the decrease of whiteness; while addition of egg white had no effect according to Benjakul et al. [22]. In our study, whiteness values were high in the surimi gels made of egg white. In general, heat induced surimi gels without additives exhibited better L* and b* values, but whiteness values were high in gel even with additives.

Samples	Lightness, L*	Redness, a*	Yellowness, b*	Whiteness, W
RS-1	71.76 ± 0.11	- 1.40 ± 0.17	7.80 ± 0.31	70.67 ± 0.59
RS-2	71.48 ± 0.70	- 1.32 ± 0.05	7.66 ± 0.15	70.39 ± 0.90
RS-3	72.43 ± 0.18	- 1.44 ± 0.08	9.58 ± 0.34	70.78 ± 0.60
RS-4	74.55 ± 1.01	- 1.02 ± 0.05	6.52 ± 0.66	74.75 ± 1.72
RS-5	71.55 ± 0.70	- 0.94 ± 0.26	8.19 ± 0.87	70.40 ± 1.83
RS-6	70.67 ± 0.50	- 1.35 ± 0.09	7.86 ± 0.54	70.52 ± 1.13
RS-7	72.01 ± 0.28	- 1.17 ± 0.49	9.04 ± 0.57	70.57 ± 1.34
RS-8	70.43 ± 0.40	- 0.86 ± 0.01	8.33 ± 0.08	69.26 ± 0.49

RS-1 Surimi, RS-2 Surimi with corn, RS-3 Surimi with casein, RS-4 Surimi with egg white, RS-5 Surimi with corn and casein, RS-6 Surimi with corn and egg white, RS-7 Surimi with casein and egg white, RS-8 Surimi with corn, Casein and egg white.

Table 3: Colour values of the different restructured surimi gel products.

Microstructure (SEM)

Microstructure analysis is also another important phenomenon to identify the change in molecular structural arrangement after high temperature processing of surimi gels. SEM images at 1000 X magnifications of the surimi gel products are shown in Figure 4. In general, the appearances of the matrices were different with regular or irregular surfaces and formation of cracks and cavities. The control surimi gel (RS -1) showed an irregular pattern and looked very smooth with deep cavities. As all the surimi gel products are prepared with 2.5% NaCl, they are expected to have more extensive solubilisation of myofibrillar proteins [29-38] with regular mesh in the matrices. But, Gomez-Guillen et al. [30] had found in gels with 1.5% NaCl, that the mesh was rather more irregular with large pore size than in the gels with 2.5% NaCl [39-43]. Apart from the salt concentration, the heating temperature at 95°C was also the cause for the haphazard protein aggregation, giving rise to a less stable structure for the gels.

The matrix of surimi gel with corn (RS-2) was more uniform with regular cavities expressing a true 3-D network [44-50]. In this gel, the corn starch that gelatinizes at 67°C to 70°C in water did not gelatinize even at 90°C, indicating that gelatinization temperature is also influenced by the presence of fish myofibrillar proteins and the available hydrodynamic water for gelation. Also, the myofibrillar proteins were not been fully solubilised leading to aggregation upon heating, accompanied by more extensive protein syneresis. As a result, the water got expelled, the muscle protein got retracted and a number of holes or cavities had appeared in the gel [51-60]. The number and size of the cavities are determined by the amount of water expelled from the matrix [15]. Large white coloured spherical clusters on the surface of the RS-2 gel matrix indicated self-aggregation of corn molecules by means of hydrophobic interactions or hydrogen bonding. This

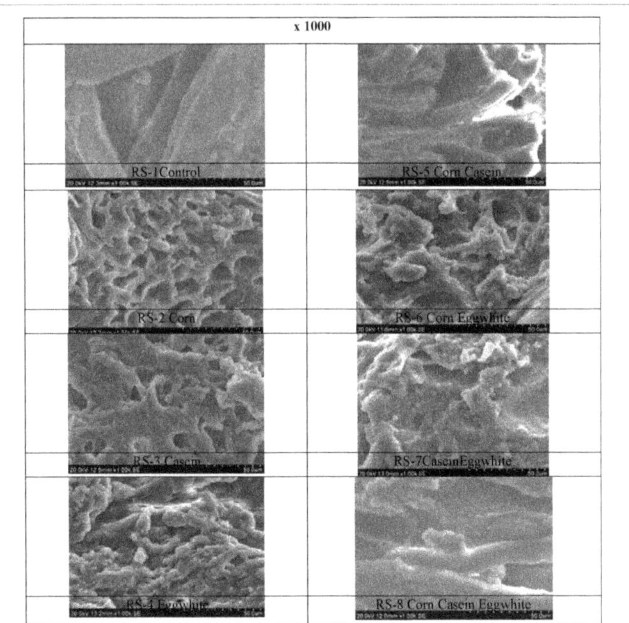

Figure 4: Electron microscopic images of the different restructured surimi gel products of short nosed white tripod fish at 1000 X magnification.

propensity of self-interaction of corn interferes with the cross-linking of myofibrillar proteins, to form an adequate gel network.

The surimi gel with casein, RS-3, more or less compact with less cavities. Casein acts as functional filler by increasing shear stress but not the shear strain values [27]. Casein and gluten were arranged in randomly distributed clusters in sardine surimi gels [30]. The aggregation of protein was such that there were deep cracks in the matrix; and there was lack of fibrillar orientation that had contributed for their very low gel strength.

The surimi gel with egg white (RS-4) had the matrix with more uniform, compact and fibrous appearance with clusters but no cavities. Egg white being a functional binder increases both the shear stress and shear strain values [27]. There was a negative relationship between the size of these cavities and gel strength. Thus, more visible cavities in the surimi gels with corn contributed for the lowest gel strength (2.68 kgF), less numbers of cavities in gels with casein had slightly lower gel strength (3.29 kgF), and no cavities in gels with egg white had the highest gel strength (6.08 kgF). Gomez-Guillen et al. [30] had reported that the sardine gels with egg white are better formed and more uniform with high gel strength and breaking force than with soy protein, casein or gluten, similar to our findings.

In other restructured gels containing egg white (RS-6, RS-7and RS-8), the matrix had a reticulate appearance with round, evenly sized pores. Egg white was visibly more aggregated in gel and appeared as small clusters throughout the matrix. According to Kuhn and Soares [31] coagulating capacity of the egg white occurred together with the starch gelation in gels leading to an increase in the viscosity of the batter, which prevents the coalescence of the air cells inside the structure thereby increasing the volume without contributing to the stability.

Conclusion

To protect network protein responsible for the gel formation, many compounds of protein nature are used to improve the gel physical properties and control proteolysis by avoiding myofibrillar cleavage. These compounds are the protease inhibitors and among them, egg white gave better effect than other proteins like potato extracts or whey proteins [31]. Use of protease inhibitors independently of fish species, but at different levels in Alaska Pollack and Pacific whiting had resulted in an increase in gel strength [32,33]. In the present study, the surimi gel with egg white, RS-4 had a uniform gel network with fewer cavities due to effective crosslinking and protein aggregation, provided good physical properties to the gel, mainly gel strength and WHC. Further, egg white is also an effective protease inhibitor that could avoid myofibrillar cleavage and provide high gel strength to the surimi gel products.

Acknowledgement

The authors would like to thank the Dean, Fisheries College and Research Institute, Thoothukudi for his encouragement and for providing the necessary facilities to undertake this work.

References

1. Benjakul S, Visessanguan W, Srivilai C (2001) Gel properties of bigeye snapper (*Priacanthus tayenus*) surimi as affected by setting and porcine plasma protein. J Food Qual 24 : 453 - 471
2. MacDonald GA, Lanier TC (1994) Actomyosin solubilization to freeze-thaw and heat denaturation by lactate salts. J Food Sci 59: 101-105
3. Zhou A, Soottawat B, Ke P, Jie G, Xin L (2006) Cryoprotective effects of trehalose and sodium lactate on tilapia (*Sarotherodon nilotica*) surimi during frozen storage. J Food Chem 96: 96-103.
4. Zimmerman PA, Bissel HM, Mclntosh GS (1998) Method of processing salmonoid fish. Arctic Alaska Seafoods Inc.
5. Yetim H, Ockerman W (1995) The effects of egg white, tumbling and storage time on proximate composition and protein fractions of restructured fish product. J Aqua Food Pro Tech 4: 65-67.
6. Baker K, Lanier T, Green D (2000) Cold restructuring of seafoods using transglutaminase mediated binding. Seafoods IFT Annual Meeting Book of Abstracts 164: 75-76.
7. RM Uresti, Tellez-Luis SJ, Ramıirez JA, Vazquez M (2004) Use of dairy proteins and microbial transglutaminase to obtain low-salt fish products from filleting waste form silver carp (*Hypophthalmichthys molitrix*). J Food Chem 86: 257-262.
8. Lee CM, Chung KH (1989) Analysis of surimi gel properties by compression and penetration tests. J Textur Studi 20: 363-377.
9. Benjakul S, Visessanguan W, Tueksuban J (2003) Changes in physico-chemical properties and gel-forming ability of lizardfish (*Saurida tumbil*) during postmortem storage in ice. J Food Chem 80: 535-544.
10. National Fisheries Institute (NFI) (1991) A manual of standard methods for measuring and specifying the properties of surimi. Technical subcommittee of the surimi and surimi seafoods committee National Fisheries Institute, Washington, DC, USA.
11. Park JW, Morrissey MT (1994) The need of developing uniform surimi standards. Oregon Sea Grant Publication, Corvallis.
12. Werasinghe VC, Morrissey MT, An H (1996) Characterization of active components in food-grade proteinase inhibitors for surimi manufacture. J Agri Food Chem 44: 2584-2590.
13. Kuhn CR, Prentice-Hernandez Carlos VJ, Soares GJ (2004) Surimi of king weak fish (Macrodon ancylodon) wastes: texture gel evaluation with protease inhibitors and transglutaminase. Brazilian Arch Biol Technol 44: 895-907.
14. Tabilo MG, Barbosa-Canovas GV (2004) Color and textural parameters of pressurized and heat treated surimi gels as affected by potato starch and egg white. Int J Food Res 37: 767-775.
15. Hermansson AM (1986) Water and fat holding. In: Mitchell JR, Ledward A (eds.) Functional properties of food macromolecules. Elsevier Applied Science Publishers, London, pp: 273-314.
16. Alvarez C, Tejada M (1997) Influence of texture of suwari gels on kamaboko gels made from sardine (*Sardine pilchardus*) surimi. Agri J Food Sci 75: 472-480.
17. Sikorski ZE, Pan BS, Shahidi E (1994) Seafood proteins. Chapman and Hall, New York.
18. Balange AK, Benjakul S (2009) Effect of oxidized phenolic compounds on the gel property of mackerel (*Rastrelliger kanagurta*) surimi. J Food Sci Technol 42: 1059-1064.
19. Balange AK, Benjakul S (2009) Effect of oxidized tannic acid on the gel properties of mackerel (*Rastrelliger kanagurta*) mince and surimi prepared by different washing processes. J Food Hydrocolloids 23: 1693-1701.
20. Niwa E (1992) Chemistry of surimi gelation, In: Lanier TC, Lee CM (eds.) Surimi Technology. Marcel Dekker, New York, pp: 389-428.
21. Lin TM, Park JW (1997) Effective washing conditions reduce water usages for surimi processing. J Aqua Food Prod Technol 6: 65-79.
22. Benjakul S, Visessanguan W, Tueksuban J, Tanaka M (2004) Effect of some protein additives on proteolysis and gel forming ability of lizardfish (*Saurida tumbil*). J Food Hydrocolloids 18: 395-401.
23. Yang H, Park JW (1998) Effects of starch properties and thermal processing conditions on surimi-starch gels. Lebens-Wiss Technol 31: 344-353.
24. Pietrowski BN, Tahergorabi R, Matak KE, Tou JC, Jaczynski J, et al. (2011) Chemical properties of surimi seafood nutrified with Omega-3 rich oils. Food Chem 129: 912-919.
25. Tammatinna A, Benjakul S, Visessanguan W, Tenaka M (2007) Gelling properties of white shrimp (*Penaeus vannamei*) meat as influenced by setting condition and microbial transglutaminase. Food Sci Technol 40: 1489-1497.
26. Shie J, Park JW (1999) Physical characteristics of surimi seafood as affected by thermal processing conditions. J Food Science 64: 287-290.

27. Park JW (1995) Surimi gel colors as affected by moisture content and physical conditions. J Food Science 60: 15-18.

28. Klesk K, Yonsawatidigul J, Park J, Viratchakul S, Virulhakul P, et al. (2000) Gel forming ability of tropical tilapia surimi as compared with Alaska Pollack and Pacific whiting surimi. J Aqua Food Prod Technol 9: 91-100.

29. Burgarella JC, Lanier TC, Hamann DD, Wu MC (1985) Gel strength development during heating of surimi in combination with egg white or whey protein concentrate. J Food Sci 50: 1595-1597.

30. Gomez-Guillen MC, Montero P (1996) Addition of hydrocolloids and non muscle proteins to sardine (*Sardina pilchardus*) mince gels- Effect of salt concentration. J Food Chem 56: 421-427.

31. Kuhn CR, Soares GJD (2002) Proteases e inhibitors no processamento de surimi. Revista Brasileira de Agrociencia 8: 5-11.

32. Seymour TA, Peters MY, Morrissey MT, An H, (1997) Surimi gel enhancement by bovine plasma proteins. J Agri Food Chem 45: 2919-2923.

33. Yongsawatdigul J, Park JW, Virulhakul P, Viratchakul S (2000) Proteolytic degradation of tropical tilapia surimi. J Food Sci 65: 129-133.

34. Boye SW, Lanier TC (1988) Effects of heat stable alkaline protease activity of Atlantic menhaden (*Brevoortia tyrannus*) on surimi gels. J Food Sci 53: 1340-1342.

35. Haard NF, Simpson BK, Pan BS (1994) Sarcoplasmic proteins and other nitrogenous compounds. Food Sci Technol.

36. Ismail MI, Kamal MM, Shikha FH, Hoque MS (2004) Effect of washing and salt concentration on the gel forming ability of two tropical fish species. Int J Agri Biol 6: 762- 766.

37. Kaewudom P, Benjakul S (2011) Properties of surimi gel as influenced by the addition of fish Gelatin and Oxidized tannic acid. The 12th Asian Food Conference.

38. Lanier TC, Hart K, Martin RE (1991) A manual of standard methods for measuring and specifying the properties of surimi. National Fisheries Institute, Washington, DC.

39. Okada M (1992) History of surimi technology in Japan. In: Tyre CL, Chong ML (eds.) Surimi technology. Marcel Dekker, New York, pp: 3-21.

40. Ramirez JA, Del Angel A, Uresti RM, Velazquez G, Vazquez M, et al. (2007) Low-salt restructured fish products using low-value fish species from the Gulf of Mexico. Int J Food Sci Technol 42: 1039-1045.

41. SPSS (2000) Statistical software package for social sciences for windows-release 10. SPSS, Chicago, IL, USA.

42. Suvanichi V, Jahncke ML, Marshall LD (2000) Changes in selected chemical quality characteristics of channel catfish frame mince during chill and frozen storage. J Food sci 65: 24-29.

43. AOAC (1995) Official methods of analysis (16thedn). Association of Official Analytical Chemists, Washington, D.C.

44. Hashimoto K, Waltabe S, Kono M, Shiro K (1979) Muscle protein composition of sardine and mackerel. Bulletin Japanese Soc Fish Sci 45: 1435-1441.

45. Lanier TC (1986) Functional properties of surimi. J Food Sci 40: 107-114.

46. Lanier TC (1992) Measurements of surimi composition and functional properties. In: Lanier TC, Lee CM (eds.) Surimi process technology. Marcel Decker Inc, New York, pp: 123-166.

47. Regenstein JM (1986) The potential for minced fish. Food Technol 40: 101-106.

48. Lin D, Morrissey MT (1995) Northern squawfish (*Ptychocheilus oregonensis*) for surimi production. J Food Sci 60: 1245-1247.

49. Saiki H, Hirata F (1994) Behaviour of fish meat compounds during manufacture of frozen surimi through processing with CaCl2-washing. Fish Sci 60: 335-339.

50. Yathavamoorthi R, Sankar TV, Ravishankar CN (2010) Effect of ice storage and washing on the protein constituents and textural properties of surimi from *Labeo Calbasu* (Hamilton, 1882). Indian J Fisheries 57: 85-91.

51. Sarma J, Srikar LN, Reddy GVS (1999) Effect of ice storage on the functional properties of pink perch and oil sardine meat. J Sci Food Agri 79: 169-172.

52. Lee CM (1992) Factors affecting physical properties of fish protein gels. In: Felik GJ, Martin RE (eds.) Advances in seafood biochemistry, composition and quality. Technomic Publishing Co, Inc, Lancaster-Basal, pp: 43-67.

53. Sankar TV, Ramachandran A (1998) Utilisation of fresh water catla (*Catla catla*) for production of myofibrillar protein concentrates. Proceedings of the Asia Pacific Fisheries Commission (APFC) Symposium, Beijing. People's Republic of China, RAP Publication FAO.

54. Chaijan M, Benjakul S, Viessanguan W, Faustman C (2004) Characteristics and gel properties of muscles from sardine (*Sardinella gibbosa*) and mackerel (*Rastrelliger Kanagurta*) caught in Thailand. Int J Food Res 37: 1021-1030.

55. Benjakul S, Visessanguan W, Srivilai C (2001) Gel properties of bigeye snapper (*Priacanthustayenus*) surimi as affected by setting and porcine plasma protein. J Food Qual 24: 453-471.

56. Hossain MI, Komal M, Sakib MN, Shikha FH, Neazuddin N, et al. (2005) Influence of ice storage on the gel forming ability, myofibrillar protein solubility and Ca2+- ATP ase activity of queen fish (*Chorinemus Kysan*). J Biol Sci 5: 519-524.

57. Lin TM, Park JW (1996) Extraction of proteins from pacific whiting mince at various washing condition. J Food Sci 61: 432-438.

58. Wu YJ, Atallah MT, Hultin HO (1991) The proteins of washed mince fish muscle have significant solubility in water. J Food Biochem 15: 209-218.

59. Reppond KD, Babbit JK (1993) Protease inhibitions affect physical properties of arrow tooth flounder and walleye Pollack surimi. J Food Sci 58: 96-98.

60. Sato S, Tsuchiya T (1992) Microstructure of surimi and surimi based products. In: Lanier TC, Lee CM (eds) Surimi Technology. Marcel Dekker Inc, New York, pp: 123-166.

Molecular Characterization of *Klebsiella* Isolates from Enteral Diets

Pereira SCL[1]* and Vanetti MCD[2]

[1]*Universidade Federal de Minas Gerais, Escola de Enfermagem, Departamento de Nutrição, Belo Horizonte-MG, Brazil*
[2]*Universidade Federal de Viçosa, Departamento de Microbiologia, Viçosa-MG, Brazil*

Abstract

To evaluate the genotyping of *Klebsiella* isolates from enteral diets in public hospitals, as high-resolution support within epidemiological surveillance for controlling nosocomial infections. Methods: *Klebsiella* isolates were obtained from enteral diets at two public hospitals in the state of Minas Gerais, Brazil. These isolates were identified and serotyped. Total DNA of *Klebsiella* was extracted and subjected to three genotyping methods, to evaluate polymorphism between the strains and genetic diversity. Results: Twenty-one isolates of *Klebsiella* were obtained from the hospital enteral formulae; fifteen were identified as *K. pneumoniae* and six as *K. oxytoca*. The results from random amplified polymorphic DNA (RAPD) and 16S-23S rDNA analysis revealed high polymorphism among the *K. pneumoniae* isolates and a low level of polymorphism among the *K. oxytoca* isolates. The 1420 base pairs of DNA fragments generated through amplification of the 16S rDNA region and digestion with eight restriction endonucleases resulted in identical restriction fragment length polymorphism (RFLP) patterns for all isolates. Conclusion: Our data demonstrate that RAPD and 16S-23S rDNA analyses provide more reliable estimates of the genetic diversity among *Klebsiella* isolates than does amplification of the 16S rDNA gene. Therefore, we recommend that RAPD typing should be the preliminary method for fast and inexpensive investigation and that 16S-23S rDNA typing should be a confirmatory method if needed.

Keywords: Enteral diets; *Klebsiella*; Genotyping

Introduction

Klebsiella is a paradigm for opportunistic pathogens among Gram-negative bacilli and is associated with nosocomial infections. Resistance of the species *K. pneumoniae* to most antibiotics is on the rise [1]. In a review on surveillance surveys, the majority of episodes of nosocomial *K. pneumoniae* bacteremia were caused by extended-spectrum β-lactamase (ESBL)-producing strains [2]. Higher frequencies have been reported in some areas, such as Brazil, where up to 50% of the isolates in bloodstream infections were ESBL-producing strains of *K. pneumonia* [3]. Moreover, *Klebsiella* spp. is one of the most frequent contaminating bacteria found in enteral diets, i.e., the feeding method used for patients under intensive care [4,5]. Little attention has been paid to the role of *Klebsiella* and other intestinal foodborne pathogens in nosocomial infections, although bacteria that colonize the intestinal tract are the main cause of infections in hospitalized patients [6].

Klebsiella has recently been correlated with chronic intestinal diseases. The results of studies and reviews support the idea that *K. pneumoniae* is the most likely triggering factor involved in the initiation and development of ankylosing spondylitis and Crohn's disease [7]. Members of the family *Enterobacteriaceae*, particularly the species *Klebsiella oxytoca*, have been found to be more abundant in patients with active celiac disease than in controls [8].

In a noteworthy recent study, rapid and widespread dissemination of an epidemic clone of *K. pneumoniae* was found to cause a large nosocomial outbreak through the food chain. To our knowledge, this was the first report to provide insight on how transmission of multiresistant. *Klebsiella* can occur through food, as the vehicle in the hospital setting [9]. Thus, identification of the source of infection in hospital environments is a strategy for reducing the infection rate. Epidemiological studies on *Klebsiella* based on genotype characterization provide a suitable and accurate approach for identification purposes, thereby allowing discrimination below the strain level [1,10,11]. Compared with phenotyping, genotyping has become widely used for bacterial strain typing because of its higher resolution. Among different genotyping methods, the time and cost-effectiveness of DNA banding pattern-based typing methods have made these methods matters of interest [10]. In this light and given that few studies have been conducted on the molecular epidemiology of *Klebsiella* spp. isolates from foods destined for critical patients, the present work aimed to screen for the presence and determine the strain types of *Klebsiella* sp. present as contaminants in enteral diets served to patients admitted to two hospital units located in the State of Minas Gerais, Brazil.

Methods

Isolation, identification and serotyping of *Klebsiella* from enteral diets

Modular industrialized (milk-based and food supplement) enteral diets (30 samples) as used in two public hospitals in the State of Minas Gerais (Brazil) were analyzed, and *Klebsiella* sp. isolates were obtained from six (20%) of them. We emphasize that the samples were collected before administration to the patients. Typical *Klebsiella* colonies were isolated from these samples on MacConkey-inositol-carbenicillin selective medium [12]. Three to five typical colonies of each sample were placed on tryptic soy agar (TSA) (Oxoid, Basingtoke, Hampshire, England), and maintained on semisolid brain heart-infusion (BHI) (Oxoid), at 4°C. The isolates were identified by the API 20E System (API, Bio Meriex, L'Etole, France) and confirmed by the Crystal System

***Corresonding author:** Pereira SCL, Universidade Federal de Minas Gerais, Escola de Enfermagem, Departamento de Nutrição, Belo Horizonte-MG, Brazil
E-mail: simoneclpereira@gmail.com

(BBL-Becton & Dickinson, Bedford, MA, USA), at the Enterobacteria Laboratory of the Bacteriology Department, Oswaldo Cruz Foundation, Brazil. Serological tests were performed with antigens K1 to K6.

RAPD analysis

Total DNA of *Klebsiella* was extracted as described by Martins et al. Amplification reactions were carried [13]. The 25 µl reaction mixture consisted of 10 mM Tris-HCl, pH 8.0, 50 mM KCl, 2.0 mM $MgCl_2$, 0.1 mM dNTP, 0.4 µM RAPD primer, 1.0 U *Taq* DNA polymerase (Promega), and 25 ng total DNA. The primers OPD03, OPD07, OPD08, OPD12, OPD18, OPD20, OPF10, and OPF13 were purchased from Operon Technologies (Inc., Alameda, CA, EUA). A thermal cycling was performed in a PTC 100 machine (MJ Research Inc., Waterton, MA, EUA) programmed for an initial denaturation step at 94 ^{0}C for 5 min, followed by 30 cycles of 1 min at 94 ^{0}C, 1 min at 40 ^{0}C, and 1.30 min at 72 ^{0}C. A final extension at 72 ^{0}C for 7 min was performed. Amplification products were analyzed by electrophoresis on a 1.5% agarose gels in order to facilitate the electrophoresis analysis. The gels were photographed under UV light using an Eagle-Eye II video system (Stratagene, La Jolla, CA, USA).

rDNA analysis

Total DNA from each *Klebsiella* isolate was amplified using 16S rDNA primers (5' GCCTAACACATGCAAGTCGA and 3' AAAGTGGTAAGCGCCCTCCC) and the 16S-23S rDNA intergenic transcribed spacer (ITS) region (5' GGTGAAGTCGTAACAAG and 5' TGCCAAGGCATCCACC). DNA templates (20 ng) were amplified in a 25 µl reaction volume containing 1 U *Taq* polymerase (Promega), 0.5 µM of each primer, 0.2 mM of each of the four dNTPs (Promega), 1X PCR buffer (10 mM Tris-HCl, pH 8.0, 50 mM KCl, 2.0 mM $MgCl_2$). PCR amplification was performed with a PTC 100 thermal cycle. After an initial denaturation step of 5 min at 94°C, a total of 40 cycles of amplification was performed using the following thermal profile: 1 min at 94°C (denaturation), 1 min at 56°C (annealing) and 1 min at 72°C (elongation) followed by a 7 min extension step at 72°C. Controls were included in each set of amplifications, namely, a reaction mixture with distilled water instead of template DNA. To assess PCR reproducibility, two independent PCR amplifications were performed for each strain included in this study. 16S rDNA-amplified products were electrophoresed on a 1.5% agarose gel or ethanol precipitated for the restriction analysis with *Bgl*I, *Hha*I, *Hind*III, *Hin*fI, *Kpn*I, and *Xho*I. Restriction fragments were visualized following gel electrophoresis on 2% agarose gel. PCR products for the ITS region Amplification products were analyzed by electrophoresis on a 1.5% agarose gels in order to facilitate the electrophoresis analysis. The gels were photographed under UV light using an Eagle-Eye II video system (Stratagene, La Jolla, CA, USA).

Analysis of molecular data

Banding patterns from RAPD and ITS analysis were scored as 1 (presence) or 0 (absence) and used for determination of the genetic diversity among *Klebsiella* isolates. The presence of the same size bands in genotypes indicates similarity, while the presence in one genotype and absence in the others indicates dissimilarity. Negative controls, amplifications reactions without DNA, were used in all primer amplifications. In addition, some RAPD reactions were also repeated. The data were analyzed by the statistical software GENES [14]. The genetic diversity values (GD_{ij}) were calculated using the Jaccard's coefficient [15].

Cluster analysis was carried out using the Unweighted Pair Group Method Using Arithmetic Averages (UPGMA) provided by Statistic version 4.2, and by multidimensional scaling biplot [16]. The Tocher optimization method was also used to aid in the characterization of the groups as described by Cruz and Carneiro [15]. Hierarchical grouping methodologies and Tocher' optimization were used to cluster the less genetically divergent isolates [15].

Results

Twenty-one isolates derived from enteral diets were identified as *Klebsiella*, six was *K. oxytoca* and eleven was *K. pneumoniae* (Table 1). Among the *K. pneumoniae*, five isolates belonged to the serological group K5 and one to group K4. Analysis of the genetic diversity of *K. pneumoniae* isolates by RAPD showed a remarkable polymorphism as 31 polymorphic DNA fragments were detected against only two monomorphic (data note shown). Strains from the serogroup K5 (numbers 8, 11, 15, 17 and 21) (Table 1) revealed different band profiles, whereas no difference could be seen between the isolates 7 and 8 that belong to different serotypes. The UPGMA analysis resulted in three *K. pneumoniae* groups with genetic diversity up to 73.9%. The isolates originated from the hospital unit A formed two groups and all isolates from the hospital unit B were grouped together (Figure 1a).

RAPD markers revealed a low level of polymorphism among *K. oxytoca* isolates and sixteen monomorphic and nine polymorphic bands were detected on the agarose gel (data note shown). These isolates were originated from the same hospital unit, but they were separated into three genetically related groups with genetic distances varying between 2.4 and 20% (Figure 1b). The amplification of the 16S-23S rDNA intergenic spacer region of *K. pneumoniae* and *K. oxytoca* generated monomorphic and polymorphic DNA fragments, respectively. The banding profile clearly differentiated each species. Three monomorphic and five polymorphic bands were visualized for *K. pneumoniae* isolates (Figure 2). The UPGMA analysis resulted in two clusters: the first cluster (Group I) was formed by isolates obtained from enteral diets derived from hospital B and the second cluster (Group II) was formed

No.	Sample	Isolates	Enteral diet	Species	Sorogroup	Hospital
1	1	P1	NI[a]	*K. oxytoca*	-[c]	A
2	1	P2	NI	*K. oxytoca*	-	A
3	1	P3	NI	*K. oxytoca*	-	A
4	1	P4	NI	*K. oxytoca*	-	A
5	1	P5	NI	*K. oxytoca*	-	A
6	3	P6	NI	*K. pneumoniae*	nd[d]	A
7	3	P7	NI	*K. pneumoniae*	nd	A
8	3	P8	NI	*K. pneumoniae*	K5	A
9	3	P9	NI	*K. oxytoca*	-	A
10	4	P11	NI	*K. pneumoniae*	nd	A
11	4	P13	NI	*K. pneumoniae*	K5	A
12	4	P14	NI	*K. pneumoniae*	nd	A
13	4	P15	NI	*K. pneumoniae*	nd	A
14	4	P17	NI	*K. pneumoniae*	nd	A
15	5	U1	NIM[b]	*K. pneumoniae*	K5	B
16	5	U2	NIM	*K. pneumoniae*	nd	B
17	5	U3	NIM	*K. pneumoniae*	K5	B
18	5	U4	NIM	*K. pneumoniae*	nd	B
19	6	U5	NIM	*K. pneumoniae*	K4	B
20	6	U7	NIM	*K. pneumoniae*	nd	B
21	6	U8	NIM	*K. pneumoniae*	K5	B

Table 1: Identification, serotyping and characteristics of Klebsiella spp isolated from enteric diets.

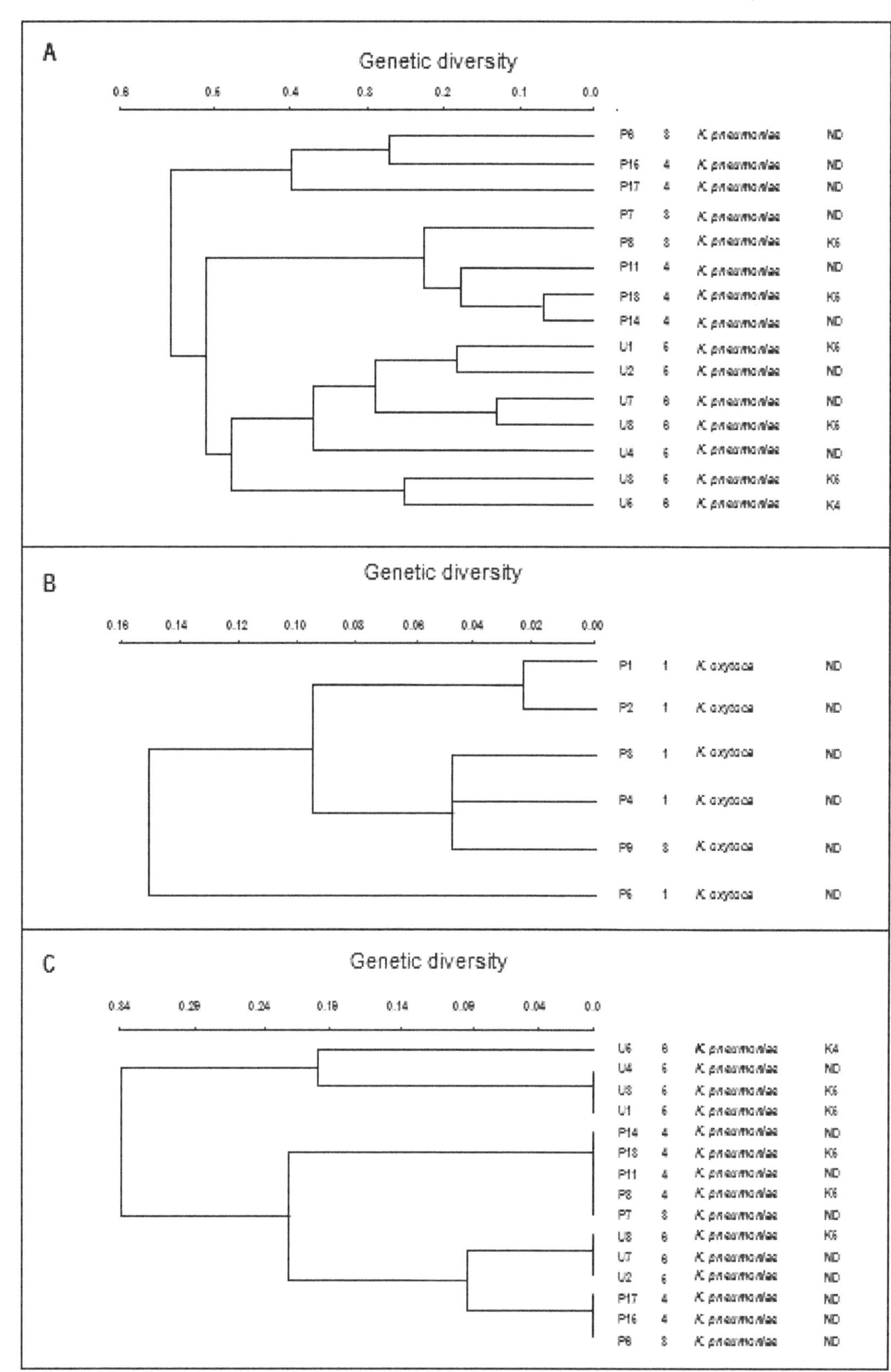

Figure 1: Dendrograms geraded by genotype grouping of isolates *K. pneumoniae* **(A and C)** and *K. oxytoca* **(B)** from enteral diets based in genetic similarity coefficient determined by amplied DNA for RAPD patterns with eight primer: OPD03, OPD07, OPD08, OPD12, OPD20, OPF10 e OPF13 **(A and B)** and based in genetic similarity coefficient determined by amplied DNA for 16S-23S rDNA spacer region patterns **(C)**. The identification of *K. pneumoniae* corresponds to isolated, sample, species, and serogroup. The identificationof *K . oxytoca* corresponds to isolate, sample, and specie nd = not determinate.

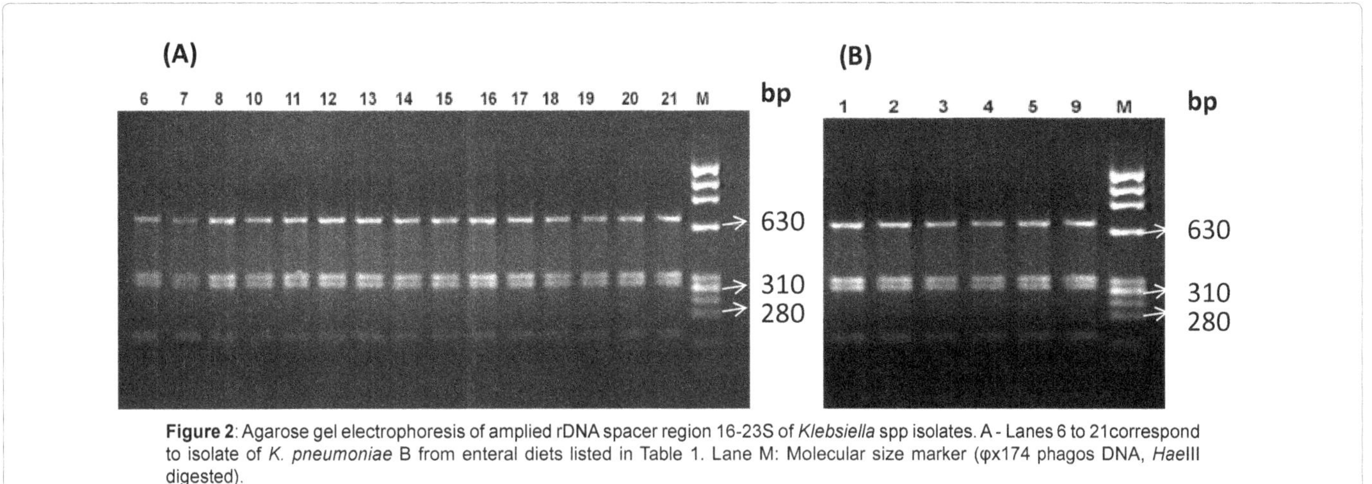

Figure 2: Agarose gel electrophoresis of amplied rDNA spacer region 16-23S of *Klebsiella* spp isolates. A - Lanes 6 to 21correspond to isolate of *K. pneumoniae* B from enteral diets listed in Table 1. Lane M: Molecular size marker (φx174 phagos DNA, *Hae*III digested).

by isolates coming from both hospitals (Figure 1c). The highest genetic diversity obtained between isolates from different hospital units and isolates from hospital B was 45%. Total similarity was only detected among isolates originated from the same hospital.

The Pearson's correlation coefficient of data generated by RAPD and ITS analysis of *K. pneumoniae* isolates was 0.34 ($p<0.01$). Amplification of *K. pneumoniae* and *K. oxytoca* total DNA with 16S rDNA gene primers produced a fragment of approximately 1420 bp. Size polymorphism was not detected among *Klebsiella* isolates (Figure 3). No restriction polymorphism was found among the isolates. The 1420 bp fragment was checked for the presence of restriction sites with the enzymes *Hinf*I, *Bgl*I, *Eco*RI, and *Hha*I. The absence of internal sites for *Eco*RV, *Hind*III, *Kpn*I, and *Xho*I was also assessed by the visualization of a single 1420 bp DNA fragment.

Discussion

The presence of *Klebsiella* in enteral diets needs to be considered as a possible source of contamination of hospitalized patients. This arises because enteral diets are especially prescribed for patients concentrated in intensive care units who are highly susceptible to nosocomial infections. These are patients who are subjects to the intrinsic risks of nosocomial infections relating to complex underlying diseases, nutritional vulnerability, extreme ages (premature babies and elderly people) and the effects of immunosuppressant and/or immunodepressive drugs and broad-spectrum antimicrobial agents. Furthermore, these patients are subject to extrinsic risks such as having a prolonged stay in the hospital environment, being subjected to a routine of invasive procedures and lacking guaranteed quality in all hospital services, given the complex nature of the organization.

Nosocomial infections are a worldwide problem because of their impact in clinical, ethical and financial terms [17]. Control programs for these infections have been developed and implemented, but the rates of such infections have not become significantly reduced, especially because of the challenge of occurrences of outbreaks that are difficult to control because of the appearance of multiresistant hospital bacteria. For many of these microbial strains, the therapeutic options are very limited and have a high cost [17-19]. In Brazil, the problem of nosocomial infections is increasing all the time. Even with the country's current legislation, nosocomial infection rates are high (15.5%). Moreover, the fact that public healthcare institutions present the highest prevalence of nosocomial infections in the country (18.4%) can be considered to be an aggravating factor [19,20].

The enteral diets analyzed were more frequently contaminated with *K. pneumoniae,* but also showed contamination with *K. oxytoca.* These species are involved in nosocomial and community infections [21,22]. Six out of the fifteen *K. pneumoniae* isolates belonged to serological groups K1 and K2, which are frequently regarded as endemic in hospital environments and associated with nosocomial infections [2].

With regard to the presence of these bacteria in the hospital dietary sector, another study conducted by the authors of the present study (data not yet published) indicated that *K. pneumoniae* was the bacterium most frequently found in all the samples analyzed (35.4% of the total), from food handlers, utensils and food processing surfaces. Recently, Ekrami et al. found a situation similar to that of the present study in seven hospital food services, with contamination caused by *S. aureus* and *K. pneumonia* [23]. There was significant presence of these microorganisms on food handlers' hands (79.5%), processing surfaces (71.4%), clothing (61.1%) and equipment (57.8%). Furthermore, this study found that Gram-negative bacteria were present on 50% of the hands that manipulated foods [24].

Corroborating these findings, the study by Calbo et al described the epidemiology of and control measures taken in relation to a foodborne nosocomial outbreak of *K. pneumoniae* in an acute care hospital [9]. This report described rapid and widespread dissemination of an epidemic clone of *K. pneumoniae,* which caused a large nosocomial outbreak through the food chain. To our knowledge, this was the first report to provide insight on how transmission of *K. pneumoniae* can occur through food as the vehicle in the hospital setting. The outbreak was stopped only after control measures were applied in the kitchen. The absence of new cases during the 14-month follow-up period suggests that these measures were effective [9]. In the present study, *Klebsiella* was isolated from enteral diets and subjected to molecular characterization as a first step towards discriminating between strains for further evaluation of their possible association with clinical isolates of this bacterium in outbreaks of nosocomial infections. These data are relevant for epidemiological surveillance because they enable identification and early intervention, thereby reducing the risk of nosocomial infections that involve the nutrition service, especially with regard to diets destined for critical patients [19]. RAPD detected genetic variation within *K. pneumoniae* isolated from enteral diets.

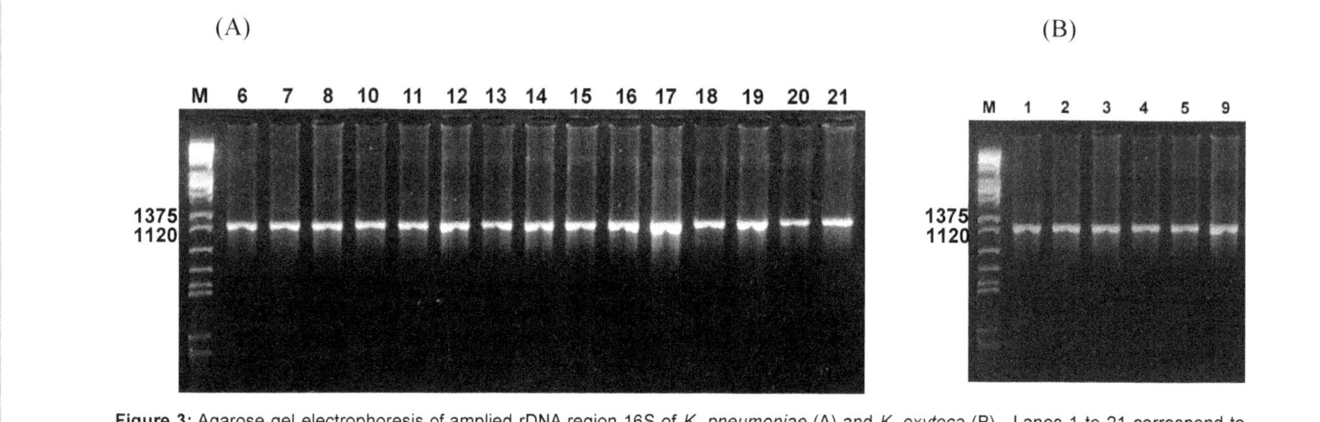

Figure 3: Agarose gel electrophoresis of amplied rDNA region 16S of *K. pneumoniae* (A) and *K. oxytoca* (B). Lanes 1 to 21 correspond to isolate listed in Table 1. Lane M: Molecular size marker (λ phagos DNA, *Hind*III-*Ban*HI-*Eco*RI digested).

Similar results were found for the clinical isolates. The ability of RAPD analysis to discriminate between strains was confirmed by our results. The amplified DNA fragments were not specific for any given serological group. In a recent study, the discriminatory power of the pulsed-field gel electrophoresis (PFGE) and random amplified polymorphic DNA (RAPD) methods for subtyping 54 clinical isolates of *Klebsiella pneumoniae* were compared. All the isolates were typable by means of RAPD, while 3.6% of them were not typable by means of PFGE.[24] PFGE is considered to be the gold standard for typing *K. pneumonia*. The repeatability of the two typing methods was 100%, with satisfactory reproducibility (95%). It was concluded that an optimized RAPD protocol is technically less demanding and time consuming, which makes it a reliable typing method that is competitive with PFGE [10,25].

Differences between *K. pneumoniae* isolates originating from the same sample or from the same place were demonstrated by the genetic distance obtained. The genetic distance was 73.9% in strains isolated from samples of enteral diets collected in hospital unit A (numbers 11 and 14). However, strain number 6, from hospital A, was 94% similar to strain number 19, derived from hospital B. These differences can be explained in terms of concomitant amplification of plasmid DNA [24]. This variation may also be due to loss of these plasmids caused by the stress conditions faced by *K. pneumoniae* strains [11]. Another possible explanation for the genetic diversity seen in this work may be the different sources of contamination in the hospital food service. High incidence of *K. pneumoniae* was found on equipment, tools, food handlers' hands and work surfaces in the nutritional and dietary sectors of the hospitals [23]. *Klebsiella* isolates were also shown to present amplification of the 16S-23S rDNA intergenic transcribed spacer region. Our data also demonstrated that RAPD and ITS analyses were more reliable for estimating genetic diversity between *Klebsiella* isolates than amplification of the 16S rDNA gene. However, the 16S region did not differentiate between the strains of *Klebsiella* in the present study. On the other hand, the importance of using this better-conserved region for the purposes of identifying the bacterial genus was highlighted. Evaluation of the 16S rDNA region would thus be applicable as a second stage as a confirmatory test for the presence of *Klebsiella*. In this regard, a recent study using this region for molecular identification found that a gene originating from V. cholera that was a novel variant of the cholera toxin gene (*ctxAB*) was present in other enterobacteria, including *Klebsiella* [11].

Final Remarks

Our results showed that ITS and optimized RAPD are equally valuable for typing *Klebsiella* isolates from enteral diets. Since RAPD is less technically demanding and time consuming, we recommend that RAPD typing should be the preliminary method for fast and inexpensive investigation and that ITS typing should be a confirmatory method if needed. Compared with phenotyping, the higher resolution of genotyping can contribute towards the basis for other studies on the distribution of nosocomial infections and their determinants that would be aimed towards preventive actions.

In this regard, the present study brings together knowledge on nosocomial infections and emphasizes the importance of measures relating to outcomes associated with the nutrition sector, with the aim of strengthening the interfaces between infection control and preventive actions and between such control and quality promotion within care services overall. It also endorses the need to direct healthcare funding policies towards adopting risk control and prevention measures in strategic sectors of the healthcare services. There is a need to reflect on all the possible strategies that might contribute towards changing the current panorama within which nosocomial infections are presented, with investment in research and updating. It should be emphasized that nosocomial infection control is grounded in results and that, from this, support within the organization can be obtained, given that its consequences for the quality of care are sufficient to justify maintaining such controls. Genotyping data support epidemiological surveillance, which thus constitutes an important instrument for planning, organizing and putting into operation healthcare services, as well as for providing standards for technical activities. This information adds knowledge on nosocomial infections and aids in detecting or preventing changes to the conditioning factors of such infections, so that timely and well-grounded recommendations can be made for appropriate and efficient measures that lead to prevention and control of these infections. It should be emphasized that dissemination of data analyses such as in the present study and other recent studies should be done routinely among all the healthcare and administrative professionals of hospital institutions. Many professionals will, through becoming aware of the results relating to their own specific services, start to rethink their practices and become more involved in prevention and control practices, with surveillance in its true sense. Thus, the information disseminated may have a significant impact on nosocomial infection rates.

References

1. Lin YT, Wang FD, Chan YJ, Fu YC, Lin CPF, et al. (2014) Clinical and microbiological characteristics of tigecycline non-susceptible *Klebsiella pneumoniae* bacteremia in Taiwan. BMC Infect.

2. Viale P, Giannella M, Lewis R, Trecarichi EM, Petrosillo N, et al. (2013) Predictors of mortality in multidrug-resistant *Klebsiella pneumoniae* bloodstream infections. Expert Review of Anti-infective Therapy 11: 1053-1063.

3. Marra AR, Wey SB, Castelo A, Gales AC, Cal RG, et al. (2006) Nosocomial bloodstream infections caused by *Klebsiella pneumoniae*: Impact of extended-spectrum beta-lactamase (ESBL) production on clinical outcome in a hospital with high ESBL prevalence. BMC Infect Dis.

4. Arias ML, Monge R, Chávez C (2003) Microbiological contamination of enteral feeding solutions used in Costa Rican hospitals. Arch Latinoam Nutr 53: 277-281.

5. Hurrell E, Kucerova E, Loughlin M, Caubilla-Barron J, Hilton A, et al. (2009) Neonatal enteral feeding tubes as loci for colonisation by members of the *Enterobacteriaceae*. BMC Infect Dis 9: 146-174.

6. Westerbeek EAM, Berg AVD, Lafeber HN, Fetter WPF, Elburg RMV, et al. (2011) The effect of enteral supplementation of a prebiotic mixture of non-human milk galacto-, fructo- and acidic oligosaccharides on intestinal permeability in preterm infants. British Journal of Nutrition 105: 268-274.

7. Rashid T, Wilson C, Ebringer A (2013) The Link between Ankylosing Spondylitis, Crohn's Disease, *Klebsiella*, and Starch Consumption. Clinical and Developmental Immunology.

8. Sánchez E, Donat E, Ribes-Koninckx C, Fernández-M urga ML, Sanz Y (2013) Duodenal-Mucosal Bacteria Associated with Celiac Disease in Children. Applied and Environmental Microbiology 79: 5472-5479.

9. Calbo E, Freixas N, Xercavins M, Riera M, Nicola´s C, et al. (2011) Foodborne nosocomial outbreak of SHV1 and CTX-M-15-producing *Klebsiella pneumoniae*: Epidemiology and control. Clin Infect Dis 52: 743-749.

10. Li W, Raoult D, Fournier PE (2009) Bacterial strain typing in the genomic era. FEMS Microbiol Rev 33: 892-916.

11. Shibata Y, Nomoto R, Vries GC, Osawa R (2013) Serendipitous Isolation of Non-*Vibrio* Bacterial Strains Carrying the Cholera Toxin Gene from Environmental Waters in Indonesia. Hindawi Publishing Corporation International Journal of Microbiology.

12. Bagley ST, Seidler RJ (1978) Primary *Klebsiella* identification with MacConkey-inositol-carbenicilin agar. App Environ Microbiol 36: 536-538.

13. Martins ML, Araújo EF, Mantovani HC, Moraes CA, Vanetti MCD, et al. (2005) Detection of the apr gene in proteolytic psychrotrophic bacteria isolated from refrigerated raw milk. International Journal of Food Microbiology 102: 203-211.

14. Cruz CD (2001) Programa Genes-versão Windows. Editora UFV, Viçosa.

15. Cruz CD, Carneiro PCS (2003) Modelos Biométricos Aplicados ao Melhoramento Genético, Viçosa.

16. Dias LAS (1998) Análises multidimensionais. Eletroforese de Isoenzimas e Proteínas Afins: Fundamentos e Aplicações em Plantas e Microrganismos. Viçosa.

17. World Health Organization (2011) Core components for infection prevention and control programs. Assessment tools for IPC programmes, Geneve.

18. Ministério da Saúde (BR) (1998) Expede na forma de anexos diretriz e normas para a prevenção e controle das infecções hospitalares: Portaria Nº 2.616, de 12 de maio de. Diário Oficial da União, República Federativa do Brasil, Brasília (DF).

19. Maruyama SAT (2008) Controle de infecção hospitalar: Histórico e papel do estado. Rev Eletr Enf 10: 775-83.

20. Prade SS (1995) Estudo Brasileiro da Magnitude das Infecões Hospitalares em Hospitais Terciários. Rev Cont Infec Hosp 2.

21. Keynan Y, Rubinstein E (2007) The changing face of *Klebsiella pneumoniae* infections in the community. International Journal of Antimicrobial Agents 30: 385-389.

22. Ifeadike CO, Ironkwe OC, Adogu POU, Nnebue CC, Emelumadu OF, et al. (2012) Prevalence and pattern of bacteria and intestinal parasites among food handlers in the Federal Capital Territory of Nigeria. Niger Med J 53: 166-171.

23. Ekrami AR, Kayedani A, Jahangir M, Kalantar E, Jalali M, et al. (2011) Isolation of common aerobic bacterial pathogens from the environment of seven hospitals, Ahvaz, Iran. Jundishapur J Microbiol 4: 75-82.

24. Mitra AP, Fereshteh E, Mobarak GM, Mahmood P, Mehdi FM, et al. (2013) Genetic profiling of Klebsiella pneumoniae: Comparison of pulsed field gel electrophoresis and random amplified polymorphic DNA. Braz J Microbiol 44: 823-828.

25. Arlet G, Rouveau M, Casin I, Bouvet PJ, Lagrange PH, et al. (1994) Molecular epidemiology of Klebsiella pneumoniae strains that produce SHV-4 beta-lactamase and which were isolated in 14 French hospitals. J Clin Microbiol 32: 2553-2558.

Gluten-Free Breakfast Cereal Prepared with Agroindustrial By-Products: Physical, Chemical, Microbiological Aspects and Sensory Acceptance

Priscila Alonso DS[1]*, Marcio Caliari[2], and Manoel Soares SJ[2]

[1]*Institute Federal Goiano Campus Rio Verde, Rodovia Sul Goiana - Rio Verde - Goiás, Brazil*

[2]*Federal University of Goiás, Food Technology Department, Goiânia, Goiás, Brazil*

Abstract

Develop foods that meet the needs of athletes and individuals with celiac disease and that serve as an alternative for the use of unconventional raw materials. Accordingly, the objective of this study was to develop and characterize a gluten-free breakfast cereal formulated with rice, passion fruit and milk by-products and to test whether its physical, chemical, and microbiological characteristics and sensory acceptance are adequate for commercialization. The experimental gluten-free cereal exhibited an expansion index of 2.56, specific volume of 1.6mL g^{-1} and chromaticity coordinate a* of 7.06. It is also a source of protein (7.55 g 100 g^{-1}), has a low lipid content (0.97 g 100 g^{-1}),and is rich in dietary fiber (6.12 g 100 g^{-1}),a third of which is soluble, providing functional value to the product. In the sensory analysis, the developed product scored average on acceptance, remaining above "I neither liked nor disliked it" and "I moderately liked it", which is considered as accepted, scoring higher than 4 in all attributes and 52% on purchase intention. The use of rice, passion fruit and milk by-products was found to be an alternative for the preparation of gluten-free extruded breakfast cereal, producing a final product with high nutritional value. The cereal met the recommended daily intake (RDI) requirements for six essential amino acids according to the FAO standards and contained 85.29, 0.78 and 39.65% of the RDI for the amino acids threonine, histidine and lysine, respectively. In addition to containing no trans-fatty acids and 20% of the mono-unsaturated fatty acid requirement, one portion of the cereal meets the Fe and Zn RDI requirements for adults.

Keywords: By-products; Extrusion; Centesimal composition; Amino acids; Fatty acids

Introduction

Food processing waste and by-products are promising sources of natural compounds, such as dietary fibers, antioxidants, essential fatty acids, antimicrobials and minerals that can be used to aggregate value to food because of their technological, nutritional and functional properties. Rice (*Oryza sativa*) is one of the three most produced and consumed cereals in the world. A portion of rice production is processed, resulting in the waste named "broken rice", which generates considerable yield loss during the production of polished grain. This by-product has been transformed into rice flour, which may partially or completely substitute for wheat flour in many foods or act as the basis for new ingredients, such as mixed flours and modified starches [1-3]. According to Molina-Infante, et al. [4], gluten sensitivity is an emerging disease with unknown prevalence. Passion fruit peel flour is considered an important source of dietary fiber [5]. Individuals with high consumption of dietary fiber apparently have lower risk of developing cardiac diseases, stroke, arterial hypertension, diabetes, obesity and certain gastrointestinal diseases.

Whey originates from cheese production, a process that involves the separation of two main proteins, casein and lactalbumin, which remain in the liquid fraction that constitutes the whey. Previous studies have demonstrated that whey proteins, such as casein, are absorbed more rapidly than other proteins. Broken rice grains (BRG), passion fruit peel and whey are gluten-free by-products, which makes them of note for the preparation of food products destined for consumption by celiac or gluten-intolerant consumers. In turn, extrusion technology allows the use of mixtures of these by-products and their transformation into industrialized food ready for consumption, such as breakfast cereals. Therefore, the objective of this study was to evaluate the physical, chemical, and microbiological characteristics and sensory acceptance of an experimental gluten-free breakfast cereal formulated with BRG, passion fruit peel flour (PFPF) and whey powder (WP)to develop and evaluate a product that meets the dietary needs of patients with celiac disease and those who practice physical activities.

Materials and Methods

Raw materials

The raw materials used for the formulation of the experimental breakfast cereal were BRG (mixture of IRGA 417 and IRGA 424 cultivars), 2013 harvest, provided by the Empresa Arroz Cristal, located at Aparecida de Goiânia, Goiás, Brazil; PFPF of the brand Natural Life, obtained in the local market of Rio Verde, Goiás, Brazil; and WP, provided by Indústria de Laticínios Italac, located in Santa Helena de Goiás, Goiás, Brazil.

Breakfast cereal formulation and processing

A gluten-free breakfast cereal was prepared with a mixture of BRG, PFPF and WP (87:03:10), homogenized in a Y-type homogenizer (*Tecnal, TE 201/05, Piracicaba, Brazil*) for 15 min, and packaged at 15g/100 g^{-1} of moisture with a manual pulverizer in low-density polyethylene (LDPE) bags and stored at 5°C for 12 h, until the time of extrusion. Thermoplastic extrusion was performed in a simple-screw extruder (*Inbramaq, PQ-30, Ribeirão Preto-SP, Brazil*). The processing conditions were defined in preliminary tests and were maintained at the following constant values: motor rotation of 250 rpm (60 Hz); circular die opening 4 mm in diameter, pre-die

***Corresonding author:** Priscila Alonso DS, Institute Federal Goiano Campus Rio Verde, Rodovia Sul Goiana - Rio Verde - Goiás, Brazil
E-mail: prialonso@yahoo.com.br

with 22 holes; screw with three entrances, 30 mm in length, and screw compression rate of 3:1; helical jacket with 335 g min^{-1} feed rate; and temperatures in the first, second and third heating zones equal to 40, 60 and 85 °C, respectively. Then, the cereals were dried in an oven at 80°C for 40 min, cooled and packaged in LDPE bags until the analyses were performed.

Grain size of the raw materials

The grain sizes of the raw materials (BRG, PFPF and WP) were determined according to the AOAC [6] method, modifying the set of sieves, using a vibrating separator (Bertel) and a set of sieves from 10 to 100 mesh.

Chemical composition of the raw materials and breakfast cereal

The moisture was measured from the mass loss of the sample heated in an oven at 105°C to constant weight. The proteins were measured using the Kjeldahl method for the determination of total nitrogen, which was converted into crude protein by a factor of 6.38 for WP, 6.25 for PFPF and 5.25 for BRG. The lipid content was measured using the Soxhlet method; the ash content, by carbonization followed by complete incineration in a muffle at 550°C; and the total, soluble and insoluble dietary fiber, using the enzyme-gravimetric method. All methods are recommended by the AOAC [6]. The carbohydrates were calculated by the differences method, subtracting from one hundred the values of the water, ash, protein and lipid contents. The total energetic value was estimated according to the Atwater conversion values, for which the carbohydrate (minus the dietary fiber content) and protein content was multiplied by four and the lipid content by nine, and the sum of the products was calculated. To determine the fatty acids, the lipids were extracted from the breakfast cereal using the hydrolysis method, adding pyrogallic acid to minimize the oxidative degradation of fatty acids. The fat was extracted using ether, and fatty acid methyl esters (FAMEs) were obtained through a reaction with boron trifluoride (BF_3) in methanol. The FAMEs were quantitatively measured by capillary gas chromatography using C 11:0 as an internal standard. The saturated and monounsaturated fats were calculated by adding their respective fatty acids and were expressed as triglyceride equivalents. Monounsaturated fats included only the cis form. For the chromatographic analysis, the following parameters were used: helium carrier gas; injector temperature of 225°C, detector temperature of 285°C and initial column temperature of 100°C (isotherm) for 4 min, and final column temperature of 240°C; and rate of 3°C/min, according to the AOAC [6] method.

The minerals (Ca^{2+}, P^{2+}, K^{+}, Mg^{2+}, Fe^{2+}, and Zn^{2+}) were quantified by atomic absorption spectrophotometry, with instrument parameters (lamp, wavelength, lamp current and gap width) specific to each nutrient, according to the official AOAC [6] method, with nitric-perchloric digestion. The amino acid profile was determined using a chromatographer, after acid hydrolysis. A sample containing approximately 25 mg of protein was processed according to the general recommendations by White, et al. [7] and Hagen, et al. [8]. All of the analyses were performed in triplicate.

Physical characteristics of the gluten-free breakfast cereal

The water activity (Aw) of the breakfast cereal was determined at 25 ± 4°C, using an Aqua Lab CX-2 Water Activity System instrument; the instrumental color parameters, according to the system CIEL L*, a* and b*, in a colorimeter (*Color Quest, XE, Reston, USA*),with fixed hue angle at 10° and standard illuminant D65, corresponding to natural daylight; the expansion index (EI),from the ratio between the diameter of the extruded products and the diameter of the extruder output hole (4 mm), measured with a paquimeter (*Digital Caliper, Messen, Danyang, China*); the volume, by the millet seed displacement method; the mass, in a semi-analytical scale; and the specific volume, from the ratio between the mean volume and the mass of cereals. The ultimate tensile strength, or hardness, was measured using a TA-XT2 texturometer (Stable Micro Systems, Surrey, England), with a 5-kg load cell, equipped with Texture Expert® software for data collection and analysis. The 2-cm-long samples were cut along a single axis using a probe with guillotine-type straight blade, using the methodology described by Chang, et al. [9], with a pre-test velocity of 2.0 mm s^{-1}, test velocity of 2 mm s^{-1}, post-test velocity of 5 mm s^{-1}, probe calibration distance of 5 mm, force threshold of 5 g and measurement under shear stress. All analyses were performed in 10 randomly selected cereals.

Microbiological risk and acceptance of the breakfast cereal

The coliform counts were performed at 35 and 45°C, according to APHA [10]. The sensory evaluation of the developed breakfast cereal was performed by 60 untrained tasters of both genders, ages 18 to 60 years. The seven-point hedonic scale acceptance test (7 = I really liked it; 6 = I liked it a lot; 5 = I moderately liked it; 4 = I neither liked nor disliked it; 3 = I moderately disliked it; 2 = I disliked it a lot; 1 = I really disliked it) was applied, and it was established that for the acceptance of the breakfast cereal, the limit score would be 4.Initially, each taster received the sample, served in a disposable plate, and a cup of cold milk (50 mL), so each taster could evaluate the breakfast cereal mixed or not with the milk, according to his or her preference. The sample was evaluated according to the attributes color, aroma, texture, flavor, and purchase intention. For this purpose, a five-point structured scale (1 = I would certainly not buy it; 5 = I would certainly buy it) was used [11]. The study was submitted to and approved by the Committee of Ethics in Research of the Goiano Federal Institute (IF Goiano) under protocol n. 040/2013. The acceptance test participants signed an informed consent form confirming they were instructed as to the study purpose, as well as the risks and benefits.

Data analysis

The data were analyzed using analysis of variance (ANOVA), and the means were compared by Tukey's test at 5% probability, using the program Assistat (2013), version 7.6.

Results and Discussion

Grain size of the raw materials

The grain size analysis of the WP indicated that all particles were smaller than 0.297 mm, with almost one fifth of the sample consisting of fine grains (<0.147 mm). The largest grain size was observed for BRG, for which almost all particles were larger than 0.841 mm because this raw material was not ground before the extrusion step to facilitate gravity feeding to the equipment [12]. The PFPF consisted of intermediate particles between 0.297 and 0.595 mm. Grain size irregularity can negatively affect physical characteristics, such as the hardness and expansion of extruded products, whereas homogeneity promotes adequate, uniform cooking of the raw material during the extrusion process, preventing hardness and partial cooking. Such problems can lead to undesirable particles with different cooking degrees, compromising the quality of the extruded product's appearance and palatability [13]. However, the grain size irregularity of the raw materials in this study did not negatively affect the expansion, texture or appearance of the breakfast cereals produced.

Chemical composition of the raw materials and breakfast cereal

The low moisture values found for the by-products are desirable (Table 1) because moisture levels lower than 14 g 100 g^{-1} avoid microbial development and increase chemical and enzymatic stability and the useful life of products [14]. The centesimal composition of the WP was similar to the values reported by the technical specification sheet of the dairy company that supplied the product, which established as standards a maximum of 3 g 100 g^{-1} of moisture and 8.5 g 100 g^{-1} of ashes, minimum of 11.0 g 100 g^{-1} of protein, and maximum of 1.5 g 100 g^{-1} of lipids.

The ash content of BRG was 142% higher than that found for rice flour by Carvalho, et al. [15], whereas the lipid content was similar, and the protein content was 13.3% lower. The moisture, ash, protein and lipid contents of PFPF were, respectively, 6%, 19%, 24% and 8.5% lower than the values found by Souza, et al. [16], whereas the ash, protein and lipid contents were, respectively, 7.3%, 14% and 58.6% higher than those found by Vernaza et al. [5]. PFPF is a food rich in fiber, of which 77.24% was insoluble and 27.76% was soluble (Table 1). The insoluble fraction is related to an increase in fecal matter, therefore ensuring intestinal peristalsis, avoiding constipation and eliminating the risk of hemorrhoids and diverticulitis. The soluble fraction, in turn, has beneficial effects on insulin metabolism and cholesterol and can be consumed by diabetics because it exerts a hyperglycemic effect by delaying gastric emptying, therefore reducing intestinal transit and glucose absorption [17]. Because the amount of soluble dietary fiber in PFPF is high, it is classified as a functional food. The WP had the highest energy value among the by-products, as it is an efficient source of calories for the biological functions of the body; further, it provides a sweet flavor to the breakfast cereal. When present in the formulation of the breakfast cereal, WP increased the moisture content after extrusion, most likely because lactose is capable of retaining water in the product due to its hygroscopic character. The high ash content of the breakfast cereal was caused by the presence of PFPF and WP, which also contain high ash contents (Table 1).

The lipid content of the breakfast cereals was low, reflecting the low content of this compound in the raw materials. Compared to the values found by Silva, et al. [18] in manioc starch and whey powder breakfast cereal (0.7574g 100 g^{-1}), the lipid content in the breakfast cereals produced in this study was only 26% higher. The gluten-free breakfast cereal made with rice, passion fruit and milk by-products may be considered a protein food (Table 1). The fortification of extruded products with proteins from selected sources can improve health and promote the quality of snacks and breakfast cereals [19]. Fibers belong to the group of biologically active compounds, and their consumption is of fundamental importance to health. The gluten-free breakfast cereal is considered to have "high fiber content" (Table 1). In turn, the incorporation of fibers in breakfast cereals may cause texture problems, thus decreasing consumer acceptance. These texture problems are partially caused by deterioration of the product's microstructure, one of the primary quality attributes of extruded breakfast cereals [20]. The carbohydrate content and the total energy value obtained in the breakfast cereal were expressive. The gluten-free breakfast cereal exhibited an adequate balance of essential amino acids when compared to the recommendation by the FAO/WHO [21] for adults (Table 2), most likely because of the addition of WP and BRG. However, contents of the amino acids threonine, histidine and lysine remained lower than the values recommended by the FAO/WHO [21] for this product. Haraguchi, et al. [22] reported that whey soluble proteins have an excellent amino acid profile, characterizing them as having high biological value because of their bioactive peptides, which confer different functional properties. The different amino acids, especially those with branched chains, favor anabolism as well as protein catabolism reduction, promoting muscle gain and reducing the loss of muscle mass during weight loss. Studies have demonstrated that only the essential amino acids, especially leucine, are necessary to stimulate protein synthesis. The gluten-free breakfast cereal had a higher content of unsaturated fatty acids than that of saturated fatty acids (Table 3). This is a positive relationship because saturated fatty acids increase cholesterolemia, reducing hepatic receptors and consequently inhibiting plasma removal of LDL, whereas unsaturated fatty acids exert protecting effects and can reduce LDL and triglyceride levels in the blood [23]. Polyunsaturated fatty acids play an important role in many physiological processes, and because they are not synthesized by the human body, they must be supplied by food. Minerals are important for many physiological functions of the human body. More than 100 mg of minerals (Na, Mg, K, Ca, P, and Cl) and less than 100 mg of microminerals (Fe, Cu, and Zn) are necessary to meet the recommended daily intake (RDI) [24]. According to the RDI [25] of minerals for adults, the necessary amount of calcium and phosphorus is 800 mg, potassium 1,950 to 5,900 mg, magnesium 300 mg, iron 14 mg, and zinc 15 mg. Thus, the amounts of iron and zinc present in the gluten-free breakfast cereal meet the daily intake value for adults, considering 100 g of cereal consumed daily. According to Lacerda, et al. [26] iron deficiency is one of the main public health

Component[1]	BRG	PFPF	WP	Breakfast Cereal
Moisture[2]	10.52 ± 0.06	5.73 ± 0.01	3.96 ± 0.02	5.38±0.02
Ashes[2]	0.58 ± 0.01	6.62 ± 0.02	7.27 ± 0.02	1.38±0.01
Protein[2]	8.96 ± 0.02	8.93 ± 0.01	13.50 ± 0.01	7.55±0.07
Lipids[2]	0.76 ± 0.01	1.45 ± 0.01	1.38 ± 0.02	0.97±0.02
TDF[2]	3.85 ± 0.01	53.94 ± 0.03	-	6.12±0.01
SDF[2]	2.95 ± 0.02	14.87 ± 0.01	-	2.2±0.01
IDF[2]	0.9 ± 0.05	39.06 ± 0.03	-	3.91±0.01
Carbohydrates[2]	79.18	77.27	73.89	84.72
Energy value[3]	359.4	357.85	361.98	353.33

[1]Mean value with standard deviation; [2]g.100 g^{-1}; [3]kcal.100 g^{-1} ;TDF (Total Dietary Fiber); SDF (Soluble Dietary Fiber); IDF (Insoluble Dietary Fiber).

Table 1: Centesimal compositions and energy values of broken rice grains (BRG), passion fruit peel flour (PFPF), whey powder (WP) and gluten-free breakfast cereal.

Amino Acid	Breakfast Cereal	FAO/WHO
Phenylalanine + Tyrosine	0.71	0.63
Leucine	0.66	0.66
Glycine	0.32	
Isoleucine	0.36	0.28
Arginine	0.58	
Alanine	0.41	
Tryptophan	0.11	0.11
Methionine + Cystine	0.28	0.25
Valine	0.49	0.35
Proline	0.36	
Serine	0.37	
Threonine	0.29	0.34
Histidine	0.15	0.19
Lysine	0.23	0.58
Taurine	<0.10	
Total amino acids	7.34	

Table 2: Amino acid profiles of the breakfast cereal (g. 100 g^{-1}) and FAO/WHO [21] essential amino acid recommendation

Fatty Acids (g.100 g^{-1})		Minerals (mg/kg)	
Polyunsaturated Fatty Acids	0.3	Calcium	80
Trans Fatty Acids	0	Phosphorus	190
Monounsaturated Fatty Acids	0.2	Potassium	370
Saturated Fatty Acids	0.31	Magnesium	60
Unsaturated Fatty Acids	0.49	Iron	33.16
		Zinc	17.86

Table 3: Fatty acid and mineral profile of the breakfast cereal.

problems in developing countries and is one of the main factors that cause anemia, affecting up to 46% and 48% of children and pregnant women, respectively, worldwide. In Brazil, anemia is the health problem most strongly related to micronutrient deficiency [27]. The calcium, potassium, magnesium and phosphorus contents in the breakfast cereal are below the recommended RDI for adults (Table 3); however, in a varied diet, such minerals might come from other food sources, thus reinforcing the need for a balanced diet.

Conclusions

New studies are necessary to accurately evaluate the effectiveness of gluten-free breakfast cereal components. The enrichment of food with whey protein would facilitate its consumption and reduce environmental damage; in addition, it is an excellent source of essential amino acids and is safe for consumption by those affected by celiac disease.

Acknowledgements

We thank the Research Foundation of the State of Goiás (Fundação de Amparo à Pesquisa do Estado de Goiás - FAPEG) and the Coordination for the Improvement of Higher Education Personnel (Coordenação de Aperfeiçoamento de Pessoal de Nível Superior - Capes) for the post-doctoral grants and the Italac and Arroz Cristal companies for the donation of raw materials.

Ethical Approval

Informed consent was obtained from all individual participants included in the study.

References

1. Carvalho WT, Reis RC, Velasco P, Soares Júnior MS, Bassinello PZ, et al. (2011) Physical and chemical characteristics of brown rice extracts, broken rice and soybean. Pesq Agropec Trop (Goiânia) 41: 422-429
2. Soares Júnior MS, Santos TPB, Pereira GF, Minafra CS, Caliari M, et al. (2011) Development extruded snacks from rice and bean fragments: Ciênc Agrárias (Londrina) 32: 189-198.
3. Tavares JS, Soares JMS, Becker FS, Eifert EC (2012) Functional changes of roasted rice flour with microwave depending on the moisture content and processing time. Ciênc Rural (Santa Maria) 42: 1102-1109.
4. Molina-Infante J, Santolaria S, Montoro M, Esteve M, Fernández-Banares F, et al. (2014) Sensibilidad al gluten no celiaca: una revisión crítica de la evidencia actual. Gastroenterol Hepatol 37: 362-371.
5. Vernaza MG, Chang YK, Steel CJ (2009) Effect of passion fruit fiber content and moisture and temperature extrusion in the development of funtional organic breakfast cereal. Braz J Food Technol (Campinas) 12: 145-154.
6. AOAC (2010) Official methods of analysis. Association of Official Analytical Chemists, Washington.
7. White JA, Hart RJ, Fry JC (1986) An evaluation of the waters pico-tag system for the amino-acid-analysis of food materials. J Autom Chem 8: 170-177.
8. Hagen SR, Frost B, Augustin J, Phenylisothiocyanate P (1989) Derivatization and liquid-chromatography of amino-acids in food. J Assoc Off Anal Chem 72: 912-916.
9. Chang YK, Hashimoto JM, Alcioli-Moura R, Martínez-Flores HE, Martínez-Bustos F, et al. (2001) Influence of extrusion conditions on cassava starch and soybean protein concentrate blend. Acta Aliment (Budapeste) 30: 189-203
10. American Public Health Association (APHA) (2002) Compendium of methods for the microbiological examination of foods. American Public Health Association, Washington DC.
11. Stone HS, Sidel JL (2005) Sensory evaluation practies. Academic Press, Zürich, Switzerland.
12. Coutinho L, Batista J, Caliari M, Soares Júnior MS (2013) Optimization of extrusion variables for the production of snacks from by-products of rice and soybean. Ciênc Tecnol Aliment (Campinas) 33: 705-712
13. Borges JTS, Ascheri JLR, Ascheri DR, Nascimento RE, Freitas AS, et al. (2003) Properties of cooking and noodle physicochemical characteristics pre-cooked meal to the entire basic quinoa (*chenopodium quinoa, wild*), rice flour (*oryza sativa*) polished in extrusion thermoplastic. Boletim do Centro de Pesquisas e Processamento de Alimentos (Curitiba) 21: 303-322.
14. Barbosa-Cánovas GV, Fontana Junior AJ, Schmidt SJ, Labuza TP (2007) Water activity in foods: fundamentals and applications. Blackwell Publishing, Oxford.
15. Carvalho AV, Bassinello PZ, Mattietto RA, Carvalho RN, Rios AO, et al. (2012) Processing and characterization of extruded snaack from broken rice flour and little band. Braz J Food Technol (Campinas) 15: 72-83.
16. Souza MWS, Ferreira TBO, Vieira IFR (2008) Centesimal properties and composition functional technological flour bark the passion. Alimentos e Nutrição (Araraquara) 19: 33-36.
17. Bernaud FSR, Rodrigues TC (2013) Dietary fiber - adequate intake and effects on metabolic health. Arquivo Brasileiro de Endocrinologia & Metabologia (São Paulo) 57: 397-405
18. Silva PA, Assis GT, Carvalho AV, Simões MG (2011) Development and characterization of breakfast cereal extruded cassava enriched with protein concentrate of whey. Braz J Food Technol (Campinas) 14: 260-266
19. Day L, Swanson BG (2013) Functionality of protein-fortified Extrudates. Compr Rev Food Sci Food Saf 12: 546-564.
20. Chassagne-Berces S, Leitner M, Melado A, Barreiro P, Correa EC, et al. (2011) Effect of fibers and whole grain content on quality attributes of extruded cereals. Procedia Food Science 1: 17-23.
21. FAO/WHO (1991) Protein quality evaluation: report of the joint FAO/WHO expert consultation. FAO, Rome.
22. Haraguchi FK, Abreu WC, Paula H (2006) Whey protein: composition, nutritional properties, applications in sports and benefits for human health. Revista Nutr (Campinas) 19: 479-488.
23. Santos KMO, Aquino RC (2008) The food pyramid: basic fundamentals of nutrition. Barueri, Manole.
24. FDA (2013) Code of federal regulations, title 21: food and drugs. Food and Drug Administration, Washington, DC.
25. Brazil (1998) Ordinance No. 33 of 13 January 1998. Diário Oficial da União, Brasília.
26. Lacerda DBCL, Soares Júnior MS, Bassinello PZ, Castro MVL, Silva-Lobo VL, Campos MRH, Siqueira BS (2010) Sharps quality of raw rice, extruded and parboiled. Pesq Agropec Trop (Goiânia) 40: 521-530.
27. Cozzolino SMF (2007) Mineral deficiencies. Estudos Avançados (São Paulo) 21: 119-126.

17

Optimization of Food Acidulant to Enhance the Organoleptic Property in Fruit Jellies

Kesava Reddy C, Sivapriya TVS, Arun Kumar U and Ramalingam C*

School of Biosciences and Technology, VIT University, Vellore, India

Abstract

Citric acid, tartaric acid and malic acids are the commonly used additives in confectionary industries. Their acidity helps in increasing the organo-leptic property of the food product. Other than the flavour, they are also used for their health benefits. They play a major role in boosting renal health, revitalizing skin, fighting free radicals etc. Every individual additive mentioned above has a wide range of health benefits. Initially these additives were used individually to obtain their source fruit's taste. Later, they were mixed to obtain unique flavours. Different level of additive mix gave different level of enhanced taste. This research work aims in appropriate optimization of these additives to provide an enhanced organoleptic property in fruit jelly. The following ratio of citric acid (CA) to malic acid (MA) to tartaric acid (TA) was used in this research work. The ratio was finalized from literature studies. (Sample-S) S1-40:35:25, S2-40:40:20, S3-33.30:33.30:33.30, S4-32.50:50:17.50. Sensory analysis was made to select the best tasting sample and mixed with stabilizers. Sodium citrate, sodium tartrate and a mix of both sodium citrate and sodium tartrate were used to make three samples using the winning proportion of (CA: MA: TA) acids. Sensory analysis was made to select the best among them and they were mixed with fruit pulp finally. Papaya pulp was used in this research work. This final sample was taken for sensory evaluation again to know the feedback on taste of the fruit jelly, which was a positive result where most of the evaluators liked it. The results were analysed using SPSS software. Thus, this research work proves that the organoleptic property of a fruit jelly can be enhanced if the additives were used in an optimized proportion as in sample S1.

Keywords: Acidulant; Citric acid; Malic acid; Tartaric acid; Optimization

Introduction

Food acidulants are commonly used to increase the acidity of the food product. Increasing the acidity of the food helps in preservation, flavour enhancement, anti-pathogenic etc. Two or more acids are combined to produce a unique flavour. Citric (E330), malic (E296) and tartaric acids (E334) are the commonly used additives in food sector. The reason behind their common use are, they are organic. They are obtained from fruits (Citrus fruits, apple, grapes, tamarind etc.) through various extraction and fermentation techniques. Hence they are easily available and economically convenient for the food sectors. Their acidity acts as a preservative. Increased acidic environment in the food kills the microbes and pathogens thus making the food safe [1]. They are Generally Regarded as Safe (GRAS) hence they act as a substitute for the chemical preservatives, since prolonged use of chemical preservative results in health hazards. Citric acid has a very long history in food industries. They are most commonly used and it is one of the conventional food additive too [2]. The health benefits obtained on using these naturally derived food additives are remarkable. Citric acid found to be a very powerful weapon against kidney stone formation. They also help in soothing sore throat, improving mineral absorption, curbing nausea etc. Similarly, tartaric acids are known for their anti-oxidant and anti-inflammatory properties. Malic acid is good for liver. It is known for its effectiveness in reducing toxicity. Initially, it was used to obtain the taste of their source fruit. Later, food sectors realized their wide range of use when they used different combinations of these additives to produce unique flavours. Tartaric acid is the least used additive when compared to citric and malic acid. It is because, it has a strong flavour which tastes absolute tangy [3]. Also, the cost price of tartaric acid is higher when compared to the other two additives. Still, they are preferred in some food product where strong tangy flavour is required. Appropriate combination of citric, malic and tartaric acid can produce a unique enhanced level of taste. [4,5]. This combination of acidulant has citric acid as a major constituent followed by malic and tartaric acid. They produce an enhanced organoleptic property to the food product and are economically convenient too. This research work focuses on optimising the combination of this additives to find the appropriate ratio of the additives which will produce enhanced organoleptic property to fruit jellies. Combination of additives were arrived from literature studies [6,7].

Initially, four combination of acidulants are arrived and using the combinations, four samples of fruit jellies are produced without any stabilizers and fruit pulp. The samples are taken for sensory evaluation and the winning sample is selected for further addition with stabilizers. Three samples are then produced along with stabilizers and sent for sensory evaluation again. The winning sample is selected and added with papaya fruit pulp. Now, the sample with winning combination of acid and stabilizers are produced with the fruit pulp and again taken for sensory evaluation to know the feedback on their taste. The sensory results are analysed using SPSS software. Papaya is used as a fruit pulp because it is easily available and blends with any type of flavour to produce the similar flavour.

Materials and Methods

Sample preparation

Fruit jellies are generally produced by heating appropriate proportion of water, sugar, liquid glucose, stabilizers until it dissolves.

***Corresponding author:** Ramalingam C, School of Biosciences and Technology, VIT University, Vellore, India, E-mail: cramalingam@vit.ac.in

It is then followed by adding fruit pulp. It is cooked up to 90°C. Then pectin and permitted colour solutions are added. The total solids will be checked by refractometer till 80-81 TS achieved. Steam is cut off after this stage and permitted flavours, acids are added and mixed. It is then allowed to cool and thus jellies are produced [8]. We followed the similar procedure to prepare our sample.

Citric (E330), malic (E296) and tartaric acid (E334) were purchased from Tirumalai chemicals. Four combinations of these additives were made based on literature studies [6,7]. Four samples of fruit jellies are produced using the arrived combination of additives in a hygienic manner. The combinations are represented in Figure 1.

Four samples were taken for sensory analysis. The winning sample's combination is used to produce fruit jellies with stabilizers. Sodium citrate (E331), sodium tartrate (E335) were used. Three samples of fruit jellies are produced using the winning additive combination and stabilizers. The three stabilizers used are represented in Figure 2. Three sample of fruit jellies were taken for sensory analysis. The winning sample's combination of acidulants, stabilizers are used to produce the final jelly with the fruit pulp. Papaya pulp is used in this research work as they are easily available and blends with any flavour to produce the desired flavour. Fresh papaya pulp is obtained by extracting from ripened fruit .750 mg of papaya pulp is used per gram of the jelly. Thus, every single fruit jelly produced, contains 60 mg of acidulants, 15 mg of stabilizers, 750 mg of pulp and other additives. The samples are assessed for physical characteristics (Like, shape, appearance, colour and moisture content of the fruit jelly), chemical characteristics (Like, water activity, reducing sugars, pH/acidity, and sucrose content), and micro characteristic (Using Total Plate Count, Yeast and Mould count, general pathogen analysis) as per FSSA procedure before sending it to sensory analysis.

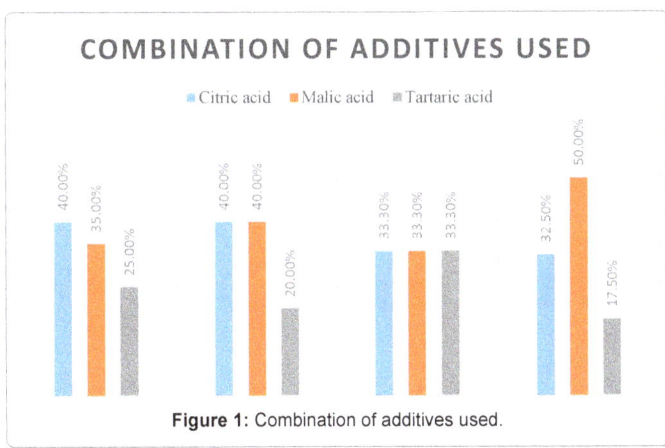

Figure 1: Combination of additives used.

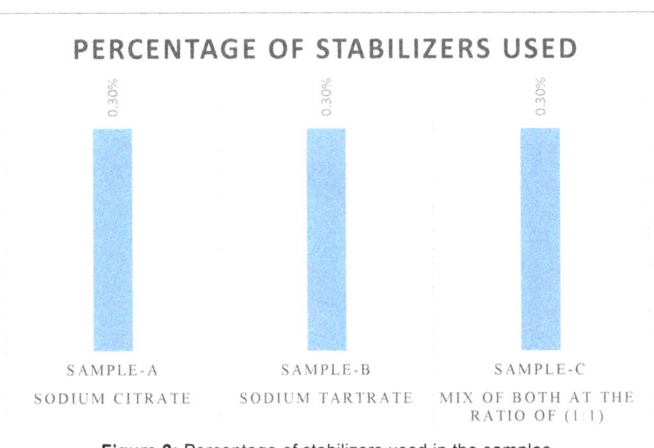

Figure 2: Percentage of stabilizers used in the samples.

Sensory analysis

Three sensory analysis was made. One for the samples with acidulants, one with acidulants +stabilizers and one with acidulant + stabilizer + papaya pulp. Sensory analysis was made by 80 volunteers from VIT University. Volunteers were from varying age group (ranging from 18-55 years old) and were asked to analyse upon their consent. They were asked to taste the samples and rate their characteristics (Like, texture, flavour and overall acceptance.) on a five point hedonic scale as shown in Table 1 [9].

Statistical analysis

The test data generated from the sensory evaluation was subjected to find the mean value, standard deviation, analysis of variance (ANOVA) and the Tukey test for multiple comparison and hence the homogenous subset from the Tukey test to find the highest value of alpha. These were performed to identify difference between the means of the samples and to find the difference at probability level $p < 0.05$ were considered to have difference between the variance in the population, also to establish a significant difference in the sample under sensory analysis. Then Tukey honesty significant difference (HSD) was calculated, to establish which sample exhibited significant difference from one and another [10]. For analysis of the data SPSS statistics 20.0 software was used.

Results and Discussion

Statistical analysis of trail-1 results over flavour

The proximate composition of the flavours in the samples for the trail test are Sample A (citric acid-40%, Malic acid-35%, Tartaric acid-25%) Sample B (Citric acid-40%, Malic acid-40%, Tartaric acid-20%) Sample C (Citric acid-33.3%, Malic acid-33.3%, Tartaric acid-33.3%) Sample D (Citric acid-32.50%, Malic acid-50%, Tartaric acid-17.5%) (Table 2). So here the *null hypothesis is rejected*, the *p*-value as Sig. is less than that of the significance value 0.05 which concludes that there is a difference between the variances in the population. It shows there is significant difference in the flavour combination of among at least three of the four formulations.

Homogeneous subsets

From the first trail, we conclude that the panellist preferred Sample A (Citric acid-40%, Malic acid-35%, Tartaric Acid-25%) over the other samples (Table 3).

Statistical analysis of trial-2 results over stabilizer from the optimized flavour sample: Stabilizers as mentioned in the literature has an influence on the sensory attribute of flavour on the fruit jellies, so three different stabilizer sodium citrate, sodium tartrate and combination of both sodium citrate and tartrate and coded as (T1, T2, T3) were used on the optimized sample from flavour trail test (Table 4).

Conclusion

From the Table 4, it can be concluded that the ANOVA result shows a higher significance value than 0.05, which shows there isn't any significance difference in the sample when the stabiliser is added. This proves that the stabilisers did not influence as of sensory and texture is concerned. And the Tukey test confirms that T3 (combination of sodium citrate and sodium tartrate) was preferred more by the population than the rest (Table 5).

Ratings	Comment
5	Excellent
4	Good
3	Moderate
2	Fair
1	Poor

Table 1: Hedonic scale rating and their interpretation.

ANOVA					
	Sum of Squares	Df	Mean Square	F	Sig.
Between Groups	7.057	3	2.352	2.679	0.047
Within Groups	242.314	276	0.878	-	-
Total	249.371	279	-	-	-

Table 2: ANOVA results for trail-1.

Tukey's HSD[a]			
Acid combinations	N	Subset for alpha = 0.05	
		1	2
Sample D	80	3.63	-
Sample C	80	3.87	3.87
Sample B	80	3.91	3.91
Sample A	80	-	4.07
Sig.		0.274	0.587

Table 3: Turkey HSD[a] result for trail-1.

ANOVA					
	Sum of Squares	Df	Mean Square	F	Sig.
Between Groups	2.155	2	1.077	1.298	0.275
Within Groups	169.275	204	0.830	-	-
Total	171.430	206	-	-	-

Table 4: ANOVA results for trail-2.

Tukey HSD[a]		
Formulation of stabilizers	N	Subset for alpha = 0.05
		1
T1	80	3.57
T2	80	3.72
T3	80	3.81
Sig.		0.253

Table 5: Turkey HSD[a] results for trail-2.

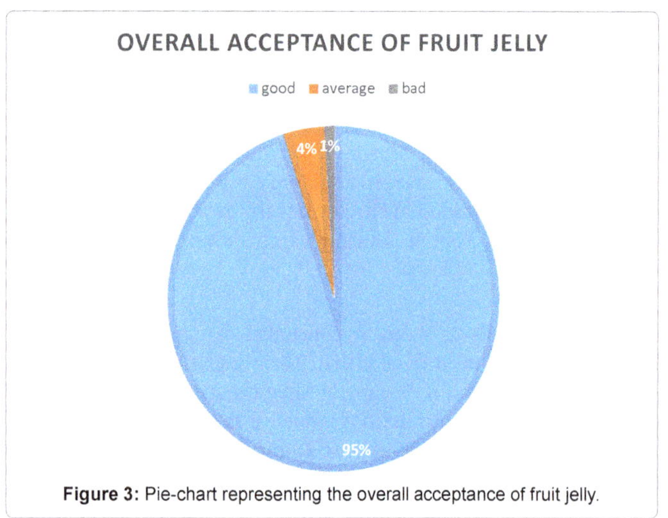

Figure 3: Pie-chart representing the overall acceptance of fruit jelly.

Trail-3 was made using the sample prepared with the composition of the winning sample of trail-1 and trail-2. Thus, it was made with 40% of citric acid, 35% of malic acid, 25% of tartaric acid and a mix of both sodium citrate and sodium tartrate as stabilizers mixed with papaya pulp. About 95% of the panellist rated the sample as good. This is represented in Figure 3.

From Figure 3, we can interpret that our sample with 40% citric acid, 35% malic acid, 25% tartaric acid, mixed stabilizers of sodium citrate and sodium tartrate along with papaya pulp has been liked by 95% of the panellist (on a total of 80 people). They rated this sample as GOOD. We have got a positive result and thus we can conclude that the above-mentioned ratio of acidulants produced enhanced organoleptic property of the fruit jelly in an economically convenient way than the generally used additives to obtain such enhanced organoleptic property.

References

1. Humayun A (2014) Effect of citric and malic acid on shelf life and characteristics of orange juice. Int J Pharmaceutic Sci 6: 117-119.
2. Tammy Nash J (2012) Acids in confections. The Manufacturing Confectioner.
3. Maria MT, Lues JR (2010) Organic acids and food preservation. CRC Press Inc, Boca Raton, USA.
4. Ahmed H, Chandan KG, Mukund M, Sumeet S, Chidambaram R (2013) The effect of citric and malic acid additives on the storage stability and sensitive parameter in lemonade. Res J Pharmaceutic Biol Chem Sci 4: 1671-1679.
5. Malundo TMM, Shewfelt RL, Ware GO, Baldwin EA (2001) Sugars and acids influence flavour properties of mango. J Amer Soc Hort Sci 126: 115-121.
6. Jai Prakash S, Satish Kumar S, Ruchika C, Gaurav P, Alok P, et al. (2014) Optimisation of common acidulant (fruitaric acid) to enhance organoleptic quality and shelf life of fruit juices. Int J Pharmaceutic Sci 46: 269-273.
7. Saha P, Jai Prakash S, Sumeet S, Ahmed H, Ramalingam C (2013) Optimisation of citric acid and malic acid to enhance flavor and shelf life of mango juice. J Chem Pharmaceutic Res 5: 90-95.
8. Catharina YW, Liu K, Yao-Wen H (1999) Asian foods: science and technology. Technomic publishing company.
9. Morten CM, Thomas BC, Gail V (1991) Sensory evaluation techniques. CRC Press Inc, Boca Raton, USA.
10. Abdi H, Williams LJ (2010) Tukey's honestly significant difference (HSD) test. Encyclopaedia of Research Design, Sage, Thousand Oaks, CA.

Sensory Characteristics, Total Polyphenol Content and *In vitro* Antioxidant Activity of Value Added Processed Barnyard Millet Flour Chapattis

Shobana Devi R and Nazni P*

Department of Food Science and Nutrition, Periyar University, Salem, Tamil Nadu, India

Abstract

Introduction: Value addition of millet is an important strategy to improve its utilization and the method of processing determines the quality characteristics of the value added products. Utilization of millets in food formulations is increasing worldwide, since they are rich sources of phytochemicals and dietary fiber which offer several health benefits.

Objective: Thus in the present study, an attempt was made to develop roasted and pressure cooked barnyard millet flour incorporated chapatti at different levels and its consumer acceptability to evaluate the total polyphenol content and *in vitro* antioxidant activity.

Materials and methods: Roasted and pressure cooked barnyard millet flour incorporated chapattis were developed at four different levels i.e., 10%, 30%, 50% and 100% millet flour to wheat flour. Sensory evaluation was done for all the developed products using 9 point hedonic scale. For the best variation, the total polyphenol content was estimated using folic ciocalteau method and the *in vitro* antioxidant activity were evaluated on the basis of measuring Ferric reducing ability power (FRAP) and scavenging activity for DPPH radicals by methanolic extracts.

Results: Roasted and pressure cooked barnyard millet flour incorporation in the preparation of chapatti at 10 per cent level was acceptable. The total polyphenol content of best variation (10%) of roasted and pressure cooked barnyard millet flour incorporated chapatti was 6.12 mg/g and 5.38 mg/g respectively. The DPPH radical scavenging activity and Ferric reducing ability power (FRAP) of the same was found to be 59% and 13.42 mg/g and 53% and 11.57 mg/g respectively. The standard chapatti prepared with wheat flour was found to have 4.02 mg/g of polyphenol content, 47% of DPPH radical scavenging activity and 9.84 mg/g of Ferric reducing ability power.

Conclusion: Thus the present study concludes that roasting and pressure cooking enhances the polyphenol content of barnyard millet grains which might also contribute significantly to the management and/or prevention of degenerative diseases associated with free radical damage due to its high polyphenol content and antioxidant activity.

Keywords: Barnyard millet; Roasting; Pressure cooking; Polyphenol; DPPH activity; FRAP assay

Introduction

Millets play very specific role in human nutrition because of their multiple qualities [1]. Barnyard (*Echinochloa frumentacea Link*) is one of the fastest growing crops of all the millets, mature in 90 to 100 days. They are an important source of vital minerals like niacin, magnesium, phosphorus, manganese, iron and potassium. They contain high amounts of protein, fiber, essential amino acid methionine, lecithin, and vitamin E [2]. Recent studies have shown that due to the high content of these nutrients, millets have therapeutic benefits such as control of asthma, migraine, blood pressure, diabetes, heart disease, atherosclerosis and heart attack. Fibre, in millet, prevents gallstones formation. Because of these benefits, millets can be used in functional foods and as nutraceuticals. Hence, they are also called as 'nutri cereals' [3].

At the household level, the common methods of food processing include wet heat treatment such as pressure cooking and dry heat treatment like roasting. These processing methods alter the nutritive value of foods. The nutrient composition and technological properties of minor millet grains offer a number of opportunities for processing and value addition to use as next generation to satisfy the consumers of different culture, location and society. Value addition of millet is an important strategy to improve its utilization and the method of processing determines the quality characteristics of the value added products. Wheat has a unique property of forming an extensible, elastic and cohesive mass when mixed with water. Millet flours lack these properties when used alone. Hence replacement of wheat flour with millet flour in wheat composite flours brings lot of innovative Ready-to-eat or Ready-to-serve minor millet based processed products [4].

Utilization of millets in food formulations is increasing worldwide, since they are rich sources of phytochemicals and dietary fiber which offer several health benefits [5]. Millets have received considerable attention in the last several decades due to the presence of unique blend of bioactive phytochemicals which are powerful antioxidants. These unique bioactive compounds such as polyphenols in whole-grains are proposed to be responsible for the health benefits of whole-grain consumption such as cardiovascular diseases, cancers, type II diabetes, neurodegenerative diseases or osteoporosis [6,7]. Fortification of diets with millets and cereals rich in phenolic acids was shown to impart many health benefit properties, and this can be exploited in developing healthy foods [8]. Chapatti as a staple food of the Indian subcontinent was selected aptly for enrichment to reach the different sections of the population [9,10]. Chapatti is a flat unleavened, hot plate baked product prepared from whole wheat flour by converting dough with water by

***Corresponding author:** Nazni P, Department of Food Science and Nutrition, Periyar University, Salem-11, Tamil Nadu, India, E-mail: naznip@gmail.com

adding other ingredients like salt and sugar according to the taste. Though wheat flour is a staple food for half of the world population; still it is not a complete diet which lacks in micronutrients [11]. Chapattis prepared with value addition of processed millet flour provide an additional dietary fiber and phytochemicals constituents that could be supportive for diabetic and obese individuals [12]. Therefore, millet polyphenols have received tremendous attention among nutritionists, food scientists, and consumers due to their roles in human health [13]. Thus in the present study, an attempt was made to develop roasted and pressure cooked barnyard millet flour incorporated chapatti at different levels and its consumer acceptability to evaluate the total polyphenol content and *in vitro* antioxidant activity.

Methods and Materials

The barnyard millet was procured from local market of Vellore district, Tamil Nadu, India. The millets were cleaned properly and stored in sealed containers till their use in different processing such as roasting and pressure cooking. Processed foods are usually less susceptible to early spoilage than fresh foods. It is widely accepted that simple and inexpensive traditional processing techniques are effective methods of achieving desirable changes in the composition of grains.

Processing techniques

Roasting: Roasting involves the application of dry heat to grains using a hot pan or at a temperature of 150 to 200°C for a short time and powdered [14].

Pressure cooking: The grains were washed, soaked and pressure cooked in tap water in the ratio of 1:3 (w/v) for two whistles in medium flame. They were solar dried for 2 days and then powdered [15].

Product development: The most acceptable product chapatti was selected for the incorporation of roasted and pressure cooked barnyard millet flour. Totally 9 variations of chapattis containing roasted or pressure cooked barnyard millet flours were developed. Four variations in roasted barnyard millet flour (10%, 30%, 50% and 100%), four variations in pressure cooked millet flour (10%, 30%, 50% and 100%) and one standard product. Wheat is an ideal grain in making chapattis. The use of millet grains as replacement in wheat composite flours seems the best method that can be used for the preparation of nutritional, healthy, safe, high quality and shelf stable food products at house hold and commercial scale to promote the utilization of millet grains. These wheat-millet composite flour blends may have the advantage of being nutritious, economical and health promoting. They have mighty potential to be included in traditional and novel foods. The value added chapatti was prepared by the addition of roasted or pressure cooked barnyard millet flour in different combination with wheat flour as mentioned below.

Preparation of chapattis from processed barnyard millet flours: Barnyard millet incorporated chapattis were prepared by incorporating roasted or pressure cooked barnyard millet flour to wheat flour at 10%, 30%, 50% and 100% levels. Standard chapattis were prepared with wheat flour without the incorporation of processed barnyard millet flour. The various levels of incorporation of roasted and pressure cooked barnyard millet flour for the development of chapattis was given in Table 1.

Preparation of chapatti

The wheat flour was mixed with roasted or pressure cooked barnyard millet flours at various levels of incorporation. The chapattis were prepared from samples according to the procedure (Table 2 and Figure 1).

Variations	Wheat Flour (g)	Processed Barnyard Millet Flour (g)	
		Roasted	Pressure cooked
Standard	100	-	-
V1	90	10	10
V2	70	30	30
V3	50	50	50
V4	-	100	100

Table 1: Variations for the preparation of chapattis from processed barnyard millet flour.

Ingredients	Weight (g)
Wheat flour	At different levels
Processed Barnyard millet flour	At different levels
Oil	1 tsp.
Salt	To taste

Table 2: Ingredients used in the preparation of chapatti.

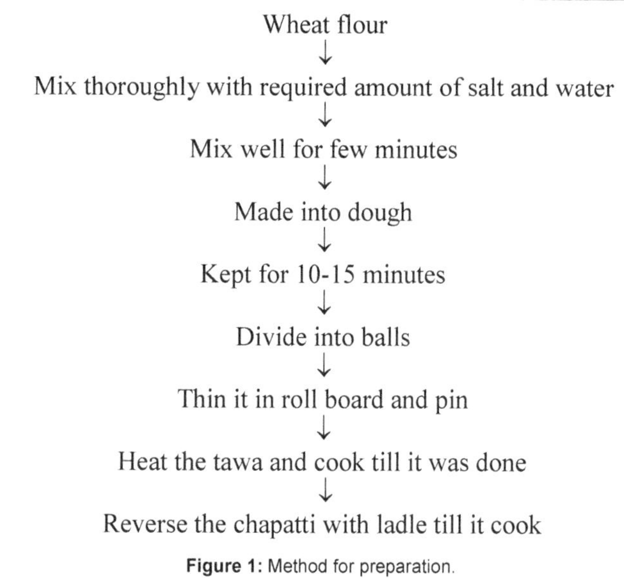
Wheat flour
Mix thoroughly with required amount of salt and water
Mix well for few minutes
Made into dough
Kept for 10-15 minutes
Divide into balls
Thin it in roll board and pin
Heat the tawa and cook till it was done
Reverse the chapatti with ladle till it cook
Figure 1: Method for preparation.

Organoleptic evaluation of the developed value added chapattis

Selection of any food products depends upon its perceived taste, nutritional labelling, family and cultural preference and prior purchase habits. Taste is generally regarded as the predominant reason for selecting a food followed by health and other considerations [16]. Organoleptic quality evaluation of the product plays an important role in the acceptance and preference of foods. The consumer acceptability of each product was carried out by panel of members through organoleptic evaluation using 9 point hedonic scale with score card of scores ranging from 9 to 1, where 1=dislike extremely, 5=neither like nor dislike and 9=like extremely was used. The parameters assessed for chapattis were colour, flavour, taste, texture, foldability, breakability and over all acceptability. A ballot sheet was prepared to evaluate sensory attributes of developed value added processed barnyard millet flour chapatti.

Selection of best variation of value added chapatti

Based on the organoleptic evaluation results, the best and highly accepted variation from each processing techniques (roasted and pressure cooked) of barnyard millet flour incorporated chapattis and the standard chapatti were selected for the determination of total polyphenol content and *in vitro* anti-oxidant activity.

Determination of total polyphenol content and *in vitro* antioxidant activity

Several studies have shown that 80% methanol is an effective solvent in extracting phenolics and other polar substances in cereals [17,18]. In this study, methanolic extract of the standard chapatti and the selected best chapatti of each processing techniques was used for the determination of total polyphenol content and antioxidant activity.

Extraction procedure

For methanolic extract of each selected product, 50g of powdered sample was macerated and extracted with 500 ml of 80% methanol in a glass jar for 48 hours at room temperature in an orbital digital agitator (Rotatest 560VIT. 15-300 T/MIN). After filtering with filter paper and the methanol solvent was removed under reduced pressure at low temperature (40°C to 45°C) using a rotary vacuum evaporator [19].

Determination of total polyphenol content

The total polyphenol content of the methanolic extract of selected products was carried out according to the Folin-Ciocalteu colorimetric method [20].

In vitro anti-oxidant activity

DPPH radical scavenging activity: The free radical-scavenging activities of extracts and ascorbic acid were measured by using 2, 2-diphenyl-1-picryl-hydrazyl (DPPH) as procedure proposed by Burits [21]. The DPPH solution (0.3 ml) was added to 2.5 ml of extracts. The mixture was shaken and left to stand at 20°C for 30 minutes at room temperature. Then, the absorbance was measured at 517 nm. Ascorbic acid was used as positive control. The inhibition percentage of the DPPH radical was calculated using the following formula,

$$\text{DPPH radical scavenging activity } (\%) = \frac{A_0 - A_1}{A_0} \times 100$$

Where,

A_0 indicate absorbance of the control and

A_1 indicate absorbance of extract sample.

Ferric reducing ability power (FRAP): The ferric ions (Fe^{3+}) reducing activity power was measured according to the method described by [22]. The method was based on the chemical reaction of Fe^{3+} to Fe^{2+}. The working solution was prepared by mixing 1 ml of extracts with 2.5 ml phosphate buffer (0.2M; pH 6.6) and 2.5 ml $N_6C_6FeK_3$ (1%; m/v). The mixing was incubated at 50°C for 30 minutes. The Fe^{2+} was monitored by measuring the formation of ferrous complex at 700 nm. The reducing power of the extracts was represented as ascorbic acid equivalent (mg AAE/g).

Statistical analysis

The final data was compiled and analyzed by using statistical methods. The results were represented as Descriptive statistics such as mean, standard deviation followed by Duncan's multiple comparison tests and correlation p values < 0.05 were considered significant.

Results and Discussion

Organoleptic evaluation of the roasted and pressure cooked barnyard millet flour incorporated chapatti

In a product development, there is a need to state the descriptive characteristics of the product along with a mere comparison of scores to assess the product quality. Descriptive characteristics of any product are very important to assess the acceptability of the product. The acceptability of the developed value added products varied with the incorporation levels [23]. The mean scores of the organoleptic evaluation of all the developed value added products using nine point hedonic scales were given in Tables 3 and 4. The pooled score of standard chapatti in all the attributes such as colour, flavour, taste, texture, foldability, breakability and overall acceptability was 8.6. Among the four variations developed, the variation 1(V1) of both roasted and pressure cooked barnyard millet flour incorporated chapatti at 10% level has got highest mean score of 8 in all the attributes respectively (Figures 2 and 3). As the level of incorporation of processed barnyard millet flour increases, the acceptability range of the developed product decreases. The chapattis prepared by the addition of 10% cereal bran showed better performance and were quite comparable with whole-wheat flour regarding the proximate components and sensory attributes [24]. Results of the Duncan's test revealed that there was significant difference for all the variations in response to all the organoleptic

Variations	Colour	Flavour	Taste	Texture	Foldability	Breakability	Overall acceptability
Standard	8.6 ± 0.54^a	8.6 ± 0.54^a	8.4 ± 0.54^a	8.6 ± 0.54^a	8.6 ± 0.54^a	8.6 ± 0.54^a	8.6 ± 0.54^a
V1	8.2 ± 0.44^a	8 ± 0.00^a	8 ± 0.00^a	7.6 ± 0.54^b	7.8 ± 0.44^b	7.8 ± 0.44^b	8 ± 0.00^a
V2	6.6 ± 0.54^b	6.4 ± 0.54^b	6 ± 0.7^b	6 ± 0.00^c	5.8 ± 0.44^c	6 ± 0.00^c	6.2 ± 0.44^b
V3	4.6 ± 0.54^c	4.4 ± 0.89^c	3.4 ± 0.54^c	3.6 ± 0.89^d	3.8 ± 0.44^d	3.4 ± 0.54^d	4 ± 0.7^c
V4	2.4 ± 0.89^d	2 ± 0.7^d	1.6 ± 0.54^d	1.8 ± 0.83^e	1.8 ± 0.44^e	1.8 ± 0.44^e	2 ± 0.00^d

a-e values in the same column with different superscripts are significantly different at (p<0.05) in Duncan's multiple range test.

Table 3: Organoleptic evaluation of roasted barnyard millet flour incorporated chapatti.

Variations	Colour	Flavour	Taste	Texture	Foldability	Breakability	Overall acceptability
Standard	8.6 ± 0.54^a	8.6 ± 0.54^a	8.4 ± 0.54^a	8.6 ± 0.54^a	8.6 ± 0.54^a	8.6 ± 0.54^a	8.6 ± 0.54^a
V1	8 ± 0.44^a	8.2 ± 0.44^a	8 ± 0.44^a	7.8 ± 0.54^b	7.6 ± 0.54^b	7.8 ± 0.54^b	8 ± 0.44^b
V2	6.2 ± 0.83^b	6.2 ± 0.44^b	5.6 ± 0.54^b	5.8 ± 0.44^c	5.6 ± 0.54^c	5.8 ± 0.44^c	5.8 ± 0.44^c
V3	4.4 ± 0.54^c	4.2 ± 0.83^c	3.4 ± 0.54^c	3.4 ± 0.54^d	3.6 ± 0.54^d	3.4 ± 0.54^d	3.6 ± 0.54^d
V4	2.4 ± 0.89^d	2 ± 0.7^d	1.6 ± 0.54^d	1.8 ± 0.83^e	1.8 ± 0.44^e	1.8 ± 0.44^e	2 ± 0.00^e

a-e values in the same column with different superscripts are significantly different at (p<0.05) in Duncan's multiple range test.

Table 4: Organoleptic evaluation of pressure cooked barnyard millet flour incorporated chapatti.

parameters evaluated for acceptability. Hence from the above results it was clear that millet incorporation in the preparation of chapatti at 10 per cent level was acceptable. Thus the variation 1 from each processing techniques of barnyard millet flour incorporated chapattis with their standard product were chosen for the estimation of total polyphenol content and antioxidant activity.

Total polyphenol content of methanolic extract of selected value added chapattis

Polyphenol is an antioxidant phytochemicals containing several hydroxyl group, they tend to prevent or neutralize the damaging effects of free radicals. It was described that alcoholic extracts of several species of millets are rich sources of phenolic compounds and show antioxidant activities, metal chelating, and reducing powers [25]. The total polyphenol content of the methanolic extract of selected value added chapattis and the standard chapatti were summarized in the Table 5.

Phenolic compounds exhibit significant pharmaceutical activities such as anticancer, anti-inflammation and anti-oxidative effects [26]. In this study, the total polyphenol content of best variation (10%) of roasted and pressure cooked barnyard millet flour incorporated chapatti was 6.12 mg/g and 5.38 mg/g respectively. The standard chapatti was found to have 4.02 mg/g of total polyphenol content. The highest levels of polyphenols were observed in roasted and pressure cooked barnyard millet flour incorporated chapatti compared to the standard chapatti. The highest content of phenolic compounds is known to have direct antioxidant property due to presence of hydroxyl groups which can function as hydrogen donor [27].

Figure 2: Roasted barnyard millet flour incorporated chapatti.

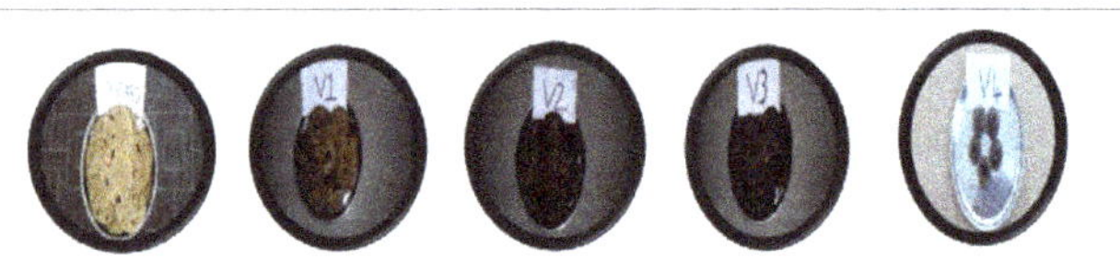

Figure 3: Pressure cooked barnyard millet flour incorporated chapatti.

Methanolic extracts of selected value added products	Total polyphenol content (mg/g)
Standard chapatti	4.02
RBMF incorporated chapatti at 10% level (V1)	6.12
PCBMF incorporated chapatti at 10% level (V1)	5.38
RBMF: Roasted barnyard millet flour PCBMF: Pressure cooked barnyard millet flour	

Table 5: Total polyphenol content of methanolic extract of selected products.

It is a well-known fact that millets are rich in polyphenols. The increase in total polyphenol content of roasted barnyard millet flour could result from release of bound polyphenols or from maillard reaction products formed during roasting that have been reported to possess scavenging activity on reactive oxygen species [28,29]. Maillard reaction involves condensation reactions between sugars and amino acids and has been found to be linked to polyphenols [30] via inhibition of polyphenol oxidase [31]. The longer the cooking time, the greater losses of the total phenolic compound measured. This could be due to the breakdown of phenolics or losses (leached out) during cooking as most of the bioactive compounds are relatively unstable to heat and easily solubilized [32]. Thus pressure cooked millets showed high total phenolic content due to its short cooking period.

In vitro antioxidant activity of methanolic extract of selected value added chapattis

Table 6 shows the finding of *in vitro* antioxidant activities of selected products. Antioxidants properties were evaluated on the basis of measuring Ferric reducing ability power (FRAP) and scavenging activity for DPPH radicals by methanolic extracts. The model of scavenging DPPH radical is a widely used method to evaluate the free radical scavenging activities of antioxidants [33]. In the DPPH assay, the colored stable DPPH radical (Purple) is reduced in the presence of an antioxidant or a hydrogen donor in to the non-radical form (Yellow). The DPPH scavenging activities of antioxidants are attributed to their hydrogen donating abilities [34]. In the present study, it should be noted that roasted and pressure cooked barnyard millet flour incorporated chapatti showed potent DPPH radical scavenging activity at 59% and 53% respectively when compared with the standard chapatti (47%).

Furthermore, FRAP assays was used to determine antioxidant activity of selected products. Methanolic extracts of roasted and pressure cooked barnyard millet flour incorporated chapatti showed better effect (13.42 mg/g and 11.57 mg/g respectively). These data marked that the roasted and pressure cooked millet flour incorporated chapatti has higher significant activities than standard (wheat flour) chapatti to reduce Fe^{3+} to Fe^{2+} by measuring the absorbance at 700 nm. Significant correlations were observed between total polyphenol content and DPPH scavenging activity (r=0.92) and ferric reducing ability power (r=0.87) indicating the role of polyphenol compounds in inhibiting free radicals and radical cations.

Conclusion

Based on the obtained results, the incorporation of 10% roasted and pressure cooked barnyard millet flour to wheat flour in the preparation of chapatti showed better sensory acceptability. This indicates that millet based products of good sensory qualities could be produced from 10% incorporation of processed millet flour. The total polyphenol content and *in vitro* antioxidant activity of the barnyard millet flour was influenced by both the processing methods. The chapatti made

Methanolic extracts of selected value added products	% inhibition of DPPH (mg/ml)	Antioxidant activity FRAP (mg AAE/g)
Standard chapatti	47%	9.84
RBMF incorporated chapatti at 10% level (V1)	59%	13.42
PCBMF incorporated chapatti at 10% level (V1)	53%	11.57
RBMF: Roasted barnyard millet flour PCBMF: Pressure cooked barnyard millet flour		

Table 6: Properties of the different antioxidant process of DPPH free radical scavenging activity and ferric reducing ability power of the selected products.

from barnyard millet flour subjected to roasting and pressure cooking exhibited higher antioxidant activity (DPPH scavenging activity and FRAP) mainly due to its high polyphenol content. Thus, the present study concludes that roasted and pressure cooked barnyard millet flour could be utilized in value addition and might also contribute significantly to the management and/or prevention of degenerative diseases associated with free radical damage due to its high polyphenol content and antioxidant activity.

Acknowledgement

The first author was supported by a Junior Research Fellowship from the University Grants Commission. The authors have no conflict of interest to report.

References

1. Verma V, Patel S (2013) Production enhancement, Nutritional security and value added products of millets of Bastar region of Chhattisgarh. Int J Res Chem Environ 3: 102-106.
2. Chandrasekhar U (2003) Appropriate technology for utilization of nutrient dense local plant foods. Proc IX Asian Cong Nutr 48: 12-16.
3. Stanly JM, Shanmugam A (2013) A study on millets based cultivation and consumption in India. Int J Mar Fin Serv Manag Res 2: 44-48.
4. Malleshi NG, Desikachar HSR, Venkat Rao S (1986) Protein quality evaluation of a weaning food based on malted ragi and green gram. Qual Plant Plant Foods Hum Nutr 36: 223-230.
5. Jones JM, Engleson J (2010) Whole grains: benefits and challenges. Annu Rev Food Sci Technol 1: 19-40.
6. Gani A, Wani SM, Masoodi FA, Hameed G (2012) Whole grain cereal bioactive compounds and their health benefits: A Review. J Food Process Technol 3: 2-10.
7. Naveena N, Bhaskarachary K (2013) Effects of soaking and germination of total and individual polyphenols content in the commonly consumed millets and legumes in India. Int J Food Nutri Sci 2: 12-19.
8. Friedman M (1997) Chemistry, biochemistry and dietary role of potato polyphenols-a review. J Agric Food Chem 45: 1523-1540.
9. Chen WL, Anderson JW, Gould MR (1994) Effect of oat bran, oat gum and pectin on lipid metabolism of cholesterol fed rats. Nutr Rep Int 24: 93-98.
10. Lee R, Manthey FA, Hall CA (2003) Effects of boiling, refrigerating and microwave heating on cooked quality and stability of lipids in macaroni containing ground flaxseed. Cereal Chem 80: 570-574.
11. Gupta RK, Sharma A, Sharma R (2007) Instrumental texture profile analysis (TPA) of shelled sunflower seed caramel snack using response surface methodology. Food Science and Technology International 47: 13455-13460.
12. Hafiz Ansar RS, Masood SB, Naureen S, Tauseef Sultan, Muhammad IC, et al. (2011) Effect of dietary fibre in lowering serum glucose and body weight in sprague dawley rats. Func Foods Health Dis 8: 261-278.
13. Tsao R (2010) Chemistry and biochemistry of dietary polyphenols. Nutrients 2: 123-146.
14. Nalaini D, Sabapathy (2006) Heat and mass transfer during cooking of chickpea - Measurements and Computational Simulation. Saskatoon.
15. Emenalom OO, Udedibie ABI (2005) Evaluation of different heat processing methods on the nutritive value of Mucuna Pruriens (Velvet Beans) seed meals for broilers. Int J Poultry Sci 4: 543-548.
16. Ballolli U (2010) Development and value addition to barnyard millet (*Echinochloa frumantacea Link*) cookies. College of Rural Home Science. University of Agricultural Sciences. Dharwad.
17. Przybylski R, Lee YC, Eskin NA (1998) Antioxidant and radical scavenging activities of buckwheat seed components. J Amer Oil Chem Soc 75: 1595-1601.
18. Zielinski H, Kozlowska H (2000) Antioxidant activity and total phenolics in selected cereal grains and their different morphological fractions. J Agri Food Chem 48: 2008-2016.
19. Venkata PS, Rajesh N, Swapna S, Chippada AR (2012) Anti hyperglycemic and antioxidant effects of Pennisetum glaucum seed extracts in STZ induced diabetic rats. J Pharm Chem 6: 36-42.
20. Sadasivam S, Manickam A (2005) Biochemical methods. New Age International Publishers.
21. Burits M, Bucar F (2000) Antioxidant activity of Nigella sativa essential oil. Phytother Res 14: 323-328.
22. Oyaizu M (1986) Studies on products of browning reaction-antioxidative activities of products of browning reaction prepared from glucosamine. Jpn J Nutr 44: 307-312.
23. Ugare R (2008) Health benefits, storage quality and value addition of barnyard millet (*Echinochloa frumentacaea Link*). College of Rural Home Science, University of Agricultural Sciences, Dharwad.
24. Butt MS, Qamar MI, Anjum FM, Aziz MA, Randhawa A (2004) Development of minerals-enriched brown flour by utilizing wheat milling by-products. Nutri Food Sci 34: 161-165.
25. Saleh ASM, Zhang Q, Chen J, Shen Q (2013) Millet grains: Nutritional quality, processing, and potential health benefits. Compre Rev Food Sci Food Saf 12: 281-295.
26. Udenigwe CC, Ramprasath VR, Aluko RE, Jones PJ (2008) Potential of resveratrol in anticancer and anti-inflammatory therapy. Nutr Rev 66: 445-454.
27. Dolai N, Karmakar I, Kumar RBS, Kar B, Bala A, et al. (2012) Free radical scavenging activity of Castanopsis indica in mediating hepatoprotective activity of carbon tetrachloride intoxicated rats. Asian Pac J Trop Biomed 2: 242-251.
28. Hayase F, Hirashima S, Okamoto G, Kato H (1990) Scavenging of active oxygens by melanoidins. Agri Biol Chem 54: 855-862.
29. Yen GC, Hsieh PP (1995) Antioxidant activity and scavenging effects on active oxygen of xylose-lysine maillard reaction products. J Sci Food Agri 67: 415-420.
30. Jokic A, Wang MC, Liu C, Frenkel AI, Huang PM, et al. (2004) Integration of the polyphenol and maillard reactions in to a unified abiotic pathway for humification in nature: The role of delta - Mno_2. Organic chem 35: 474-762.
31. Lee MK, Park I (2005) Inhibition of potato polyphenol oxidase by maillard reaction products. Food chem 91: 57-61.
32. Zhang D, Hamauzu Y (2004) Phenolics, ascorbic acid, carotenoids and antioxidant activity of broccoli and their changes during conventional and microwave cooking. Food Chem 88: 503-509.
33. Su K, Dinesh K, Manjusa KS, Nidhan S, Boodev V (2008) Antioxidant and free radical scavenging potential of Citrulus colycynths Linn schrad methanolic fruit extract. Acta Pharmacol 58: 215-220.
34. Soare JR, Dinis TCP, Cunha AP, Almeida L (1997) Antioxidant activities of some extracts of Thymus zygis. Free Radical Res 26: 469-478.

Price Flexibility and Seasonal Variations of Major Vegetables in Sindh Pakistan

Sanaullah Noonari[1]*, Irfana NM[1], Raiz AB[1], Muhammad IK[1] and Shahbaz Ali[2]

[1]*Assistant Professor, Faculty of Agricultural Social Sciences, Sindh Agriculture University, Tandojam, Pakistan*
[2]*Research Assistant, Department of Agricultural Economics, Faculty of Agricultural Social Sciences,*

Abstract

Vegetable cultivation is the most important strategy to reduce poverty as well as to overcome food security problems due to small landholdings and sufficient labour force availability in the rural areas of Pakistan. The results showed that the demand was almost elastic for potato, tomato and onions while there was flexible price trend appeared in the production. Prices on agricultural products are mostly determined by supply and demand. The results of the study showed that erratic price fluctuations both cyclical and seasonal are observed every year. The effect of over-all economic fluctuations are overlaid on a pattern of good and bad harvests, so that an analysis of the effect of a fall in demand on price and output must take account of variations in crop size due to solely the weather. Price fluctuation of these four vegetables is its seasonal character. In the post-harvest period the prices are considerably at lower side whereas in the lean season these are quite high. Thus, from the farmers' point of view they are denied of reasonable prices for their produce during post-harvest period on the consumer's side they are to pay high prices during lean season. Hence, while making a policy towards prices of the vegetables Government should increase the supply in the market by import that commodity from other markets or neighboring countries in non-harvesting seasons as well as the area and production may also increase by using new technology, high yielding seed varieties.

Keywords: Vegetable; Prices; Flexibility; Seasonal variations; Commodity; Sindh

Introduction

In Pakistan, more than 63 varieties of vegetables distributed in 44 genera, are grown on large scale and consumed as summer and winter vegetables comprising mainly potatoes, gourds, tomatoes, cucumbers, ladyfingers, turnips, cabbages, brinjal, cauliflowers etc. These vegetables are popular for their freshness, taste and nutritive value. Mostly, vegetable cultivation is concerted around the populous cities due to no difficulty of input and output market and accessibility of unskilled labour force for performing various farm practices, such as weeding, hoeing etc. Pakistan includes of five provinces namely Sindh, Punjab, Baluchistan, Khyber Pakhtunkhwa and Gilgit-Baltistan. Out of five provinces, Punjab province is not only the most densely inhabited but it has also the productive lands appropriate for cultivation of big varieties of fruits, crops and vegetables. It has a geographical area of 79.61 million acres and takes up 30.96 million acres of cultivated area. It has a total cropped area of 57.34 million acres, making 71.53 percent of the total cropped area of Pakistan. As far as vegetable cultivation is fretful, area under vegetable cultivation is 252000 hectares [1].

After growing at a steady rate in the last decade, Pakistan's vegetable exports have suffered volumetric year-on-year decrease of 40.4% in 2011-12. The drop in vegetable exports is a consequence of natural disasters, unfair profiteering by middlemen and a change in supply and demand dynamics in the foreign markets. The significant drop in vegetable exports is mainly because the onion crop was destroyed by floods. Pakistan fetched $180.2 million by exporting edible vegetables in fiscal 2011-12. Their exports increased at 39% annually between 2007 and 2011, as per the World Trade Organization (WTO). Notably, the rise in the country's vegetable exports between 2010 and 2011 alone was a staggering 122%. Export of vegetables from Pakistan. Vegetables offer good value in terms of nutrients and therefore, less developed countries, especially South Asian States have vegetable dietary habits. Hence these poor countries grow and consume much more vegetables for their main food requirements. Pakistan has greater opportunities, being a centre for vegetable production and can export fresh and canned vegetables in most of the Asian countries to earn foreign exchange. At present, mostly the growers depends on imported seeds, but it is true that many jobs farm of labourers, could be created by growing vegetables for seed production, seed trade and export business may also increase, which reduce annual import costs on vegetable seeds. Farmers prefer to grow vegetables due to short plantation duration and it is considered as the low delta crop. The vegetables can play great role in boosting the economy of the country, due to the fact that this sector has not been explored to earn more income through exports to other countries.

Pakistan has a potential to export these products with trade liberalization under the system of World Trade Organization. Production of vegetables is beneficial; nonetheless, it labor demanding. Thus, it provides income support especially to small farmers and employment opportunity for landless laborers in rural areas. Production functions of onion, tomato and chilies are quite complex since different inputs with different combinations are used. The differences across farms in use of various factors of production and various combinations of factors of production cause changes in crop yields. The input use level and combinations are different across farms and regions resulting in different yields. Furthermore, there is a wide gap in yields of experimental stations and farmer fields indicating the suboptimal use of inputs. Vegetables produced in different zones by

***Corresonding author:** Sanaullah Noonari, Assistant Professor, Department of Agricultural Economics, Faculty of Agricultural Social Sciences, Sindh Agriculture University, Tandojam, Pakistan
E-mail: sanaullahnoonari@gmail.com

using different production technologies during different seasons are traded across regional markets of Pakistan in order to meet consumer demand across the country. Eighty percent of vegetable production in Pakistan is marketable [2-7] (Table 1).

In agriculture the demand for any crop does not play important role in determining the price. The demand for any crop remains inelastic while supply is highly elastic. That is why high variations in the supply of the Cropwell create variability in its price. The price flexibility analysis reveals that the relation of price ad production of gram and mung was positive it may be because these are that major pulses of their respective season also having the high demand. While of masoor and mash it was negative. The supply (Production) of gram increases its current year's price was also increases [6,8-13]. A 10 percent increase in the production of gram increase the price of gram the 9.8 percent in the same production season. Hence price response elasticity in case of gram was elastic closed to unitary elastic. In case of own price elasticity of masoor is highly elastic. A 10 percent increase in the in the production of masoor decrease will cause a decline of 17.7 percent in its price. In case of mung the own price elasticity is highly elastic because due to 10 percent increase in the production in mung that will bring increase in its price by 12.4 percent. Price response elasticity in case of mash is inelastic [3] (Table 2).

Objectives

I. To estimate the price flexibility of different vegetables.

II. To observe the part of various factors towards the price flexibility at different productivity levels.

III. To determine the simple average approach seasonality of different vegetables.

IV. To suggest some policy measures based on the results of the study.

Methodology

Primary purpose of this chapter is to explain various tools and techniques in the selection of sample, collection, analysis and interpretation of data relating to research. Intend of this study was to investigate the existing price flexibility and seasonal variations in prices of vegetables in District Hyderabad Sindh [14-19]. Planned strategy was used to study the area, type and number of respondents without which it would be an ineffective effort. Therefore, it is essential to define variables included in the research to make it more scientific and objective.

Selection of the research study

To measure the price flexibility and the seasonal variation in prices of vegetables, the analysis covers major city of Hyderabad Sindh [20-22]. The area was selected due to their major contribution in the economy of Sindh. Hyderabad city has a larger contribution in the production of Onion, Potato and Tomato has the biggest markets in the Sindh province.

Data collection

For this study, secondary data of monthly prices and quantity data was collected from market of onion, tomato and potato of last two years (2012 and 2013). Data were collected from various sources including, vegetable market committee Hyderabad, Government of Sindh and vegetable wholesale market Hyderabad Sindh [21, 23].

Theoretical framework

This section is devoted to the theoretical description of the different price flexibility approaches and seasonality methods with special reference.

Particulars	Onion		Tomato		Potato	
	Area	Production	Area	Production	Area	Production
	Hectares	Tonnes	Hectares	Tonnes	Hectares	Tonnes
Pakistan	147.6	1939.6	52.3	529.6	159.4	3491.7
Sindh	63.2	861.5	14.6	114.8	0.4	3.9
Punjab	44.7	367.9	6.7	87.8	148.1	3339.9
KPK	11.0	181.3	12.6	113.2	8.9	118.2
Baluchistan	28.7	328.9	18.4	213.8	2.0	29.7

Source: Fruit, vegetable and condiments statistics Government of Pakistan for the year 2013

Table 1: Area and production of different vegetables in Pakistan.

Months	Unit	Year 2012			Year 2013		
		Potato	Onion	Tomato	Potato	Onion	Tomato
January	100Kg	1054.00	2939.00	4902.00	2100.00	1700.00	5700.00
February	100Kg	1066.00	1955.00	2625.00	1900.00	2400.00	5100.00
March	100Kg	1098.00	1194.00	2515.00	1800.00	4100.00	4700.00
April	100Kg	1450.00	1094.00	1912.00	1900.00	4800.00	4000.00
May	100Kg	2700.00	1012.00	788.00	1900.00	4500.00	5000.00
June	100Kg	2325.00	877.00	1643.00	2000.50	3600.00	7200.00
July	100Kg	2734.00	1373.00	1643.00	2500.00	3800.00	7700.00
August	100Kg	2996.00	1905.00	3033.00	2100.00	5500.00	7500.00
September	100Kg	2670.00	2435.00	3856.00	2400.00	4800.00	6600.00
October	100Kg	2677.00	3495.00	5585.00	3500.00	4600.00	5900.00
November	100Kg	2370.00	3871.00	5247.00	5700.00	5000.00	5500.00
December	100Kg	1128.00	2687.00	4159.00	3700.00	3700.00	3300.00

Source: Secondary data on price for Hyderabad Wholesale Market

Table 2: Average monthly wholesale prices of different vegetables prices in Rs. Per 100 Kg in Hyderabad 2012-13.

Price flexibility

Although the relationships among estimated demand and supply coefficients have been examined at length, the link between the direct price flexibility and elasticity of demand has not been discussed explicitly in the literature although often mentioned in passing; this particular relationship remains a source of confusion. In order to clarify it, only a little matrix algebra and some economic theory are needed. It is shown here that, under rather general conditions, the reciprocal of the direct price flexibility (often estimated in econometric work) is the lower absolute limit of the corresponding direct price elasticity [24,25]. Price elasticity of demand is concerned with the responsiveness of consumers in the quantities they will produce in response to a price change. But the price forecaster is frequently concerned with the volatility that might be expected in prices as result of a change in the quantity of product made available for sale. Since elasticity measures the quantity response to a price change, the inverse of elasticity would measure the responsiveness of price to a quantity change. Price flexibility is the term used to describe the inverse of the elasticity relationship (Table 3).

$$\textit{Price flexibility (PF)} = 1/\frac{\%\,\text{change in Q}}{\%\,\text{change in P}} = \frac{1}{\text{E}}$$

Seasonal Variations

Crop prices tend to follow a general season pattern, which is a function of relative changes in supply and demand as the marketing year progresses. Generally, crop prices set their seasonal low at harvest followed by a post-harvest rally. Postharvest rallies occur because the supply of the crop is fixed and consumption gradually uses up that supply, causing prices to rise. Seasonality is a phenomenon that occurs over one production cycle for crops this is generally twelve months. Seasonal forces are different from cyclical or trend forces. A seasonal is one special type of cycle [10,16]. A cycle is a continuous and self-sustaining price pattern which can occur over any length of time. Although there is some evidence of cycles influencing livestock markets, there is little evidence other than of a "technical analysis" nature of other cycles affecting crops. Major market shocks (droughts, embargoes, dramatic policy events, etc.,) can cause crop prices to behave in a "contra-seasonal" manner. Consequently, some analysts separate out years that had a special "condition" and build seasonal that consist only of those years.

The simple average approach

To calculate a seasonal price indexed based on a simple average calculation; we would simply array the prices by months for each of the crop years, calculating a two year average price for each month and for the entire period. Thus using the two year annual average price as a base, we could construct our index by dividing each two year average monthly price by the overall two year average [3,7]. The resulting index would be interpreted as the expected monthly percentage deviation that might be expected from the average price. Thus, an index value of any for January would suggest that we would normally expect January prices in any vegetable year to be 3 percent higher than the season average price.

In order to glean some estimate how dependable our seasonal index might be, we could construct a range or "zone" of seasonal instability. This range of instability is calculated in the case of the simple average index by dividing the price for each month of a marketing year by the annual average price for that year, and then selecting the high and low values for each month as the limits of the range of seasonal instability.

The simple average approach to measuring price seasonality can be modified to at least partially compensate for the limitations observed simply by calculating the real price for any given month during a crop year as a percentage of the season average price for the year. Once the individual monthly percentages have been calculated, the simple average percentage for each month can be calculated as an index of price seasonality.

Results

The production of any agricultural crop/vegetable is subject to the congenial soil structure, climatic conditions, social organization, availability of resources and favorable marketing condition both in factor and product markets. The general objective of the study was to identify the price flexibility with the changes in the prices of potato, onion, tomato and how seasonal can be used to identify the timing of a market's major turning points and to predict the magnitude of price changes at the sample farms in District Hyderabad Sindh.

Price flexibility of potato

Table 3 shows the price elasticity and price flexibility coefficient for potato. In the month of February if there is 4 percent increase in the quantity made available would be associated with a 9.5 percent reduction in price. There is a lot of variation in the prices of potato

Months	**Monthly Average Price** (Rs.per 100 KG)	**Monthly Average Supplied Quantity** (Tones)	**Price Elasticity**	**Price Flexibility Coefficient**
Jan.	950	4588	-	-
Feb.	917	4710	-0.76	-1.30
Mar.	1000	3964	-1.74	-0.57
Apr.	983	4025	-0.90	-1.10
May	1700	3740	-0.097	-10.30
Jun	2142	3250	-0.50	-1.98
July	2567	3020	-0.35	-2.80
Aug	3917	2775	-0.15	-6.48
Sept	2725	2930	-0.18	-5.44
Oct	2458	3095	-0.57	-1.73
Nov	2483	3155	1.9	0.52
Dec	1283	3445	-0.19	-5.24

Source: Secondary data on price for Hyderabad market

Table 3: Price elasticity and price flexibility coefficient for potato in Hyderabad 2012.

due to larger fluctuation in the prices. Price flexibility coefficient is at his higher level at -10.30 where there is a big difference in the prices of April and May. In the month of May there is highest price flexibility with a value of 10.30. There is only one point when the prices of potato are inflexible in the month of November

Figure 1 explains that in the second month of 2013 there is some price flexibility in the prices of onion again the same results related to the last year but in the early month of year 2013 the price flexibility shows less variation. After the month of February the variation is higher. According to the results the price flexibility in the month of May is -9.7 which is the highest value in the whole year but then suddenly in the month of June the value of price flexibility is much less.

Price flexibility of onion

Table 4 shows the price elasticity and price flexibility coefficient for onion. In the early month of the prices of onion are not much higher but as time passes the prices go higher because of the non-harvesting time of the onion. The export to different countries brings significant impact on prices in local market. The augmented supplies tend to keep prices in the domestic market low thereby offering an opportunity for export. Therefore, onion exports are mainly undertaken during this period, predominantly from Sindh crop.

Figure 2 explains that in the second month of 2013 there is some price flexibility in the prices of onion. The highest variation is seen in the price of onion is between the month of October and November which is Rs.3292/100 Kg to Rs.5883/100 Kg because of the non-harvesting season of the onion. The value of price flexibility coefficient is higher during this time period with a value of -3.01 which means that if 1 percent change in the quantity supplied is available may cause 3.01 percent change or reduction in the price. The value of coefficient is lower in the month of June which is the peak time of harvesting. Only in the month of December the value of coefficient is positive which means the prices are inflexible in that month.

Price flexibility of tomato

Table 5 shows the price elasticity and price flexibility coefficient for Tomato. In growing season of tomatoes the prices are less but as the season passed the prices are going higher and higher gradually. From January to May the prices are slowly increases but in the month of June the prices increased rapidly but in the whole year highest average price of tomato are seen in the month of August. So as the price is higher in August the quantity supplied in the market of tomato is less in the whole year. It is observed that prices are lowest during May when Tomato is supplied from Punjab and highest during August and September when it is supplied from N.W.F.P.

The prices tend to normal when supply starts from Sindh. Figure 3 explains that in the second month of 2013 there is some price flexibility in the prices of tomato. Price elasticity of demand for tomato for whole year is inelastic which means the price flexibility coefficient in flexible in the whole year. The interpretation for the price flexibility in the month of February is that if there is 1 percent increase in the quantity made

Months	**Monthly Average Price** (Rs.per 100 KG)	**Monthly Average Supplied Quantity** (Tones)	Price Elasticity	Price Flexibility Coefficient
Jan.	1883	75	-	-
Feb.	1733	68	-2.25	-0.44
Mar.	1967	71	-1.80	-0.55
Apr.	2367	58	-0.54	-1.84
May	1858	63	-0.23	-4.16
Jun	1467	72	-3.39	-0.29
July	1550	53	-1.00	-0.99
Aug	2150	39	-1.04	-0.95
Sept	3000	41	-0.51	-1.93
Oct	3292	48	-1.18	-0.84
Nov	5883	60	-0.33	-3.01
Dec	4167	40	0.53	1.85

Source: Secondary data on price for Hyderabad market

Table 4: Price elasticity and price flexibility coefficient for onion in Hyderabad 2012.

Months	**Monthly Average Price** (Rs.per 100 Kg)	**Monthly Average Supplied Quantity** (Tones)	Price Elasticity	Price Flexibility Coefficient
Jan.	1217	75	-	-
Feb.	1417	68	-0.56	-1.76
Mar.	1358	71	-1.05	-0.94
Apr.	1774	58	-0.59	-1.67
May	1508	63	-0.57	-1.73
Jun	1433	72	-2.86	-0.34
July	2975	53	-0.24	-4.07
Aug	4683	39	-0.46	-2.17
Sept	4033	41	-0.36	-2.70
Oct	3883	48	-4.59	-0.21
Nov	2933	60	-1.02	-0.97
Dec	4208	40	-0.76	-1.30

Source: Secondary data on price for Hyderabad market

Table 5: Price elasticity and price flexibility coefficient for tomato in Hyderabad for 2012.

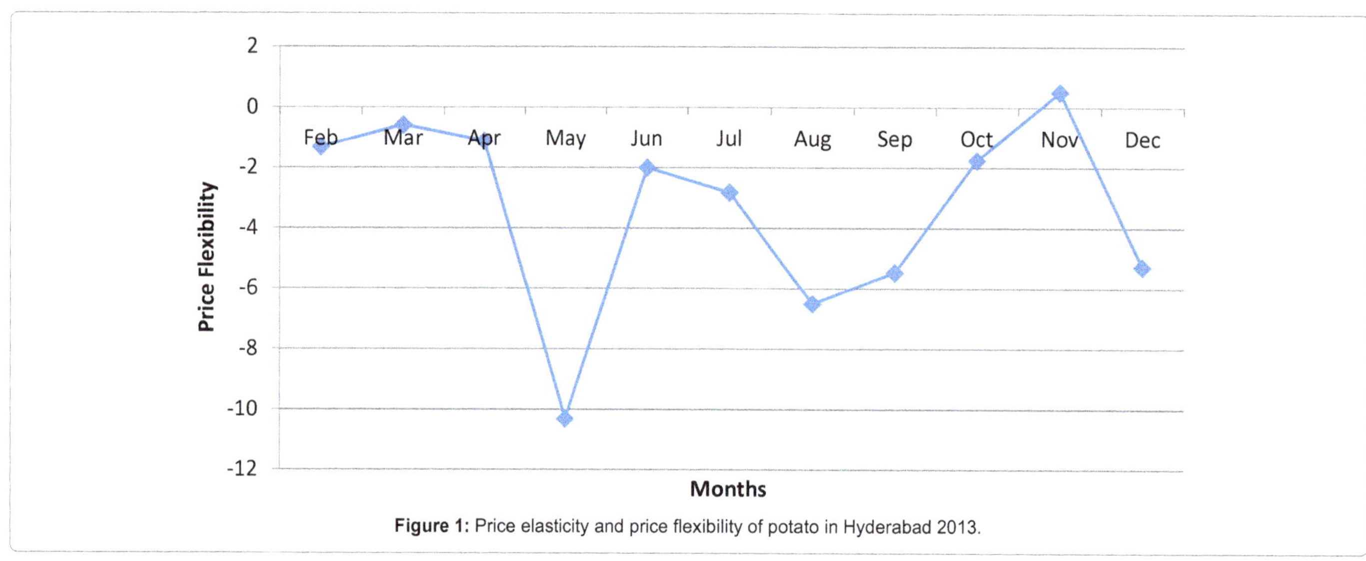

Figure 1: Price elasticity and price flexibility of potato in Hyderabad 2013.

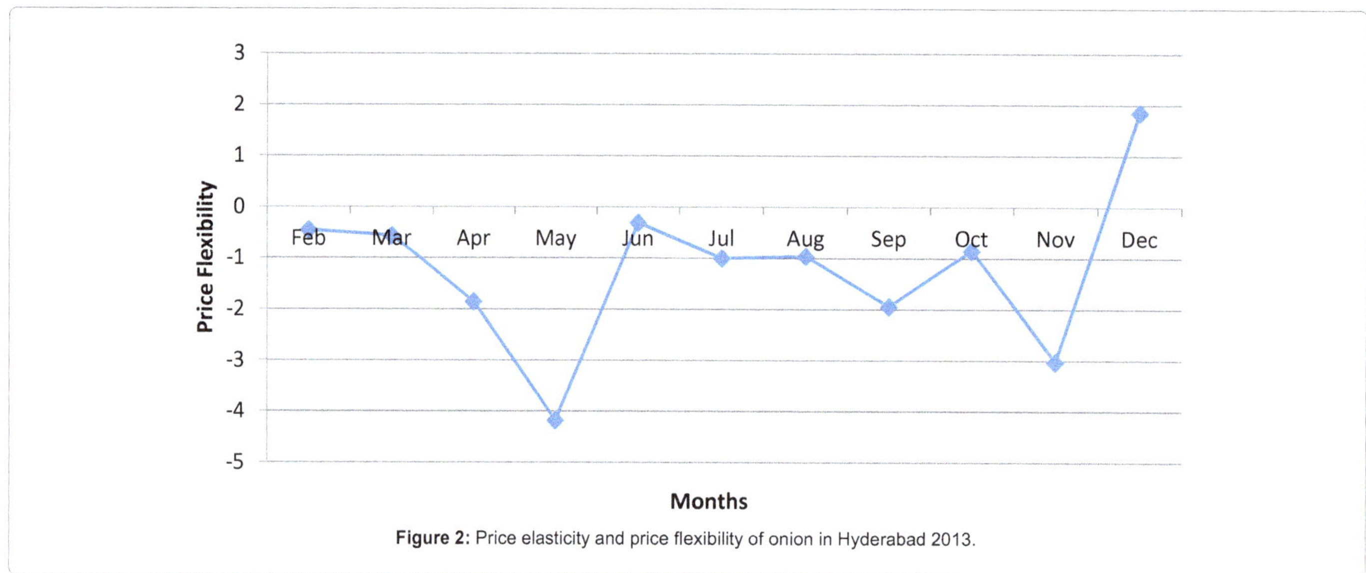

Figure 2: Price elasticity and price flexibility of onion in Hyderabad 2013.

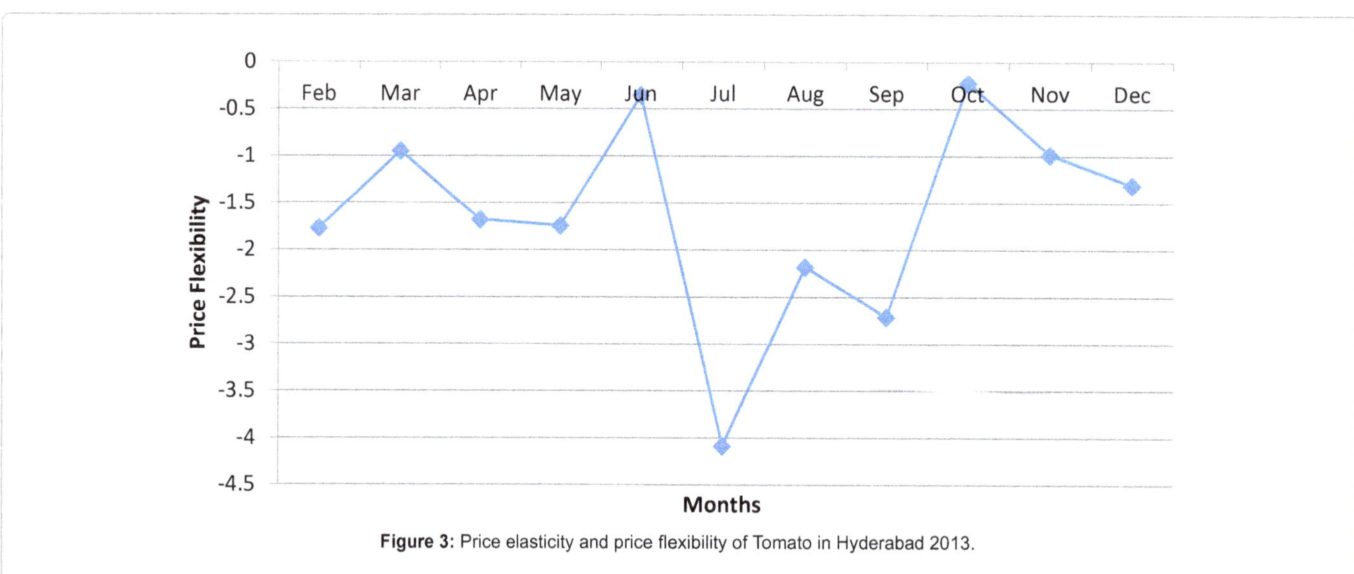

Figure 3: Price elasticity and price flexibility of Tomato in Hyderabad 2013.

available would be associated with a 1.76 percent reduction in price. The highest variation in the price flexibility is between the month of July and August. In the month of July the price flexibility coefficient has the highest value of -4.07 which means if there is 1 percent increase in the quantity made available than it would cause 4.07 percent reduction in the price. The lowest value is between the month of June and July where there is less difference between the prices of tomatoes.

Seasonality of vegetables

Table 6 shows the monthly average price data over the period of 2012 and 2013. The result shows in the same table. The prices of potato vary in both years which cause the variation in the seasonal index. The average price of both years is lowest in the month of February because of the harvesting season of this crop while the prices at its peak with a value of Rs. 4091 in the month of Nov when there is a non-harvesting season.

The simple average approach for onion

Table 6 shows the monthly average price data over the period of 2012 and 2013. The result shows in the same table. The prices of onion vary in both years which cause the variation in the seasonal index. The average price of both years is lowest in the month of February because of the harvesting season of this crop while the prices at its peak with a value of Rs. 4435 in the month of Nov when there is a non-harvesting season.

The simple average approach for tomato

Table 6 shows the monthly average price data over the period of 2012 and 2013. The result shows in the same table. The prices of tomato vary in both years which cause the variation in the seasonal index. The average price of both years is lowest in the month of May because of the harvesting season of this crop while the prices at its peak with a value of Rs. 5742 in the month of Oct when there is a non-harvesting season (Tables 7 and 8).

Conclusion and Suggestions

In Agriculture, any analysis of output and prices must take into account the special role played by variations in harvests. The effect of over-all economic fluctuation is overlaid on a pattern of good and bad harvests, so that an analysis of the effect of a fall in demand on price and output must take account of variations in crop size due solely to weather. The introduction of tunnel technology and hybrid seed are required to enhance yield and lengthen the production period. The value addition in vegetables can help to even out its prices. Harvest and post-harvest management can also bring positive changes. The vegetable marketing system in Sindh in particular and Pakistan in general is not efficient. Poor marketing system causes considerable losses. These losses become more prominent in perishable vegetables. Suitable standards and latest technologies are needed at various levels of marketing to improve functioning of the system. Vegetable export potential to different countries causes of unstable vegetable prices, along with ways to reduce price instability and economic assessment of the potential for the processing of vegetables should be explored.

Month	Potato Prices		2 year average price for each month.	2 year annual average price	Seasonal Index
	2012	2013			
Jan.	950	2100	1525	2212	76.25
Feb.	917	1900	1408	2212	70.40
Mar.	1000	1800	1400	2212	70.00
Apr.	983	1900	1441	2212	70.70
May	1700	1900	1800	2212	90.00
Jun	2142	2000	2071	2212	103.55
July	2567	2500	2573	2212	128.65
Aug	3917	2100	2508	2212	125.40
Sept	2725	2400	2562	2212	128.10
Oct	2458	3500	2676	2212	133.80
Nov	2483	5700	4091	2212	204.55
Dec	1283	3700	2491	2212	124.55

Table 6: Simple average seasonal index for potato.

Month	Onion Prices		2 year average price for each month.	2 year annual average price	Seasonal Index
	2012	2013			
Jan.	2939	1700	2319	3057	65.95
Feb.	1955	2400	2177	3057	108.85
Mar.	1194	4100	2647	3057	132.35
Apr.	1094	4800	2947	3057	147.35
May	1012	4500	2756	3057	137.80
Jun	877	3600	2238	3057	111.90
July	1373	3800	2586	3057	129.30
Aug	1905	5500	3725	3057	186.25
Sept	2435	4800	3617	3057	180.85
Oct	3495	4600	4047	3057	202.35
Nov	3870	5000	4435	3057	221.75
Dec	2687	3700	3193	3057	159.65

Table 7: Simple average seasonal index for onion.

Month	Tomato Prices		2 year average price for each month.	2 year annual average price	Seasonal Index
	2012	2013			
Jan.	4902	5700	5301	4420	265.05
Feb.	2625	5100	3867	4420	193.35
Mar.	2515	4700	3607	4420	180.35
Apr.	1912	4000	2956	4420	148.25
May	788	5000	2894	4420	144.70
Jun	1643	7200	4421	4420	221.05
July	1643	7700	4671	4420	233.55
Aug	3033	7500	5266	4420	263.30
Sept	3856	6600	5228	4420	261.40
Oct	5585	5900	5742	4420	287.10
Nov	5247	5500	5368	4420	268.40
Dec	4159	3300	3729	4420	186.45

Table 8: Simple average seasonal index for tomato.

In this study, an attempt has been made to develop relationship between prices and the quantity supplied to the market. The present study estimated the price flexibility and the seasonal variation in the prices of potato, tomato and onion in Hyderabad market which tells us the trend of prices during the two years monthly 2012 and 2013 data of price. Weather conditions, as an example cans cause wide variations in the quantities of agricultural products produced from year to year. Changes in the level of supply in a market characterized by an inelastic demand lead to disproportionate price level changes. Thus, price levels in agriculture tend to be unstable over the short-term. Fluctuations in the demand for specific foods and fluctuations in the demand for different quality levels of food shift in relation to price level changes related to supply changes in an inelastic market. Similarly, supply conditions in foreign markets where demand also is inelastic affect the demand for agricultural products produced in a domestic market. In many countries, agricultural policies are based on quantitative analysis of agricultural production system. Different types of quantitative analyses are performed and these include measuring scale economies, producers' responsiveness to product and input price variation and the relative efficiency of resource use. Therefore, there is a need to conduct such kinds of studies focusing on issues relating to policy options. However, the sector needing dire attention is the vegetable sector, since this sector possesses a lot of opportunities to flourish. The present study was planned to estimate price flexibility and the seasonal variation in the prices vegetables on the basis of their price s and the quantity supplied to the market in Lahore district. On the basis of this study, following suggestions are made for the selected vegetables in Lahore to boost up vegetable production for the decrease in the prices.

If there is price hike due to the failure of the targeted crop or have low crops than in short run import these vegetables from the neighboring countries; India and Iran or increase the supply from the other markets of different provinces. If the prices are low in the harvesting season in this area than I purposed to enhance the time period or to produce the short day varieties. Per acre yield of these vegetables in this area is very less as compared to the other regions of the country for this purpose I suggest to improve the farm management practices leading to better quality and more yields. For this new high yielding varieties may be evolved and also use the balanced fertilizers along with micronutrients.

In the case of high price fluctuations we would have to collect and disseminate the market information. For this purpose the policy makers must forecast the prices of vegetables according to the estimates. Agriculture marketing crops reporting service timely release of area and the production estimation and be prepare for the future values.

If the production is low than I suggest three ways to increase vegetable production in Pakistan

a) By increasing cropped area under vegetable cultivation

b) By developing new technology and

c) By using available resources efficiently.

Third option is the most suitable because this option does not require more area and development of new technology. Balanced use of inorganic nutrients in vegetable production needs the attention of vegetable growers to enhance per acre yield substantially. In spite of the application of large quantities of pesticides on vegetables, insect pests were not well controlled and they were causing lot of damages to the vegetable crops. This shows that vegetable growers have poor knowledge about vegetable pests and diseases. Application of irrigation is another production aspect requiring the immediate attention. Growers should irrigate their vegetable fields appropriately, since these vegetables are water loving and quality of output depends on irrigation. Canal water deficiencies forced the vegetable growers to irrigate their vegetables with sewage water resulting in low and poor quality yield. When the production is high than definitely supply of these vegetables is high in Lahore than ultimately the prices are low in the market.

Small farmers were ignored by the extension services. Since the vegetable growers possess small land holdings, therefore, they are not provided the extension services. This suggests that the extension services should be expanded to vegetable growers especially the Onion, potato and tomato growers and this will help in increasing in the production. Thus it is the time for the extension workers to revise their policies and to focus on minor crops as well including vegetables. Training of extension staff in vegetable growing practices is the most important step needing careful attention of the concerned authorities. According to Fraser and Cordina, "it is all very well having best practice described by an extension officer on a farm visit, but being able to observe directly best-practice farming techniques will enhance the learning experience".

Vegetable export potential to different countries causes of unstable vegetable prices, along with ways to reduce price instability and economic assessment of the potential for the processing of vegetables should be explored.

References

1. GOP (2012) Economic Survey of Pakistan, 2011-12. Government of Pakistan, Economic Affairs Division, Ministry of Finance, Islamabad.

2. Mukhtar MM (2004) Agricultural Marketing System and Trade Enhancement:

Issues and Policies. Pakistan Journal of Agricultural Economics. Agricultural Prices Commission, Islmabad 5: 17-26.

3. Rani S, Shah H, Ali A, Rehman B (2012) Growth, Instability and Price Flexibility of Major Pulses in Pakistan. Asian Journal of Agriculture and Rural Development 2: 107-112.
4. Ali, M, Abedullah (2002) Nutritional and economic benefits of enhanced vegetable production and consumption. Journal of Crop Production 6: 145-176.
5. Ali M, Tsou S (2000) Combating micronutrient deficiencies through vegetables a neglected food frontier in Asia. The Journal of Food Policy 22: 17-38.
6. AMIS (2011) Agriculture Marketing Information Service (AMIS) Punjab Agriculture Department, Government of Punjab, Pakistan.
7. Adjemian MK, Smith A (2012) Using USDA Forecasts to Estimate the Price Flexibility of Demand for Agricultural Commodities American journal of Agricultural Economics 94: 978-995.
8. Athar M, Bokhari T (2006) Ethnobotany and Production Constraints of Traditional and Commonly Used Vegetables of Pakistan. J. Vegetable Sci 12: 27- 38.
9. Alan B, Huang J, Rozelle S (2000) Responsiveness, Flexibility and Market Liberalization in China's Agriculture. American journal of Agricultural Economics 82: 1133-1139.
10. Chaudhry GM, Ahmad B (2000) Dynamics of vegetable production, distribution and consumption in Asian Vegetable Research and Development Center, Tainan, Taiwan. AVRDC 498: 271-302.
11. Charnsirisakskul KP, Griffin, Keskinocak P (2004) Pricing and Scheduling Decisions with Leadtime Flexibility. School of Industrial and Systems Engineering, Georgia Institute of Technology,Atlanta,USA. 171: 153-169.
12. Dutta SM, Levy D (2002) Price flexibility in channels of distribution: Evidence from scanner data. Journal of Economic Dynamics & Control 26: 1245-1900.
13. Dong D, Lin BH (2009) Fruit and Vegetable Consumption by Low-Income Americans would a Price Reduction Make a Difference? United States Department of Agriculture, Economic Research Report Number 70.
14. FAO (2004) Fertilizer use by crop in Pakistan. Land and Plant Nutrition Management Service, Land and Water Development Division, FAO, Rome.
15. Francis NS, Chiu, Bjornsson H (2002) Quantifying price flexibility in material procurement as a real option CIFE Seed Project - Real Options in Material Procurement Contracts October.
16. Huang KS (2000) Forecasting Consumer Price Indexes for Food: A Demand Model Approach. Food and Rural Economics Division, Economic Research Service, U.S. Department of Agriculture.
17. Huang KS (2006) A Look at Food Price Elasticities and Flexibilities. Economic Research Service, U.S. Department of Agriculture, Washington, DC.
18. James PH (2008) The Relationship of Direct Price Flexibilities to Direct Price Elasticities.
19. Nicholls WH (2012) Price Flexibility and Concentration in the Agricultural Processing Industries. Journal of Political Economy 48: 883-888.
20. Richard MA (2012) Statistical Measurement of Price Flexibility. The Quarterly Journal of Economics 56: 497-502.
21. Schnepf R (2006) Price Determination in Agricultural Commodity Markets: A Primer. Congressional Research Service, The Library of Congress.
22. Senay O, Sutherland A (2006) Can Endogenous Changes in Price Flexibility Alter the Relative Welfare Performance of Exchange Rate Regimes? National Bureau of Economic Research.
23. Taussig F W (2012) Price-Fixing as Seen by a Price-Fixer. The Quarterly Journal of Economics 33: 205-241.
24. Toaha M (1974) Estimation of Marketing Margins and Measurement of Seasonal Price Variations of Selected Agricultural Commodities in Sindh Province of Pakistan. Sindh Agriculture College, Tando Jam, Pakistan.
25. Thornsbury S, Jerardo A (2012) Low Price Continue for Fresh-Market Vegetables, Vegetables and Pulses Outlook. Economic Research Service, U.S. Department of Agriculture Washington, DC.

Safety Assessment of Polycyclic Aromatic Hydrocarbons (PAHs) in Cold Smoked Fish (*Mugil Cephalus*) Using GC-MS

Hafez NE[1]*, Awad AM[1], Ibrahim SM[2] and Mohamed HR[2]

[1]*Department of Food Science and Technology, Faculty of Agriculture, El-Fayoum University, Egypt*

[2]*Fish Processing and Technology Laboratory, Fisheries Division, National Institute of Oceanography and Fisheries, Egypt*

Abstract

This work was planned to determine the safety level of polycyclic aromatic hydrocarbons (PAHs) in cold smoked mullet fish obtained from two fish farms (A and B) localized at El-Fayoum Governorate, Egypt during August 2015. PAHs components were determined by GC-MS. Results showed that total of PAHs components was 28.8 and 5 ppb in both smoked fish products (A and B), respectively. Also, low molecular weight (LMW) of PAHs was found especially in smoked fish (A), followed by medium molecular weight (MMW) and however high molecular weight (HMW) was not detectable. Levels of Benzo [a] Pyrene (B {a} P) equivalent were 0.051 and 0.005 in products (A and B), respectively. However, PAH4 (BaP+CHR+BaA+BbF) and PAH8 (Benzo (a) anthracene, Chrysene, Dibenzo (a, h) anthracene, Benzo (g, h, i) perylene, Benzo (b) Fluoranthene, Benzo (k) fluoranthene, Benzo (a) Pyrene, Indeno (1, 2, 3-c, d) Pyrene) were not detectable. Also, categories of PAHs concentration are considered a minimally contaminated (10 to 99 µg/kg) and not contaminated (<10 ppb) compared with the maximum recommended levels. Based on our results, it could be concluded that Benzo (a) pyrene compound was not detectable in all smoked samples which are considered as a safe product for human consumption.

Keywords: Fish; Smoking; PAHs; GC-MS

Introduction

Contamination of food by polycyclic aromatic hydrocarbons (PAHs) can be resulted from food processing at high temperature such as frying, smoking or roasting Codex [1-3]. There are many parameters limited the amounts of PAHs in food such as composition of the smoke, technology used in smoking, combustion temperature, type of wood, and exposure of the edible parts to the smoke [3-5]. The ranges of MW could be divided into three classes; 152 g/mol to 178 g/mol, 202 and 228 g/mol to 278 g/mol of low, medium and high MW, respectively [6,7]. Seven of the PAHs have been considered human carcnogens; benzo (a) anthracene, benzo (b) fluoranthene, benzo (k) fluoranthene, chrysene, benzo (a) pyrene, dibenzo (a, h) anthracene, and indeno (1, 2, 3-c, d) pyrene. Therefore, this study was designed to determine the safety level of PAHs in cold smoked mullet fish samples that were obtained during August 2015 from two fish farms localized at El-Fayoum governorate, Egypt (Figure 1).

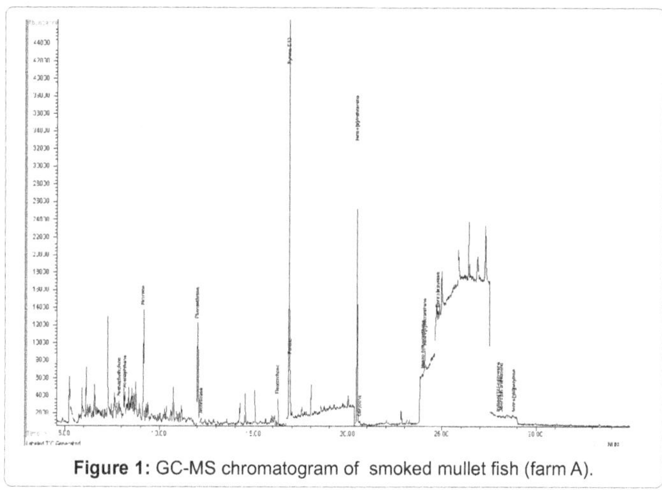

Figure 1: GC-MS chromatogram of smoked mullet fish (farm A).

Materials and Methods

Fish samples

Mullet fish (*Mugil cephalus)* samples were obtained after directly catch from two fish farms (A and B). The main resources of irrigation water were industrial for A and agricultural discharge for B during August 2015 at El-Fayoum governorate. They were trans within two hours using ice box to Fish Processing and Technology Lab, Shakshouk Station for Water Resource, National Institute of Oceanography and Fisheries (NIOF), Egypt. Average of weight 453.3 gm ± 51.7 gm and length 35 gm ± 3 cm for raw samples from Farm A (Industrial discharge) also, average weight and length of raw mullet samples from Farm B (Agricultural discharge) are 526.6 gm ± 18 gm and 38 cm ± 2 cm respectively.

Smoking process and source of PAHs

After that, chilled whole fish samples were washed carefully with tap water, soaked in 10% brined solution for 2 hrs, rinsed with tap water for 1 min, semi-dried at 250°C for two hrs, traditional cold smoked at 35°C to 45°C for 8 to 10 hrs, using sawdust and finally cooled under ambient temperature.

Analytical methods

The edible of smoked mullet fish products was manually separated, homogenized, packed in polyethylene bags and then stored in a freezer at -20°C till analysis.

PAHs determination: PAHs were determined at Central Laboratory of Residue Analysis of Pesticides and Heavy Metals in Food

***Corresponding author:** Hafez NE, Department of Food Science and Technology, Faculty of Agriculture, El-Fayoum University, Egypt
E-mail: m_fathy8789@yahoo.com

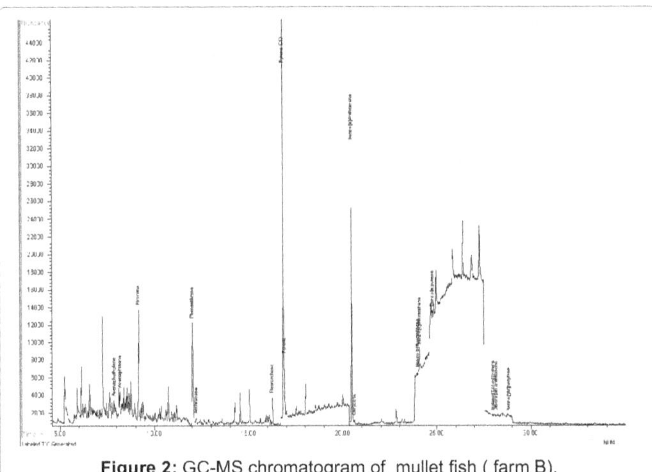
Figure 2: GC-MS chromatogram of mullet fish (farm B).

Compound	Abbrev.	Mw	Rings	Concentration (ppb)	
				Farm (A)	Farm (B)
Chrysene	CHR	228	4	ND	ND
Anthracene	ANT	178	3	2.8	ND
Acenaphthene	ACE	154	3	ND	ND
Benzo(b)fluoranthene	BbF	252	5	ND	ND
Benzo(k)fluoranthene	BkF	252	5	ND	ND
Dibenzo(a,h)anthracene	DahA	278	5	ND	ND
Fluorene	FLU	166	3	6.6	ND
Naphthalene	NA	128	3	ND	ND
Benzo(a)pyrene	BaP	252	5	ND	ND
Benzo(g,h,i)perylene	BghiP	276	6	ND	ND
Indeno(1,2,3,cd)pyrene	IcdP	276	6	ND	ND
Acenaphthylene	ACY	152	3	ND	ND
Fluoranthene	FLA	202	4	3.3	2.1
Pyrene	PYR	202	4	4.6	2.9
Benzo(a)anthracene	BaA	228	4	ND	ND
Phenanthrene	PHE	178	3	11.5	ND
Σ 16PAHs				28.8	5

Table 1: Molecular weight (MW), number of rings and concentration of PAHs in smoked mullet samples.

(QCAP), Agricultural Research Centre. Cairo, Egypt as described by Forsberg et al. [8], Smoker et al. [9] and Khorshid et al. [10].

B{a}P equivalent: The $B\{a\}P_{eq}$ was calculated as the $\Sigma\, B\{a\}P_{eqi}$ value for individual PAHs. The $B\{a\}P_{eqi}$ calculated as the following equation [11]:

$$BaP_{eq} = \Sigma\left(BaP_{eqi}\right) = \Sigma\left(C_{PAHi} \times TEF_{PAHi}\right)$$

C_{PAHi}: Concentration of each PAH in the sample,

TEF_{PAHi}: Toxic equivalency factor for each individual PAH.

Statistical analysis: The results obtained were analyzed statistically using the least significant difference test (LSD) at ($P \leq 0.05$) and were expressed as Mean ± SD using SPSS 16 for windows.

Results and Discussion

As previous mentioned that raw mullet samples were obtained from different fish farms based on irrigation main resources; farm (A) was dependent on industrial discharge and localized at eastern Fayoum governorate farms whereas farm (B) was dependent on agriculture discharge and localized at Qarun lake farms (Figure 2).

Polycyclic aromatic hydrocarbons (PAHs)

Table 1 shows the PAHs concentration of cold smoked mullet fish flesh. 16 components of PAHs were detected in edible part of investigated products. Concentrations of ANT, FLA, PYR, FLU and PHE were 2.8, 3.3, 4.6, 6.6 and 11.5 ppb, respectively in smoked fish farm (A).

Farm (A): Industrial discharge. Farm (B): Agricultural discharge. Mw: Molecular weight.

In the other batch, FLA and PYR levels were 2.1 and 2.9 ppb, respectively in smoked fish farms (B). The PAHs have been grouped according to its molecular weights, we found that the high MW components of PAHs (228 g/mol to 278 g/mol), were Benzo (a) anthracene, Benzo (b) fluoranthene, Benzo (k) fluoranthene, Benzo (j) fluoranthene, Benzo (e) pyrene, Benzo (a) Pyrene, Benzo (ghi) perylene, dibenzo (h) anthracene, chrysene, cyclopenta (cd) pyrene, indo (1, 2, 3-c, d) Pyrene, and anthranthrene. The medium MW components (202 g/mol) were fluoranthene and pyrene. The low molecular weights (152 g/mol to 178 g/mol) included naphthalene, Acenaphthylene, acenaphthene, fluorene, anthracene and phenanthrene. Total of PAHs were 28.8 and 5 ppb in both smoked fish farms (A) and (B), respectively. In addition, rings number of PAHs in smoked fish farm (A) ranged from 3 to 4 rings and its MW ranged from 166 to 178 whereas in case of farm (B) only 4 rings and 202 as MW were found. Comparison of fish species smoked as fillets and as whole fish illustrated that the level of PAHs was higher for smoked fillets in comparison with the same fish species smoked as a whole fish.

Category of PAH concentration

Categories of concentration of PAH are considered a minimally contaminated (10 to 99 ppb) and not contaminated (<10 ppb) compared with recommended levels as set by Soares-Gomes et al. [12]. Concentrations of PAH were 28.8 and 5.0 μg/kg in smoked fish farms (A) and (B), respectively. Based on these results, Farm A (Industrial discharge) and B (Agricultural discharge) classified as minimally contaminated and Not contaminated respectively.

Molecular weight of PAHs in smoked fish

Also, Table 2 exhibits the molecular weight (MW) of PAHs in smoked mullet fish. The total concentration of the low molecular weights (LWM) of PAHs was higher than the medium molecular weights (MMW) in smoked fish farm (A). The high concentration of LWM in smoked fish farm (A) was 20.9 KDa. In other side, total concentration of medium molecular weights of PAHs was 5 KDa. Lipophilic nature of the PAHs and fish's skin may be make better protection from the HMW-PAHs than LMW as reported by Mohammadi et al. [13]. In addition, results showed that there is no found a HMW-PAHs in different smoked fish farms either (A) or (B). In all products, HMW-PAHs were below the limit of quantification or not detectable.

Toxic equivalent factors (TEFs) and B {a} P equivalent of PAHs

TEF is an estimate of the relative toxicity of individual PAH fraction

Fish farm	HMW	MMW	LMW
A	-	7.9	20.9
B	-	5	-
Farm (A): Industrial discharge. Farm (B): Agricultural discharge			

Table 2: Molecular weight (MW) of PAHs in cold smoked fish.

compared to benzo(a) pyrene. The toxic equivalent factors (TEFs) and B [a] P Equivalent of PAHs in smoked mullet fish are present in Table 3. Concentrations of phenanthrene, flourene, pyrene, fluoranthene and anthracene in smoked fish farm (A) were 11.5, 6.6, 4.6, 3.3 and 2.5 ppb, respectively and total B [a] P Equivalent was 0.051. On the other side, concentrations of pyrene and fluoranthene in smoked fish farm (B) were 2.9 and 2.1 ppb, respectively and total B [a] P Equivalent was 0.005.

Benzo [a] pyrene (BaP), PAH4 and PAH8

In this study the PAH4 (the sum of BaP, chrysene, benz [a] anthracene and benzo [b] fluoranthene) and PAH8 {Benzo (a) anthracene, Chrysene, Dibenzo (a, h) anthracene, Benzo (g, h, i) perylene, Benzo (b) Fluoranthene, Benzo (k) fluoranthene, Benzo (a) Pyrene, Indeno (1, 2, 3-c, d) Pyrene} were not detected in smoked fish samples too. So, in the current study, cold smoked products are safety for human consumption because BaP did not detect comparing to the maximum permissible limit 2 ppb as set by the European Commission Regulation (OJEU, 835/2011) [14]. These results are in accordance to the findings by El-Lahamy et al. [15] who reported that BaP not detected in cold and hot smoked catfish fillets.

Sources and assessment of PAHs

Table 4 shows that source characterization and assessment of PAHs comparing with the previous references. It is well known that the sawdust wood was used as source of PAHs in this work as mentioned above (material and methods part). Results showed that the ratio of anthrancene to anthracene plus phenanthrene [An/(An+Phen)] was 0.19 in case of product obtained from farm A while it was not detected in farm B. This indicates that the wood combustion is the main source of PAHs compared with the mass 178. [An/(An+Phen)] ratio <0.10 usually is referred to petroleum while a ratio >0.10 indicates dominance of combustion [16-18]. The [Fl/(Fl+Py)] ratio also ranged from 0.41 to 0.42 in farm A and B products, respectively. The ratios ranged between 0.4 and 0.5 the refers to some amount of fossil fuel combustion sources (vehicular, fat and crude oil) of PAHs [19-22].

Compound	TEF	Farm (A)		Farm (B)	
		PAHs (μg/kg)	BaP_{eqi}	PAHs (μg/kg)	BaP_{eqi}
Naphthalene	0.001	ND	-	ND	-
Acenaphthylene	0.001	ND	-	ND	-
Acenaphthene	0.001	ND	-	ND	-
Fluorene	0.001	6.6	0.0066	ND	-
Phenanthrene	0.001	11.5	0.0115	ND	-
Anthracene	0.01	2.5	0.025	ND	-
Fluoranthene	0.001	3.3	0.0033	2.1	0.0021
Pyrene	0.001	4.6	0.0046	2.9	0.0029
Benzo(a)anthracene	0.1	ND	-	ND	-
Chrysene	0.01	ND	-	ND	-
Benzo(b)fluoranthene	0.1	ND	-	ND	-
Benzo(k)fluoranthene	0.1	ND	-	ND	-
Benzo(a)pyrene	1	ND	-	ND	-
Indeno(1,2,3,c)pyrene	0.1	ND	-	ND	-
Dibenzo(a,h) anthracene	1	ND	-	ND	-
Benzo(g,h,i)perylene	0.01	ND	-	ND	-
$\sum (BaP_{eqi})$			0.051		0.005

TEF: Toxic equivalent factor; BaP_{eqi}: B[a] P equivalent; Farm (A): Industrial discharge; Farm (B): Agricultural discharge.

Table 3: Toxic Equivalent factors (TEFs) and B [a] P Equivalent of PAHs in smoked mullet fish.

Source of PAHs		PAH Ratios	
		[An/(An + Phen)] 178	[Fl /(Fl + Py)] 202
Wood combustion		>0.10	>0.5
Petroleum		<0.10	0.40
Saw dust	Farm A	0.19	0.41
	Farm B	ND	0.42

[An/(An+Phen)]: anthracene plus phenanthrene. [Fl/(Fl+Py)]: fluoranthene to fluoranthene plus pyrene. Farm (A): Industrial discharge. Farm (B): Agricultural discharge.

Table 4: Sources and assessment of PAHs comparing with the previous reference.

Conclusion

In conclusion, the safety of smoked fish has been controlled by measuring benzo (a) pyrene level, which is one of the most carcinogenic PAHs. European Commission has limited the maximum acceptable concentrations of benzo (a) pyrene at 2 ppb for smoked fish and smoked fishery products, excluding bivalve molluscs. In addition, the categories of concentration of PAH are considered a minimally contaminated and not contaminated compared with international recommended levels.

References

1. CCFAC (2005) Discussion paper on polycyclic aromatic hydrocarbons contamination. The Hague, The Netherlands.
2. Moret S, Purcaro G, Conte LS (2005) Polycyclic aromatic hydrocarbons in vegetable oils from canned foods. Eur J Lipid Sci Technol 107: 488-496.
3. Yurchenko S, Mölder U (2005) The determination of polycyclic aromatic hydrocarbons in smoked fish by gas chromatography mass spectrometry with positive-ion chemical ionization. J Food Comp Anal 18: 857-869.
4. Duedahl-Olesen L, White S, Binderup ML (2006) Polycyclic aromatic hydrocarbons (PAH) in Danish smoked fish and meat products. Polycycl Aromat Comp 26: 163-184.
5. Stumpe-Viksna I, Bartkevics V, Kuka´ re A, Morozovs A (2008) Polycyclic aromatic hydrocarbons in meat smoked with different types of wood. Food Chem 110: 794-797.
6. European Food Safety Authority EFSA (2002) Scientific Committee on Food. Opinion on the risks to human health of polycyclic aromatic hydrocarbons in food, Italy.
7. ATSDR (1995) Toxicological profile for polyaromatic hydrocarbons-update. US Department of Health and Human Services, Atlanta, GA.
8. Forsberg ND, Wilson GR, Anderson KA (2011) Determination of parent and substituted polycyclic aromatic hydrocarbons in high-fat salmon using a modified QuEChERS extraction, dispersive SPE and GC-MS. J Agric Food Chem 10: 8108-8116.
9. Smoker M, Tran K, Smith RE (2010) Determination of polycyclic aromatic hydrocarbons (pahs) in shrimp. J Agric Food Chem 58: 12101-12104.
10. Khorshid M, Souaya ER, Hamzawy AH, Mohammed MN (2015) QuEChERS method followed by solid phase extraction method for gas chromatographic mass spectrometric determination of polycyclic aromatic hydrocarbons in fish. Int J Analytical Chem 7: 205-217.
11. Nisbet ICT, LaGoy PK (1992) Toxic equivalency factor (TEFs) for polycyclic aromatic hydrocarbons (PAHs). Regulat Toxicol Pharmacol 16: 290-300.
12. Soares-Gomes A, Neves RL, Aucélio R, Van Der Ven PH, Pitombo FB, et al. (2010) Changes and variations of polycyclic aromatic hydrocarbon concentrations in fish, barnacles and crabs following an oil spill in a mangrove of Guanabara bay Southeast Brazil. Mar Pollut Bull 60: 1359-1363.
13. Mohammadi A, Ghasemzadeh-Mohammadi V, Haratian P, Khaksar R, Chaichi M (2013) Determination of polycyclic aromatic hydrocarbons in smoked fish samples by a new microextraction technique and method optimisation using response surface methodology. J Food Chem 141: 2459-2465
14. OJEU (2011) Commission regulation (EU) No 835/2011 of 19 August 2011.

Amending Regulation (EC) No 1881/2006 as regards maximum levels for polycyclic aromatic hydrocarbons in foodstuffs.

15. El-Lahamy AA, Khalil IK, El-Sherif SA, Awad AM (2016) The influence of smoking method on the levels of poly cyclic aromatic hydrocarbons (PAHs) in smoked catfish (*Clarias gariepinus*) fillets. Int J Adv Res 4: 1529-1538.

16. Budzinski H, Jones I, Bellocq J, Pierard C, Garrigues P (1997) Evaluation of sediment contamination by polycyclic aromatic hydrocarbons in the Gironde estuary. Marine Chem 58: 85-97.

17. Zhang HB, Luo YM, Wong MH, Zhao QG, Zhang GL (2006) Distributions and concentrations of PAHs in Hong Kong Soils. Environ Pollut 141: 107-114.

18. Pies C, Hoffmann B, Petrowsky J, Yang Y, Ternes TA (2008) Characterization and source identification of polycyclic aromatic hydrocarbons (PAHs) in river bank soils. Chemosphere 72: 1594-1601.

19. Placha D, Raclavska H, Matysek D, Rummeli MH (2009) The polycyclic aromatichydrocarbon concentrations in soils in the region of Valasske Mezirici, the Czech Republic. Geochem Transact 10: 12.

20. Phillips DH (1999) Polycyclic aromatic hydrocarbons in the diet. Mutation Res 443: 139-147

21. Kazerouni N, Sinha R, Hsu CH, Greenberg A, Rothman N (2001) Analysis of 200 food items for benzo(a)pyrene and estimation of its intake in an epidemiologic study. Food Chem Toxicol 39: 423-436.

22. Yin CQ, Jiang X, Yang XL, Bian YR, Wang F (2008) Polycyclic aromatic hydrocarbons in soils in the vicinity of Nanjing, China. Chemospher 73: 389-394.

Selective Removal of Phenylalanine Impurities from Commercial κ-Casein Glycomacropeptide by Anion Exchange Chromatography

Nakano T[1]* and Ozimek L[1]

[1]*Department of Agricultural, Food and Nutritional Science, University of Alberta, Canada*

Abstract

Bovine κ-casein glycomacropeptide (GMP) found in sweet whey is a 64 amino acid residue phosphorylated glycopeptide. Because it lacks aromatic amino acids including phenylalanine, GMP is thought to be an important dietary source of amino acids for patients suffering from phenylketonuria. There is, however, very little information available concerning preparation of phenylalanine-free GMP for human consumption. This study was, therefore, undertaken to remove phenylalanine containing impurities from commercially available crude GMP by anion exchange chromatography on diethylaminoethyl (DEAE)-Sephacel. The results demonstrated that phenylalanine containing proteins or peptides do not bind to the column, while most GMP accounting for 93% of total recovered sialic acid can bind to the column. The purified GMP, which accounted for average 43% of dry weight of crude GMP, contained undetectable level of phenylalanine. Carbohydrate analyses and cellulose acetate electrophoresis showed that the purified GMP is a product with high sialic acid content (average 15.5% dry weight). Gel filtration chromatography on Sephacryl S-100 and size exclusion HPLC on Superdex 75 confirmed our previous findings that GMP monomers form aggregates and elute as a single peak with its elution volume close to the elution volume of dimeric β-lactoglobulin (36.6 kDa). It was concluded that the crude preparation of GMP can be highly refined by selectively removing phenylalanine impurities using DEAE-Sephacel chromatography.

Keywords: κ-Casein glycomacropeptide; Caseinomacropeptide; Sialic acid; Anion exchange chromatography; DEAE-Sephacel

Introduction

Glycomacropeptide (GMP) found in sweet whey (or cheese whey) from cow's milk is a 64 amino acid residue C-terminal phosphorylated glycopeptide (residues 106-169) released from κ-casein by the action of chymosin, which catalyzes cleavage between residues 105 (phenylalanine) and 106 (methionine) of κ-casein during cheese making [1-5]. GMP contains varying amounts of carbohydrates including N-acetylgalactosamine, galactose and N-acetylneuraminic acid (sialic acid) [6]. GMP is known to have various biological activities (e.g. protection against toxins, bacteria, and viruses, and modulation of immune responses [3], and is thought to be a potential ingredient for functional foods and pharmaceuticals. GMP, which lacks aromatic amino acids (phenylalanine, tyrosine, and tryptophan), is thought to be suitable for the source of dietary amino acids for patients suffering from phenylketonuria (PKU), a hereditary disorder of phenylalanine metabolism causing mental retardation [7,8]. Thus, much attention has been given to the development of techniques to prepare high purity GMP.

Olieman and van Riel [9] by using trichloroacetic acid treatment and reversed-phase HPLC isolated GMP from sweet whey, and reported absence of phenylalanine in their final preparation. Nakano et al. [10] also reported that GMP fraction from sweet whey prepared by trichloroacetic acid treatment followed by gel filtration chromatography contained undetectable level of phenylalanine. However, the use of trichloroacetic acid is not suitable for production of GMP for human consumption. For preparation of GMP as a food, ion exchange is one of the common techniques, in that GMP having an isoelectric point (pI) < 3.8 [11], which is lower than the pI (> 4.1) of major whey proteins (β-lactoglobulin, α-lactalubumin, serum albumin, immunoglobulins etc.) [1], can be separated from whey proteins by the difference of pI. GMP fractions obtained by ion exchange techniques contain traces of aromatic amino acids as contaminants. For example, Nakano and Ozimek [12] reported that GMP isolated from sweet whey using anion exchange chromatography on diethylaminoethyl (DEAE)-Sephacel contained less than 1 residue of each of phenylalanine, histidine and arginine per peptide. Similarly, it has been reported that GMP isolated from sweet whey or whey protein concentrate by anion exchange techniques contained low but detectable levels of aromatic amino acids including phenylalanine, and other amino acids not found in GMP [11,13,14]. More recently, LaClair et al. [7], in their experiment of PKU diet preparation, reported that the phenylalanine concentration (5 mg/g of product) in a commercially available GMP (Davisco Foods International, Inc. USA.) is too high, and thus, they refined the product by using cation exchange chromatography to reduce phenyalanine level to 2.7 mg/g protein equivalent (43% reduction). This suggests that purification of GMP by ion exchange chromatography without having contaminating amino acids including phenylalanine is very difficult, although the following information is available in the literature. Léonil and Mollé [15] isolated GMP from sweet whey using cation exchange HPLC on a Mono-S column, whereas Saito et al. [16] purified GMP by using ethanol precipitation followed by DEAE-Toyopearl anion exchange chromatography. Both groups of researchers reported preparation of GMP with no contaminating amino acids including histidine, tyrosine, arginine, and phenylalanine. However, these results must be interpreted carefully. Amino acid analysis in either study showed recovery of less than 64 residues/peptide [i.e. 59 and 60 residues/peptide, each calculated from the data of Léonil and Mollé

***Corresponding author:** Takuo Nakano, Department of Agricultural, Food and Nutritional Science, University of Alberta, Edmonton, Alberta, Canada
E-mail: takuonakano@hotmail.com

[15] and Saito et al. [16], respectively], suggesting that the assay was not sensitive enough to rule out the occurrence of contaminating amino acids in the purified product. The present study was, therefore, undertaken to determine whether commercially available GMP can be refined by anion exchange chromatography with high reproducibility to provide GMP with no contaminating amino acids.

Materials and Methods

Materials

A commercial sample of crude GMP prepared using ion exchange technique was obtained from Davisco Foods International, Inc., Eden Prairie, MN., USA. DEAE-Sephacel and Superdex G-75 were products of GE Healthcare, Baie d'Urfé, PQ, Canada. Sialic acid (N-acetylneuraminic acid from sheep submaxilally glands), galactose, galactosamine-HCl, and Sephacryl S-100-HR were obtained from Sigma-Aldrich, Canada Ltd., Mississauga, ON, Canada. Cellulose acetate strips (Sepraphore III, 2.5 cm × 15.2 cm) were obtained from Pall Corporation, Ann Arbor MI, USA. Malachite green was obtained from Difco Laboratories, Detroit, MI., USA.

Anion exchange chromatography

To a sample (~0.2 g) of crude GMP, 40 ml of water was added. This amount of water was used for the convenience of pH measurement and centrifugation. The mixture was adjusted to pH 3.0 with 1 M HCl. A small amount of precipitate formed was removed by centrifugation at 20,000 × g and 21°C for 20 min, and a supernatant obtained was applied to a 1.5 cm × 6.2 cm column of DEAE-Sephacel equilibrated with water adjusted to pH 3.0 with 1 M HCl. Immediately after application of sample, the column was washed with approximately 40 ml of water adjusted to pH 3.0, and then eluted with a linear gradient formed from 45 ml of water and 45 ml of 1 M NaCl, both adjusted to pH 3.0. Factions (~2 ml) collected at a flow rate of 23 ml/h were monitored for ultra violet (UV) absorbance at 210 and 230 nm for peptide amide bond, and 280 nm for protein, which is dependent mainly on the amounts of tyrosine and tryptophan [17]. Fractions were also monitored for carbohydrates (sialic acid, galactose, and galactosamine), and phosphorus. The recovery of GMP was estimated by monitoring sialic acid, a marker compound specific to sialylated GMP [18]. The major sialic acid containing fractions and those containing components unadsorbed to DEAE-Sephacel were separately pooled, exhaustively dialyzed in water, and freeze-dried for further studies. This experiment was repeated six times.

Analytical methods

The sialic acid content was determined by the modified thiobarbituric acid reaction [19,20], in that 1-propanol instead of cyclohexanone [21] was used to extract chromophore formed during reaction. Cyclohexanone used by Warren [21] in the original thiobarbituric acid reaction is a hazardous chemical too difficult to handle in assaying many samples. The absorbance of chromophore was read at 549 nm, and the concentration of sialic acid in the crude or purified preparation of GMP was calculated by comparison of its absorbance with the absorbance of a known concentration of sialic acid. Galactose was determined by the anthrone reaction [22] using galactose as a standard, and galactosamine was determined by the indole reaction [23] after hydrolysis of samples in 4 M HCl at 100°C for 4 h. The galactosamine value obtained was corrected for hydrolytic loss of 10% determined on standard galactosamine-HCl. Phosphorus concentrations in crude and purified GMP were determined by using the molybdate-vanadate reagent [24] with potassium dihydrogen phosphate as a standard phosphate. Fractions obtained after anion exchange chromatography or cellulose acetate electrophoresis was also monitored for phosphorus by using the malachite green dye binding method [25].

For analysis of amino acids except tryptophan, a sample (~3 mg) was hydrolyzed in 3 ml of 6 M HCl in the presence of nitrogen at 110°C for 24 h. For tryptophan analysis, a sample (~3 mg) was first dissolved in 3 ml of 4.2 M NaOH, to which 0.75 ml of 2 M pyrogallol was added, and the mixture was hydrolyzed in the presence of nitrogen at 110°C for 20 h. Amino acids in both the acid and alkali hydrolysates were derivatized using the o-phthaldialdehyde method [26,27] using a fluoraldehyde reagent prepared by dissolving 0.25 g of o-phthaldialdehyde in 6 ml of methanol followed by the addition of 56 ml of 0.04 M sodium borate buffer, pH 9.5, 0.25 ml of 2-mercaptoethanol, and 2 ml of Brij 35. Chromatographic analysis of the derivatized amino acids was carried out using a Supelcosil 3 micron LC-18 reverse phase column (4.6 mm × 150 mm, Supelco) with a Varian Fluorichrom fluorescence detector (excitation 340 nm and emission 450 nm).

Size exclusion HPLC

A 20 μL of sample containing 100 μg of purified GMP, crude GMP or the product unadsorbed on DEAE-Sephacel was applied to 1cm × 30 cm column of Superdex 75 10/300 GL equilibrated and eluted with 0.05 M sodium phosphate-0.15 M NaCl, pH 7.0. The eluate was monitored for peptide by measuring absorbance at 214 nm.

Cellulose acetate electrophoresis

Electrophoresis of GMP on cellulose acetate strips was carried out in 0.1 M pyridine-1.2 M acetic acid, pH 3.5 [20]. After electrophoresis, GMP was located by monitoring sialic acid and phosphorus in serial fractions obtained from the cellulose acetate strip [20].

Gel filtration chromatography

Molecular size of GMP purified using DEAE-Sephacel anion exchange chromatography was examined using gel filtration chromatography on Sephacryl S-100-HR. Two columns with similar size were prepared. The first one (0.9 cm × 57 cm) was equilibrated and eluted with 0.05 M phosphate-0.15 M NaCl, pH 7.0, and the second one (0.9 cm × 58.5 cm) was equilibrated and eluted with 6 M guanidine hydrochloride-0.1 M sodium acetate, pH 7.0. In each case, approximately 2.5 mg portion of GMP sample was chromatographed at a flow rate of 9 ml/h, and elution patterns of GMP were compared between the two columns.

Results and Discussion

The crude GMP suspension showed an average pH value of 6.4 with relatively low turbidity [0.052 ± 0.001 (standard deviation, SD, n =6)] against water at wavelength 500 nm]. Its turbidity increased 2.3 times (0.120 ± 0.006) when the pH was adjusted to 3.0. After centrifugation, the supernatant collected was applied to the anion exchange column of DEAE-Sephacel, whereas the precipitate obtained, which accounted for 1.6 ± 0.4% dry weight of crude GMP, was discarded.

A representative DEAE-Sephacel chromatogram for the crude GMP is shown in Figure 1. A small proportion of sialic acid (corresponding to 6.8 ± 5.7% of total recovered sialic acid) failed to bind to the anion exchanger, whereas most (93.2 ± 5.7%) of the recovered sialic acid was adsorbed on the column, and eluted at 0.3-0.6 M NaCl (Figure 1A). The sialic acid found in the unadsorbed component (fractions 9-33, Figure 1A) was confirmed to be GMP sialic acid (but unlikely sialic acid from whey proteins including α-lactalbumin and immunoglobulin) by cellulose acetate electrophoresis (data not shown). The major sialic acid peak fractions (57-68) (Figure 1A) also contained peptide, galactose, galactosamine, and phosphorus, each showing its elution position identical to that of sialic acid (Figure 1B), reflecting the structure of GMP as phosphorylated peptide to which sialylated oligosaccharides are covalently attached [1-5]. The product adsorbed on the column (referred to as purified GMP), and the product unadsorbed on the column (unadsorbed product) accounted for 42.6 ± 5.9% and 36.0 ± 10.0%, respectively, of dry weight of crude GMP sample applied to the column. The apparently larger peak area of UV absorbance (at 210 nm and 230 nm) seen in the unadsorbed than in the adsorbed product (Figure 1A), which apparently does not reflect the difference in the recovered dry weight between the two products as reported above, is likely due to the higher concentration of non-peptide component (i.e. carbohydrate) in the latter (see below).

Amino acid analysis (Table 1) showed that the small amount of phenylalanine, present in the crude GMP, was also found in the unadsorbed product, but not in the purified GMP. This indicated that phenylalanine containing protein or peptide did not bind to the anion exchanger. The amino acid composition of the purified GMP was similar among the six experiments, suggesting that the separation of GMP with undetectable level of phenylalanine is highly reproducible. The purified product, however, still contained small amounts of tryptophan and arginine (amino acids not present in GMP). The source of these amino acids is unknown. The relatively low but positive absorbance at 280 nm seen in the GMP peak (Figure 1A) as well as in the purified GMP (Table 1) may be due to the presence of tryptophan. The absorbance value for the purified GMP was 4 and 5 times lower compared to the absorbance values for the crude GMP and the unadsorbed product, respectively (Table 1). The molar ratio of amino acid calculated for the purified GMP in this study was in general comparable to the theoretical value for the bovine GMP [1].

	Crude GMP	Product adsorbed on the column (purified GMP)	Product unadsorbed on the column
Asx	0.50[a]	0.47 ± 0.05[b]	0.62 ± 0.04[b]
Glx	1.10	1.05 ± 0.08	1.34 ± 0.10
Ser	0.66	0.57 ± 0.08	0.72 ± 0.10
Gly	0.12	0.11 ± 0.01	0.13 ± 0.01
Thr	1.17	1.17 ± 0.10	1.28 ± 0.07
Arg	0.02	0.01 ± 0.00	0.03 ± 0.01
Ala	0.53	0.52 ± 0.04	0.74 ± 0.25
Val	0.50	0.54 ± 0.02	0.64 ± 0.05
Ile	0.58	0.60 ± 0.04	0.69 ± 0.08
Leu	0.15	0.12 ± 0.01	0.24 ± 0.03
Lys	0.27	0.29 ± 0.02	0.40 ± 0.02
Tyr	0.06	nd	0.02 ± 0.02
Phe	0.03	nd	0.04 ± 0.01
Trp	< 0.01	< 0.01	< 0.01
Absorbance at 280 nm for 1 mg/mL solution[c]	0.061 ± 0.00	0.014 ± 0.001	0.072 ± 0.030

Proline, methionine, and cysteine were not determined.
[a]Average of two determinations.
[b]Mean ± SD (n = 6).
nd: Not detected.
[c]Each sample was filtered through 0.22 μm membrane before measurement of absorbance, and absorvance values are shown as mean ± SD (n = 6).

Table 1: Amino acid concentrations (μmole/mg) and UV absorbance determined in the crude GMP and its fractions obtained by anion exchange chromatography (Figure 1).

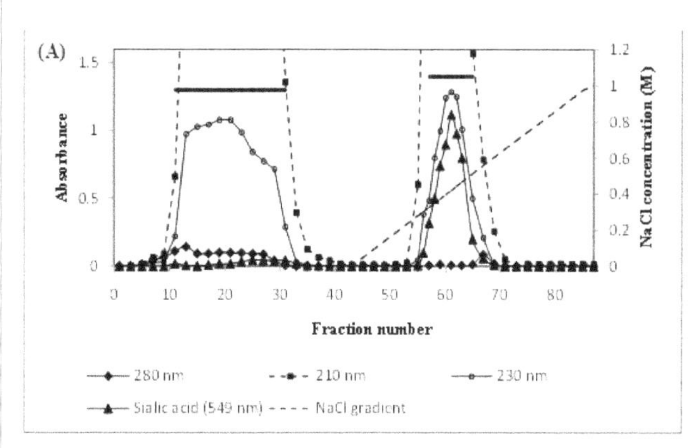

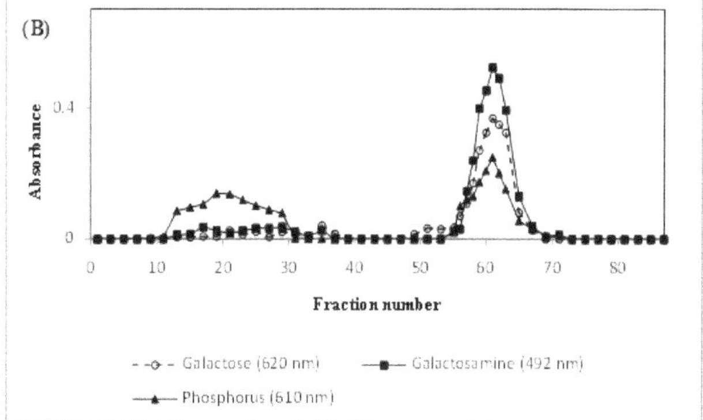

Figure 1: Anion exchange chromatography of the crude GMP on a 1.5 × 5.7 cm column of DEAE-Sephacel. (A): Fractions collected were monitored for UV absorbance and sialic acid. (B): Fractions collected were also monitored for galactose, galactosamine, and phosphorus. Horizontal bars denote the fractions that were pooled for further study.

Carbohydrate analysis (Table 2) showed that concentrations of sialic acid, galactose and galactosamine were 18.9, 7.3 and 6.5 times higher in the purified GMP than in the unadsorbed product, suggesting that most of sialylated GMP was adsorbed and eluted from the column. The sialic acid, galactose and galactosamine concentrations were, respectively, 1.8, 1.6 and 1.6 times higher in the purified than in the crude GMP as expected. In contrast to carbohydrate concentrations, phosphorus concentrations (Table 2) were 1.4 times higher in the unadsorbed product compared to the purified GMP, but similar between the purified and crude GMP.

Carbohydrate and phosphorus concentrations in the purified GMP (Table 1) were in general within the range of values reported for bovine GMP [6]. To PKU infants, GMP with high sialic acid content as seen in the present study (average 15.5%) may be an important nutrient for brain growth [28] as well as dietary amino acid source.

Samples of the crude GMP, purified GMP and unadsorbed product were then examined using size exclusion HPLC on Superdex-75 (Figure 2). The retention time was the least in the purified GMP (27.96 ± 0.17 min, n = 5), and less in the crude GMP (28.78 ± 0.04 min, n = 3) than in the unadsorbed product (29.04 ± 0.17 min, n = 5) (Figure 2A). This difference appears to be due to the difference in carbohydrate contents (Table 2). The retention time for the purified GMP was close to that (28.34 ± 0.09 min, n = 3) of dimeric β-lactoglobulin (36.6 kDa) (Figure 2B), confirming the similarity of molecular size between GMP aggregate and dimeric β-lactoglobulin shown by gel filtration chromatography (see below). In addition to the major peak, the purified GMP also had a small peak at ~43 min (Figure 2A), which is likely related to the small peak that eluted near the total column volume on Sephacryl S-100 chromatography (see below).

The purified GMP was further studied using cellulose acetate electrophoresis and gel filtration chromatography on Sephacryl S-100. Cellulose acetate electrophoresis (Figure 3) showed a single but relatively broad peak of GMP sialic acid in fractions 8–10, which were seen to contain phosphorus, confirming that the purified product contained sialylated phosphorylated glycopeptide as shown above by DEAE-Sephacel chromatography (Figure 1).

Elution patterns of the purified GMP on Sephacryl S-100 gel filtration chromatography are given in Figue 4. With 0.05 M sodium

	Crude GMP	Product adsorbed on the column (purified GMP)	Product unadsorbed on the column
Sialic acid	8.44[a]	15.51 ± 1.75[b]	0.82 ± 0.47[b]
Galactose	3.25	5.10 ± 0.85	0.70 ± 0.11
Galactosamine	3.85	6.14 ± 1.74	0.94 ± 0.26
Phosphorus	0.42	0.43 ± 0.02	0.62 ± 0.05

[a]Average of two determinations.
[b]Mean ± SD (n = 6).

Table 2: Concentrations (g/100g) of carbohydrate and phosphorus in the crude GMP and products obtained by anion exchange chromatography.

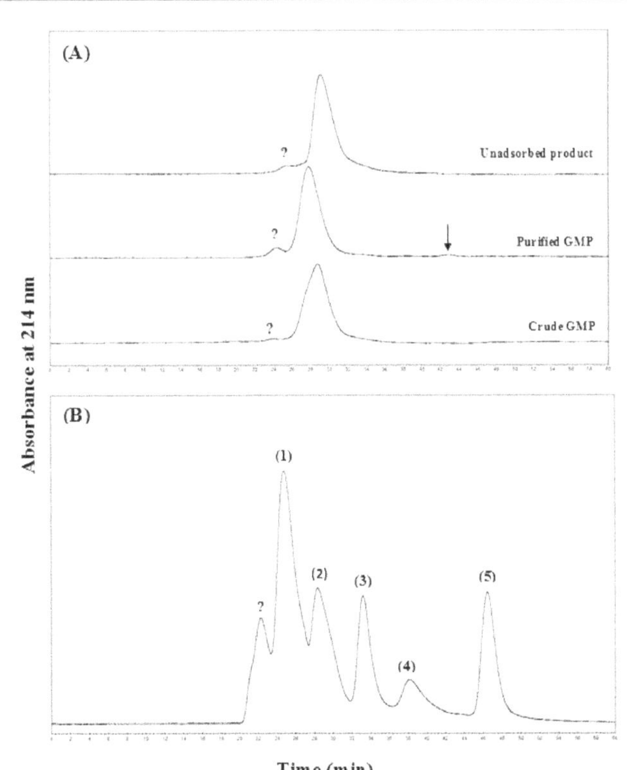

Figure 2: Size exclusion HPLC on Superdex-75. (A): Elution pattens for the crude GMP, purified GMP, and unadsorbed product are shown. The purified GMP had a small peak eluting at ~43 min shown by an arrow. (B): The chromatogram shows elution patterns of molecular size markers including: (1) bovine serum albumin (67 kDa); (2) dimeric β-lactoglobulin (36.6 kDa); (3) cytochrome-C (13.6 kDa); (4) aprotinin (6.512 kDa); and (5) vitamin B_{12} (1.355 kDa).

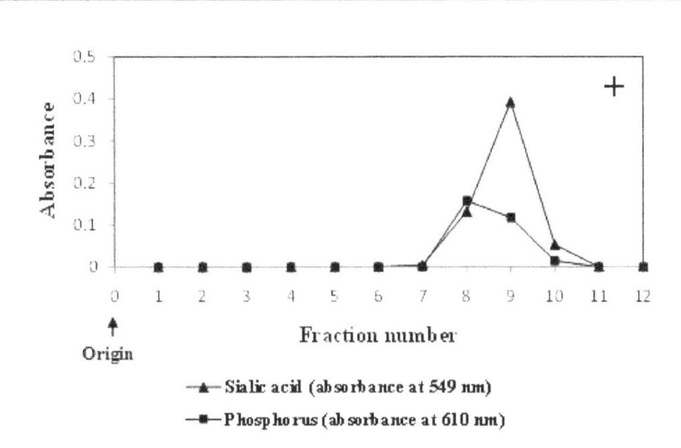

Figure 3: Electrophoresis of the purified GMP on a cellulose acetate strip in 0.1 M pyridine-1.2 M acetic acid, pH 3.5.

phosphate-0.15 M NaCl, pH 7.0 as an eluent, greater than 99.5% of the purified GMP, monitored by sialic acid assay and UV absorbance measurement, eluted as a single peak with its elution volume close to the elution volume of dimeric β-lactoglobulin (36.6 kDa) (Figure 4A). A small peak corresponding to <0.5% of total recovered sialic acid or peptide appeared near the total column volume (33.0 ml determined using NaCl), which might have been a mixture of degradation products of GMP. This was not investigated further. In contrast to the results obtained above, when chromatographed with 6 M guanidine hydrochloride-0.1 M sodium acetate, pH 7.0, as an eluent, the GMP eluted at the position close to the elution position of α-lactalbumin (14.2 kDa) (Figure 4B). This indicated that the purified GMP was in aggregated form with 0.05 M sodium phosphate-0.15 M NaCl, but in disaggregated form in the presence of 6 M guanidine hydrochloride (dissociating agent). These results confirm the finding of Nakano and Ozimek [29] who reported that GMP aggregate is comprised of approximately three monomers.

In this study, most of non-sialylated GMP (not quantified in this study) or low-sialylated GMP was unlikely adsorbed to the anion exchanger, and thus not included in the purified GMP. This caused the apparently low recovery of purified GMP (average 43%). However, for preparation of foods for PKU patients, the level of phenylalanine in GMP is very important. We need a reliable method for constant supply of GMP with undetectable level of phenylalanine, which helps dietitians to estimate GMP content to be safely used in foods for PKU patients. The purified GMP can also be used as a research chemical. It is interesting to know whether the laboratory scale technique in this study can be scaled up (preferably with batch method) without losing its high reproducibility.

No attempt was made in this study to further separate impurities from the non-sialylated and low-sialylated GMP. Since the isoelectric point (pI) of GMP peptide (calculated to be 4.04 and 4.14 for its genetic variants A and B, respectively [11]) is close to the pI of α-lactalbumin (4.2-4.5) [1], it is uncertain whether ion exchange method is an efficient technique for selective separation of phenylalanine containing impurities from non-sialylated or low-sialylated GMP. Further research is needed to develop inexpensive methods to separate non-sialylated or low-sialylated GMP from sweet whey proteins.

Conclusion

Results obtained in this study suggest that DEAE-Sephacel

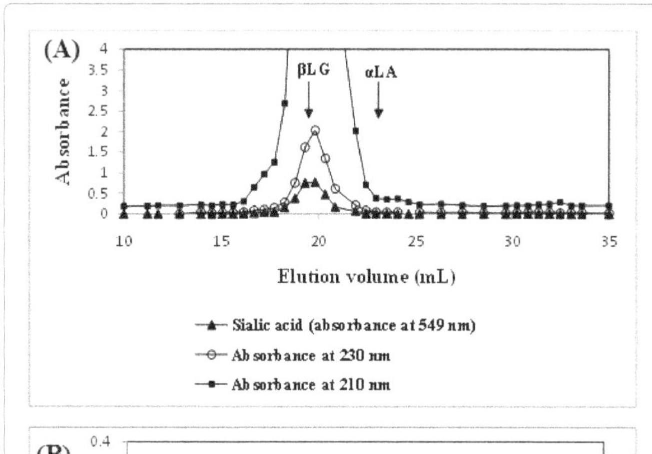

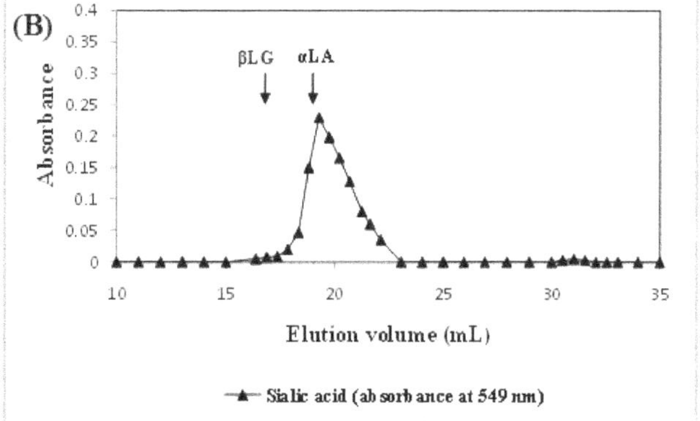

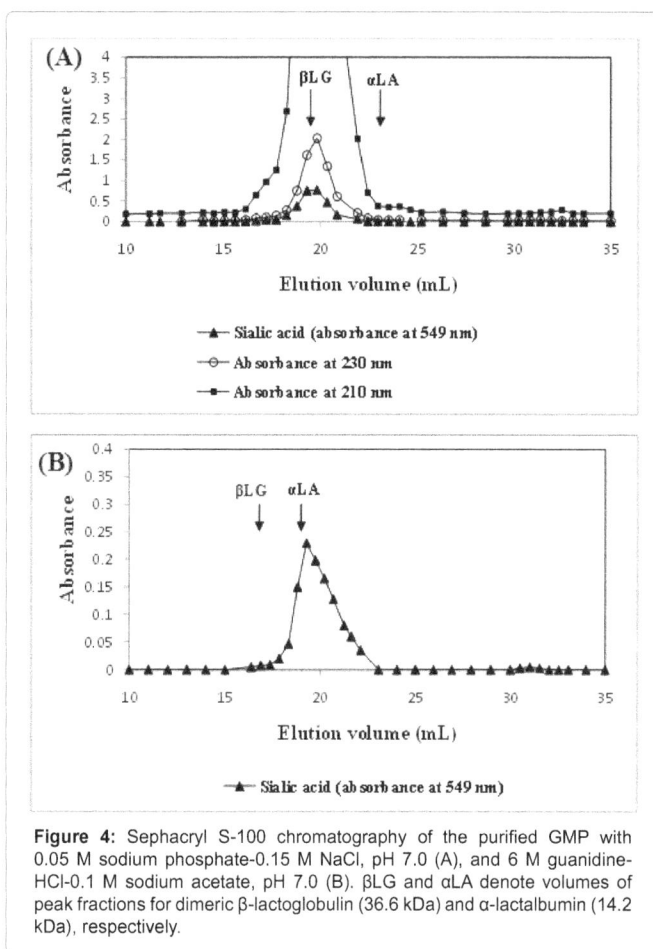

Figure 4: Sephacryl S-100 chromatography of the purified GMP with 0.05 M sodium phosphate-0.15 M NaCl, pH 7.0 (A), and 6 M guanidine-HCl-0.1 M sodium acetate, pH 7.0 (B). βLG and αLA denote volumes of peak fractions for dimeric β-lactoglobulin (36.6 kDa) and α-lactalbumin (14.2 kDa), respectively.

chromatography is a relatively simple reproducible method to selectively remove phenylalanine impurities from crude GMP. The purified product is a GMP with high sialic acid content. To our knowledge, this is the first report of food grade GMP preparation without detectable level of phenylalanine. Further research is needed to scale up the method for industrial production of GMP as a food for PKU patients. It is also important to develop methods to recover non-sialylated and low-sialylated GMP which were removed with phenylalanine impurities during anion exchange chromatography, and thus not included in the purified GMP fraction in the present study.

Acknowledgments

This study was financially supported by the grant from the Alberta Livestock and Meat Agency Ltd. We thank Gary Sedgwick for amino acid analysis and size exclusion chromatography.

References

1. Eigel WN, Butler JE, Ernstrom CA, Farrel HM Jr, Harwalker VR, et al. (1984) Nomenclature of proteins of cow's milk. Fifth revision. Journal of Dairy Science 67: 1599-1631.

2. Abd El-Salam HM, Elshibiny S, Buchheim W (1996) Characteristics and potential uses of the casein macropeptide. International Dairy Journal 6: 327-341.

3. Brody EP (2000) Biological activities of bovine glycomacropeptide. British Journal of Nutrition 84 Suppl 1: 39-46.

4. Thomä-Worringer C, Sørensen J, López-Fandiño R (2006) Health effects and technological features of caseinomacropeptide. International Dairy Journal 16: 1324-1333.

5. Neelima, Sharma R, Rajput YS, Mann B (2013) Chemical and functional properties of glycomacropeptide (GMP) and its role in the detection of cheese whey adulteration in milk: a review. Dairy Science and Technology 93: 21-43.

6. Kawakami H, Kawasaki Y, Dosako S, Tanimoto M, Nakajima I (1992) Determination of κ-casein glycomacropeptide by high performance liquid chromatography without trichloroacetic acid pretreatment. Milchwissenschaft 47: 688-693.

7. Laclair CE, Ney DM, MacLeod EL, Etzel MR (2009) Purification and use of glycomacropeptide for nutritional management of phenylketonuria. Journal of Food Science 74: 199-206.

8. van Calcar SC, Ney DM (2012) Food products made with glycomacropeptide, a low-phenylalanine whey protein, provide a new alternative to amino Acid-based medical foods for nutrition management of phenylketonuria. Journal of Academy of Nutrition and Dietetics 112: 1201-1210.

9. Olieman C, van Riel JAM (1989) Detection of rennet whey solids in skim milk powder and buttermilk powder with reversed-phase HPLC. Netherland Milk Dairy Journal 43: 171-184.

10. Nakano T, Silva-Hernandez ER, Ikawa N, Ozimek L (2002) Purification of k-casien glycomacropeptide from sweet whey with undetectable level of phenylalanine. Biotechnology Progress 18: 409-412.

11. Nakano T, Ozimek L (2000) Purification of glycomacropeptide from dialyzed and non- dialyzed sweet whey by anion-exchange chromatography at different pH values. Biotechnology Letters 22: 1081-1086.

12. Nakano T, Ozimek L (1999) Purification of glycomacropeptide from non-dialyzable fraction of sweet whey by anion-exchange chromatography. Biotechnology Techniques, 13: 739-742.

13. Ayers JS, Coolbear KP, Elgar DF, Pritchard M (1998) Process for isolatin glycomacropeptide from dairy products with a phenylalanine impurity of 0.5% w/w. POT/WO 98/14071.

14. Nakano T, Ikawa N, Ozimek L (2004) Use of epichlorohydrin-treated chitosan resin as an adsorbent to isolate k-casein glycomacropeptide from sweet whey. Journal of Agricultural and Food Chemistry 52: 7555-7560.

15. Léonil J, Mollé D (1991) A method for determination of macropeptide by cation-exchange fast protein liquid chromatography and its use for following the action of chymosin in milk. Journal of Dairy Research 58: 321-328.

16. Saito T, Yamaji A, Ito T (1991) A new isolation method of caseinoglycopeptide from sweet cheese whey. Journal of Dairy Science 74: 2831-2837.

17. Peterson GL (1983) Determination of total protein. Methods in Enzymology 91: 95-119.

18. Nakano T, Ozimek L (2014) A sialic acid assay in isolation and purification of bovine k-casein glycomacropeptide: a review. Recent Patents on Food Nutrition and Agriculture 6: 38-44.

19. Nakano K, Nakano T, Ahn DU, Sim JS (1994) Sialic acid contents in chicken eggs and tissues. Canadian Journal of Animal Science 74: 601-606.

20. Nakano T, Ikawa N, Ozimek L (2007) Detection of sialylated phosphorylated κ-casein glycomacropeptide electrophoresed on polyacrylamide gels and cellulose acetate strips by the thiobarbituric acid and malachite green dye reactions. Journal of Agricultural and Food Chemistry 55: 2714-2726.

21. Warren L (1959) The thiobarbituric acid assay of sialic acids. Journal of Biological Chemistry 234: 1971-1975.

22. Trevelyan W, Harrison JS (1952) Studies on yeast metabolism. I. Fractionation and microdetermination of cell carbohydrates. Biochemical Journal 50: 298-303.

23. Dische Z, Borenfrend EA (1950) A spectrophotometric method for the micro determination of hexosamines. Journal of Biological Chemistry 184: 517-522.

24. AOAC (1998) Official methods of analysis: Association of Official Analytical Chemists, USA.

25. Van Veldhoven PP, Mannaerts GP (1987) Inorganic and organic phosphate measurements in the nanomolar range. Analytical Biochemistry 161: 45-48.

26. Jones BN, Gilligan J (1983) O-Phthalaldehyde precolumn derivatization and reversed-phase high-performance liquid chromatography of polypeptide hydrolysates and physiological fluids. Journal of Chromatography 266: 471-482.

27. Sedgwick GW, Fenton TW, Thompson JR (1991) Effect of protein precipitating agents on the recovery of plasma free amino acids. Canadian Journal of Animal Science 71: 953-957.

Phytochemical Profiling of Conventional and Supercritical Ginger Extract Based Baked Bars

Tanweer S[1,3]*, Shehzad A[1], Sadiq Butt M[1] and Shahid M[2]

[1]*National Institute of Food Science and Technology, Faculty of Food, Nutrition and Home Sciences, University of Agriculture, Faisalabad, Pakistan*
[2]*Department of Biochemistry, Faculty of Basic Sciences, University of Agriculture, Faisalabad, Pakistan*
[3]*Bioactive Natural Product Laboratory, Plant Soil and Science Building, Michigan State University, MI, USA*

Abstract

Contemporary, nutraceutics have attracted the consumers owing to their therapeutical potential aligned with metabolic arrays. In this ambiance, ginger is a famous herb that has the ability to mitigate various health related disorders due to its unique photochemistry with special reference to gingerol and shagoal. To evaluate the health boosting ability of ginger, product namely ginger bars were prepared by the addition of 3% ginger conventional nutraceutical (CSE) as well as 0.3% of supercritical nutraceutic (SFE). The product was observed for color tonality in the form of L*, a*, b*, Chroma and Hue. The antioxidant potential of ginger bars was assessed by different antioxidant tests *i.e.* TPC, DPPH, Antioxidant activity, FRAP, ABTS and metal chelating ranges from 67.45 ± 2.29 to 112.28 ± 3.81 mg GAE/100g for TPC, 8.28 ± 0.28 to 30.72 ± 1.05% for DPPH, 13.27 ± 0.45 to 33.61 ± 1.14% for antioxidant activity, 22.15 ± 0.75 to 48.81 ± 1.66 µmole TE/g for FRAP assay, 5.94 ± 0.20 to 19.05 ± 0.65 µmole TE/g for ABTS and for metal chelating it varied from 16.41 ± 0.56 to 21.22 ± 0.72 by the addendum of ginger extracts. Furthermore the ginger bars were marked by hedonic response in terms of color, crispiness, taste, flavor and overall acceptability.

Keywords: Designer product; Physico-chemical analysis; TPC; DPPH; FRAP; ABTS; FRAP; Antioxidant activity; Metal chelating

Introduction

Diet along with its constituents contributes improved state of health other than reduced risk of diseases to enhance the quality of life. These concepts motivated to the addition of functional foods in routine diet that are health boosting foods processed with biologically active ingredient in precise quantity having both qualitative and quantitative influence on health. Hence, in modern age these healthy foods are important source in management and prevention of chronic disorders [1].

In present era, the consumption rate of designer foods is increasing day by day because of its health benefits beyond to nutritional value along with enhanced shelf life owing to the addition of antioxidants that lowers the process of rancidity [2-4]. A few epochs ago, the trend of cereal based food product moved towards designer foods by the addendum of phytoceutics that improves the health stratum along with enhanced shelf life [2]. Innately, the marked chances resulted due to the oxidation reactions that transpire slowly during storage [3]. The food recipes that are modified by the supplementation of spices have improved stability against oxidation. These spices are more often used for oxidation stability to enhance shelf life in addition to providing flavor [5].

Baked products have been recognized as best vehicles for amalgamation of ginger, although there are positive effects on the physicochemical properties of baked products after the addition of ginger along with health benefits [6]. Baking is a complex process and results in manifold physical and biochemical effects including structure formation, taste development, color formation and synthesis of health promoting and health impairing constituents [7]. The bars are well known as a cradle of carbohydrates in food pyramid that ensures that a person is taking sufficient amount of nutrients in balance to require by the body. In formulation of bars, the ingredients provide its characteristics including color, flavor, taste, texture along with calories. Other parameters that have impact on attributes of bar are replacement of sugar and fat replacement over and above to addition of spices [8].

Incorporation of antioxidants such as bioactive ingredients in food products *viz.*, baked bars have been grown rapidly because of improved health status awareness [9]. These natural moieties also act as mold inhibitors that delay the production and growth of mold on baked products and help in improved shelf life. The other method to get the interest of consumer is to develop the formation of chemical free product by replacing undesired ingredients augmented by antioxidants and enzymes [8].

Ginger owing to be a rich source of aromatic and pleasant flavoring properties is commonly used in the preparation of baked products, condiments and curries [10]. It has strong antioxidant potential that has been verified to be effectual in lipid oxidation inhibition as well as declining the level of oxidation in baked products. Although, nutraceutics from ginger have desirable characteristics such as being natural, non-GMO food and clean label ingredient as it can be labeled as a food ingredient in the label of food product [11].

Materials and Methods

Three types of bars were prepared using best treatment of each $\text{nutraceutical}_{CSE}$ and $\text{nutraceutical}_{SFE}$ as described in AACC (2000) method no. 10-50D. The first (T1) contained $\text{nutraceutical}_{CSE}$ whilst other (T2) enriched with $\text{nutraceutical}_{SFE}$ along with control (T0) for comparison purpose Table 1.

Physico-chemical analysis

The prepared bars were analyzed for the color, texture and

***Corresponding author:** Tanweer S, National Institute of Food Science and Technology, Faculty of Food, Nutrition and Home Sciences, University of Agriculture, Faisalabad, 38000, Pakistan, E-mail: sairatanweer1116@gmail.com

Treatments	Description
T_0	Control
T_1	Ginger bars with 3% nutraceutical $_{CSE}$
T_2	Ginger bars with 0.3% nutraceutical $_{SFE}$

Table 1: Treatments used in product development.

antioxidant potential during the storage period. The color and texture parameters of bars were measured using the methods of Parn et al. [12].

Color analysis

The color analysis was performed by using CIE-Lab Color Meter (CIELAB SPACE, Color Tech-PCM, USA). Prior to analysis, the colorimeter was calibrated using the zero and white calibration plates, respectively. Samples were also analyzed to find out their hue and chroma values.

$$\text{Chroma } (C^*) = \left[(a^*)^2 + (b^*)^2\right]^{1/2}$$

$$\text{Hue angle } (h) = \tan^{-1}\left(\frac{b^*}{a^*}\right)$$

Texture analysis

Texture analysis was performed using texture analyzer (single arm texture analyzer TA-XT Plus, Stable Micro Systems, Surrey, UK) with a load cell of 2 kg weight. A force versus time curve for a two-cycle compression was measured, with a disk probe of 35 mM diameter and at a displacement speed of 10 mM/min. Built in software of the texture analyzer was used for analyzing the data generated.

Antioxidant potential

Antioxidant potential of ginger bar was determined by the protocols described by Sharma and Gujral [13].

Total Phenolic Content (TPC)

Total phenolic contents (TPC) in ginger bars extract were measured using Folin-Ciocalteu method that was based on the reduction of phosphotungstic acid to phosphotungstic blue and as result absorbance increased due to rise in number of aromatic phenolic groups. For the purpose, 50 µL of ginger bar extract was separately added to test tube containing 250 µL of Folin-Ciocalteu's reagent, 750 µL of 20% sodium carbonate solution and volume was made up to 5mL with distilled water. After two hours, absorbance was measured at 765 nm using UV/visible light Spectrophotometer (CECIL CE7200) against control that has all reaction reagents except sample extract. Total polyphenols was estimated and values were verbalized as gallic acid equivalent (mg gallic acid/100 g).

Total phenolic compounds of each extract in gallic acid equivalents (GAE) was calculated by following formula:

$$C = c \times \frac{V}{M}$$

C = Total phenolic contents (mg/g plant extract, in GAE)

c = Concentration of gallic acid (mg/mL)

V = Volume of extract (mL)

M = Weight of ginger extract (g)

Free Radical Scavenging Activity (DPPH assay)

Sample solution of ginger bar extract was prepared by dissolving 0.025 mL of sample extract in 10 mL of respective solvent with 3 mL of freshly prepared DPPH solution in respective solvent that was mixed with 77 µL sample extract. Each sample was kept in dark place for about 15 minutes at room temperature and decrease in absorbance was measured at 517 nm on UV/visible light spectrophotometer. Similarly, blank sample absorbance having the same amount of solvent and DPPH solution except extract was prepared and absorbance was estimated at same wavelength on UV/visible light spectrophotometer. The free radical-scavenging activity of each ginger extract can be presented as percentage reduction in DPPH due to given amount of each extract.

$$\text{Reduction of absorbance } (\%) = \left[\frac{(AB - AA)}{AB}\right] \times 100$$

AB = Absorbance of blank sample at t = 0 minute

AA = Absorbance of tested extract solution at t = 15 minutes

Antioxidant Activity (AA)

Antioxidant activity of ginger bar extracts was based on coupled oxidation of ß-carotene as well as linoleic acid. In this method, 2 mg of β-carotene was dissolved in 20 mL of chloroform. A 3 mL of aliquot was taken in flask containing 40 mg linoleic acid along with 400 mg Tween 20 and the mixture was then evaporated at 40°C for 10 min using rotary evaporator to remove chloroform. This mixture was diluted with 100 mL distilled water and was mixed properly by vortex mixer to prepare emulsion. 3 mL of β-carotene emulsion as well as 0.12 mL phenolic extracts were taken in test tubes and were thoroughly mixed. Afterward, test tubes were incubated at 50°C in a water bath for time duration of 30 minutes. Absorbance of each sample was measured at 470 nm on UV/visible light spectrophotometer. The degradation rate of the extracts was also calculated according to the first order kinetic reaction using following expression.

$$\text{Sample degradation rate} = \ln\left(a/b\right) \times 1/t$$

ln = Natural log

a = Initial absorbance on 470 nm at time zero

b = Absorbance on 470 nm after 30 min

t = Time in minutes

The antioxidant activity was expressed as percentage inhibition (%) relative to the control by following equation.

$$AA\ (\%) = \frac{\text{Degradation rate of control} - \text{Degradation rate of sample}}{\text{Degradation rate of control}} \times 100$$

Ferric Reducing Antioxidant Power (FRAP) assay

The reducing power of ginger bar extracts was determined by measuring capability of extracts to reduce ferric tripyridyltriazine into blue colored ferrous that can be detected at 593 nm. FRAP reagent was prepared by mixing 25 mL acetate buffer (0.1 M at pH 3.6), 2.5 mL TPTZ (10 mM), and 2.5 mL ferric chloride (20 mM) and was incubated at 30°C for 10 minutes. To determine reducing power of ginger extract immediately 1.5 mL of FRAP reagent was mixed with 100 µL of ginger extract or standard and 100 µL of distilled water. Then absorbance was taken at 593 nm on UV/visible light spectrophotometer. A calibration curve was drawn using trolox (0-500 µmol/mL) and was expressed as µmol trolox equivalent per gram of sample.

ABTS (2,2-Azino-Bis, 3-Ethylbenzothiazoline-6-Sulphonic Acid) Assay

ABTS assay is a decolorizing method, the ABTS radical was freshly prepared by adding 5 mL of a 4.9 mM potassium persulfate solution to 5 mL of a 14 mM ABTS solution and keeping the mixture in the dark for 16 hr. This solution was diluted further with respective solvent to yield an absorbance of 0.7 ± 0.02 at 734 nm and was used for antioxidant assay. The final reaction mixture (1 mL) comprised of 950 μL of ABTS solution and 50 μL of the extract or water was mixed for 30 seconds and allowed to stay for 5 min at ambient temperature. After the absorbance was recorded at 734 nm using a UV-visible spectrophotometer (Shimadzu UV-160A, Kyoto, Japan) and compared with the control ABTS solution. A calibration curve was made by making various concentration of Trolox (780-1000 μL/mL). ABTS radical scavenging activity was expressed as μmol trolox equivalent antioxidant capacity (TEAC) per gram of sample.

Metal chelating potential

Ferrous ions chelating activity of extracts was estimated in which ginger bar extracts (0.1 mL) were added to a solution of 2 mM $FeCl_2$ (0.05 mL). The reaction was initiated by the addition of 5 mM ferrozine (0.1 mL) and 2.75 mL of distilled water. The mixture was shaken vigorously and left at room temperature for 10 min. The absorbance of the solution was then measured at 562 nm. The scavenging activity was calculated as follows:

$$MC\ (\%) = \frac{A_{blank} - A_{sample}}{A_{blank}} \times 100$$

Where,

A_{blank} = absorbance of the control reaction

A_{sample} = absorbance in the presence of plant extract

Samples were analyzed in triplicate.

Hedonic response

The resultant bars were evaluated by a trained panel of judges using 9 point hedonic scale as described by Parn et al. in 2015. Attributes to be tested on the products included various quality parameters such as that of aroma, taste, color, texture, overall acceptability, which were based on a nine point hedonic test scale.

Results and Discussion

Physico-chemical analysis of bars

The baked bars prepared by using conventional nutraceutical (3%) and nutraceutical supercritical (0.3% ginger extract) concentrations were analyzed for color tonality, texture and antioxidant potential to assess the impact of treatment as well as 60 days storage interval.

Color: The mark of consumer acceptance of food product is principally centered on color. The scrutiny of color tonality is mainly carried out with CIELAB (Commission International de l'Eclairage (CIE) *L∗* (lightness), *a∗* (redness), and *b∗* (yellowness)) color operating system that gives its interpretation as L*, a* and b* traits where L* displays brightness, a* points greenish to reddish tonality, whilst b* indicates bluish to yellowish color. Values concerning L* values of bars are represented in Figure 1. The L* values for bars T_0 (control), T_1 (bars containing 3% ginger nutraceutical$_{CSE}$) and T_2 (bars containing 0.3% ginger nutraceutical$_{SFE}$ extract) were 60.21 ± 2.05, 58.08 ± 1.97 and 59.42 ± 2.02, correspondingly. During sixty days storage the L* value of bars gradually decreased from 61.02 ± 2.07 to 56.80 ± 2.04. The Values (Figure 1) for a* value of T_2 was maximum 7.03 ± 0.24 followed by T_0 and T_1 of bars (6.87 ± 0.23 and 3.69 ± 0.23, respectively). During storage, maximum decrease in a* value was observed in T_1

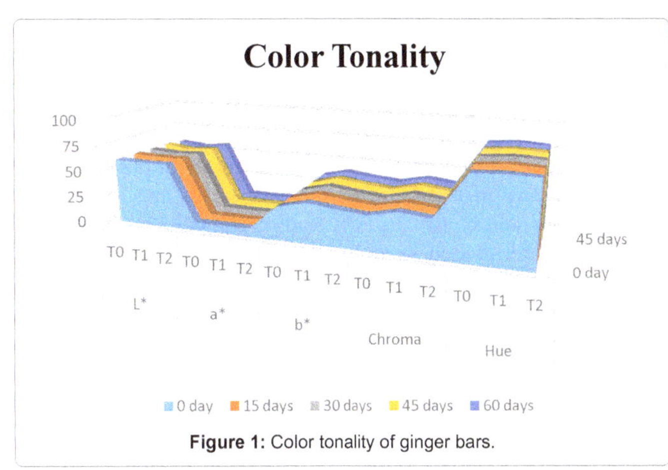

Figure 1: Color tonality of ginger bars.

ranging from 7.86 ± 0.27 to 5.77 ± 0.21. It was noticed that a* value of bars decreased gradually as a function of storage intervals from 7.75 ± 0.26 at 0 day to 6.15 ± 0.22 at 60th day. According to Figure 1 it is noticeable that b* value increased as color of product changes to yellowish with the passage of time. As it was observed from present study that b* value for T_1 was maximum 38.68 ± 1.32 due to yellowish color of ginger conventional extract while minimum for T_0 (32.42 ± 1.10). Collaborative effect of treatment and storage has exposed that highest b* value was recorded in T_1 (37.68 ± 1.28) at beginning that increased to (39.52 ± 1.42) at end of storage. However, 60 days storage has increased b* values from 34.65 ± 1.18 to 36.44 ± 1.31. Values in Figure 1 revealed that addition of ginger conventional extract in T_1 and ginger supercritical extract in T_2 bars produced significant change in chroma value *i.e.* 39.25 ± 1.34 and 36.97 ± 1.33, respectively however, for T_0 it was minimum 33.15 ± 1.13 that changed as function of time from 35.51 ± 1.21 to 36.96 ± 1.33. For hue angle, values shown in Figure 1 specified maximum value for T_1 (80.17 ± 2.73) followed by T_2 and T_0 (78.81 ± 2.68 and 78.02 ± 2.65, consistently). Values concerning hue angle reduced during storage from 77.34 ± 2.63 to 80.36 ± 2.89. So, we can interpret that with the increase in concentration of ginger conventional extract L* and a* values decreased while b*, chroma and hue was increased.

Color parameters: The results of current research work are comparable with the findings of Abdel-Samie et al. [14] observed the color parameters of ginger enriched cookies and suggested that L* value of cookies was 65.3 ± 0.6 for control cookies prepared from wheat flour and 65.9 ± 0.9 to 59.9 ± 0.2 by the addendum of various concentration of ginger. Similarly the a* was 8.9 ± 0.5 in control and changed to 7.1 ± 0.2 by the addition of ginger powder. In the case of b* values, it changed between 36.6 ± 0.2 to 37.5 ± 0.5 by the adjunct of ginger powder that was 38.6 ± 0.5 in control.

Furthermore, another group of scientists Ashoush and Gadallah, [15] prepared the biscuits by the augmentation of mango kernel peel and mango kernel powder along with control wheat flour biscuits. In their research work the color intensity of control wheat flour was 62.48 ± 1.90 for L* at start that decreased to 51.22 ± 0.27 at end of storage, 8.97 ± 0.93 for a* that decreased to 2.44 ± 0.31 while for b* value increased from 31.64 ± 0.39 to 33.69 ± 2.20 in the end of storage interval. The chroma intensity increased from 32.90 ± 0.43 to 34.04 ± 2.17 however for hue it changes from 70.62 ± 1.70 to 60.00 ± 0.39. Similarly, Haase et al. [7] concluded that by changing baking temperature from 180 to 240°C the L* values of baked products changes from 79.9 ± 4.27 to 72.7 ± 4.42.

Nonetheless, Sharma and Gujral, [13] who prepared wheat flour chapatties by the addition of barely flour. In their research work the L* of wheat flour chapatties decreased from 84.7 ± 0.3 to 80.4 ± 0.2 whilst the a* value decreased from 13.44 ± 0.53 to 12.41 ± 0.14 during storage. The L* value decreased due to the production of some melanoids that decrease L* value however the a* decreased because of the product of these intermediate maillard reaction compounds that moves it towards greenish shade and bluish shade instead of yellow in the case of b*. Although the decrease in L* and a* are due to the baking in which brown pigments produced during baking [16]. Moreover, Pasqalone et al. [17] prepared the wheat flour biscuits by the enrichment of grape mac and concluded that the L*, a* and b* values of biscuits was 41.14 ± 4.20, 9.74 ± 2.14 and 32.18 ± 0.40, respectively.

Texture: Figure 2 has represented the effect of treatment as well storage on hardness of bars. It was observed that hardness for T_1 (3% ginger conventional extract) was minimum 0.62 ± 0.02 in contrast to T_2 (0.3% ginger supercritical extract) 0.65 ± 0.02 and T_0 (control) 0.72 ± 0.02 kg force. Interactive effect of treatment as well as storage has depicted that hardness of bars containing 3% ginger conventional extract (T_1) was less effected from 0 to 60 days storage period as compared to supercritical treatment. Storage of bars caused significant decrease in hardness from 0.75 ± 0.03 to 0.56 ± 0.02 kg. It is clear from the Figure 2 that the bars became soft with the passage of time due to which the force gradually decreased from 1st day to 60th day of storage.

The outcomes were in line with the findings of Abdel-Samie et al. [14] that prepared ginger based cookies and evaluate texture of cookies as the force required breaking it. They depicted that the force required for control cookies was 4.7 ± 1.0 kg force whilst after the supplementation of ginger this force changed from 4.2 ± 0.1 to 3.4 ± 0 that was 4.7 ± 1.0 kg for control cookies.

Antioxidant activity of bars

Total phenolic content: In baked products the main problem is the rancidity that reduces the attention of people. So, some attempt was employed in present research work by incorporating ginger powder and its extract in bars having bioactive moieties of ginger that were examined for their antioxidant perspective during four days storage. It is cleared from means Table 2 TPC of bars that T_2 (0.3% ginger nutraceutical$_{SFE}$) has maximum phenolic contents 112.28 ± 3.81 mg GAE/100g as compared to T_1 (3% ginger nutraceutical$_{CSE}$) 87.12 ± 2.96 and T_0 (control) 67.45 ± 2.29 mg GAE/100g. During storage of bars, total phenolics gradually decreased from 92.54 ± 3.15 to 84.97 ± 3.06 mg GAE/100g while minimum reduction was noted in T_2 from 115.08 ± 3.91 at 0 day to 84.97 ± 3.06 mg GAE/100g at 60 days.

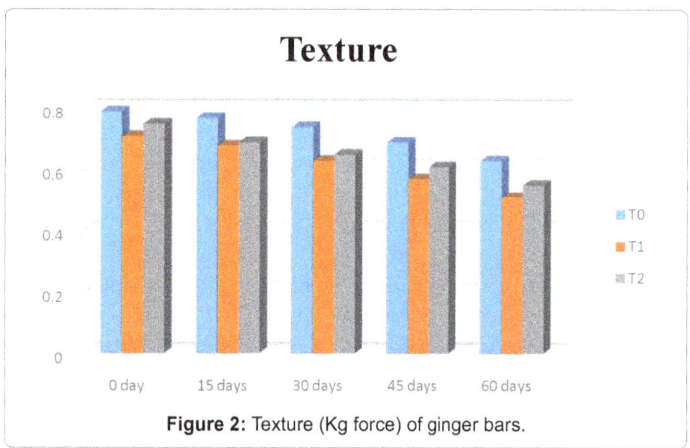

Figure 2: Texture (Kg force) of ginger bars.

Storage intervals (days)	Treatments			Means
	T_0	T_1	T_2	
0	71.23 ± 2.42	91.30 ± 3.10	115.08 ± 3.91	92.54 ± 3.15ᵃ
15	69.78 ± 2.23	88.94 ± 2.85	114.22 ± 3.66	90.98 ± 2.91ᵃᵇ
30	68.04 ± 2.04	86.53 ± 2.60	112.95 ± 3.39	89.17 ± 2.68ᵇ
45	65.70 ± 2.50	85.06 ± 3.23	110.54 ± 4.20	87.10 ± 3.31ᶜ
60	562.51 ± 2.25	85.06 ± 3.23	108.6 ± 3.91	84.97 ± 3.06ᵈ
Means	67.45 ± 2.29ᶜ	87.12 ± 2.96ᵇ	112.28 ± 3.81ᵃ	-

T_0= (control bars)
T_1= (bars containing 3% ginger CSE)
T_2= (bars containing 0.3% ginger SFE)

Table 2: Effect of treatments and storage on TPC (mg GAE/100 g) of bars.

Storage intervals (days)	Treatments			Means
	T_0	T_1	T_2	
0	7.61 ± 0.26	15.92 ± 0.54	29.38 ± 1.00	17.64 ± 0.60ᶜ
15	7.98 ± 0.26	16.19 ± 0.52	30.57 ± 0.98	18.25 ± 0.58ᵇ
30	8.24 ± 0.25	16.44 ± 0.49	30.86 ± 0.93	18.51 ± 0.56ᵇ
45	8.59 ± 0.33	17.70 ± 0.67	31.21 ± 1.19	19.17 ± 0.73ᵃ
60	8.98 ± 0.32	18.13 ± 0.65	31.56 ± 1.14	19.56 ± 0.70ᵃ
Means	8.28 ± 0.28ᶜ	16.88 ± 0.58ᵇ	30.72 ± 1.05ᵃ	-

T_0= (control bars)
T_1= (bars containing 3% ginger CSE)
T_2= (bars containing 0.3% ginger SFE)

Table 3: Effect of treatments and storage on DPPH (%) of bars.

Storage intervals (days)	Treatments			Means
	T_0	T_1	T_2	
0	12.39 ± 0.42	19.52 ± 0.66	32.43 ± 1.10	21.45 ± 0.73ᶜ
15	12.81 ± 0.41	20.35 ± 0.65	32.97 ± 1.06	22.04 ± 0.71ᵇ
30	13.28 ± 0.40	20.93 ± 0.63	33.65 ± 1.01	22.62 ± 0.68ᵇ
45	13.76 ± 0.52	21.48 ± 0.82	34.12 ± 1.30	23.12 ± 0.88ᵃ
60	14.1 ± 0.51	21.73 ± 0.78	34.86 ± 1.25	23.56 ± 0.85ᵃ
Means	13.27 ± 0.45ᶜ	20.80 ± 0.71ᵇ	33.61 ± 1.14ᵃ	-

T_0= (control bars)
T_1= (bars containing 3% ginger CSE)
T_2= (bars containing 0.3% ginger SFE)

Table 4: Effect of treatments and storage on antioxidant activity (%) of bars.

DPPH: It is mostly used to assess the antioxidant potential that valued the antioxidant indices through free radical scavenging. Means for DPPH Table 3 demonstrated that free radical scavenging activity of T_2 was maximum 30.72 ± 1.05% followed by T_1 and T_0 16.88 ± 0.58 and 8.28 ± 0.28% respectively. Throughout storage interval the DPPH assay increased from 17.64 ± 0.60 to 19.56 ± 0.70% whilst, maximum increase was observed in bars prepared by ginger supercritical extract ranging from 29.38 ± 1.00 to 31.56 ± 1.14%.

Antioxidant activity (AA): Mean antioxidant potential Table 4 regarding three treatments *i.e.* T_0 (control), T_1 (3% conventional ginger extract) and T_2 (0.3% supercritical ginger extract) has revealed maximum activity (33.61 ± 1.14%) was observed in T_2 followed by T_1 (20.80 ± 0.71) and T_0 (13.27 ± 0.45%). In the same way, storage factor has also influenced ß- carotene bleaching rate of each treatment that was highest at 60th day 23.56 ± 0.85% and lowest on first day *i.e.* 21.45 ± 0.73%.

Ferric Reducing Antioxidant Potential (FRAP)

Results Table 5 have illustrated that T_2 extract has maximum ferric reducing power 48.81 ± 1.66 that was low in T_1 35.60 ± 1.21 μmole trolox equivalents/g ginger bar and 22.15 ± 0.75 in control bars. In the same way, significant effect was noted in storage time factor for each treatment that was higher 36.34 ± 1.24 at 0 day while lower 34.64 ± 1.25 μmole trolox equivalents/g bar at 60th day.

ABTS assay

From means Table 6, it was observed that maximum ABTS value was recorded in T_2 19.05 ± 0.65 followed by T_1 11.29 ± 0.38 and lowest was in control 5.94 ± 0.20 μmol trolox equivalents/g. Furthermore, it was also predicted as function of storage duration that maximum ABTS value 12.51 ± 0.43 μmol trolox equivalents/g was measured for ginger bar extract at 0 day while at 60th day it was lowest 11.65 ± 0.42 μmol trolox equivalents/g.

Metal chelating potential

Values for effect of solvent and time Table 7 have shown highest chelating potential in T_2 21.22 ± 0.72% followed by 17.88 ± 0.61% in T_1 and 16.41 ± 0.56% in T_0. Storage time also effected chelating potential as maximum amount 19.36 ± 0.70% was observed at 60th day while minimum 17.66 ± 0.60% at 0 day.

Antioxidant potential

The results of current research were in harmony with the findings of Abdel-Samie [14] with his colleagues evaluated the effect of ginger as antioxidant on the dough mixing properties and quality of cookies and concluded that the total phenolic content of control cookies that were prepared by wheat flour alone were 78.5 ± 1.1 mg GAE/100 g of cookies that increased from 90.8 ± 0.8 to 109.8 ± 2.7 mg GAE/100 g of cookies by the gradually supplementation of ginger. Similarly, the antioxidant assay of ginger based cookies increased from 45.8 ± 1.8 to 64.6 ± 1.0% by increasing the concentration of ginger that was 41.0 ± 0.6% in control cookies.

Storage intervals (days)	Treatments			Means
	T_0	T_1	T_2	
0	23.05 ± 0.78	36.20 ± 1.23	49.76 ± 1.69	36.34 ± 1.24^c
15	22.76 ± 0.73	36.04 ± 1.15	49.24 ± 1.58	36.01 ± 1.15^c
30	22.18 ± 0.67	35.73 ± 1.07	48.83 ± 1.46	35.58 ± 1.07^b
45	21.63 ± 0.82	35.19 ± 1.34	48.31 ± 1.84	35.04 ± 1.33^b
60	21.14 ± 0.76	34.86 ± 1.25	47.92 ± 1.73	34.64 ± 1.25^a
Means	22.15 ± 0.75^c	35.60 ± 1.21^b	48.81 ± 1.66^a	-

T_0= (control bars)
T_1= (bars containing 3% ginger CSE)
T_2= (bars containing 0.3% ginger SFE)

Table 5: Effect of treatments and storage on FRAP (μmole TE/g) of bars.

Storage intervals (days)	Treatments			Means
	T_0	T_1	T_2	
0	6.37 ± 0.22	11.72 ± 0.40	19.43 ± 0.66	12.51 ± 0.43^b
15	6.14 ± 0.20	11.59 ± 0.37	19.26 ± 0.62	12.33 ± 0.39^b
30	5.96 ± 0.18	11.26 ± 0.34	19.07 ± 0.57	12.10 ± 0.36^b
45	5.72 ± 0.22	11.03 ± 0.42	18.85 ± 0.72	11.87 ± 0.45^a
60	5.49 ± 0.20	10.85 ± 0.39	18.62 ± 0.67	11.65 ± 0.42^a
Means	5.94 ± 0.20^c	11.29 ± 0.38^b	19.05 ± 0.65^a	-

T_0= (control bars)
T_1= (bars containing 3% ginger CSE)
T_2= (bars containing 0.3% ginger SFE)

Table 6: Effect of treatments and storage on ABTS (μmole TE/g) of bars.

Storage intervals (days)	Treatments			Means
	T_0	T_1	T_2	
0	15.37 ± 0.52	17.06 ± 0.58	20.54 ± 0.70	17.66 ± 0.60^c
15	15.94 ± 0.51	17.53 ± 0.56	20.8 ± 0.67	18.09 ± 0.58^b
30	16.45 ± 0.49	17.92 ± 0.54	21.18 ± 0.64	18.52 ± 0.56^b
45	16.82 ± 0.64	18.28 ± 0.69	21.64 ± 0.82	18.91 ± 0.72^{ab}
60	17.49 ± 0.63	18.61 ± 0.67	21.97 ± 0.79	19.36 ± 0.70^a
Means	16.41 ± 0.56^c	17.88 ± 0.61^b	21.23 ± 0.72^a	-

T_0= (control bars)
T_1= (bars containing 3% ginger CSE)
T_2= (bars containing 0.3% ginger SFE)

Table 7: Effect of treatments and storage on metal chelating potential (%) of bars.

Furthermore, Ashoush and Gadallah [15] prepared wheat flour biscuits ad concluded that the total phenolic contents on wheat flour were 1.59 ± 0.05 mg GAE/g of wheat flour biscuit that increased to 7.08 ± 0.07 mg GAE/g by the addition of mango kernel powder as well as the DPPH assay of control wheat flour biscuits were 26.13 ± 0.05% that increased to 91.57 ± 0.11% by the enrichment of mango kernel powder. At the same moment, Zhu et al. [18] assessed the antioxidant potential of defatted wheat germ and resulted that the total phenolic content in wheat germ was 14.63 ± 0.04 mg GAE/g and DPPH assay was 75%. For ABTS radical scavenging the value was 9.37 ± 0.05 mg/mL as IC_{50} β carotene based antioxidant activity was 35.90% and for metal chelating potential the value of wheat germ was 25.7%.

Additionally, an alternative group of researchers, Haase et al. [7] evaluated the ABTS and FRAP assay of wheat flour biscuits and clinched that after baking the ABTS assay of wheat flour based biscuits were 7.12 ± 2.06 to 7.68 ± 1.91 mmol TE/kg wheat flour though for FRAP assay the value was 3.10 ± 0.98 to 3.84 ± 1.01 mmol TE/kg wheat flour that varied by changing the baking temperature from 210°C to 240°C. Moreover, Ahmad et al. [19] who prepared tiger nut enriched biscuits and assessed for nutritional and sensory aspects. They concluded that control biscuits without tiger nut supplementation have the total phenolic content of 2.11 mg/g of wheat flour and DPPH assay of 6.51% that gradually increased by the supplementation of tiger nut flour.

Another group of scientist Sharma and Gujral [13] prepared the wheat chapatties by the addition of barely flour and concluded that total phenolic content of wheat flour based chapatties were 2062 ± 36 μg/g in flour which increased by the incorporation of barely flour but decreased during baking (2016 ± 22 μg/g of chapatti) due to the decomposition of molecules at higher temperature beyond to 80°C. Similarly, the antioxidant of wheat chapatties was 16.1 ± 1.1% in the start that increased up to 30.6 ± 0.40% during storage. During the processing of baking the antioxidant activity of baked products increased as compared to flour due to the maillard reaction that takes place in the availability of sugars and proteins. Some dark compounds normally brown colored are produced due to the thermal processing of baked products. These melanoidins (brown pigments) are briefly known to possess antioxidant properties [14]. Likewise, they determined the metal chelating power of wheat chapatties that was 27.4 ± 0.5% and increased to 30.9 ± 1.0% during baking and storage. In the meanwhile, they observed reducing power of wheat flour chapatties that was 29.1 ± 1.2 μmole ascorbic acid at the start and decreased slightly during storage.

Recently, Parn et al. [12] evaluated the antioxidant potential of wheat based fruit bars by utilizing date paste and concluded that the total phenolic content of bar ranges in 240.33 ± 6.35 to 224.33 ± 1.15 mg GAE/100 g although, the DPPH scavenging varied from 30.69 ± 1.06 to 32.75 ± 0.46.

Sensory evaluation of bars

For sensory evaluation, bars were ranked using 9 point hedonic scale for their color, flavor, taste, crispiness and overall acceptability. Color being the most important character is the key of success of any product. If color does not affect then consumer would not like to even taste it. Means color marks for outcome of treatment has Figure 3 elucidated non-significant effect on color of bars; maximum 7.27 ± 0.25 were assigned to T_1 (3% ginger extract) followed by T_0 (7.26 ± 0.25) while minimum 7.23 ± 0.25 to T_2 (0.3% ginger nutraceutical$_{SFE}$). Color scores for bars significantly decreased as a function of storage from 7.43 ± 0.25 to 7.07 ± 0.25 during sixty days. It is obvious from the Figure 3 that the color was approximately same for all the treatments and hormonally the score of color decreased with time. Flavor is one of the characteristics which make product liked or disliked by the consumers. The flavor showed various results for treatments as well as storage intervals. It is obvious from the Figure 3 that T_2 got higher marks for flavor 7.33 ± 0.25 in contrast to T_1 (7.05 ± 0.24) and T_0 (6.66 ± 0.23) as ginger extracts in T_1 and T_2 caused pleasant flavor in bars. Figure 3 showed that the flavor of supercritical extract based bars got maximum marks among all the treatments followed by conventional extract based and control bars. Similarly, storage study has revealed that flavor of bars also changed significantly ranging from 7.27 ± 0.25 to 6.75 ± 0.24. Values for taste (Figure 3) showed that maximum score for taste was assigned to T_2 (7.29 ± 0.25) while minimum to T_0 (6.97 ± 0.24). Likewise, storage also decreased the taste marks from 7.38 ± 0.25 to 6.95 ± 0.25. Crispiness specifies the crusty expertise of the food products. Same as the flavor and taste of nutraceutical$_{SFE}$ extract based bars was best from all three treatments as depicted by Figure 3. For bars crispiness Figure 4, maximum scores 7.41 ± 0.25 was noted for T_2 while minimum for T_0 and T_1 7.25 ± 0.25 and 7.15 ± 0.24, correspondingly. Storage intervals also showed significant reduction from 7.42 ± 0.25 to 6.98 ± 0.25 in bars crispiness. Figure 4 proved that the control bars have good crispiness as compared to nutraceutical based bars. In view of the overall acceptability (Figure 4), T_2 was considered best with allocated marks 7.34 ± 0.25, whereas T_0 at the lower level with marks 7.04 ± 0.24. Overall acceptability also decreased with time from 7.47 ± 0.25 to 6.98 ± 0.25 during sixty days storage of bars however remained highest for T_2. Making an allowance for hedonic scale response, Figure 4 concluded that the bars containing 0.3% ginger supercritical extract were rated higher marks.

The results of current research work were in accordance to the finding of Abdel-Samie et al. [14] who observed the sensory profile of ginger based cookies. They concluded that the appearance color score of control cookies was 8.0 ± 1.21 that changed from 8.0 ± 0.9 to 7.5 ± 1.1 by the addition of ginger. Similarly the texture of ginger based cookies was marked as 7.3 ± 1.3 to 6.9 ± 1.1 however it was 7.4 ± 1.7

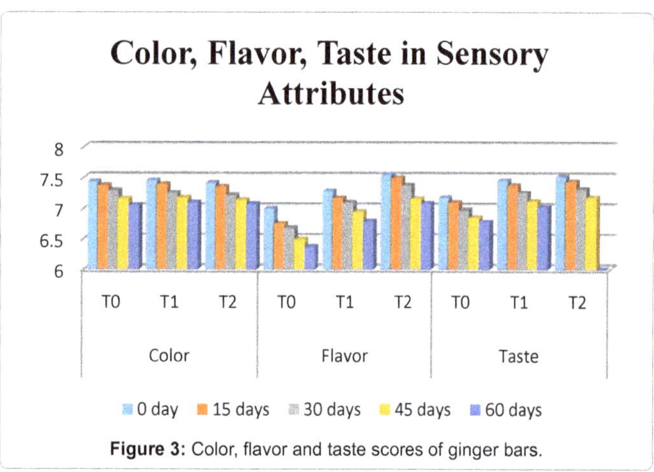

Figure 3: Color, flavor and taste scores of ginger bars.

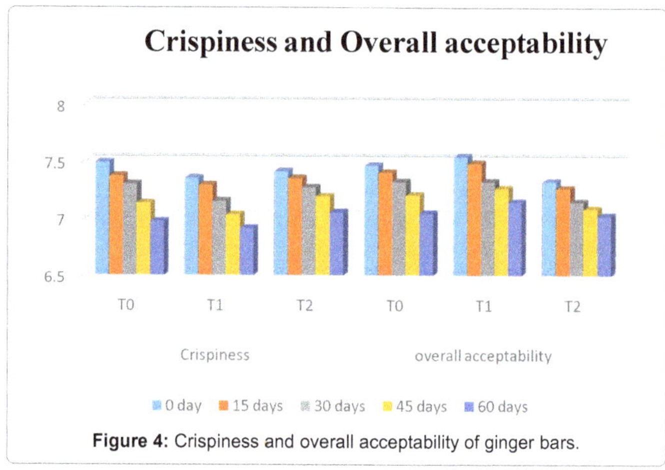

Figure 4: Crispiness and overall acceptability of ginger bars.

for control cookies. Furthermore, for the flavor grades were 7.5 ± 1.4 for control and 7.0 ± 1.7 to 6.4 ± 1.4 after the augmentation of ginger powder. Nonetheless, for overall acceptability best count was 8.1 ± 0.8 to 7.0 ± 1.5 for ginger cookies that was 7.4 ± 1.4 for control cookies.

Similarly, Oluwamukomi et al. [20] prepared wheat-cassava composite biscuits by the addition of soy flour. In their research work they concluded that the crispiness of wheat flour biscuits was marked as 8.0 while taste, aroma, shape, color and overall acceptability was scored as 8.0, 7.5, 6.0, 7.0 and 8.5 correspondingly.

Conclusion

During research work, two types of nutraceutical bars were prepared after supplementing with ginger enriched fractions against control. In the case of bars the T_1 contained 3% ginger nutraceutical$_{CSE}$ and T_2 supplemented with 0.3% ginger supercritical extract. The treatments and storage exhibited significant variations in color tonality and texture that decreased from 61.02 ± 2.07 at 0 day to 56.80 ± 2.04 at 60th day for L*, 7.75 ± 0.26 to 6.15 ± 0.22 for a* however the values of b*, chroma and hue increased during storage from 34.65 ± 1.18 to 36.44 ± 1.31 for b*, 35.51 ± 1.21 to 36.96 ± 1.33 for chroma and for hue the value was 77.34 ± 2.63 to 80.36 ± 2.89. Among antioxidant perspectives, T_2 Showed maximum values for all tests such as TPC (112.28 ± 3.81 mg GAE/100g), DPPH (30.72 ± 1.05%), antioxidant activity (33.61 ± 1.14), FRAP (48.81 ± 1.66 μ mole TE/g), ABTS (19.05 ± 0.65 μ mole TE/g) and metal chelating (21.22 ± 0.72%). T_2 was followed by T_1 with values 87.12 ± 2.96 mg GAE/100g TPC, 16.88 ± 0.58% DPPH, 20.80 ± 0.71% antioxidant activity, 35.60 ± 1.21 μ mole TE/g FRAP, 11.29 ± 0.38 μ mole TE/g ABTS and 11.29 ± 0.38% for metal chelating. During storage the antioxidant potential decreased from 92.54 ± 3.15 to 84.97 ± 3.06 mg GAE/100 g in TPC, 36.34 ± 1.24 to 34.64 ± 1.25 μ mole TE/g for FRAP and 12.51 ± 0.43 to 11.65 ± 0. μ mole TE/g for ABTS although, in DPPH it increased from 17.64 ± 0.60 to 19.56 ± 0.70%, 21.45 ± 0.73 to 23.56 ± 0.86% antioxidant activity and metal chelating potential increased from 17.66 ± 0.60 to 19.36 ± 0.70 ± . Hedonic response was also assessed using 9-point hedonic scale for the estimation of color, flavor, crispiness, taste and overall acceptability of ginger bars. The maximum scores for color was 7.27 ± 0.25 (T_1), 7.33 ± 0.28 (T_2) for flavor, 7.29 ± 0.25 for taste (T_2), 7.25 ± 0.25 for control and 7.34 ± 0.25 (T_2) for overall acceptability.

Acknowledgement

This work was carried out under Pak-US Science and Technology project for establishment of Functional and Nutraceutical Research Section at University of

Agriculture, Faisalabad-Pakistan. The financial and technical assistance under the framework of this project is highly acknowledged.

References

1. Martirosyan DM (2011) Introduction to functional food science In: Functional foods and chronic diseases. Danik M, Martirosyan (eds), Food Science Publisher: Richardson Texas USA. Pp: 173-204.
2. Fischer UA, Dettmann, JS, Carle R, Kammerer DR (2011) Impact of processing and storage on the phenolic profiles and contents of pomegranate (*Punica granatum L.*) juices. Europ Food Res Technol 233: 797-816.
3. Jensen S, Oestdal H, Clausen MR, Andersen ML, Skibsted LH (2011) Oxidative stability of whole wheat bread during storage. LWT Food Sci Technol 44: 637-642.
4. Patil S, Valdramidis VP, Tiwari BK, Cullen PJ, Bourke P (2011) Quantitative assessment of the shelf life of ozonated juice. Europ. Food Res Technol: 469-477.
5. Zielinski H, Del Castillo MD, Przygodzka M, Ciesarova Z (2012) Changes in chemical composition and antioxidative properties of rye ginger cakes during their shelf life. Food Chem 135: 2965-2973.
6. Tuncel NB, Yilmaz N, Kocabiyik H, Uygur A (2014a) The effect of infrared stabilized rice bran substitution on physicochemical and sensory properties of pan bread: part I. J Cereal Sci 59: 155-161.
7. Haase NU, Grothe KH, Mattaus B, Vosmann K, Lindhauer MG (2012) Acrylamide formation and antioxidant level in biscuits related to recipe and baking. Food Add Contam A 29:1230-1238.
8. Ibrahim UK, Salleh RM, Maqsood-ul-haque SNS (2015) Bread towards functional food: An overview. Int J Food Eng 1: 39-43.
9. Sivam AS, Sun-waterhouse D, Siew YQ, Perera CO (2010) Properties of bread dough with added fiber polysaccharides and phenolic antioxidants: A review. J Food Sci 75: 163-174.
10. Malipatili NB, Manjunath S, Shruthi DP (2015) Evaluation of effect of aqueous extract of Zingiber officinale Roscoe (ginger) on acute and chronic inflammation in adult albino rats. Asian J Pharma Clin Res 8: 113-116.
11. Embuscado ME (2015) Spices and herbs: natural sources of antioxidants-a mini review. J Func Foods.
12. Parn OJ, Bhat R, Yeoh TK, Hassan AA (2015) Development of novel fruit bars by utilizing date paste. Food Biosci 9: 20-27.
13. Sharma P, Gujral HS (2014) Antioxidant potential of wheat flour chapattis as affected by incorporation of barely flour. LWT- Food Sci Technol 56: 118-123.
14. Abdel-samie MAS, Wan J, Haung W, Chung OK, Xu B, et al. (2014) Effects of cumin and ginger as antioxidants on dough mixing properties and cookie quality. Cereal Chem 87: 454-460.
15. Ashoush IS, Gadallah MGE (2011) Utilization of mango peels and seed kernels powders as source of phytochemicals in biscuits. World J Dairy Food Sci 6: 35-42.
16. Sharma P, Gujral HS, Singh B (2012) Antioxidant activity of barley as affected by extrusion cooking. Food Chem 131: 1406-1413.
17. Pasqalone A, Bianco AM, Paradiso VM, Summo C, Gambactora G, et al. (2014) Physico-chemical sensory and volatile profiles of biscuits enriched with grape marc extract. Food Res Int 65: 385-393.
18. Zhu KX, Lian CX, Guo XN, Peng W, Zhou HM, et al. (2011) Antioxidant activities and total phenolic contents of various extracts from defatted wheat germ. Food Chem 126: 1122-1126.
19. Ahmad ZS, Abozed SS, Nagum MS (2014) Nutritional value and sensory profile of gluten free tiger nut enriched biscuits. World J Dairy Food Sci 9: 127-134.
20. Oluwamukomi MO, Oluwalana IB, Akinbowale OF (2011) Physicochemical and sensory properties of wheat-cassava composite biscuit enriched with soy flour. African J Food Sci 5: 50-56.

Spent Brewer's Yeast Autolysates as a New and Valuable Component of Functional Food and Dietary Supplements

Podpora B, Świderski F, Sadowska A*, Piotrowska A and Rakowska R

Department of Functional and Organic Food and Commodities, Faculty of Human Nutrition and Consumer Sciences, Warsaw University of Life Sciences, Warsaw, Poland

Abstract

The aim of the work was to obtain autolysates derived from spent brewer's yeast and demonstrate their potential capabilities as a natural and valuable ingredients intended for functional food and dietary supplements production. The research material consisted of yeast *Saccharomyces cerevisiae* which was the remains after the beer production process. In these autolysates the following analyses were performed - protein, dry matter and amino acids content, determination of molecular weight of proteins, antioxidant properties and sensory quality. It was found that the tested autolysates were characterised by a high content of essential amino acids, exciding the amount present in the reference protein developed by FAO/WHO. The sensory quality of the tested autolysates was dependent on the degree of autolysis process of the yeast protein which further determined their usage in a food production. Precise time control of the autolysis process allows obtaining a product with the designed functional properties, characterised by the desired content of free amino acids, peptides with a specific molecular weight and high antioxidant properties. Based on the obtained results it can be concluded that the post-fermentation yeast may be a valuable raw material for the preparation of yeast autolysates which may be new and valuable component in functional food and dietary supplements production.

Keywords: Spent brewer's yeast; Yeast's autolysates; Functional food; Supplements

Introduction

Brewer's yeast is increasingly finding application in the food industry, due to its high functional properties. Most post-fermentation brewer's yeast, which is a waste product, was used as a source of B vitamins, protein and minerals in animal feed production [1]. Nowadays, more and more companies are involved in the processing of post-fermentation yeast products with bioactive properties to be used in the production of functional food. Some bioactive compounds, such as β-glucans, monooligosaccharides, significant amounts of B vitamins, minerals and also yeast extracts can be obtained from the post-fermentation yeast. Yeast extracts are more widely used as a natural flavouring preparation of meat profile which replaces protein hydrolysates obtained by acid hydrolysis. Yeast extracts are produced by breaking the cell walls with the use of endogenous or exogenous enzymes [2-4]. The methods of the manufacturing preparation of extracts are divided into hydrolysis and autolysis. Hydrolysis is the most efficient method of yeast solubilisation. Yeast extracts prepared by the use of acids or the addition of enzymes is known as 'yeast hydrolysates'. Despite a high production yield, acid hydrolysis is less attractive to the manufacturers because of relatively high salt content and high probability of carcinogenic compounds content, such as monochloropropanol and dichloropropanol [5]. Autolysis with using the endogenous enzymes occurs in yeast when the cell growth cycle is completed and the death phase is initiated [3,6]. Intracellular enzymes are activated by appropriate process conditions, such as temperature and time, which results in a partial degradation of the cell wall structures. This allows extracting valuable proteins, carbohydrates and vitamins from the cells, preserving their native structure. Autolysates are the purest product group, as hydrolysates contain large amounts of salt or sodium used during plasmolysis process or to neutralise the acids used in the hydrolysis process, respectively [7,8]. Currently, yeast autolysates are used in the fermentation industry, especially as starting materials and in the food industry mainly as flavour enhancers [9-11].

In the literature, there are only few reports on the possibility of the application of the post-fermentation yeast for the production of yeast extracts, which are characterised by the preferred nutritional properties, that may be used in functional food and dietary supplements production. The aim of this study was to investigate the possibility of using post-fermentation yeast in order to obtain yeast autolysates with the properties valuable from a nutritional point of view that can be used as a natural ingredient in functional food and dietary supplements production.

Material and Methods

Material

Material consisted of yeast autolysates obtained from the fermented yeast slurry of *Saccharomyces cerevisiae*, which is a byproduct of the brewing industry. The slurry was subjected to a filtration process (to remove physical impurities) and centrifugation (to remove beer residual). The feedstock was then subjected to a standardisation process to obtain the following parameters: dry matter content: 15, 5-16, 5%, protein content: min. 42%, pH (5,2-6,2), suitable flocculation and content of viable yeast cells. Selected yeast slurry in an amount of 3000 g was transferred to the autolytic reactor placed in an oil bath filled with polyethylene glycol. The slurry was subjected to pre-incubation at 38°C for 4 hours to completely fermentate sugars remaining in the yeast slurry. Autolysis was carried out at 47°C during 48 h. After starting the reactor, the stirrer was working during the whole process. During the

***Corresonding author:** Anna Sadowska, Department of Functional and Organic Food and Commodities, Faculty of Human Nutrition and Consumer Sciences, Warsaw University of Life Sciences, Nowoursynowska, Warsaw, Poland
Email: anna_sadowska@sggw.pl

trial the samples were taken for testing at specified intervals. In order to inactivate the lytic enzymes obtained samples were heated at 95°C for 30 min. Tested yeast autolysates were in liquid form, they were not subjected to compaction nor drying process. The protein and dry matter content in autolysates used for the study was respectively 7.53 ± 0.34% and14.45 ± 0.03%.

Methods

Chemical composition of autolysates was conducted referring to the method of AOAC [12]. In the autolysates the dry matter content was determined using the gravimetric method. The content of the nitrogen was analysed by the Kjeldahl method using a conversion factor of 6.25 for nitrogen quantities of protein content.

Determination of amino acids content in tested extracts was followed by acid hydrolysis in order to carry out the peptide bonds hydrolysis until they reach the free amino acids. After hydrolysis the derivatization using dansyl chloride was done, and then amino acids were separated and determined using high performance liquid chromatography UV/VIS [13]. For determination of tryptophan alkaline hydrolysis was used. The tryptophan was determined by using HPLC with fluorescence detection [14]. On the basis of the amino acid composition Chemical Score (CS) [15] and Essential Amino Acid Index (EAAI) were calculated [16]. As the pattern of the protein amino acid composition the standards of whole egg and these developed by FAO/WHO [17] were adopted.

Determination of molecular weights of proteins by mass spectrometry was performed using a mass spectrometer equipped with a MALDI ion source type. For the analysis, the mass spectrometer MALDI-TOF Reflex IV (Bruker-Daltonics, Bremen, Germany) was used. The analysis was performed at a linear mode operation mass spectrometer in the range 500-5500 m/z, 2000-19000 m/z 10000-105000 m/z. As a matrix the sinapic acid (SA) and the α-cyano-4-hydroxycinnamic acid (HCCA) (dissolved in a solution of 2:1 acetonitrile in water to saturate the solution) were used. Each sample was thoroughly mixed and centrifuged (10000 RCF, 10 min, 4°C). Then, the supernatant was diluted five times with 0.1% TFA. On the plate 0.5 ml of the prepared sample with 0.5 mL of the matrix was loaded and then the plate was allowed to dry. After drying, the sample was introduced into the mass spectrometer ion source and ionized. There was 1 μg of sample used for a single analysis. The nominal sensitivity of the spectrometer for the peptides of 1000-5000 m/z was 1 fmol.

Determination of peptides with a specific molecular weight range using LC-MS was performed using high performance liquid chromatography coupled with a mass detector LTQ Thermo Finnigan. Each sample was thoroughly mixed, collected in tubes of Falcone type in an amount of 10 ml and then centrifuged (15 min, 4000 RCF, 4°C). To clean tubes 5 ml of the supernatant was collected and was acidified with 10% TFA to pH<3. Thus prepared samples were loaded on a C18 SPE column (200 mg, Merck), after activation the bed (3 mL 0.1% TFA in 80% acetonitrile, 3 ml 0.1% TFA). Then the column was washed with 3 mL of 0.1% TFA, dried and eluted with 3 ml of 0.1% TFA in 40% acetonitrile. The eluent was freeze-dried for over 12h, the dry residue was taken up in 100 μl of 0.1% formic acid and injected on LC-MS. Spectra MS-MS was performed based on the calculation program OMSSA using the Swissprot database, for peptides derived from proteins of yeast (*Saccharomyces cerevisiae*) without the specified proteases.

The antioxidant properties were determined by spectrophotometric method, "in vitro", using the synthetic ABTS radicals according to Pellegrini et al. [18] method. The content of polyphenols was determined by the method of Singleton and Rossi [19], using the Folin-Ciocolteu reagent.

Detailed sensory characteristics of extracts samples was performed by a quantitative descriptive analysis (Quantitative Descriptive Analysis - QDA), using the analytical procedure described in BS EN ISO 13299:2010 [20] 24 quality parameters were selected to analyse the profile of the extracts. The sensory evaluation of the samples was conducted by 8-person team of evaluators who are qualified assessors - experts, with theoretical and practical preparation in the field of sensory methods. The sensory evaluation was performed in two independent replications. Ratings were performed at the Laboratory of Sensory Analysis, which meets all the requirements specified in BS EN ISO 8589:2010 standard [21]. For planning sessions with ratings scaling method, generation of random numbers for coding samples, records of individual results and their pre-treatment a computerised support system of sensory analysis ANALSENS NT was used.

Statistical analysis of the results was performed using STATISTICA 10 software. The significance of differences in chemical research of yeast aytolysates was verified using one-way analysis of variance (ANOVA). In order to study the differences between groups, a Duncan test (α = 0.05) was used. The graphic projections of distribution of tested aytolysates in the coordinate system formed by two main factors were prepared in the ANALSENS based on the covariance matrices.

Results and Discussion

The sum of amino acid of yeast autolysates obtained by varying autolysis time is shown in Table 1. Along with prolonged autolysis time, there was an increase of free amino acids - it reached a level of 11.2% after 2h and 77.5% after 48 hours. The obtained results of amino acid content of yeast autolysates obtained in the present work is different from the amino acid composition of pure brewer's yeast, *Saccharomyces cerevisiae.* Brewer's yeast contain substantially lower amount of glutamic acid, proline and arginine [22,23]. This is due to the fact that the autolysates investigated in the present work were not prepared from pure brewer's yeast but from their suspension in the wort remaining in the brewing process, which is rich in proteins, peptides and free amino acids derived from barley malt and other additives used in beer production. Consequently, the amino acid profile of the obtained autolysates is much more developed and is characterised by a much higher concentration of amino acids that are normally present in yeast cells in minor amounts. The amino acid profile in such case is dependent on the wort composition [24,25].

On the basis of the obtained data the amino acid content in autolysates was converted to gram of protein and the results were compared to a WHO/FAO [17] standard (Table 2). Among egzogenous amino acids there was a large number of phenylalanine, threonine, tryptophan and valine. These amino acids are not typical for the

	AUTOLYSATES							
	2h	6h	8h	10h	12h	16h	20h	48h
Sum of the free amino acids [g/100 g of dry weight]	4.69	8.44	9.29	9.77	10.17	12.45	16.06	32.38
Percentage of free amino acids [%]	11.2	20.2	22.2	23.4	24.3	29.8	38.4	77.5

Table 1: The amino acid composition of yeast autolysates after the autolysis process, per g/100 g of dry weight.

Amino Acids	Autolysate	Reference Protein According To Fao/Who	CS**)
	mg/g of protein	mg/g of protein	%
Isoleucine (EG.)*)	23.7	28	84.6
Leucine + Norleucine (EG.)	65.1	66	98.7
Lysine (EG.)	47.2	58	81.4
Methionine (EG.) + Cysteine	23.8	25	92.0
Phenylalanine (EG.) + Tyrosine	90.6	63	143.8
Threonine (EG.)	61.3	34	180.3
Tryptophan (EG.)	11.9	11	108.2
Valine (EG.)	69.7	35	199.2
Sum of exogenous amino acids	393.3	320	-
EAAI***)	117	100	-

*) EG. - essential amino acids
**) CS - chemical score
***) EAAI - essential amino acids index

Table 2: Comparison of the content of essential amino acids in the yeast's autolysate after 48 h of autolysis in relation to the content of amino acids in protein standard established by the FAO/WHO [17].

yeast amino acids but more particularly for plant proteins, especially for barley's proteins [26]. It was found that the limiting amino acids in the obtained autolysates was primarily lysine and in lesser extent isoleucine, where the rate of limiting amino acid content to a reference protein developed by FAO/WHO is 81.4% for lysine and 84.6% for isoleucine. The data indicate that the obtained autolysates are very valuable source of series of amino acids, such as tryptophan, valine, threonine and phenylalanine+tyrosine. The content of these amino acids, as compared to the reference protein developed by FAO/WHO, was significantly higher, ranging from 143.8% for valine + tyrosine to 199.2% for valine. It is worth emphasizing that the tested autolysates were characterised by a high sum of the exogenous amino acids of about 303.3 mg/g of protein. This value is much higher than the total amino acids present in the reference protein developed by FAO/WHO. The integrated ratio of essential amino acids (EAAI), as an indicator of the maximum potential nutritional value of the protein [27] reached 117% for the tested autolysates in relation to a reference protein by FAO/WHO. It can be concluded that the autolysates obtained from post-fermentation yeast are a valuable source of amino acids, which is important from a nutritional point of view. These amino acids come not only from yeast but also from the components used for the beer production (e.g., malt and hops). A high total content of amino acids in yeast autolysates, superior to that present in the reference protein developed by the FAO/WHO, indicates that these autolysates can be a valuable component of several products from the group of functional food and dietary supplements. The obtained autolysates are characterised by a different degree of hydrolysis, resulting in obtaining a product with a high and low content of free amino acids which allows expanding their use depending upon of the desired functional characteristics of the proposed application.

MALDI-TOF-MS analysis of yeast autolysates showed the presence of small amounts of the compounds with masses in the range of 1000-6000 Da. The peaks were characterised by a small intensity and mainly limited to the range of 1000-2500 m/z (Figure 1). The most intense peak in all the investigated autolysates was found in the vicinity of 2414 m/z. In all the autolysates there were no peaks in the range of 10000-15000 m/z or substances of high molecular weight (up to 200000 m/z). Given the fact that the average molecular weight of amino acid is approximately 110 Da, it can be assumed that the peptides formed in the process of autolysis contain 20-30 amino acid residues [28]. Literature data indicates that in the composition of the autolysates obtained on industrial scale free amino acids represent 35-45%, di-, tri- and tetrapeptides of a molecular weight of less than 600 Da constitutes 10-15%, oligopeptides of 2000-3000 Da weight constitute 40-45%, whereas those with a mass of 3000-100000 Da represent only 2-5%. These values refer to the total protein present in autolysates [3,11,29]. Table 3 shows the results of the molecular masses of the peptides present in the obtained yeast autolysates, according to the duration of the autolysis process. The molecular masses were defined on the basis of the calculation resulting from the molecular ion (m/z), charge and the areas under the peak for the identified peptides. The presence of peptides with masses of 1000-2000 Da was revealed in the analysed autolysates. Based on these results it can be concluded that the amount of peptide in each sample decreases with lenghtening the autolysis duration, due to the fact that they are decomposed into free amino acids. There are known applications of protein hydrolysates with a high degree of hydrolysis, in which the majority are the peptides with a molecular weight less than 2500 Da in the feeding of young children or people with allergies. In the literature there are studies showing that the loss of immunogenic properties is due to the loss of tertiary structure of proteins as a result of their hydrolysis. It was demonstrated that peptides of molecular weight less than 10 kDa are the tolerance limits but they may have little allergenic properties.

Currently, products of different composition can be found on the market. In these products, depending on the application, differences in the content of peptides of certain molecular weights can be determined. In addition to molecular weight of peptides, an additional important

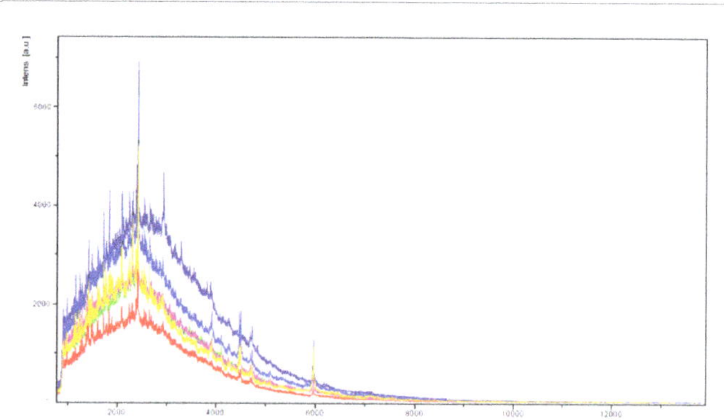

Figure 1: Comparison of the spectra obtained for the tested autolysate samples at different autolysis time (6H, 8H, 10H, 12H, 16H, 20H) in the range of 1000-15000 m/z.

Autolysis Time	Number Of Identified Peptides	Percentage Of Identified Peptides In The Mass Range		
		<1000 Da	1000-2000 Da	>2000 Da
2 h	64	0,27	96,10	3,63
6 h	43	8,18	87,20	4,62
8 h	34	3,72	89,94	6,34
10 h	37	4,01	95,20	0,79
12 h	37	0,73	94,62	4,65
16 h	35	1,04	89,08	9,88
20 h	29	1,10	86,25	12,65
48 h	22	2,52	74,28	23,20

Table 3: The percentage of the peptides identified in the individual ranges of the masses.

aspect is the source and the way of their reception as well as the ratio of branched chain amino acids to aromatic peptides [30-32].

The results of the antioxidant activity and the content of polyphenolic compounds found in the tested yeast autolysates are shown in Table 4. The obtained results indicate that the test autolysates are characterised by a high antioxidant activity, ranging from 22.18 to 32.73 mMol TEAC/100 ml. The content of polyphenolic compounds ranged from 228.3 to 336.1 mgGAE/100 ml. High antioxidant activity is likely to be related to the presence grade of polyphenolic compounds, which are derived from hops and malt used for beer production and the grade of proteins degradation to amino acids. Yeast autolysates may further contain Maillard reaction products and glutathione. It has been shown that the antioxidant activity of Maillard reaction products is dependent on the amino acid sequence of the dipeptides (the constituents of the yeast autolysates). Fresh baker's yeast contains about 0.65% of glutathione and after processing this content may increase up to 5%. It has been shown that the addition of glutathione to butter in the amount of 0.01-0.02% significantly slowed down the process of rancidity [11,33]. Studies have shown that whey protein and soy hydrolysates have antioxidant properties. Short peptides are probably responsible for this action [34]. The strongest antioxidant properties have soy hydrolysates containing mainly peptides of a size 0.7-1.4 kDa [34]. During analysing of the obtained results it was noted that the antioxidant activity increased with the increase in duration of the autolysis process wherein the autolysates obtained after 16h and 20h of autolysis process did not differ statistically significantly both in terms of its antioxidant properties and the content of polyphenolic compounds. In the context of this study, the sensory evaluation of autolysates solutions was performed depending both on the duration of the autolysis process (compared to samples obtained after 6, 8, 10 and 20 hours of autolysis process) and autolysate concentration in the aqueous solution (0.7%, 3.6%, and 7%). Principal Components Analysis (Figure 2) showed that the autolysate solutions differed one from another considering their concentration (0.7, 3.6 and 7%). The vast majority of variation (87.38%) was assigned to the first component and set out the main changes in the sensory profile of the tested autolysates. Location of autolysate solutions next to each other proves their significant similarity. The first group consisted of samples representing the lowest concentration of autolysate, which showed a much lower intensity parameters in both a positive and negative, and overall quality. Location of autolysates of higher concentrations on the opposite side than the samples (0.7%) after 6h, 8h, 10h and 20h of autolysis process confirmed their differences. The odour of the samples obtained in the various duration of the autolysis process and prepared in a concentration of 3.6% was primarily cereal, bouillon, yeast and acidic. In the samples representing the highest concentration of autolysate (7%) such sensory trials as fermentation, beer, yeast and tart

Autolysis Time	Antioxidant activity Xav. (SD)		Polyphenolic compounds Xav. (SD)	
	mMol TEAC/100 ml		mg GAE/100 ml	
6 h	$22,18^{a}$	(0,24)	$228,34^{a}$	(1,59)
8 h	$24,78^{b}$	(0,26)	$274,82^{b}$	(1,35)
10 h	$27,36^{c}$	(0,12)	$300,30^{c}$	(2,10)
12 h	$30,04^{d}$	(0,33)	$307,02^{d}$	(1,51)
16 h	$32,73^{e}$	(0,22)	$329,98^{e}$	(1,28)
20 h	$32,71^{e}$	(0,19)	$336,14^{e}$	(1,28)

Xav-the average content; SD-the standard deviation; $^{a-e}$the mean values in columns indicated with different letters differed significantly (α=0,05)

Table 4: The antioxidant activity and the content of polyphenolic compounds of yeast autolysates obtained during the autolysis process.

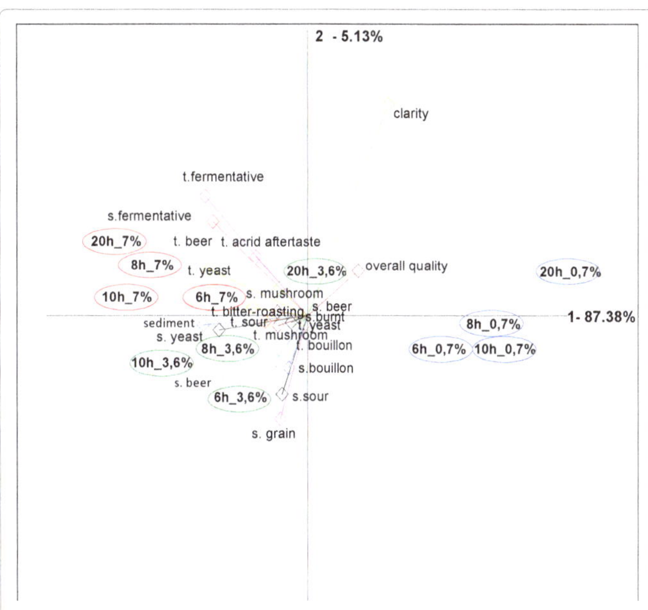

Figure 2: Similarities and differences in the sensory characteristics of the tested yeast autolysates.

predominated. Analysing the effect of concentration on the sensory quality of yeast autolysates it can be concluded that the intensity of quality parameters increased with the increase in concentration of the autolysate in the solution. Sensory quality of yeast autolysates solutions was complex and depended on the presence and the changing the intensity of many flavour traits. The sensory profile of the tested yeast autolysates was clearly dependent on such notes as yeast, fermentation, "beer", bitter-astringent and roasting. It was observed that autolysates obtained after 20 hours of the autolysis process were characterised by the least intensive aroma, such as yeast, fermentation, beer, cereal and acid as compared to other autolysates. These autolysates were also characterised by the highest clarity and lowest amount of sediment and represented the lowest rate of fermentation and beer taste. The intensity of the other quality parameters was comparable, irrespective to the duration of the autolysis process. Overall quality remained at the highest level for the autolysates obtained after 20 hours of the autolysis process.

Conclusions

I. The post-fermentation brewing yeast may be a valuable material for the preparation of yeast autolysates which can be then used in the production of functional food and dietary supplements. The functional properties of obtained autolysates were affected by the use of yeast slurry rich in amino acids and the bioactive compounds derived from barley and hops used for beer production.

II. The obtained yeast autolysates were characterised by a high content of amino acids, exciding the amount present in the reference protein developed by FAO/WHO. Varied content of free amino acids and high content of low weight molecular peptides in autolysates is very advantageous and may indicate the possibility of their broader use in functional food and dietary supplements.

III. The obtained autolysates were characterised by a high antioxidant properties, which were affected, among others, by polyphenolic compounds derived from hops and malt present in the post-fermentation yeast slurry used for autolysate production.

IV. The tested autolysates were characterised by a different sensory profile with a clearly noticeable bitter aftertaste derived from bitter substances extracted from hops. This bitter aftertaste changes favourably with the increase in duration of the autolysis process and growth of free amino acids content, particularly glutamic acid.

V. The sensory quality and functional properties of autolysates derived from post-fermentation yeasts are dependent on the degree of the yeast protein autolysis which determines their usage in a food production.

VI. Precise time control of the autolysis process allows obtaining a product with designed composition and functional properties. Such product is characterised by a desired content of free amino acids, peptides with a specific molecular weight and antioxidant properties.

References

1. Waszkiewicz-Robak B (2013) Spent Brewer's Yeast and Beta-Glucans Isolated from Them as Diet Components Modifying Blood Lipid Metabolism Disturbed by an Atherogenic Diet: Lipid Metabolizm. InTech Pub, Rijeka, Croatia.
2. Breddam K, Beenfeld T (1991) Acceleration of yeast autolysis by chemical methods for production of intracellular enzymes. Applied Biochemistry and Microbiology 35: 323-329.
3. Chae HJ, Joo H, In MJ (2001) Utilization of brewer's yeast cells for the production of food-grade yeast extract, Part 1: Effects of different enzymatic treatments on solid and protein recovery and flavor characteristics. Bioresource Technology 76: 253-258.
4. Choi SJ, Chung BH (1998) Simultaneous production of invertase and yeast extract from baker's yeast. Biotechnology and Bioengineering 13: 308-311.
5. Nagodawithana T (1992) Yeast-derived flavors and flavor enhancers and their probable mode of action. Food Technology 46: 138-144
6. Belousova NI, Gordienko SV, Eroshin VK (1995) Influence of autolysis conditions on the properties of amino-acid mixtures produced by ethanol-assimilating yeast. Applied Biochemistry and Microbiology 31: 391-395.
7. Peppler HJ (1982) Yeast extracts. Economic Microbiology. Academic Press London.
8. Stone CW (1998) Yeast Products in the Feed Industry: A Practical Guide for Feed Professionals. Diamond V Mills, Cedar Rapids, Iowa.
9. In MJ, Kim DC, Chae HJ (2005) Downstream process fort the production of yeast extract using brewer's yeast cells. Biotechnology and bioprocess engineering 10: 85-90.
10. Lee C, Tsang SK, Urakabe R, Rha CK (1979) Disintegration of dried yeast cells andist effect on protein extractability, sedimentation property and viscosity of the cell suspension. Biotechnology and Bioengineering 21: 1-17.
11. Sommer R (1998) Yeast Extracts: Production, Properties and Components. Food Australia 50: 181-183.
12. AOAC (1995) Official methods of analysis of the Association of Official Analytical Chemists. Arlington, Virginia, USA.
13. Stephens K (1986) Amino Acid Analysis by Dansylation: A Revised Method. Ball State University, Muncie, Indiana.
14. Bech-Andersen S (1991) Determination of Tryptophan with HPLC after Alkaline Hydrolysis in Autoclave using α-methyl-tryptophan as Internal Standard. Acta Agriculturae Scandinavica 41: 305-309.
15. Block RJ , Mitchell HH (1946) The Correlation of the Amino Acid Composition of Proteins with their Nutritive Value. Nutrition Abstracts and Reviews 16: 249-278.
16. Oser BL (1951) Methods for the integrating essential amino acid content in the nutritional evaluation of protein. J Am Dietetic Assoc 27: 399-404.
17. FAO/WHO (1991) Protein quality evaluation. Report of the Joint FAO/WHO Expert Consultation. FAO Food and Nutrition Paper 51. Food and Agriculture Organization of the United Nations, Rome.
18. Pellegrini N, Serafini M, Colombi B, Del Rio D, Salvatore S, et al. (2003) Total Antioxidant Capacity of Plant Foods, Beverages and Oils Consumed in Italy Assessed by Three Different In Vitro Assays. American Society of Nutritional Sciences 6: 2812-2819.
19. Singleton VL, Rossi JA (1965) Colorimetry of total phenolics with phosphomolybdic-phosphotungstic acid reagents. American Journal of Enology and Viticulture 16: 144-158.
20. BS EN ISO 13299:2010 Sensory analysis. Methodology. General guidance for establishing a sensory profile.
21. BSEN ISO (2010) Sensory analysis. General guidance for the design of test rooms.
22. Martini Vaughan AE, Miller MW, Martini A (1979) Amino Acid Composition of Whole Cells of Different Yeasts. J Agric Food Chem 27: 982-984.
23. Yamada EA, Sgarbieri VC (2005) Yeast (Saccharomyces cerevisiae) Protein Concentrate: Preparation, Chemical Composition, and Nutritional and Functional Properties. J Agric Food Chem 53: 3931-3936.
24. Mez-alonso SG, Rrez IH, Garciaa-romero E (2007) Simultaneous HPLC Analysis of Biogenic Amines, Amino Acids, and Ammonium Ion as Aminoenone Derivatives in Wine and Beer Samples. J Agric Food Chem 55: 608-613.
25. Voss K, Galensa R (2000) Determination of L- and D-amino acids in foodstuffs by coupling of high-performance liquid chromatography with enzyme reactors. Amino Acids 18: 339-352.
26. Barczak B, Nowak K (2008) Skład aminokwasowy białka biomasy jęczmienia ozimego (*hordeum vulgare* l.) w zależności od stadium rozwoju rośliny i nawożenia azotem. Acta Sci. Pol. Agricultura 7: 3-15.
27. Gawęcki J, Hryniewiecki L (2005) Białka: Żywienie człowieka, Gawęcki J, Polskie Wydawnictwo Naukowe.
28. Stryer L (1997) Biochemia. Wydawnictwo Naukowe PWN: 17-44.
29. Rose AH, Harrison JS (1970) The Yeasts, Yeast Technology. Academic Press, London.
30. Clemente A (2000) Enzymatic protein hydrolysates in human nutrition. Trends in Food Science & Technology 11: 254-262.
31. Maldonado J, Gil A, Narbona E, Molina JA (1998) Special formulas in infant nutrition: a review. Early Human Development Suppl 53: 23-32.
32. Netto FM, Galeazzi MA (1998) Production and characterization of enzymatic hydrolysate from soy protein isolate. Lebensm-Wiss u-Technol 31: 624-631.
33. Risch SJ, Ho CT (2000) Flavor Chemistry. Industrial and academic research; Studies on potent aroma compounds generated in Maillard-type reactions using using the odor-activity-value concept. American Chemical Society, Washington DC: 133-150.
34. Moure A, Dominguez H, Parajo JC (2006) Antioxidant properties of ultrafiltrationrecovered soy protein fractions from industrial effluents and their hydrolysates. Process Biochemistry 41: 447-456.

Seasonal Variation of Chemical and Fatty Acids Composition in Atlantic Mackerel from the Tunisian Northern-East Coast

Guizani SEO* and Moujahed N

National Institute of Agricultural Sciences of Tunisia, 43 Av. Charles Nicolle 1082 Tunis Mahrajene, Tunisia Laboratory of animal and food resources, Tunisia

Abstract

Seasonal variation of chemical and fatty acids composition of Atlantic Mackerel populations from the Northern-East coast of Tunisia were investigated in order to assess the best period for its consumption. Fresh specimens were purchased monthly from Bizerte fishing port located in the north-east coast of Tunisia. It was noted that total lipids and protein contents levels varied significantly (P<0.05) according to the season. The highest protein level was registered in autumn (27.6%), while the lowest was in spring (18.7%). Total fat content decreased significantly (P<0.05) from (11.1%) in spring to (4.5%) in summer. The major fatty acids in each season were 16:0 and18:0 (palmitic and stearic) acids as saturated; 16:1 and 18:1 (palmitoleic and oleic) acids as monounsaturated; and 20:4 (arachidonic acid), 20:5 Eicosapentanoic (EPA), 22:6 Docosahexanoic (DHA)as polyunsaturated. These components varied significantly (P<0.05) during the sampling period.

It was shown that high PUFA values are related to the high n-3 PUFA proportions, mainly represented by EPA and DHA. Docosahexanoic acid was the most abundant (P<0.05) during the study period which highest rate (P>0.05) was observed in autumn (40.1%). It was noted that n-6 PUFA proportions were very low comparatively to those of n-3 PUFA series. The highest level of n-6 PUFA series corresponded to arachidonic acid in autumn (3.9%). The best (n-3)/(n-6) ratio was registered in spring (10.81).

Keywords: Atlantic mackerel; Fatty acids; Lipids; Seasonal variation

Introduction

In many countries fish species are important sources of food in human diet composition. Besides its high level of protein content and its moderate cost, Mackerel species seems to contain substantial amounts of polyunsaturated fatty acids (PUFA) which have benefic effects on human health. These fatty acids are suspected to prevent some diseases such as cardiovascular, rheumatoid, arthritis, mental illness [1,2] and several types of cancer [3,4].

PUFA found in fish oils can be divided into two biochemical families, n-3 and n-6, with different biological effects [5] those of n-3 family mainly represented byeicosapentaenoic acid (EPA, 20:5), docosapentaenoic acid (DPA, 22:5)and docosahexaenoic acid (DHA, 22:6). While those of n-6 family generally present at very low quantity [6] are represented by (arachidonic, 20:4) eicosadienoic (20:2) and linoleic acids (18:2). Moreover, n-3 PUFA series play an important role in the improvement of visual function and in the prevention of atherosclerosis and thrombosis development [7]. Also, studies on newborns indicate that DHA is essential for the normal functional development of the retina and brain, particularly in premature infants [8].

Scomberscombrus is a small pelagic fish species well distributed through the Mediterranean Sea, the Black sea and on the both sides of Atlantic Ocean [9]. Belonging to the Scombridae family, *Scomber scombrus*is characterized by a blue back and silver abdomen, it is one of the most popular fish in Tunisian gastronomy, its annual production was about 196 tones [10]. This species is known for its high fat content and seems to be one of the highest sources of long chain polyunsaturated fatty acids LC-PUFA [11].

In Tunisia, previous studies examined mackerel biometry and biological reproduction, it is characterized by a carnivorous diet and its spawning period was in winter [12]. However, there are few or no available details on biochemical studies of its flesh composition. Moreover, this species is caught during several periods of the year [12] consumer could request data about its seasonal nutritive variation.

The aim of the present study was to evaluate the seasonal variation of chemical and fatty acids composition of Atlantic Mackerel populations from the Northern-East coast of Tunisia in order to determine the most favorable periods for its consumption.

Materials and Methods

Fishes sampling and conditioning

The analyzed specimens were collected monthly from the fishery off northern east of Tunisia. A case of Mackerel population was sampled randomly and purchased each month at the hall fish port of Bizerte during the period extending from May 2010 to April 2011. At landing, fishes were transferred in polystyrene boxes containing ice and quickly transported at the laboratory were somato metric measurements were taken; Samples length and weight ranged from 26 cm to 35 cm and 160 g to 310 g. Then fishes were beheaded, washed, filleted, bag packed and frozen at -20°C.

Chemical analysis and fatty acid characterization

Moisture: Moisture content was determined by drying to constant weight in an oven at 105°C (n=6) according to the AOAC, 2000. Results were expressed as percentage of dry weight.

Ash: Ash content was determined after ignition at 550°C (n=6) for 4 h (AOAC, 2000). Results were expressed as percentage of dry weight.

***Corresonding author:** Guizani SEO, National Institute of Agricultural Sciences of Tunisia, 43 Av. Charles Nicolle 1082 Tunis Mahrajene, Tunisia Laboratory of animal and food resources, Tunisia, E-mail: salma.inat@yahoo.fr

Protein: Crude protein (N×6.25) was determined in triplicate by the Kjeldahl method (AOAC, 2000). Results were expressed as percentage of dry weight.

Lipid extraction: Total lipid extraction was carried out by Soxhlet method (AOCS, Ba 3-38) in triplicate using 5 g (dry weight) of flesh powder with 200 ml of petroleum ether for 6 h. The extracted oil was evaporated under vacuum at 65°C using a rotaryevaporator, and then placed in an oven at 45°C for 1 h before being transferred into desiccators and reweighing; results are expressed as a percentage of dry tissue.

Fatty acids analysis

Methyl esters were prepared by direct trans esterification method using 200 µl acetyl chlorides according to Mosers [13]. The extracted lipid was dissolved in 2 ml (methanol/methylene chloride (3:1)); followed by the addition of an internal standard. The mixture was vortexed and heated at 75°C for one hour. After cooling, 4 ml of potassium carbonate (7% W/V), hexane and acetonitrile were added and the mixture was centrifuged for 10 mn; the hexane layer (1 µl) was taken for GC analysis. Fatty acid analysis was performed using a gas chromatograph (HP, 6890), with a split/split less injector and a flame ionization detector. The device includes a 30 m long HP Innowax capillary column with an internal diameter of 250 µm and a 0.25 µm film, the stationary polar phase of the column being polyethylene glycol. Comparison of the retention times of the fatty acids under study and those of standard fatty acid methyl esters (PUFA-3) allowed to identify the different fatty acids contained in mackerel oil extract. Results are expressed as percentage of total fatty acids.

Statistical analysis

For the different biochemical parameters, statistical analysis was performed using analysis of variance according to ANOVA procedure (SAS software, version 9.1). Season effect was analyzed and month sampling in each season was considered as replication. Means were compared using Student Newman and Kull tests and differences were considered significant when $P<0.05$.

Results

Chemical composition

The chemical composition of *Scomber scombrus* is presented in Table 1. The highest moisture level was obtained in summer (72.25%). Ash proportions ranged from 1.7% in winter to 2.6% in autumn. Moisture and Ash content vary significantly according to the season ($P<0.05$). Protein content reached in autumn 27.6%, lowest value was obtained in spring 18.7%. Lipid content ranged from 11.1% in spring to 4.5% in summer. Protein and lipids seasonal differences were significantly high ($P<0.001$).

Fatty acids contents

Fatty acids composition of the analyzed specimens is presented in Table 2. PUFA constitute the majority of the fatty acids pool, followed by saturated (SFA) and monounsaturated fatty acids (MUFA).

Total Saturated fatty acid fraction ranged between 32.35% and 38.22% (Figure 1) with palmitic acid (16:0) as the most important fatty acid, followed by stearic acid (18:0) and myristic acid (14:0) in each season. It is noted that high percentages of (16:0) and (18:0) are obtained in autumn with about 24.97% and 12.01%, however seasonal variation of these components was not statistically significant ($p>0.05$); whereas (14:0) vary significantly with season ($p>0.05$); peak is obtained in spring 3.5% and its minimal value was registered in autumn1.24%.

Total monounsaturated fatty acids mainly represented by (16:1) and (18:1) increasing in winter 16.58% (Figure 1). The most important MUFA corresponds to (18:1) reaching in winter 13.81%; whereas (16:1) high percentage is obtained in spring 4.13%. These two components (16:1 and 18:1) vary significantly with seasons ($p<0.05$) in fact it is noted that their minimal values were obtained respectively in autumn 2.63% and 7.09%.

Polyunsaturated fatty acids content constitute the biggest part in Mackerel lipid. It was noted that high PUFA levels are related to n-3 PUFA family mainly represented by EPA (20:5) and DHA (22:6) increasing respectively in spring 7.8% and in autumn 40.10%, and which seasonal variation was statistically significant ($p<0.05$); however minimal values were obtained in autumn 4.54% and 35.66%.

The high proportion of n-6 PUFA series is linked to that of arachidonic acid (20:4) increasing in autumn (3.91%) and the eicosadienoic acid (20:2) maximum rate was obtained in spring (0.34%). Seasonal variation of these two components was statistically significant ($P<0.05$). The highest n-3/n-6 ratio is observed in spring (10.81).

Discussion

Proximate composition

Proximate composition of mackerel flesh was investigated; results showed that moisture contentsin studied samples were the highest when lipid was the lowest leading to deduce an inverse relation between thesetwo components. This finding is in agreement with Orban et al.[14] study on *Boops boops* from the Southern Adriatic coast of Italy. Ash content range is between 1.70 - 2.68%. These values were close to those found on *Sardina pilichardus* by Caponio et al. [15]. Moisture and ash contents seasonal variation was statistically significant (p<0, 05) as observed on other studied fish species by Kacem et al. [16].

Scomber scombrus protein and lipid contents vary significantly ($P<0.01$) according to the seasons, this trend is in accordance with Orban et al. [14] study on horse mackerel from the Southern Adriatic coast of Italy; moreover, according to Wallace [17], mackerel fat fillets content vary with season with a maximum level ranging from 25 to 30% in December and a minimum of around 5% in May when the fish spawns.

The observed seasonal variations in mackerel flesh composition may be explained by the impact of some environmental parameters such as temperature, salinity and fluctuations in food availability and fish life cycle. In addition, during the reproductive periods, lipids and proteins are mobilized from muscles and transferred to the gonads [18] influencing fish nutritional value and particularly flesh lipid composition.

Fatty acids

Our results show that palmitic acid 16:0 is the predominant fatty acid in SFA family [19-21]; in fact, palmitic acid is the key metabolite in fish species [22] its rate seemed to be constant through the sampling

Parameters	Autumn	Winter	Spring	Summer	P-value
Moisture (%)	71.1 ± 0.8^b	70.15 ± 0.7^b	71,6 ± 0.9^b	72.25 ± 2.8^a	0.04
Ash (%)	2.6±0.9^a	1.70±0.3^b	1.8±0.2^b	2.05±0.7^a	0.01
Protein (%)	27.6±1.82^b	20.6±2.23^a	18.7±1.09^b	22.7±1.53^a	0.001
Lipid (%)	6.5±0.51^c	6±0.22^b	11.1±1.51^a	4.5±0.80^d	0.001

Means with the same letter in the raw are not significantly different.

Table 1: Proximate composition of studied samples.

Nutrients	Autumn	Winter	Spring	Summer	P-value
		Saturated fattya cids (SFA)			
C14:0	1.24±0.52[a]	1.57±1.03[a]	3.5±2.55[b]	1.79±0.54[a]	**0.04**
C16:0	24.97±6.15[a]	20.94±5.61[a]	21.76±1.31[a]	22.67±3.28[a]	**0.11**
C18:0	12.01±3.03[a]	9.84±2.15[a]	7.85±3.14[a]	9.69±0.92[a]	**0.07**
		Mono unsaturated fatty acids (MUFA)			
C16:1	2.63±0.77[a]	2.77±1.39[a]	4.13±1.16[b]	2.69±0.69[a]	**0.001**
C18:1	7.09±3.92[b]	13.81±6.60[a]	12.23±3.73[a]	11.93±4.21[a]	**0.01**
		Poly unsaturated fatty acids : n-6 series			
C18:2n-6	1.34±0.40[a]	1.26±0.31[a]	1.46±0.51[a]	1.42±0.23[a]	**0.02**
C18:3 n-6	0.15±0.10[a]	0.12±0.05[a]	0.10±0.03[a]	0.19±0.09[a]	**0.53**
C20:2n-6	0.15±0.06[a]	0.26±0.04[a]	0.34±0.13[a]	0.17±0.16[a]	**0.03**
C20:4n-6	3.91±0.57[a]	2.81±0.79[b]	2.37±0.79[b]	2.72±0.70[b]	**0.004**
		Poly unsaturated fatty acids : n-3 series			
C18:3 n-3	0.29±0.05[c]	0.50±0.18[a]	0.73±0.38[b]	0.47±0.17[a]	**0.001**
C20:3n-3	0.35±0.17[a]	0.51±0.29[a]	0.26±0.14[b]	0.38±0.05[a]	**0.02**
C20:5n-3	4.57±0.91[a]	5.79±0.77[a]	7.80±0.97[b]	6.08±1.10[b]	**0.001**
C22:5n-3	1.05±0.23[a]	1.63±0.68[a]	1.75±0.42[a]	1.46±0.24[a]	**0.009**
C22:6n-3	40.10±7.93[b]	38±8.91[b]	35.66±4.45[a]	38.28±5.8[b]	**0.04**
PUFA/SFA	1.36	1.56	1.52	1.49	
ω3/ ω6	8.36	10.40	10.81	10.37	

Means (n = 6) with the same letter in the raw are not significantly different (p>0.05).

Table 2: Fatty acids composition of Atlantic mackerel samples (% TFA).

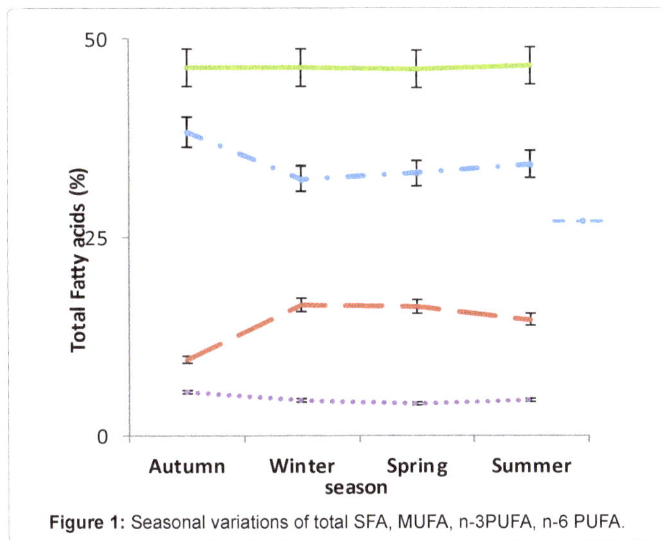

Figure 1: Seasonal variations of total SFA, MUFA, n-3PUFA, n-6 PUFA.

period and which seasonal variation was not statistically significant (p>0.05) leading to deduce that this compound is not influenced by diet and environmental condition variations.

Total MUFA mainly represented by oleic (18:1) and palmitoleic (16:1) acids varied significantly (P<0.05) with fishing season increasing in winter 16.5% which is in accordance with Soriguer et al.[23] finding on Atlantic mackerel from the south of Spain. Moreover, high MUFA level is an indicator of high degree of carnivory as signaled by Dalsgaard et al. [24], in addition the analyzed specimens diet composition show that this fish species is typically carnivorous feeding essentially by small fish species as (*Sardina pilchardus*, *Engraulis encrasicolus*) crustacean mainly represented by (mysidacea and euphosiacea) and mollusks as (*Sepia officinalis*, *Loligo vulgaris*, *Octopus vulgaris*) personal unpublished.

Comparatively with SFA and MUFA, PUFA constitute the highest proportion in mackerel extracted lipid as indicated by Ozogul et al. [25]. Moreover, n-3 PUFA family was dominant comparatively with n-6 series present at very low quantity [6]. Also, it was noted that highest total PUFA level was obtained in autumn 52% and the lowest in spring 50.5% (Figure 1). These high percentages are linked to n-3 PUFA mainly represented by EPA and DHA, where the latter is present at an abundant amount during the study period ranging from 35.7% in spring to 40.2% in autumn which is in accordance with Ozogul et al. [25] finding in Mackerel mussels from Turkey(35.2%).

In comparison with other fish species DHA in mackerel mussels are close to those of *Boops boops* captured in March from the Southern Adriatic coast of Italy in Orban et al. [14] finding (31.18%), whereas it is much higher than those of Horse mackerel in the same study. DHA in our study is higher than *Scomber scombrus* value from Marmara Sea (25.2%), but it is close to *Merangius merlangus* value (40.5%) from the Black Sea in Tanakol et al. [26] finding. In this study DHA high values may be explained by the impact of many factors influencing Mackerel flesh composition essentially *n-3* PUFA series as:location, environmental parameters, and food availability with a large variety of prey items such as fish (*Engraulis encrasicolus, Sardina pilchardus....*); crustacean (mysidacea and euphosiace)and mollusks present respectively with 67%, 20.17%, 12% of frequency of occurrence (F%) (El Oudiani unpublished).

Regarding n-6 PUFA, high values are linked to arachidonic acid (20:4) reaching in autumn (3.91%), the obtained results are close to Tanakol et al. [26] finding on mackerel from Marmara sea (2.8%),and *Merangiu smerlangus* from the Black Sea(3.5%). The n-3/n-6 PUFA ratio is an essential element of human nutritional evaluation and an indicator of the lipid quality [14]. Its high level was obtained in spring 10.81 which is in agreement with Ozogul et al. [25] finding on mackerel from Turkey. Indeed, the n-3 PUFA are known to decrease the rate of triacylglycerols, cholesterol, mainly low density lipoproteins in the human serum, and to inhibit the aggregation of blood platelets and the damage of blood vessels [27-29]. In addition, they are suggested to play a role in the prevention of some cancers [30].

In human diet, seafood products are the most important significant source of n-3 PUFA, that's why dietary guidelines recommend more and more their consumption. The period from May to November where *n-3* PUFA values range from (36% to 46%) is considered as the most important for mackerel consumption due to the optimum n-3 PUFA level; although the consumption of this fish is recommended throughout the year. However, from nutritional point of view *Scomber scombrus* is classified among the fatty fish species. The present data shows that its flesh composition is characterized by a high protein and lipid contents and an important n-3 PUFA level mainly DHA.

Conclusion

This study dealing with seasonal variation of the mackerel flesh composition is very important and may provide some useful information in the production of higher value-added products such as oil extraction in view of commercial exploitation.

These variations are essentially due to the impact of many factors such as sex, size, age, reproduction, water temperature, degree of pollution, nutritional condition, geographical location, species genetic differences, processing, storage, and distribution which could affect the quality of the value-added product. The advantage of oil from mackerel flesh is its higher level of EPA and DHA. These applications include essentially the production of nutraceutical foods with high level of PUFAs that could potentially decrease cholesterol levels and prevent heart diseases.

References

1. Ruxton CH, Calder PC, Reed SC, Simpson MJ (2005) The impact of long chain n-3 polyunsaturated fatty acids on human health. Nut Res Rev 18: 113-129.
2. Song C, Zhao S (2007) Omega-3 fatty acid eicosapentaenoic acid. A new treatment for psychiatric and neurodegenerative diseases: A review of clinical investigations. Exp Opin Inv Dru 16: 1627-1638.
3. Chen YQ, Edwards IJ, Kridel SJ, Thornburg T, Berquin IM (2007) Dietary fat-gene interaction in cancer. Can Meta Rev 26: 535-551.
4. Calviello G, Serini S, Piccioni E (2007) N-3 polyunsaturated fatty acids and the prevention of colorectal cancer: molecular mechanisms involved. Curr Medi Chem 14: 3059-3069.
5. James MJ, Cleland LG, Gibson RA (2000) Dietary polyunsaturated fatty acids and inflammatory mediator production. American Journal of Clinical Nutrition 71: 343-348.
6. Sargent JR (1997) Fish oils and human diet. Brit J Nut 78: S5 -13.
7. Calder PC (2003) New evidence in support of the cardiovascular benefits of long-chain n-3 fatty acids. Ital Heart J 4: 427-429.
8. Connor WE (2000) Importance of n-3 fatty acids in health and disease. Ame J Clin Nutr 71: 171S-175S.
9. MacKay KT (1967) An ecological study of *mackerel Scomberscombrus* (Linnaeus), in the coastal waters of Canada. Can Fish Res Board Tech Rep.
10. Tunisian general direction of fishing and aquaculture (TGDFA), statistics of the year 2013.
11. Ackman RG (1990) Seafood lipids and fatty acids. Food Rev Int 6: 617-646.
12. Hattour A (2000) Contribution à l'étude des poissons pélagiques des eaux tunisiennes. Thèse de Doctorat d'Etat de la Faculté des Sciences de Tunis.
13. Moser HW, Moser AB (1991) Measurement of very long fatty acids in plasma. Techniques in diagnostic human biochemical genetics: A Laboratory Manual. Willey-Liss, Inc: 177-191.
14. Orban E, Dilena G, Nevigato T, Masci M, Casini I, et al. (2011) Proximate unsaponifiable lipid and fatty acid composition of bogue (*Boops boops*) and *horse mackerel* (*Trachurus trachurus*) from the Italian trawl fishery. J Food Comp Anal 24: 1110-1116.
15. Caponio F, Lestingi A, Summo CMT, Bilancia MT, Laudadio V (2004) Chemical characteristics and lipid fraction quality of sardines (*Sardina pilchardus* W.): Influence of sex and length. J Appl Icht 20: 530-535.
16. Kacem M, Sellami M, Kammoun W, Frikha F, Miled N, et al. (2011) Seasonal variations in proximate and fatty acid composition of Viscera of *Sardinella aurita*, *Sarpasalpa*, and *Sepia officinalis* from Tunisia:J Aqua Food Prod Techn 20:233-246.
17. Wallace PD (1991) Seasonal variations in fat content of mackerel (*Scomber scombrus* L.) caught in the English Channel. Fish Res Techn Rep.
18. Love RM (1997) Biochemical dynamics and the quality of fresh and frozen fish. Fish Proc Techn 1-31.
19. Osman H, Suriah AR, Law EC (2001) Fatty acid composition and cholesterol content of selected marine fish in Malaysian waters. Food Chem 73: 55-60.
20. Aidos I, Van Der Padt A, Luten JB, Boom RM (2002) Seasonal changes in crude and lipid composition of herring fillets, by-products, and respective produced oils. J Agric Food Chem 16: 4589-4599.
21. Passi S, Cataudella S, Marco DP, Simone DF, Rastrelli L (2002) Fatty acid composition and antioxidant levels in muscle tissue of different Mediterranean marine species of fish and shellfish. J Agr Food Chem 50: 7314-7322.
22. Andrade AD, Rubira AF, Matsushita M, Souza NE (1997) Omega-3 fatty acids in baked freshwater fish from South Brazil. Arch Latinoam Nutr 47: 73-76.
23. Soriguer F, Serna S, Valverde E, Hernando J, Martin-Reyes HA, et al. (1997) Lipid, protein and calorie content of different Atlantic and Mediterranean fish, shellfish and molluscs commonly eaten in the south of Spain. Eur J Epi 13: 451-463.
24. Dalsgaard J, St John M, Kattner G, Muller-Navarra D, Hagen W (2003) Fatty acid trophic markers in the pelagic marine environment. Adv Mar Biol 46: 225-340.
25. Özogul Y, Özogul F, Alagoz S (2007) Fatty acid profiles and fat contents of commercially important seawater and freshwater fish species of Turkey: A comparative study. Food Chem 103: 217-223.
26. Tanakol R, Yazıcı Z, Sener E, Sencer E (1999) Fatty acid composition of 19 species of fish from the Black Sea and the Marmara Sea. Lipids 34: 291-297.
27. Kromhout D, Bosschieter EB, Coulander de Lezenne C (1985) The inverse relation between fish consumption and 20-year mortality from coronary heart disease. New Eng J Med 312: 1205-1209.
28. Herold PM, Kinsella JE (1986) Fish oil consumption and decreased risk of cardiovascular disease: A comparison of findings from animal and human feeding trials. Am J Clin Nutr 43: 566-598.
29. Simopoulos AP (1991) Omega-3 fatty acids in health and disease and in growth and development. Ame J Clin Nutr 54: 438-463.
30. Larsson SC, Kumlin M, Ingelman-Sundberg M, Wolk A (2004) Dietary long-chain n-3 fatty acids for the prevention of cancer: A review of potential mechanisms. Am J Clin Nutr 79: 935-945.

Studies on Preparation and Quality of Nutritious Noodles by Incorporation of Defatted Rice Bran and Soy Flour

Pakhare KN[1], Dagadkhair AC[1*], Udachan IS[2] and Andhale RA[3]

[1]Department of Technology, Food Technology Division, Shivaji University, Kolhapur, India
[2]MIT College of Food Technology, Rajbaugh, Loni Kalbhor, Pune, India
[3]College of Food Technology, VNMKV, Parbhani, India

Abstract

The protein and fiber rich noodles were prepared by incorporation of defatted soy flour (DSF) and defatted rice bran (DRB). The quantity DSF and DRB was used at 10% and 6% respectively for incorporation. The chemical composition of incorporated noodles utters the information such as, moisture 8.43%, total carbohydrate 68.30%, crude protein 14.29%, crude fat 4.98%, crude fiber 4.02%, ash 1.54% and calcium 498 mg/100 g which reveals it's a good source of nutrients. The cooking quality was also examined and conclusions are; cooking time 7.30 min, cooking loss 1.25 g and water uptake 10.5 g. The DSF and DRB greatly affect the chemical composition and cooking quality of noodles, as soy and rice bran are the rich sources of protein and fiber respectively. The shelf life study was conducted by undertaking microbial study of noodles packed in different packaging material's viz. HDPE and LDPE. Though TPC of HDPE packed noodles was between 0.1×10^2 and 0.35×10^2 CFU/gm during whole tenure of study. The yeast and mold count was ranged from 0.06×10^2 to 0.12×10^2 CFU/gm. Simultaneously LDPE packed noodles showed TPC from 0.1×10^2 to 0.42×10^2 CFU/gm and yeast and mold was between 0.06×10^2 and 0.26×10^2 CFU/gm while study. The comprehensive study of shelf life demonstrates that HDPE packed noodles have better shelf life than LDPE packed noodles as it having less microbial growth.

Keywords: Noodles; DSF and DRB; Incorporation; Protein and fiber; Packaging; Shelf life

Introduction

Noodles are one of the staple foods consumed in many Asian countries. Instant noodles have become internationally recognized food, and worldwide consumption is on the rise. Many researchers are exploring the potential of noodle fortification as an effective public health intervention and improve its nutritional properties. The properties of instant noodles like taste, nutrition, convenience, safety, longer shelf life, and reasonable price have made them popular. Quality factors important for instant noodles are color, flavor, and texture, cooking quality, rehydration rates during final preparation, and the presence or absence of rancid taste after extended storage Gulia et al. [1]. Instant noodles are widely consumed throughout the world and it is a fast-growing sector of the noodle industry. Global consumption of the noodles second only to bread. Noodles are a staple food in many cultures made from unleavened dough which is stretched, extruded, or rolled flat and cut into one of a variety of shapes Okoye et al. [2]. According to Lee et al. [3] noodles came to China as early as 5000 BC, and then spread to other Asian countries such as Japan, Thailand, Korea and Malaysia, and now days it has become one of the fastest growing sectors in the world with the compound annual growth rate (CAGR) reaching 4%. Worldwide, China ranks first in the consumption of noodles followed by Indonesia, Japan, and Vietnam according to the world instant noodle association [4] Instant noodles are made from wheat flour, starch, water, salt or kansui (an alkaline salt mixture of sodium carbonate, potassium carbonate, and sodium phosphate), and other ingredients that improve the texture and flavor of noodles, partially cooked by steaming and further cooked and dehydrated by a deep-frying process [5].

Nowadays Consumers all around the world, all are more at the risk of many diseases such as diabetes due to obesity, high cholesterol, cardiovascular diseases, high blood pressure and irregular blood sugar levels. These risk factors are because of the unfit diet which is low in essential nutrients like dietary fiber, phytochemical and antioxidants. Functional foods provide health benefits and help in the avoidance of diseases by incorporating nutraceutical ingredients and other essential nutrients. In the world now days the market of instant noodles gaining popularity. Usually, wheat flour is preferred to prepare instant noodles with low protein and dietary fiber content [6].

Wheat flour noodles can be supplemented with a range of materials that can boost fiber, protein etc. [7]. Recently, food manufacturers have responded to consumer demands for foods with higher fiber content by developing products in which high-fiber ingredients are used dietary fiber can also impart some functional properties to foods, e.g. increase water holding capacity, oil holding capacity, emulsification and/or gel formation. Traditionally, consumers have chosen foods such as whole grains, fruits and vegetables as sources of dietary fiber. Flour of hard wheat (*Triticumaestivum L.*) is the main primary ingredient which is usually used to make instant noodles is low in fiber and protein contents but also poor in essential amino acid, lysine [8]. Noodles also can be made from other flours like rice, buckwheat, and starches derived from potato, sweet potato, and pulses [9]. Most of the essential nutrients are lack in traditional noodles such as dietary fiber, vitamins and minerals, which are lost during wheat flour refinement. Thus, noodle products which represent a major end-use of wheat, are suitable for enhancing health after incorporating sources of fiber and essential nutrients. The essential amino acids are not present in Instant noodles mainly consist of wheat flour whose protein quality is not sufficient [10].

***Corresponding author:** Dagadkhair AC, Department of Technology, Food Technology Division, Shivaji University, Kolhapur, MH, India
E-mail: amoldagadkhair007@gmail.com

The soybean, a grain legume, is one of the richest and cheapest sources of plant protein that can be used to improve the diet of millions of people, especially the poor and low income earners in developing countries. The main ingredients of noodles are wheat, which is having deficiency of essential amino acid lysine, whereas soybean is richer in lysine and can be complement to wheat in noodles. Soybean protein is more economical than high priced meat protein and so they are considered as best source of protein especially in vegetarian diet. It increases nutritional status of vulnerable groups like pregnant woman, nursing mother, school going and young children. High protein soya product reduces incidence of malnutrition and encourage the farmers to grow more soybeans due to increasing demand in the market [11].

Rice bran is a byproduct of rice milling industry and constitutes around 10% of the total weight of rough rice. It is primarily composed of aleurone, pericarp, sub aleurone layer and germ. Rice bran is a rich source of vitamins, minerals, essential fatty acids, dietary fiber and other sterols. There is a widespread scientific agreement on various health benefits associated with consumption of dietary fiber. Consumer attitude towards health foods is promising and the scope of functional foods is growing in the world markets. Rice bran is finding increased applications in food, nutraceutical and pharmaceutical industries. However, potential applications of rice bran in food industry are limited by its instability owing to rancidity caused by exposure of oil to lipases during milling [12].

Though, the study was carried to produce a fiber and protein rich noodles by incorporating defatted soy flour and defatted rice bran in noodles with its quality analysis and shelf life study in various packaging materials.

Materials and Methods

Preparation of defatted soy flour (DSF) and defatted rice bran (DRB) incorporated noodles

The DSF and DRB incorporated noodles were prepared by using standard formulation, which was finalized by comprehensive process of individual sensory evaluation of DSF and DRB incorporated noodles by semi trained panels. The individual DSF incorporated noodles were prepared by addition of 5, 10, 15, and 20 percent of DSF in comparison with refined wheat flour. Simultaneously, the DRB incorporated noodles were prepared by taking various trials of DRB incorporation in 2, 4, 6, 8, 10, 12, and 15 percent in comparison with refined wheat flour. The comprehensive process of sensory evaluation of incorporated noodles was leads to standardize recipe such as; refined wheat flour 84 g, DSF 10 g, DRB 6 g, water 40 ml, oil 5 g, salt 2 g, Kansui solution 1 g, gluten 2 g and guar gum 0.2 g. (Figure 1) [13].

Chemical analysis of noodles samples

The chemical analysis of control noodle sample and noodles incorporated with DSF and DRB were analyzed for moisture, crude protein, crude fat, crude fiber, total ash and calcium by the methods demonstrated by Ranganna [14].

Cooking quality evaluation of noodles: The cooking qualities of the dried noodles were evaluated with respect to cooking time, cooking loss and water uptake Gatade and Sahoo [15].

***Cooking time*:** Optimal cooking time was evaluated by observing the time of disappearance of the core of the noodle strand during cooking (every 20 sec) by squeezing the noodles between two transparent glass slides.

***Cooking loss*:** The cooking loss was determined by measuring the amount of solid substance lost to cooking water. 10 gm sample of noodles was placed into 100 ml of boiling water in a 500 ml beaker. Cooking water was collected in a pre-weighed glass dish and was placed in a hot air oven at 105°C and evaporated to dryness. The dry residue was weighed and reported as a percentage.

***Water uptake*:** The water uptake was calculated by getting the difference between weight of cooked noodles and weight of dried noodles. The cooked noodles were placed on filter paper for 5 min before weighing, to blot the excess adhered water.

Sensory evaluation

The sensory evaluation of DSF and DRB incorporated noodles was undertaken by semi trained panel members of department of food technology, department of technology, Shivaji University, Kolhapur. The sensory attributes like appearance, color, taste, flavor, texture and overall acceptability were evaluated by semi trained judges using 9 point hedonic score system. The panelist gives score 9-1 to the product, ranging from 'like extremely' to 'disliked extremely' to find out the most suitable composition of noodles [16].

Effect of storage (Ambient temperature) and packaging material on shelf life of incorporated noodles

Shelf life study was carried out to determine the keeping quality of product. The effect of packaging material and storage condition on microbial study was determined. The shelf life of any food product is depending on various factors like processing, method of preservation, packaging and microbial count so it founds obligatory to go for microbial examination.

Packaging material used for incorporated noodles

The high-density polyethylene and low density polyethylene were used for packaging of incorporated noodles.

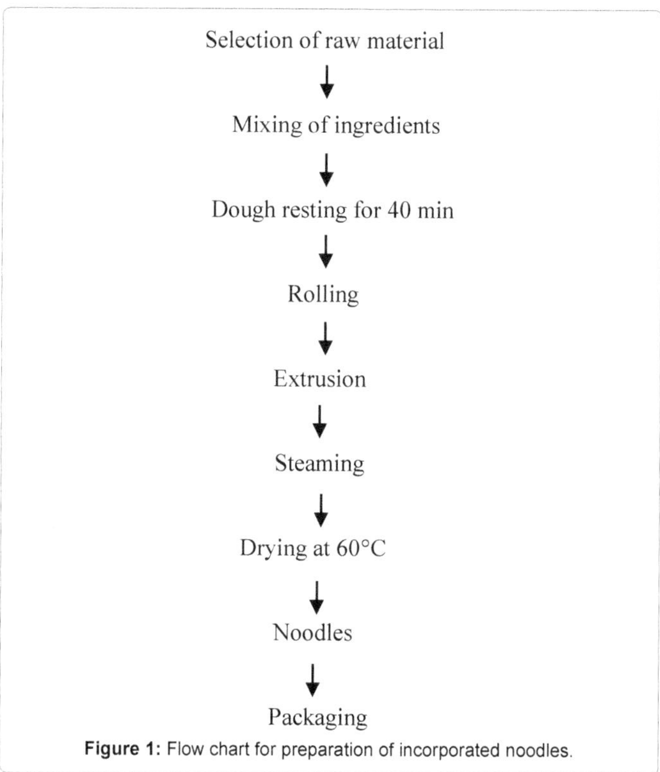

Figure 1: Flow chart for preparation of incorporated noodles.

Microbial study of incorporated noodles: Microbial study of incorporated noodles was carried by determining Total Plate Count and Yeast Mold count. The methods used for determination of TPC and yeast and mold are explained below.

Total plate count (TPC): The total plate count of incorporated noodles was determined by using simply a total plate count agar (Nutrient agar) method. The pour plate technique was used for the Isolation of microorganisms from the incorporated noodles. The dilutions were made up to 10^{-2} by using 1 g of sample and 0.1 ml of aliquot was used for the isolation. All process was carried out in a strictly sterile area with the help of laminar air flow. Plates were incubated at 37°C for 48 hrs, and results noted in cfu/ml.

Total plate count (TPC) of incorporated noodles was examined and the effect of storage and packaging material on microbial quality (TPC) of incorporated noodles was examined on fresh, 15, 30, 45, 60, 75 and 90 days.

Yeast and mold: The yeast and mold count of incorporated noodles was determined by using potato dextrose agar (PDA), the streak plate and pour plate technique were used for the Isolation. The media was sterilized and poured into plates. The dilutions of sample were made up to 10^{-2} as like TPC and then 0.1 ml of aliquot was used for streaking. Plates were incubated at 37°C for 48-72 hrs. and results noted in cfu/ml.

The yeast and mold count of incorporated noodles was examined and the effect of on storage and packaging material microbial quality (Yeast and Mold) of fish sauce was examined on fresh, 15, 30, 45, 60, 75 and 90 days.

Result and Discussion

Chemical composition and cooking quality of incorporated noodles with DSF and DRB

Proximate analysis and cooking quality of incorporated noodles has been studied and presented in below Table 1. The following parameters were checked like moisture, carbohydrate, protein, crude fat, crude fiber, ash and calcium content. In cooking quality evaluation cooking time, cooking loss and water uptake was studied.

The information regarding chemical composition of incorporated noodles emerges from the Table 1 and results found that moisture content was 7.43%, total carbohydrate 68.30%, protein 14.29%, crude fat 4.98%, crude fiber 4.02%, ash 1.54% and calcium was 498 mg. It could be visualized from the give result that noodles found rich source of carbohydrate and protein along with dietary fiber. As incorporated noodles are rich source of dietary fibers it encompasses various health benefits. The protein rich noodles could be prepared by incorporation of defatted soy flour Shogren et al. [17] and Amudha et al. [18]. Cooking time of incorporated noodles was 7.30 min which was slightly higher than control sample (7 min). The cooking loss was 1.25 g which was also quite more than control sample (1.2 g). Water uptake was also determined and founds 10.5 g which was also higher than control sample (10.2 g). The reason behind increase in this cooking quality parameter may be due to increase in protein concentration of incorporated noodles, which directly effects on textural and ultimately cooking quality of noodles.

Parameters	Value
Moisture (%)	7.43
Total carbohydrate (%)	68.30
Protein (%)	14.29
Crude fat (%)	4.98
Crude fiber (%)	4.02
Ash (%)	1.54
Calcium (mg)	498
Cooking Time (min)	7.30
Cooking Loss (g)	1.19
Water Uptake (g)	10.2
*each result is of three determinations	

Table 1: Chemical composition and cooking quality of noodles incorporated noodles with DSF and DRB.

Effect of storage (Ambient temperature) and packaging material on shelf life of incorporated noodles

The study was conducted to determine the microbial growth and chemical composition of incorporated noodles in different packaging materials such as high density polyethylene (HDPE) and low density polyethylene (LDPE). As microbial evaluation founds prime factor to determine the shelf life of any food products so, the major emphasis was on microbial study of incorporated noodles. The microbial growth was measured at 15 days' interval up to 90 days; simultaneously the chemical composition of noodles was also analyzed and results are narrated in Table 2. The microbial study was conducted by undertaking total plate count (TPC) and yeast and mold count of incorporated noodles.

The microbial study was conducted by undertaking TPC and yeast and mold count of incorporated noodles. Results were narrated in Tables 2 and 3.

The TPC of incorporated noodles in HDPE was found within 0.1 × 10^2 to 0.35 × 10^2. From the data in Table 2, it expresses that TPC of fresh sample was 0.1 × 10^2 which was gradually increased from 15 days (0.22 × 10^2) to 30 days (0.25 × 10^2), 45 days (0.30 × 10^2), 60 days (0.31 × 10^2), 75 days (0.33 × 10^2) and (0.35 × 10^2) at 90 days of study. The Yeast and Mold count was also increased gradually and founds a dominant microflora in incorporated noodles. The yeast and mold count of incorporated noodles was ranged from 0.06 × 10^2 to 0.12 × 10^2. On the first day, it was 0.06 × 10^2 while it was steadily increased from 15 days (0.08 × 10^2) to 30 days (0.09 × 10^2), 45 days (0.1 × 10^2), 60 days (0.12 × 10^2), 75 days (0.12 × 10^2) and (0.13 × 10^2) at 90 days of study. The similar results regarding microbial characteristics of noodles were discussed by Akhigbemidu et al. [18].

The consistent increased growth in microbes especially Yeast and Mold in incorporated noodles shows that noodles was containing enough moisture and other nutrients required for microbial growth. The reason behind growth of Yeast and Mold could be the lower water activity of incorporated noodles. The TPC of incorporated noodles in LDPE was ranged from 0.1 × 10^2 to 0.42 × 10^2. From the Table 3 founds that the fresh sample TPC was 0.1 × 10^2which was exponentially increased from 15 days (0.23 × 10^2) to 30 days (0.25 × 10^2), 45 days (0.31 × 10^2), 60 days (0.34 × 10^2), 75days (0.38 × 10^2) and (0.42 × 10^2) at 90 days of study. Simultaneously LDPE packed incorporated noodles were evaluated for yeast and mold. The given data utters that yeast and mold count of incorporated noodles was ranged from 0.06 × 10^2 to 0.26 × 10^2. On the first day, it was 0.06 × 10^2 while it was gradually increased from 15 days (0.11 × 10^2) to 30 days (0.15 × 10^2), 45 days (0.18 × 10^2), 60 days (0.19 × 10^2), 75 days (0.21 × 10^2) and (0.26 × 10^2) at 90 days of study. The similar results were discussed by [18] ref.

From discussion of Tables 2 and 3 it could be realized that the microbial growth of HDPE packed incorporated noodles was less

Sr. No.	Microbial study parameter	Colony count cfu/gm						
		Fresh	15 days	30 days	45 days	60 days	75 days	90 days
1	TPC	0.1×10^2	0.22×10^2	0.25×10^2	0.30×10^2	0.31×10^2	0.33×10^2	0.35×10^2
2	Yeast and mold	0.06×10^2	0.08×10^2	0.09×10^2	0.1×10^2	0.12×10^2	0.12×10^2	0.13×10^2

Table 2: Microbial study of incorporated noodles packed in high density polyethylene (HDPE).

Sr. No.	Microbial study parameter	Colony count cfu/gm						
		Fresh	15 days	30 days	45 days	60 days	75 days	90 days
1	TPC	0.1×10^2	0.23×10^2	0.25×10^2	0.31×10^2	0.34×10^2	0.38×10^2	0.42×10^2
2	Yeast and mold	0.06×10^2	0.11×10^2	0.15×10^2	0.18×10^2	0.19×10^2	0.21×10^2	0.26×10^2

Table 3: Microbial study of incorporated noodles packed in low density polyethylene (LDPE).

than LDPE packed noodles. It could be the result of good packaging functionalities of HDPE as compare to LDPE.

Conclusion

A nutritious noodle could be prepared by incorporation of DSF and DRB in such a way that, it could increase the protein and fiber content up to required optimum level. Soy founds rich source of protein and rice bran founds rich source of fibers, so it could be incorporated in noodles. The noodles packed and stored in HDPE having better storability than LDPE without affecting much quality.

Acknowledgment

At the outset, I am glad to say thank you to an enthusiastic person Dr. Sahoo A. Director Department of Technology, Shivaji University, Kolhapur for his kind support and guidance. Also, cordially thanks to my research guide Prof. Udachanlranna for their valuable guidance while study.

References

1. Gulia N, Dhaka V, Khatkar BS (2014) Instant noodles: Processing, quality and nutritional aspects. Food Sci Nutri 54: 1386-1399.
2. Okoye JI, Nkwocha AC, Ogbonnaya AE (2008) Production, proximate composition and consumer acceptability of biscuits from wheat/soybean flour blends. Continent J Food Sci Technol 2: 6-13.
3. Lee SJ, Kim JJ, Moon HI, Ahn JK, Chun SC, et al. (2008) Analysis of isoflavones and phenolic compounds in Korean soybean [*Glycine max* (L.) *Merrill*] seeds of different seed weights. J Agri Food Chem 56: 2751-2758.
4. Anonymous (2011) WINA (World instant noodle association): Expanding market.
5. Kim SG (1996) Instant noodles, pasta and noodle technology. American Association of Cereal Chemistry, St Paul MN.
6. Yadav S, Gupta RK (2015) Formulation of noodles using apple pomace and evaluation of its phytochemicals and antioxidant activity. J Pharmacog Phytochem 4: 99-106.
7. Ross AS (2013) Fiber enriched and whole wheat noodles. Food Sci Technol Nutri 3: 291-360.
8. Fu BX (2008) Asian noodles History classification raw materials and processing. Food Res Int 41: 888-902.
9. Tan HZ, Li ZG, Tan B (2009) Starch noodles history classification materials processing structure nutrition quality evaluating and improving. Food Res Int 42: 551-576.
10. Neha M, Ramesh C (2014) Development of functional biscuit from soy flour & rice bran. Int J Agri Food Sci 1: 27-32.
11. Khalid B, Vidhu A, Lubna M (2012) Physio-chemical and sensory characteristics of pasta fortified with chickpea flour and defatted soy flour. J Environment Sci Toxicol Food Technol 1: 34-39.
12. Omeire GC, Umeji OF, Obasi NE (2014) Acceptability of noodles produced from blends of wheat acha and soybean composite flours. Nigeria Food J 32:31-37.
13. Ranganna S (1986) Handbook of analysis and quality control for fruit and vegetable products (2nd edn). Tata McGraw Hill Publ Co Ltd, New Delhi.
14. Gatade AA, Sahoo AK (2015) Effect of additives and steaming on quality of air dried noodles. J Food Sci Technol 13: 197-215.
15. Taneya MLJ, Biswas MMH, Shams-Ud-Din M (2014) The studies on the preparation of instant noodles from wheat flour. J Bangladesh Agril Univ 12: 135-142.
16. Shogren RL, Hareland GA, Wu YV (2006) Sensory evaluation and composition of spaghetti fortified with soy flour. J Food Sci 71: 1750-3841.
17. Amudha S, Ravi R, Bhat KK, Seethalakshmi RM (2002) Studies on the quality of fried snacks based on blends of wheat flour and soya flour. Elsevier Sci.
18. Akhigbemidu W, Musa A, Kuforiji O (2015) Assessment of the microbial qualities of noodles and the accompanying seasonings. Nigeria Food J 33: 48-53.

26

Research on Some Questions of Inter-Esterification of Oils and Fats

Oltiev AT* and Majidov KH

Bukhara Engineering and Technological Institute, Bukhara City, Uzbekistan

Abstract

Development of methods of reception and the organization of industrial production of food plastic fats with balanced fatty-acid structure lowered to natural level by the maintenance of trance-isomerized acids at the optimum maintenance of irreplaceable linoleic acid is one of the actual problems which decision provides the further scientific and technical progress of fat-processing industries.

Keywords: Cotton oil and products of its processing; Cotton palmitin; Inter-esterification fats; Technology of their manufacture

Introduction

The scientific essence of the problem consists in research of advanced methods of modification of natural fats and oils, allowing to keep basic biologically important components of food fats in native condition and to ensure plastic mixtures of triglycerides in which the basic structuring components are triglycerides of saturated fat acids. Scientific work is directed on creation of universal technology of food plastic fats reception with adjustable acid and triglyceride structure and optimum physical and chemical indicators on the basis of purposeful application of statistical interesterification of fats [1].

Materials and Methods

Interesterification of oils, fats and modeling mixtures have been carried out in closed temperature-controlled reactor with volume of 0.1 dm^3, supplied with intensive mixer. The general concentration of sodium, concentration of active sodium alcoholate and sodium soaps in oil suspension estimated by methods of titration analysis. Activity of catalysts estimated on degree of interesterification of modeling mixtures. For research of chemical compound of the catalyst disintegration products were used selective extraction, gas-liquid and thin-layer chromatography and also nuclear magnetic resonance method. The maintenance of trans-isomershas been defined by the IR-spectroscopy method; division of triglycerides on molecular weight has been defined by the method of high-temperature gas-liquid chromatography; Research of distribution of fat acids in triglycerideshas been defined by the method of enzymatic hydrolysis with the subsequent calculation triglyceride structure [2-4]. Maintenance of monoesters of fat acids, mono- di- and triglycerideshas been defined by method of thin-layer and gas-liquid chromatography.

The polymorphic crystal structure of fat was analyzed by methodsX-ray diffraction and differential-thermal analysis.

Results and Discussion

Dynamics of redistribution of fat acids in molecules and between molecules triglycerides analyzed on modeling mixtures of liquid vegetable oils, hydrogenated fats with fusion temperature of 31°C to 65°C, grease, palm-oil and palm stearin, differing by the various maintenance of saturated acids CT_9-CO_9. isomerized mono-non-saturated acids and linoleic acids.

By method of lipase hydrolysis it is established that distribution of fat acids between extreme and average positions of glycerides changes gradually in process of interesterification, and ends with achievement offatty-acid structure identical to all positions equalto fatty-acid structure of an initial mixture of fats [3]. Fatty-acid structure of mono, di and triglycerides, monoester of fat acids and free fat acids present in interesterificated fat also answers to structure of initial fat that testifies to an equivalence of fat acids in process of interesterification, irrespective of their structure and an initial arrangement in molecules of triglycerides. The parity of symmetric and asymmetrical mono- and di-saturatedtriglycerides in mixture in process of interesterification continuously changes and in interesterificated fat reaches the quantity of 1:2 [5]. At interesterification of mixtures with identical fatty-acid structure the identical molecular structure of statistically interesterificated fat is reached, and change of triglyceride structure occurs gradually in process of interesterification and ends with formation of glyceride structure corresponding to statistical distribution of fat acids (Figure 1).

Designations

Π_3, Π_2H - mole rate of trisaturated and disaturated glycerides correspondingly,%;

1, 2, 3-mixtures of cotton oil with the beef fat (60:40), palm stearin (58:42) and deeply hydrogenated fat (78:22) accordingly;

N_{01}, N_{02}, N_{03}-triglyceride structure of initial fatty mixtures 1, 2, 3 accordingly;

Apparently on Figure 1, during the selection of mixtures of defined fatty-acid structure and spending of interesterification to certain depth, it is possible to receive any triglyceride structure between structure of an initial mixture of fats and statistical distribution.

Points designate the structures of fat received by interesterification; the continuous line designates the structures received by mixture of initial and interesterificated fat. In practice reproducibility of partial interesterification is insufficient because of fluctuations of concentration of the catalyst, qualityof preparation of fatty raw materials, rather high speed of process and absence of express methods of the direct control

***Corresponding author:** Oltiev AT, Bukhara Engineering and Technological Institute, Bukhara City, Uzbekistan, E-mail: kafedra-03@mail.ru

triglyceride structure. It is established that mixturesof triglycerides, close on molecular structure to partially interesterificated fats, can be received by mixture of initial and interesterificated fatty mixture in certain parities. This variant of process is more favorable from the economic point of view too, because waste and fat losses in process interesterification are decreased [6]. It is shown that the major advantage of statistical interesterification of fats is possibility of reception of relatives on physical and chemical indicators interesterificated fats from the various fatty mixtures having identical group of fatty-acid structure. The basic physical and chemical indicators of these fats coincide within accuracy of methods of the analysis that is reasoned by close properties of mix-acidic triglycerides of palmitic and stearin acids, and also of oleic and linoleic acids.

For the first time is obtained systematized data characterizingdependence of physical and chemical indicators of interesterificated fats from their group fatty-acid and triglyceride of structure.

Influence of fatty-acid structure of interesterificated fat on fusion temperature, hardness and the maintenance of a firm phase at 15°C is shown on Figure 2. With increase of the maintenance of the high-molecular saturated fat acids (palmitic and especially stearin acids)

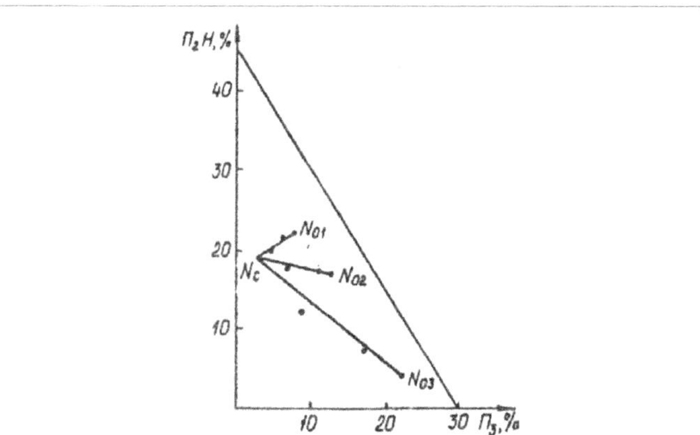

Figure 1: Statistical interesterification of mixtures of fats and oils with identical fatty-acid and various triglyceride structure (mole rate of saturated acids $C_{16} - C_{22}$ in initial fatty mixtures $N_{01} = N_{02} = N_{03} = 30\%$).

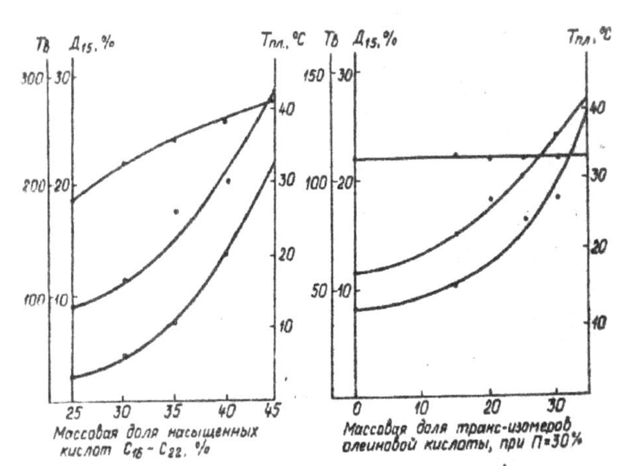

Figure 2: Dependence of physical and chemicalindicators ofinteresterificated fats from concentration of saturated fat acids C_{16} - C_{22} and trance-isomerized acids.

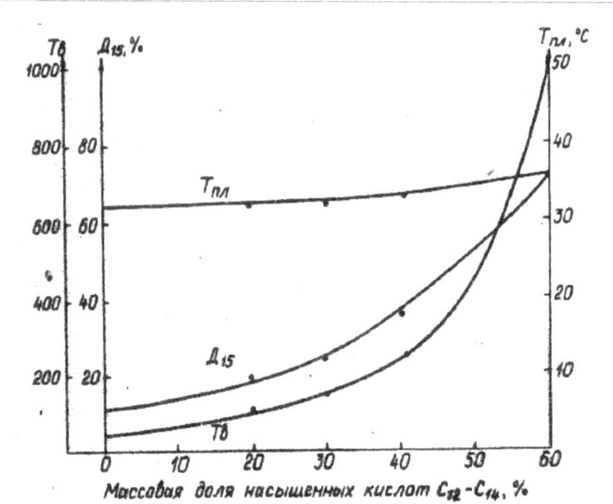

Figure 3: Dependence of physical and chemical indicators of interesterificated fats from concentration of saturated fat acids C_{12}-C_{14} (at П = 30%).

from 25% to 45% the fusion temperature of interesterificated fat increases from 28°C to 41°C. Hardness of interesterificated fat from 20 to 220 g/cm and mass fraction of firm phase simultaneously rises at 15°C from 8% to 28%. Parallel character of change of hardness and $П_{15}$ shows to possible interdependence of these indicators.

At increase of the maintenance of trans-isomersof oleic acids from 0% to 35% against the constant maintenance of saturated fat acids C_{16}-C_{22} (П = 30%) fusion temperature of interesterificated fat increases slightly, however essentially increases hardness (from 40 g/cm to 120 g/cm) and mass fraction of a firm phase at 15°C.

Especially strong influence on properties of interesterificated fat is rendered by replacement of part of non-saturated fat acids with average-molecularsaturated acids C_{12}-C_{14}. In the fats, containing 30% of saturated fat acids C_{20}-C_{22}, at the increase maintenance average-molecular acids from 0 to 60% the fusion temperature increases from 32°C to 36°C, hardness from 40 g/cm to 1000 g/cm, mass fraction of firm phase at 15°C from 11% to 71% (Mass concentration of saturated acids C12-C14 % (Figure 3)).

For the purpose of the quantitative descriptionof dependences between fatty-acid and triglyceride structure of interesterificated fats and their major physical and chemical indicators we have representation about structuring triglycerides. Fat as a first approximation represents the heterogeneous diphasic system which liquid phase is formed fused triglycerides. The firm phase of this system is formedby triglycerides which at the given temperature are insoluble or partially soluble in a liquid phase (structuring glycerides).

At each temperature structure of structuring triglycerides is different, depending on temperature of their fusion and solubility in a liquid phase of system. Compounding of interesterificated fats for physiological tests on fatty-acid structure intended as for foodof young, growing organism, and for food of patients with infringements lipidic metabolism [7]. Liquid vegetable oils of linolic-oleic groups with close to statistical distribution of fatty acids were entered into compounding, that provided similar distribution of linoleic acids in initial fatty mixture. Because of this after interesterification was reached not only preservation of the general level of linoleic acids, but also its concentration in physiologically important 2nd position of triglycerides (Table 1).

Structure of initial mix of fats and oils	Concentration of linoleic acids in 2nd position of triglycerides, %	
	Before inter-esterification	After inter-esterification
Mutton fat	3.5	3.5
Cotton oil	39.0	33.6
Mutton fat + Cotton oil (50:50)	19.0	17.7
The hydrogenated and natural cotton oil (50:50)	12.2	11.6
The hydrogenated cotton oil + mutton fat + natural cotton oil (55:15:30)	11.5	11.8

Table 1: Influence of statistical interesterification of mixtures of fats and of oils on concentration of linoleic acids in 2nd position of triglycerides.

Soft temperature conditions of interesterification of fats on sodium alkoxydes (60°C to 90°C) and rather low dosages of the catalyst (0.2% to 0.3%) allow keeping biologically important components of food fats in native condition. It is established that in process of interesterification free sterols almost completely etherified byfat acids of triglycerides, but it does not reducebiological value of sterols. There is also considerable decrease in concentration of tocopherols (50% from initial level), however this decrease much lower than at hydrogenation of fats [4]. At the contact of fat with sodium alkoxydes there will be destruction hydroperoxide connections which maintenance is decrease in 5-6 times. Deep clearing of fatty raw materials of oxidation products provides stability increase of interesterificated fats at the storage.

Conclusion

Analysis of the literary information and industrial production of margarine has shown that for radical improvement of quality and biological valueof mass kinds of margarine production, expansions of its raw-material base, the organization of manufacture of dietary margarine, products of a children's food, soft and bar margarine of sandwich-type are necessary plastic food fats with balanced fatty-acid structure, lowered to natural level by the maintenance of trans-isomerized acids at the optimum maintenance of irreplaceable linoleic acid.

In work is offered advanced technology of updating of naturalfats and oils, allowing to keep in native condition the basicbiologically important components of fatty raw materials and to ensure some plastic food fats in which the basic structuring components are triglycerides of the saturated fat acids.

On the basis of consideration of plastic fats as solution of high-meltingtriglycerides in liquid fraction laws of change of physical and chemical indicators of plastic mixtures triglycerides at change of their molecular structure are theoretically proved. Correlation between group of triglyceride structure and the major physical and chemical indicators (fusion temperature, hardness, a mass fraction of a firm phase in the range of temperature 0°C to 40°C) of statistically interesterificated fats and their mixtures is experimentally confirmed.

The obtained data was theoretical base for formationof fatty bases of margarine production with the set properties and substantiations of chemical-technological requirements to interesterificated fats for manufacture of margarine, culinary, baking fats and other food stuff.

Methods of calculation of componential structure of fatty basesreceived by mixture of fats of known glyceride structure and partial interesterification of fatty raw materials are developed. Process of interesterification of triglycerides at presence of sodium alkoxydes is analyzed and influence of technology factorson speed of intramolecular and intermolecular migration of acyls of fat acids in triglycerides is estimated. The treatment of process of interesterification of fats as chain non-ramified heterolytic reaction is offered, in which the active center conducting the chain of interesterification is sodium glyceroxyde.

The phasic mechanism of chain process is analyzed. It is established that the stage of chain origin represents reversible reaction of initial sodium alkoxyde with triglycerides, kinetic laws are analyzed, and the constant of balance and energy of activation of this reaction is defined. It is shown that distribution of the active centers of sodium glyceroxyde in mixture of triglycerides carries the statistical property. Kinetic laws are analyzed and the mechanisms of collateral nucleophilic reactions of splitting of sodium glyceroxyde are offered. It is established that the stage of interesterification represents the complex of reversible reactions of 2nd order. Principles of selectionof models and condition of carrying out of the reaction are offered, allowing to provide a quasi-stationary state of chain process and to receive the kineticequation of the pseudo-first order. Kinetic laws of reaction of interesterification are analyzed and optimum parameters of process at production conditions are proved.

References

1. Arutyunyan NS, Kornena EP, Yanova LI (1999) Technology of processing of fats. Pishepromizdat.
2. Akramov OA (2008) Modification of cotton oil on effective catalysts. Tashkent, TCTI.
3. Minzdrav (2002) Hygienic requirements of safety and food value of foodstuff. Sanitary and prophylactic instructions and norms.
4. Stopskiy VS, Klyuchkin VV, Andreev NV (1992) Chemistry of fats and products of processing of fatty raw materials. Kolos.
5. Tyutyunnikov BN, Gladkiy FF (1992) Chemistry of fats. Kolos.
6. Perkel RL (2000) Modern problems of interesterification of oils and fats. Development of fat-and-oil complex of Russia in the conditions of market economy. All-Russian conference, Moscow.
7. Shenfeld T (2002) New methods of fat refining. Texizdat.

Physico-Chemical Characteristics of Cookies Prepared with Tomato Pomace Powder

Mudasir Ahmad Bhat* **and Hafiza Ahsan**

Department of Post-Harvest Technology, SKUAST-K, Shalimar, Srinagar

Abstract

This study was conducted to investigate the physico-chemical characteristics of cookies added with different levels of tomato pomace powder. Cookies were prepared with six different levels of tomato pomace powder (0, 5, 10, 15, 20 and 25%) and the physicochemical properties were examined. Crude protein and ash content of cookies containing 20, 25% of tomato pomace powders were significantly higher than those of control and rest of treatments ($p<0.05$). The spread factor of control cookie was higher than that of cookies containing 5, 10, 15, 20 and 25% of tomato pomace powder. The incorporation of tomato pomace powder in cookie lowered the lightness values but increased redness and yellowness values. Sensory evaluation revealed that overall desirability scores were not significantly different between control and tomato pomace powder incorporated cookies. The cookies up to 5% substitution of powder were found acceptable by the consumers.

Keywords: Tomato pomace powder; Cookies; Physico-chemical characteristics

Introduction

Today foods are not only intended to satisfy hunger and to provide necessary nutrients for humans but also to prevent nutrition-related diseases and improve physical and mental well-being [1]. Functional foods are quiet remedy. Cookies are widely consumed baked products which can be served from breakfast to bedtime. Cookies are appreciated for their taste, aroma, convenience, and long shelf stability due to low moisture content. Recently, increasing consumer demand for healthier foods has triggered the development of cookies made with natural ingredients exhibiting functional properties and providing specific health benefits beyond those to be gained from traditional nutrients. Tomatoes are an integral part of the human diet rich in several nutrients including vitamin A, vitamin C, potassium, calcium and lycopene. Most tomatoes produced worldwide are used as tomato juice, paste, ketchup and sauce, although a number of tomatoes are commonly consumed fresh [2]. Recent studies have indicated that a diet rich in tomatoes and tomato products is associated with a reduced risk of certain cancers [3]. The consensus seems to be that lycopene; a major carotenoid present in red tomatoes having antioxidant capacity is a natural cancer-fighting agent. It is also suggested that complex interactions among multiple nutrients present in tomatoes contribute to its anticancer properties [4].

Tomato pomace is highly perishable in its fresh state because it contains about 95% moisture and its storage is difficult [5]. Therefore, the waste is just dumped and allowed to decay which increases landfill costs and societal concerns about solid waste. These generated interest in finding an economical outlet for this byproduct. Bakery product cookies are fast and convenient food based on wheat. Cookies, which constitute an important item of bakery industry have now become a common item of consumption among all classes of people with tea or coffee, Different variety of cookies are used as one of the tasty and nutritious snacks. Cookies are more popular as a convenient food.

In present scenario, there is an increasing demand for conversion of fruit and vegetable wastes into useful products as well as to minimize environmental impact of these by-products. Successful incorporation of tomato pomace into bakery products that deliver physiologically active components represents a major opportunity for food processors providing the consumer a healthy wheat based product to choose from which is currently lacking in the marketplace. Keeping in view the bioactive potential and health benefits of tomato pomace, the proposed study shall be undertaken to investigate the utilization of tomato pomace for development of wheat based cookies with the objective to evaluate the effects of tomato powder on the quality characteristics of cookies. Cookies were prepared with tomato powder which was substituted for 0%, 5%, 10%, 15%, 20% and 25 % of wheat flour and the physicochemical and sensory properties were investigated.

Materials and Methods

Materials

Ingredients for cookies such as wheat flour, milk, sugar, butter, egg white and baking powder were purchased from a local market. Tomato pomace prepared from local tomato variety was dried in a cabinet dryer at 60°C and then grinded in a grinder-mixer to obtain tomato pomace powder with following proximate composition: moisture 14.89%, protein 20.36%, fat 0.35%, and ash 8.97%.

Preparation of cookies

The cookie recipe used in this study is shown in Table 1. Tomato pomace powder was used to replace part of the whole wheat flour (0%, 5%, 10%, 15%, 20% and 25%) in a standard cookie recipe. Egg white, butter and sugar were creamed in a hand mixer for 3 min. Flour, tomato pomace powder and baking powder were sifted and added to the liquid ingredients and mixed for 30 seconds. The firm dough was then sheeted to a thickness of 5 mm thick with roller and cut with a cookie cutter of

***Corresponding author:** Mudasir Ahmad Bhat, Department of Post-Harvest Technology, SKUAST-K, Shalimar, Srinagar, E-mail: mudasagar@gmail.com

50 mm diameter. The cookies were then transferred to a lightly greased baking tray and baked in an oven at an upper heating temperature of 160°C for 15 min. The baked cookies were removed from the baking pan, cooled to room temperature for 1 hr before analysis.

Proximate composition analysis

Moisture, crude protein, crude fat and ash content were determined according to the AOAC method [6].

Physical characteristics

Cookie diameter (D) and thickness (T) of six cookies were measured as described in the AACC methods [7]. The spread factor was calculated as D/T. The volume of cookies were determined by seed displacement method

Color measurement

Surface color of cookies was determined by measuring tristimulus L (lightness), a (redness), b (yellowness) values with a colorimeter calibrated with a white standard plate (L=+98.5, a=+0.07, b=-0.40).

Sensory evaluation

Sensory evaluation was conducted after cooling the cookies for 1hr at room temperature. Cookies were placed on a plastic dish coded by a three-digit random number and offered to 8 trained panelists in an individual booth with lighting. Surface color, taste, texture, appearance and overall acceptability were evaluated using the five-point scale with 5 indicating strong attributes.

Statistical analysis

All data were recorded as means ± standard deviation of at least triplicate measurements. Means were compared with Duncan's multiple range tests with $\alpha=0.05$ using SAS (Statistical Analysis System, version 8.12).

Results and Discussion

Proximate composition

The results of the proximate composition of cookies prepared with tomato powder are shown in Table 2. Moisture content of control was 2.49% and that of cookies containing tomato powder ranged from 2.24 to 3.39%. Crude protein and ash content values of the cookies containing tomato powder of 5.0, 7.5 and 10.0% were significantly higher than those of control and 2.5% added sample ($p<0.05$). This is attributable to the higher protein and ash content of tomato powder than wheat flour. Crude fat contents were 21.71-21.85% and no significant differences among cookies were observed.

Physical characteristics

Cookie diameter, thickness and spread factor are shown in Table 3. There were no significant differences in the diameter and thickness of the cookies between those containing up to 5.0% and the control. However, significant differences were found with higher percentages of tomato powder ($p<0.05$). Larger diameter and lower thickness values were observed in the formulations containing 7.5 and 10.0% of tomato powder. Thus, it could be stated that the incorporation of tomato powder in the cookies at the level of 7.5 and 10.0% affected cookie expansion by lowering gas retaining power below that of wheat flour [8]. The spread factor of cookies made with 5.0, 7.5 and 10.0% tomato powder was significantly higher than that of control ($p<0.05$), but a trend with the varying levels of tomato powder substitution for flour was not found. Chung and Kwon [8] reported that cookies made with yam powder exhibited a reduction in size and thickness as the content of yam powder increased. According to Kim et al. [9] barley germ increased cookie diameter and spread ratio, whereas decreased thickness and weight as the barley germ substitution level increased, and they have proposed that such increases in diameter and spread ratio may be due to the decrease in cookie dough consistency with increasing level of barley germ in the cookie formula. Dissimilar results have been

Ingredients %	Treatment combinations					
	D1T1	D1T2	D1T3	D1T4	D1T5	D1T6
Flour (g)	200	190	180	170	160	150
Tomato pomace powder (%)	0	5	10	15	20	25
Butter (g)	50	50	50	50	50	50
Milk powder (g)	50	50	50	50	50	50
Sugar (g)	100	100	100	100	100	100
Baking powder (g)	2	2	2	2	2	2
Egg white (No.s)	1	1	1	1	1	1

Table 1: Cookie recipe combinations used in the study.

	Moisture (%)	Crude protein (%)	Ash (%)	Crude fat (%)
Wheat flour	14.45 ± 0.19	9.50 ± 0.83	0.48 ± 0.01	0.87 ± 0.01
Tomato pomace powder	10.20 ± 0.75	18.25 ± 0.25	7.25 ± 0.05	0.65 ± 0.09
D1T1	2.49 ± 0.10^{a}	5.186 ± 0.25^{a}	4.883 ± 0.61^{a}	21.71 ± 0.01^{a}
D1T2	2.40 ± 0.20^{a}	5.503 ± 0.20^{b}	4.844 ± 0.22^{a}	21.80 ± 0.02^{b}
D1T3	3.39 ± 0.10^{b}	6.25 ± 0.22^{c}	4.916 ± 0.10^{b}	21.81 ± 0.55^{b}
D1T4	2.24 ± 0.01^{c}	6.406 ± 0.01^{d}	4.920 ± 0.01^{b}	21.82 ± 0.61^{b}
D1T5	2.97 ± 0.01^{d}	6.446 ± 0.01^{d}	4.912 ± 0.01^{b}	21.83 ± 0.10^{b}
D1T6	3.24 ± 0.02^{ab}	6.870 ± 0.01^{ab}	4.927 ± 0.02^{b}	21.85 ± 0.01^{c}

Each value is mean ± standard deviation (SD).

Means with different letters within a column are significantly different from each other at $\alpha=0.05$ as determined by Duncan's multiple range tests.

Table 2: Proximate compositions of cookies prepared with tomato powder.

Groups	D1T1	D1T2	D1T3	D1T4	D1T5	D1T6
Diameter(cm)	19.26 ± 0.10^{a}	19.30 ± 0.10^{b}	19.33 ± 0.05^{c}	19.75 ± 0.02^{d}	19.92 ± 0.25^{ab}	19.96 ± 0.10^{ac}
Thickness (cm)	3.10 ± 0.10^{a}	2.85 ± 0.08^{b}	2.82 ± 0.12^{c}	2.75 ± 0.25^{d}	2.64 ± 0.54^{ab}	2.55 ± 0.09^{ac}
Spread factor	6.21 ± 0.06^{a}	6.77 ± 0.12^{b}	6.85 ± 0.25^{c}	7.18 ± 0.12^{d}	7.54 ± 0.25^{ab}	7.82 ± 0.01^{ac}

Each value is mean ± SD.

Means with different letters within a row are significantly different from each other at α=0.05 as determined by Duncan's multiple range test.

Table 3: Diameter, thickness and spread factor of cookies prepared with tomato powder.

Group	L	a	b
D1T1	52.33 ± 0.88^{a}	3.31 ± 0.16^{a}	30.47 ± 1.16^{a}
D1T2	50.35 ± 0.22^{b}	5.68 ± 0.01^{b}	37.63 ± 0.10^{b}
D1T3	48.22 ± 0.30^{c}	6.74 ± 0.25^{c}	36.66 ± 2.45^{c}
D1T4	48.20 ± 2.88^{c}	7.49 ± 0.29^{d}	37.85 ± 0.36^{b}
D1T5	47.55 ± 0.55^{d}	8.33 ± 0.90^{ab}	41.59 ± 0.12^{d}
D1T6	47.45 ± 0.10^{d}	8.56 ± 0.01^{ab}	41.66 ± 0.01^{d}

Each value is mean ± SD.

Means with different letters within a column are significantly different from each other at α=0.05 as determined by Duncan's multiple range tests.

Table 4: Colorimetric characteristics of cookies prepared with tomato powder.

Group	Taste	Texture	Color	Appearance	Overall acceptability
D1T1	4.22 ± 0.85^{a}	3.55 ± 0.80^{a}	3.44 ± 0.10^{a}	3.62 ± 0.01^{a}	3.70 ± 0.01^{a}
D1T2	4.25 ± 0.02^{a}	3.85 ± 0.25^{b}	4.56 ± 1.41^{b}	4.50 ± 1.60^{b}	4.29 ± 1.24^{b}
D1T3	3.24 ± 1.21^{b}	3.65 ± 0.68^{b}	3.28 ± 0.98^{a}	3.68 ± 0.27^{a}	3.46 ± 0.32^{a}
D1T4	3.29 ± 0.58^{b}	3.22 ± 0.01^{a}	3.3 ± 0.10^{a}	3.25 ± 0.01^{a}	3.26 ± 0.10^{a}
D1T5	3.88 ± 0.25^{c}	3.26 ± 0.05^{a}	3.35 ± 0.02^{a}	2.66 ± 0.01^{c}	3.28 ± 0.05^{a}
D1T6	3.10 ± 1.25^{b}	3.45 ± 2.15^{a}	2.25 ± 0.05^{c}	2.22 ± 0.01^{c}	2.75 ± 0.01^{c}

Each value is mean ± SD.

Means with different letters within a column are significantly different from each other at α=0.05 as determined by Duncan's multiple range tests.

Table 5: Sensory scores of cookies prepared with tomato powder.

observed by other researchers, who reported that the spread ratio of cookies prepared with bamboo leaf powder decreased with increasing amount of bamboo leaves powder [10]. Cho et al. [11] also reported that the addition of sea tangle powder lowered the spread ratio of cookies. From these results, it is suggested that the incorporation of different ingredients into cookie products result in different physical properties. In general, cookie spread or diameter is used as an indicator of cookie quality and cookies with larger spread or diameter are considered more desirable [12].

Color measurement

The results of Hunter Lab color value of cookies are shown in Table 4. The L value (lightness) of control cookies was 52.33 and those of tomato powder cookies ranged from 50.35 to 47.45, indicating that lightness decreased with the reduction in the proportion of wheat flour because of the loss of white color of the flour ($p < 0.05$). Therefore, it could be expected that cookies would become darker with increasing amount of tomato powder level. The 'a' value (redness) of control cookies was 3.31 and those of tomato powder groups were 5.68-8.56, showing more reddish color than control. This is attributable to the reddish color of tomato powder. The b value (yellowness) of control group was 30.47 and cookies substituted with different levels of tomato powder ranged from 37.63 to 41.66, having more yellowish color than control. Similar results were obtained by Kim et al. [9] who reported that the color of sugar- snap cookies became darker, more red and yellow with increasing amounts of barley germ. Singh et al. [13] reported that lightness value decreased while redness and yellowness values increased as the substitution.

Sensory evaluation

Results of sensory evaluation of cookies with added tomato powder are shown in Table 5. The color was evaluated to become darker as the tomato powder level increased, which is due to the reddish color of tomato powder. This is in agreement with the result of lightness values shown in Table 4. Scores of tomato taste were higher in cookies added with tomato powder than those of control cookies because tomato powder gave a characteristic taste. The hardness score was higher in 5% substitution than those of cookies substituted with different levels of tomato powder. The appearance score was found higher in 5% substitution level than rest of the treatments. Appearance score was judged on the basis of number of cracks on the surface of cookies. The higher the cracks with uniformity better will be the cookie. There were no significant differences in the overall desirability scores between control cookies and tomato powder added cookies. Therefore, replacing up to 5% wheat flour with tomato powder would not result in significant differences in the acceptability of cookies.

Conclusion

In the experiment it was concluded that the cookies incorporated with the tomato pomace powder were found acceptable up to 5% substitution. The color of the cookies becomes more and darker as the level of pomace powder increased. Spread factor of the cookies was increased due to the pomace incorporation.

Acknowledgement

This work was supported by the Post-Harvest Technology, Division of Sher-e-Kashmir University of Agricultural Sciences and Technology, Shalimar Srinagar.

References

1. Takachi R, Manami I, Junko I, Norie K, Motoki I, et al. (2008) Fruit and vegetable intake and risk of total cancer and cardiovascular disease: Japan public health center-based prospective study. American Journal of Epidemiology 167: 59-70.

2. Sancez MC, Valencia C, Ciruelos A, Latorre A, Gallegos C, et al. (2002) Rheological properties of tomato paste: Influence of the addition of tomato slurry. J Food Sci 68: 551-554.

3. Giovannucci E (1999) Tomatoes, tomato-based products, lycopene and cancer: Review of the epidemiologic literature. J Nat Cancer Inst 91: 317-331.

4. Rao AV, Agarwal S (1999) Role of lycopene as antioxidant carotenoid in the prevention of chronic diseases: a review. Nutr Research 19: 305-323.

5. Akanbi CTR, Adeyemi S, Ojo A (2006) Drying characteristics and sorption isotherm of tomato slices. Journal of Food Engineering 73: 157-163.

6. AACC (1995) Approved Methods of the AACC. American Association of Cereal Chemists, St Paul, MN.

7. AOAC (1995) Official Methods of Analysis. Association of Official Analytical Chemists, Washington DC.

8. Chung, KM, Kwon CS (1999) Properties of cookies added of flour and Chinese yam powder. Food Sci Biotechnol 8: 341-343.

9. Kim IS, Lee YT, Seog HM (2002) Effects of barley germ on sugar-snap cookie quality. Food Sci Biotechnol 11: 515-519.

10. Lee JY, Ju JC, Park H J, Heu ES, Choi SY, et al. (2006) Quality characteristics of cookies with bamboo leave powder. Korean J Food and Nutr 19: 1-7.

11. Cho HS, Park BH, Kim KH, Kim HA (2006) Antioxidative effect and quality characteristics of cookies made with sea tangle powder. Korean J Food Culture 21: 541-549.

12. Fimney KF, Morris VH, Yamazaki WT (1950) Micro versus macro cookie baking procedures for evaluating the cookie quality of wheat varieties. Cereal Chem 27: 42-49.

13. Singh J, Singh N, Sharma TR, Saxena SK (2003) Physicochemical, rheological and cookie making properties of corn and potato flours. Food Chem 83: 387-393.

28

Study the Sensory Attributes and Shelf Life of Developed Digestive Pills from Makoi (*Solanum nigrum*)

Bimal Bibhuti* and Yadav AK

Department of Food Process Engineering, Vaugh School of Agricultural Engineering and Technology, Sam Higginbottom Institute of Agriculture, Technology and Sciences, Allahabad, India

Abstract

Solanum nigrum belongs to family Solanaceae. Blacknight-shade and Makoi are the common name for it. The chemical constituents commonly found in *Solanum nigrum* are glycoalkaloids, glycoproteins, polysaccharides, polyphenolic compounds such as gallic acid, cathechin, protocatechuic acid, caffeic acid, epicatechin, rutin etc. *Solanum nigrum* has very much importance as a medicinal plant. Root, whole plant and leaves are used but fruits of black colour are not used as they possess toxicity, therefore they are not used for medicinal purposes. Reddish- brown coloured fruits are used for edible purpose. It has been also extensively used in traditional medicine in India and other parts of world to cure liver disorders, digestion, chronic skin ailments, inflammatory conditions, painful periods, fevers, diarrhoea, eye disease, hydrophobia etc. Extracts prepared with using spices condiments exhibited stomach disorders and also improved digestion activity as compared to other digestive pills.

Keywords: *Solanum nigrum*; Medicinal; Liver disorders; Digestion; Condiments; Digestive pills

Introduction

India is the largest producer of medicinal plants and brightly called as the "Botanical garden of the world". Plant derived drugs even today remain important resource especially in developing countries to combat serious diseases. Approximately 62% to 80% of world's population still relies on traditional medicines for the treatment of common illness [1]. *S. nigrum* is an important ingredient in traditional Indian medicines. Infusions are used in dysentery, stomach complaints, and fever. In India, the berries are casually grown and eaten, but not cultivated for commercial use. The quest to manufacture foods for healthy benefits is an underpinning goal of the modern food industry. Food processing has evolved to carryout steps for the controlled destruction of natural food structures. These steps facilitate separation of valuable components from the original matrix in which they are embedded [2-7]. The separated ingredients are then converted into recognizable processed foods with desirable's textural and sensorial properties by application of one or more processing steps. Recent evidence indicates that how the food structure breaks down during gastric digestion significantly affects the rate of uptake of nutrients in the gastrointestinal (GI) tract. The digestion process has been well studied in terms of secretion of gastric fluids, enzymatic breakdown of fats, proteins and carbohydrates, and molecular and ionic transport across the intestinal epithelium. However, there remains a notable lack of understanding about the food disintegration kinetics and the extraction of small molecules from complex food structures in the gastric environment. The rate of food disintegration in stomach is a key factor influencing emptying rate and subsequently affecting absorption of nutrients in the intestine. Faster disintegration and emptying of drug tablets is responsible for the faster absorption of drug ingredients in the intestine. The stomach contraction, particularly terminal antral contraction, imposes a considerable mechanical destructive force on food particulates and thus is crucial on the disintegration of solids. Researchers have measured contraction forces present in the stomach ranging from 0.2 N to 2 N [8-11].

Findings from this research will provide an improved understanding of the interaction of the food matrix and active ingredients during gastric digestion. The computational modeling of the human stomach will predict the kinetics of disintegration of a food matrix under known physiological conditions of the stomach. These findings should provide new information for the food processing industry to develop structured foods for healthful benefits and develop strategies for controlled release of food nutrients at desired sites in the GI tract [12-15]. The anticipated information will enhance understanding of the stomach emptying of foods to develop approaches to control it. Control of gastric emptying is essential for ensuring optimal digestion. The rate of food disintegration in the stomach appears to be a key factor influencing emptying rate and subsequently affecting absorption of nutrients in the intestine [16-23]. The potential for modulation of the rate of gastric emptying to control obesity and diabetic patients is now being explored vigorously by the pharmaceutical industry. Study of gastric disintegration of foods should also help our understanding of the interactions between food and drugs during digestion [5,8,9]. The disintegration activity of a drug is substantially affected by the presence of food components. Thus the understanding of food disintegration should help improve the control of pill dissolution in stomach.

Kakamachi (*Solanum nigrum*) is widely described in the Ayurvedic classics & also have references in Vedas. This herb has its own ertho-medical importance since it plays a significant role in the treatment of various diseases. It is having both curative & nutritive value [21,24-28]. It is used as single drug & in compound formulations. In this research, we use different ingredients along with *Solanum nigrum* to increase their pharmacological properties, therapeutic or dietarg utility indigestion. Hence the project is carried out under the above objectives.

Materials and Methods

This chapter deals with the description of various materials and

***Corresponding author:** Bimal Bibhuti, Department of Food Process Engineering, Vaugh School of Agricultural Engineering and Technology, Sam Higginbottom Institute of Agriculture, Technology and Sciences, Allahabad, India
E-mail: bimalbibhuti@rediffmail.com

methods which was used to accomplish the experimental work done to attain the desired objectives of the work entitled, Study the sensory attributes and shelf life of developed digestive pills from Makoi (*Solanum nigrum*). The experimental technique, materials which was used in this study and the associated methodology for product development and their sensory analysis was elaborated in this chapter. The experimental plan was shown in Table 1 with different variables, levels and description [27-31].

From the above formula, the development of product from different treatments were done and in these treatments the treatment T_1 was widely accepted by the judgers and the further experiments were carried-out by using this formula [32-35] (Table 2). According to the judger's confirmation for treatment T_1, the alteration in moisture were calculated and by this alteration the total effect on the Physico-chemical properties of the developed product was estimated (as R_1, R_2 and R_3) and the mean value of the data for the product was selected which were given as follows. All the Physico-chemical values were calculated as per 100 g of the developed product (Table 3).

Methods

After the development of the product, the pills were further brought for study the sensory attributes and storage conditions from the following methods which were explained follows:

Organoleptic evaluation: Sensory attributes including colour, aroma, taste and overall acceptability was determined by hedonic rating tastes as recommended by Karmakar [18]. Hedonic rating taste was used for evaluation of sensory characteristics. This test was used for acceptability by consumer for the product. The detail methodology was presented below. A panel of 5 expert judges of different age group having different habit was be selected and samples were served to them [36-39]. The expert panellist was be asked to rate the acceptability of the product through sense organs on scale of nine (9) points ranging from like extremely to dislike extremely.

Packaging material: The protection offered by a package was determined by the nature of the packaging materials and by the type of package condition. A great variety of materials were used in packaging, as LDPE, HDPE and Glass bottles.

Results and Discussion

This chapter deals with the findings of the study and effect of developed pills with different spices condiments, on the parameters such as sensory attributes and shelf life of developed digestive pills from Makoi (*Solanum nigrum*).

Sensory evaluation

Sensory attributes including colour, taste, aroma, and overall

S.no.	Variables/ Parameters	Levels	Description
1	Product	1	Digestive pills
2	Ingredients	8	Makoi (*Solanum nigrum*), *Hing*, *Piper longum* and Tailed pepper, Ajwain, Jeera, Black Salt, Imli Paste and Artificial sweetner.
3	Treatment	1 × 5	T_0, T_1, T_2, T_3, T_4
4	Replication	3	R_1, R_2, R_3
5	Physico-chemical Analysis	7	Moisture, Protein, Fat, Total ash, Crude fibre, Carbohydrate and Calories.
6	Storage condition	1	Ambient temperature
7	Storage period	1	Two months
8	Packaging material	3	LDPE, HDPE, Glass bottle
9	Sample size	1	100 gms

Table 1: Experimental plan.

Ingredients	Treatments				
	T_0	T_1	T_2	T_3	T_4
Makoi	12%	20%	24%	25%	26%
Piper longum and Tailed pepper	5%	4%	3%	3%	3%
Hing	1%	1%	1%	1%	1%
Artificial sweetner	12%	12%	14%	15%	15%
Ajwain	20%	8%	8%	8%	7%
Jeera	15%	10%	10%	8%	8%
Black salt	10%	15%	15%	15%	15%
Imli paste	25%	25%	25%	25%	25%

Table 2: Experimental design.

Physico-chemical properties	Replication			Mean Value
	R_1	R_2	R_3	
Moisture	22.43%	16.23%	15.02%	17.89%
Protein	0.82%	0.85%	0.86%	0.84%
Fat	1.95%	1.97%	1.98%	1.97%
Total Ash	2.40%	2.60%	2.80%	2.60%
Crude Fibre	1.62%	1.60%	1.59%	1.60%
Carbohydrate	72.40%	78.35%	79.34%	76.70%
Calories	310.43 Kcal	334.53 Kcal	338.62 Kcal	327.89 Kcal

Table 3: Experimental design for the developed product.

acceptability was determined by hedonic rating tastes recommended by Hedonic rating taste which was used for evaluation of sensory characteristics. This test was used for acceptability by the consumer for the developed product [40]. The colour, taste, aroma and overall acceptability of developed digestive pills were accepted at different storage periods at ambient temperature by the judgers (Tables 4-6). Effect of packaging materials and ambient storage period on sensory evaluation of developed digestive pills [41,42] (Figures 1-3).

Summary

This present study on sensory attributes and shelf life of developed digestive pills from Makoi (*Solanum nigrum*) was undertaken in the Department of Food Process and Engineering, Vaugh school of Agricultural Engineering and Technology, Sam Higgin bottom Institute of Agriculture, Technology and Sciences Allahabad. In the present investigation efforts were made to develop the digestive pills

Packaging Material	Colour	Taste	Aroma	Overall Acceptability
LDPE	8.5	8.6	8.42	8.22
HDPE	8.5	8.6	8.42	8.22
Glass Bottle	8.5	8.6	8.42	8.22

Table 4: Sensory scores of digestive pills on day one.

Packaging Material	Colour	Taste	Aroma	Overall Acceptability
LDPE	8.5	8.5	8.40	8.20
HDPE	8.5	8.5	8.42	8.21
Glass Bottle	8.5	8.6	8.42	8.22

Table 5: Sensory scores of digestive pills on 30th day.

Packaging Material	Colour	Taste	Aroma	Overall Acceptability
LDPE	8.5	8.4	8.30	8.0
HDPE	8.5	8.5	8.41	8.20
Glass Bottle	8.5	8.5	8.42	8.21

Table 6: Sensory scores of digestive pills on 60th day.

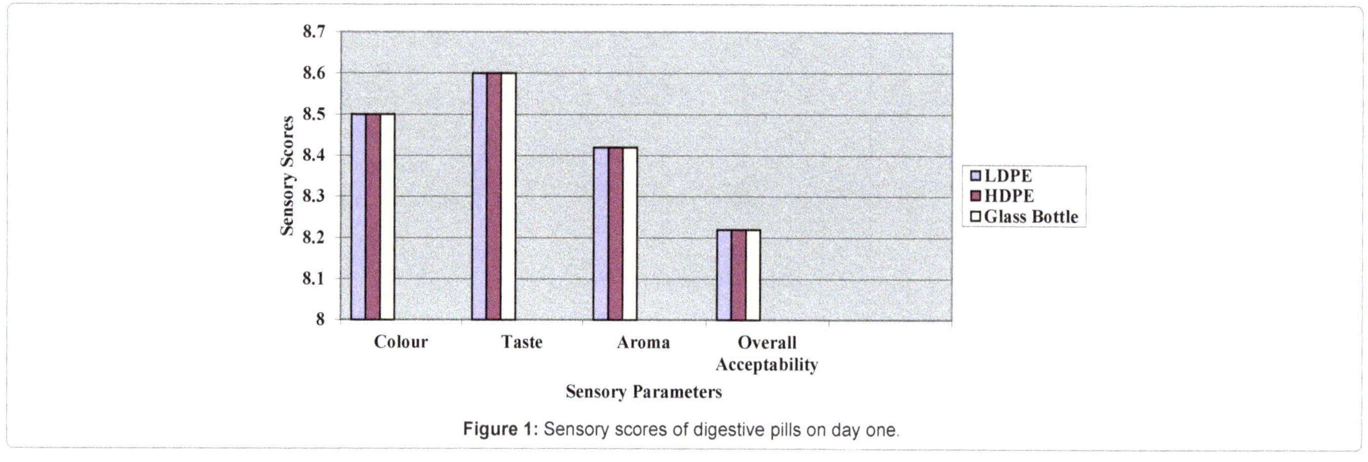

Figure 1: Sensory scores of digestive pills on day one.

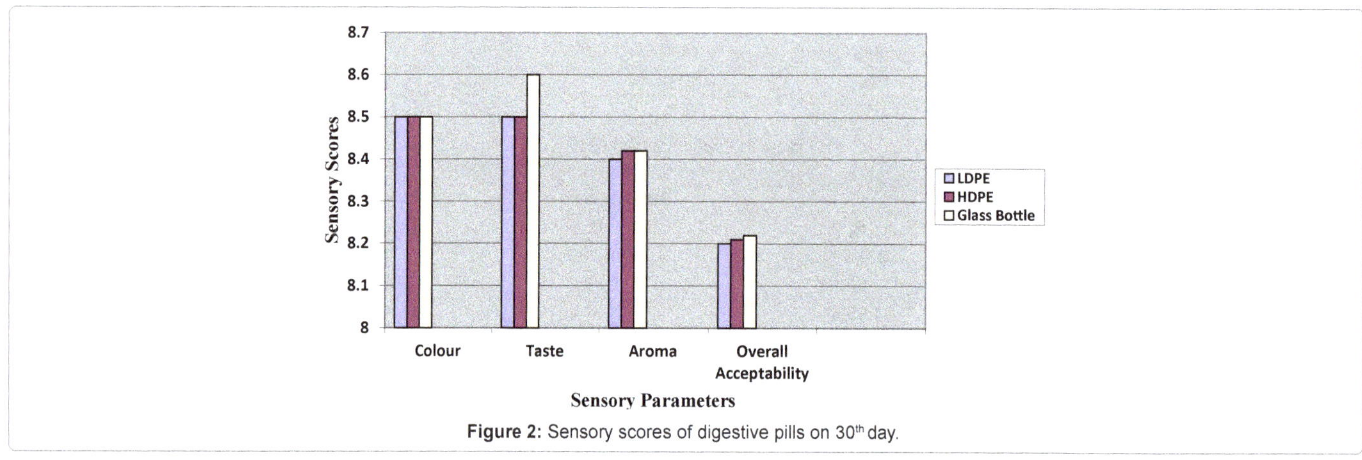

Figure 2: Sensory scores of digestive pills on 30th day.

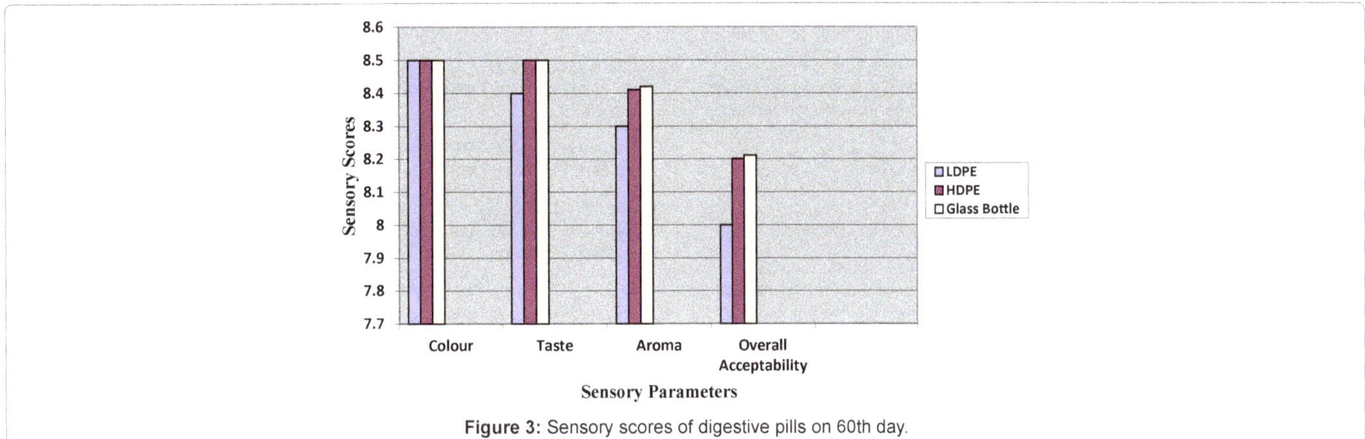

Figure 3: Sensory scores of digestive pills on 60th day.

with incorporation of other digestive ingredients which also helped the sensory characteristics to the developed digestive pills [43-45]. The sensory attributes such as colour, taste, aroma and overall acceptability was higher at initial packaging period i.e. at day one in LDPE, HDPE and Glass bottle and it was slightly decreases with increase in storage period at ambient temperature but in case of Glass bottle, it was not much affected. It was also found from Annova that the variation between the product treatments was non-significant and the variation within the product replications was significant.

Conclusion

From this preliminary investigation and research, it has been concluded that the development and Physico-chemical properties of developed digestive pills have good source of protein, fat, crude fibre, carbohydrate and calories which also lies under the dietary limit of daily intake. The developed digestive pills have found satisfactory in the stomach digestion after testing their physic-chemical analysis like moisture, ash, protein, fat, crude fibre, and also depended upon different sensory attributes like colour, taste, aroma, and overall acceptability during shelf life study in two months were not much effected from controlled sample i.e. Physico-chemical and Organoleptic characteristics were good and was acceptable. Thus, we concluded that the developed digestive pills were gave good response in HDPE and Glass bottle, storage at maximum storage period at ambient temperature compared than LDPE; while the extract mechanism of action remains to be elucidated in many cases of disease. The Makoi (*Solanum nigrum*) has wide-ranging Therapeutic properties needs to be investigated in well-designed studies and further research is in process to find more uses of *Solanum nigrum*.

References

1. Aguilera JM, Stanley DW (1999) Microstructural principles of food processing and engineering. Aspen Publishers Inc, Maryland, USA.
2. Kelly K, O'Mahony B, Lindsay B, Jones T, Grattan TJ, et al. (2004) Comparison of the rates of disintegration, gastric emptying, and drug absorption following administration of a new and a conventional paracetamol formulation, using γ scintigraphy. Pharm Res 20: 1668-1673.
3. Vassallo MJ, Camilleri M, Prather CM, Hanson RB, Thomforde GM, et al. (1992) Measurement of axial forces during emptying from the human stomach. Am J Physiol Gastrointest Liver Physiol 263: 230-239.
4. Kambizi L, Afolayan AJ (2001) An ethnobotanical study of plants used for the treatment of sexual transmitted diseases (njovhera) in Guruve District, Zimbabwe. J Ethnopharmacol 77: 5-9.
5. Rayner CK, Samsom M, Jones KL, Horowitz M (2001) Relationships of upper gastrointestinal motor and sensory function with glycaemic control. Diabetes Care 24: 371-381.
6. Agrawal KR, Lucas PW, Prinz JF, Bruce IC (1997) Mechanical properties of foods responsible for resisting food breakdown in the human mouth. Arch Oral Biol 42: 1-9.
7. Kojia M, Popovia R, Karadyia B (1997) Vaskularne biljke Srbiyekaoin dikatoristanista, Institut zaistrayivanja u polyoprivedi "Srbija" and Institut zabioloskai strayivanja "sinisa Stankovia", Beograd.
8. Lee KR, Kozukue N, Han JS (2004) Glycoalkaloids and metabolites inhibit the growth of human colon (HT29) and liver (HepG2) cancer cells. J Agric Food Chem 52: 2832-2839.
9. Ali NS, Singh K, Khan MI, Rani S (2010) Protective effect of ethanolic extracts of *Solanum nigrum* on the blood sugar of albino rats. IJPSR 1: 97-99.
10. Sridhar TM, Josthna P, Naidu CV (2011) Antifungal activity, phytochemical analysis of *Solanum nigrum* (L.)-An important antiulcer medicinal plant. J Ecobiotechnol 3: 11-15.
11. Sultana S, Perwaiz S, Iqbal M, Athar M (1995) Crude extracts of hepatoprotective plants, *Solanum nigrum* and Cichoriuminty businhibit free radical-mediated DNA damage. J Ethnopharmacol, 45: 189-192.
12. Atanu FO, Ebiloma UG, Ajayi EI (2011) A review of the pharmacological aspects of *Solanum nigrum* Linn. Biotechnol Molecul Biol Rev 6: 1-7.
13. Chifundera K (1998) Livestock diseases and the traditional medicine in the Bushi area, Kivu province, Democratic Republic of Congo. Afr Study Monogr 19: 13-33.
14. Sharma YK, Singh H, Mehra BL (2004) Hepatoprotective effect of few Ayurvedic herbs in patients receiving antituberculus treatment. Indian J Traditional Knowledge 3: 391-396.
15. Sikdar M, Dutta U (2008) Traditional phytotherapy among the Nath people of Assam. Entno- Med 2: 39-45.
16. Cooper MR, Johnson AW (1984) Poisonous plants in Britain and their effects on animals and man. London HMSO.
17. Dhellot JR, Matouba E, Maloumbi MG, Nzikou JM, Dzondo MG, et al. (2006) Extraction and nutritional properties of *Solanum nigrum* L. Seed oil. Africa J Biotechnol 5: 987-991.
18. Karmakar UK, Tarafder UK, Sadhu SK, Biswas NN, Shill MC (2010) Biological investigations of dried fruit of *Solanum nigrum* Linn. S J Pharm Sci 3: 38-45.
19. Kaushik D, Jogpal V, Kaushik P (2009) Evaluation of activities of *Solanum nigrum*, fruit extract. Arch Appl Sci Res 1: 43-50.
20. Elhag RAM, Badwi MAE, Bakhiet AO, Galal M (2011) Hepatoprotective activity of *Solanum nigrum* extracts on chemically induced liver damage in rats. J Vet Med Ani Heal 3: 45-50.
21. Atanu U, Ebiloma, Ajayi EI (2010) A Review of the Pharmacological Aspects of *Solanum nigrum* Linn. Biotechnology and molecular Biology Review 6: 1-7.
22. Hamil FA, Apio S, Mubiru NK, Bukenya-Ziruba R, Mosanyo M, et al. (2003) Traditional herbal drugs of southern Uganda, II. Literature analysis and antimicrobial assays. J Ethnopharmacol 84: 57-78.
23. Parveen, Upadhyay B, Roy S, Kumar A (2007) Traditional uses of medicinal plants among the rural communities of Churu district in the!ar Desert, India. J Ethnopharmacol, 113: 387-399.

24. Patels S, Gheewala N, Suthar A, Shah A (2009) *In-vitro* cytotoxicity activity of *Solanum nigrum* extracts against Hela cell lines and Vero cell lines. Int J Pharm Pharmaceutic Sci 1: 38-46.

25. Purohit SS, Vyas SP (2004) Medicinal plants cultivation a scientific approach including processing and financial guidelines. Agrobios Publishers, Jodhpur, India.

26. Harborne JB (1984) Phytochemical methods: A guide to modern techniques of plant analysis. Chapman and Hall, London.

27. Hawkes JG, Edmonds JM (1972) *Solanum* L. Flora European University Press, Cambridge 3: 197-199.

28. Sweta P, Ashok Jain K (2011) Antifungal activity and preliminary photochemical studies of leaf extracts of *Solanum nigrum* Linn. Int J Pharma Pharmaceutic Sci.

29. Venkatesan D, Karrunakaran CM (2010) Antimicrobial activity of selected Indian medicinal plants. J Phytol 2: 44-48.

30. Heo KS, Lee SJ, Ko JH (2004) Glycoprotein isolated from *Solanum nigrum* L. inhibits the DNA-binding activities of NF-kappaB and AP-1, and increases the production of nitric oxide in TPA-stimulated MCF-7 cells. Toxicol *In Vitro* 18: 755-763.

31. Hussain AOD, Virmani, Pople SP (1992) Dictionary of Indian Medicinal Plants. Central Institute of Medical and Aromatic plants, Lucknow.

32. Jabar ZK, Khattak (2012) *Solanum nigrum* as potent therapy: A review. British J pharmacol Toxicol 3: 185-189.

33. Jain R, Sharma A, Gupta S, Sarethy IP, Gabrani R, et al. (2011) *Solanum nigrum*: Current perspectives on therapeutic properties. Alternat Med Rev 16: 78-85.

34. Amarsingham RD, Bisset NG, Millard AH, Woods MC (1964) A phytochemical survey of Malaya-III, Alkaloids and saponins. Economic Botany 18: 270-278.

35. Arunachalam G, Subramanian N, Pazhani GP, Karunanithi M, Ravichandran V, et al. (2009) Evaluation of anti-inflammatory activity of methanolic extract of *Solanum nigrum* (Solanaceae). Iran J Pharmaceutic Sci Sum 5: 151-156.

36. Jainu M, Devi CSS (2004) Antioxidant effect of methanolic extract of *Solanum nigrum* berries on aspirin induced gastric mucosal injury. Indian J Clin Biochem 19: 65-70.

37. Ramya J, Anjali S (2011) *Solanum nigrum*: Current perspectives on therapeutic properties. Alternat Med review 16: 78-85.

38. Joo HY, Lim K, Lim KT (2009) Phytoglycoprotein (150 kDa) isolated from *Solanum nigrum* L has a preventive effect on dextran sodium sulfate-induced colitis in A/J mouse. J Appl Toxicol 29: 207-213.

39. Li J, Li QW, Gao DW (2009) Antitumor and immune modulating effects of polysaccharides isolated from *Solanum nigrum* L. Phytother Res 23: 1524-1530.

40. Paul Singh R (2012) Role of food material properties and disintegration kinetics in gastric digestion- A quest for foods for healthy benefits.

41. Rajani C, Ruby KM (2012) *Solanum nigrum* with dynamic therapeutic role; A Review. Int J Pharmaceutic Sci Rev Res 15: 65-71.

42. Ravi V, Saleem TM, Maiti PP, Gauthaman K, Ramamurthy J, et al. (2009) Phytochemical and pharmacological evaluation of *Solanum nigrum* Linn. Afri J Pharm Pharmacol 3: 454-457.

43. Rawani A, Ghosh A, Chandra G (2010) Mosquito larvicidal activities of *Solanum nigrum* L. leaf extract against *Culex quinquefasciatus*. Parasitol Res 107: 1235-1240.

44. Schilling EE, Ma QS, Andersen RN (1992) Common names and species identification in black nightshades, Solanum sect. *Solanum* (Solanaceae) Econ Bot 46: 223-225.

45. Sridhar TM, Josthna P, Naidu CV (2011) *In vitro* antibacterial activity and phytochemical analysis of *Solanum nigrum* (Linn.): An important antiulcer medicinal plant. J Experiment Sciences. 2: 24-29.

Rosmarinus officinalis L. and *Myrtus communis* L. Essential Oils Treatments by Vapor Contact to Control *Penicillium digitatum*

Ladu G[1]*, Cubaiu L[1], d'Hallewin G[1], Pintore G[2], Petretto GL[2] and Venditti T[1]

[1]*CNR - ISPA, Trav. La Crucca, 3 - 07040 Sassari, Italy*

[2]*Department of Chemistry and Pharmacy, University of Sassari, via Muroni, 23 / A - 07100 Sassari, Italy*

Abstract

Background: The antimicrobial activity of Essential Oils (EOs) has been used for centuries and nowadays the efforts to develop natural preservatives in postharvest management have augmented interest in their possible applications.

Materials and Methods: *Rosmarinus officinalis* L. and *Myrtus communis* L. EOs, and two of their components, *α*- and *β*-pinene, have been tested *in vitro* against *Penicillium digitatum*, with the aim to assess their antifungal effects when applied as fumigation. The pathogen, inoculated on PDA dishes, was treated by EO-vapors contact and the fungal growth inhibition was recorded in order to evaluate the EOs antifungal activity.

Results: The exposure to the EOs vapors shows different ability in the control of fungal growth related to the EO concentration used and the elapsed time between the fungal inoculum and EO vapor contact. The greatest antifungal activity was observed for rosemary EO, while a less control was found for the myrtle one. Treatment performed with *α*-pinene showed a control of the pathogen that was similar to the myrtle, whereas control with *β*-pinene was very poor.

Highlight: Our finding revealed that plant EOs could be successful in controlling fungal postharvest disease in a dose and compound dependent manner, but deeper researches are needed about treatment parameters because the effectiveness seems to be affected by treatment modalities.

Keywords: Antifungal; Post-harvest disease; Plant extracts; SEM

Introduction

Microbial spoilage is one of the several reactions that alter the safety and the organoleptic properties of perishable products. These reactions decrease the value of food because of the destruction of functional substances, the production of off-flavors and the production of food borne illness related to unsafe food intake [1]. In the field and during postharvest several infections that occur are related to fungal diseases. Fungi are universal biological agents able to colonize fruits and vegetables, causing pathologic disorders because of their potential to synthesize a broad range of hydrolytic enzymes [2]. The high losses in quality, nutrient content and monetary value recorded during postharvest are related to the settlement of molds toxigenic strains [3,4]. Furthermore many mold species can also synthesize mycotoxin, hazardous compounds to animal and human health. Thus inhibition of fungal growth is an effective way to prevent mycotoxin accumulation [5-7]. Furthermore consumers have become more aware of potential health risks associated to synthetic fungicide residues and are increasingly interested in commodities treated with compounds from natural sources [8]. For these reasons new safe technologies, with no risk for consumers and low impact for the environment are needed.

Essential oils are secondary metabolites of plants, which have been traditionally used due to their antimicrobial/antifungal activity [9], and many of them, such as clove, oregano, thyme, basil or cinnamon are categorized as GRAS (Generally Recognized as Safe) by the US-FDA (Food and Drug Administration) [10]. Indeed, essential oils and their active compounds could potentially serve as an effective alternative to conventional antimicrobial agents. The antimicrobial activity depends on the constituents and their concentration: phenolic compounds possess the highest antimicrobial properties followed by alcohols aldehydes ketones ethers and hydrocarbons [11]. The strong antimicrobial activity could be correlated to the high percentage of monoterpenes, phenols and ketones [12]. It is suggested that antimicrobial activity of EOs could be the results of disturbance in several enzymatic systems involved in energy production and structural components synthesis; also they affect and disturb genetic material functionality [13,14]. Presence of phenols with hydroxyl group (-OH) are responsible of antimicrobial properties, the interaction with the membrane enzymes and protein can cause an opposite flow of protons, affecting cellular activity [15]. Nowadays, the efforts to develop natural preservatives in postharvest management have augmented interest in their applications. The most widely used methods to determine *in vitro* the antimicrobial properties of EOs are agar diffusion and broth dilution, while in our study we have performed the EOs application by fumigation. Due to their bioactivity in the vapor phase [16] EOs could be used as fumigant during postharvest protection. The aim of the present investigation was to evaluate the *in vitro* fungi toxic activity of *Rosmarinus officinalis* L. and *Myrtus communis* L. EOs, and two of their ccomponents, α- and β-pinene, against *Penicillium digitatum* when applied as vapor contact.

Materials and Methods

Plant material and EO extraction

Aerial parts of wild *Myrtus communis* L. and *Rosmarinus officinalis* L. were collected in April (2013) from the Sassari area, Sardinia, Italy. A sample of leaves, weighing 50 g, was subjected to hydro-distillation using a Clevenger type apparatus for 2h, according to the European

***Corresonding author:** Ladu G, CNR - ISPA, Trav. La Crucca, 3-07040 Sassari, Italy, E-mail: gfladu@gmail.com

Pharmacopoeia protocol (2002). The EO isolation was carried out in triplicate, the obtained EOs were collected separately, dried over anhydrous sodium sulfate (Na_2SO_4) and stored under nitrogen atmosphere at 4°C, in amber glass vials, until analysis.

Gas chromatography-mass spectrometry analysis

The GC-MS analysis was carried out using a Hewlett Packard 5890 GC (Palo Alto, USA) equipped with a Hewlett Packard 5971 (Palo Alto, USA) mass selective detector (MSD, operating in the EI mode at 70 eV). The GC capillary column was a HP5 (30 m × 0.25 mm, film thickness 0.17 µm), the following temperature program was used: 60°C hold for 3 min, then increased to 210°C at a rate of 4°C/min, then held at 210°C for 15 min, then increased to 300°C at a rate of 10°C/min, and finally held at 300°C for 15 min. Helium was used as carrier gas at a constant flow of 1 mL/min for both columns. Identification of the individual components was performed by comparison with the co-injected pure compounds or, when pure standard was not available, by matching the MS fragmentation patterns and retention indices with the built in libraries or literature data and commercial mass spectral libraries (NIST/EPA/NIH 1999; HP1607 purchased from Agilent Technologies).

Gas chromatography – flame ionization detector analysis

The GC analysis of the EOs was carried out using an Agilent 6890 N instrument equipped with a FID and an HP-5 capillary column (30 m × 0.25 mm, film thickness 0.17 µm). The column temperature program was the same described in GC-MS section. Injector and detector temperatures were 280°C. Helium was used as carrier gas at a flow rate of 1 mL/min. The relative proportions percentages of the HS constituents were obtained by FID peak area normalization without the use of any correction factor.

A hydrocarbon mixture of n-alkanes (C9-C22) was analyzed separately under the same chromatographic conditions used on the HP-5MS capillary columns to calculate the retention indexes with the generalized equation by Vandel Dool, Kartz [17].

As regards α- and β-pinene, these compounds were purchased from Sigma Aldrich.

Antifungal activity assay

The plant EOs as well as the α- and β-pinene activity was evaluated against a wild strain of *Penicillium digitatum* (Ispa-Pd1) isolated from orange fruit. PDA dishes were inoculated with 20 µL of a conidial suspension at the concentration of 1×10^4 conidia/mL. The dishes were then treated by vapor contact, after 0 (t 0), 24 (t 24) and 48 (t 48) hours from inoculation, with 0, 50 or 100 ppm of each compound.

Fumigations were performed at 20°C inside suitable boxes (Figure 1), equipped with a circulation fan (12 V - 0, 16 A) and a rubber septum for EOs applications. In order to obtain a gradual and controlled evaporation a heating system was constructed in our laboratory. Heat was delivered by a couple of electric resistances (10 ohms each) placed inside the chambers, connected to a stabilized power supply located outside. The power supply was provided with a voltage regulator (1-15 V of output) that allows controlling the temperature of the resistances. The established amounts of the compounds were injected through the rubber septum into a heatproof glass vessel, placed on the electric resistances. Once each dose was injected inside the vessels, the circulation fans were turned on for 45 min to circulate the compound within the chambers. The experimental design was used to examine the inhibiting impact on the pathogen of the two concentrations of each compound used. After treatments the dishes were incubated for 7 days at 25°C and the radial growths were daily recorded. Moreover, at the fourth day after treatment, some plugs from the PDA plates were taken for Scan Electron Microscopy (SEM) observations.

Data analysis

Data were reported as mean ± SD of the radial growth of nine dishes, for each concentration and compound. Analysis of Variance (ANOVA) of all data from analytical determinations was performed using the MSTAT-C software (Michigan State University, East Lansing, 1995).

Results and Discussion

The EOs of rosemary and myrtle used in this study were obtained by steam distillation of wild plants collected in Sardinia, Italy. The qualitative analysis was performed by Gas Chromatography-Mass Spectrometry (GC-MS), whereas their percentage composition (semi-quantitative analysis) was achieved by GC-FID technique. The chemical composition of rosemary and myrtle EOs are reported in Tables 1 and 2, respectively and data are consistent with those reported in previous papers [18-20]. The GC-MS was also used to perform a dynamic study of the system, aimed to determine the atmosphere composition inside the boxes during the experiment (data not shown).

The antifungal activity of the compounds was screened by means of vapor contact as described under Material and Methods. Pure essential oils of rosemary and myrtle, and two of their components, α-pinene and β-pinene, were evaluated for their *in vitro* activity against the postharvest fruit-decaying agent *P. digitatum*.

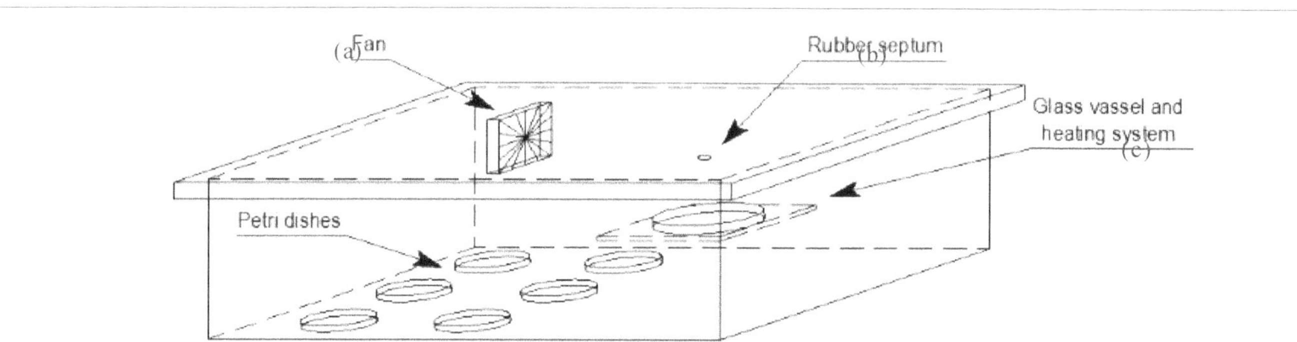

Figure 1: Fumigation box designed to obtain a gradual and controlled evaporation: (a) Circulation fan attached in the lid inner side; (b) Rubber septum for EOs application; (c) Heatproof glass vessel disposed over a heating system.

Compound	Composition (%)	SD	RI	ID
tricyclene	0,34	0,03	927	MS, RI
α-thujene	0,23	0,01	928	MS, RI
α-pinene	30,89	0,15	938	MS, RI, STD
camphene	6,76	0,51	951	MS, RI
verbenene	0,97	0,11	963	MS, RI, STD
β-pinene	2,61	0,21	974	MS, RI, STD
myrcene	1,79	0,07	993	MS, RI, STD
α-phellandrene	0,42	0,00	999	MS, RI, STD
α-terpinene	0,47	0,02	1014	MS, RI, STD
p-cymene	1,78	0,05	1026	MS, RI, STD
3-carene	0,45	0,03	1031	MS, RI, STD
limonene	4,19	0,08	1035	MS, RI, STD
cineole 1,8	13,26	0,03	1040	MS, RI, STD
ocimene beta (z)	0,17	0,00	1050	MS, RI
γ–terpinene	0,78	0,02	1087	MS, RI, STD
terpinolene	1,01	0,00	1108	MS, RI, STD
linalool	0,88	0,04	1138	MS, RI, STD
chrysantenone	0,28	0,03	1122	MS, RI
capholenal	0,36	0,01	1128	MS, RI
camphor	7,19	0,34	1142	MS, RI, STD
pinocamphone trans	0,16	0,00	1163	MS, RI
pinocarvone	0,18	0,01	1168	MS, RI
borneol	3,80	0,02	1171	MS, RI, STD
terpinen 4-ol	0,81	0,02	1192	MS, RI, STD
α-terpineol	0,61	0,75	1197	MS, RI, STD
myrtenol	0,69	0,63	1197	MS, RI, STD
verbenone	2,78	3,65	1201	MS, RI, STD
trans carveol	2,86	3,92	1215	MS, RI
bornyl acetate	8,94	0,38	1280	MS, RI, STD
pinocarvyl acetate	0,08	0,01	1300	MS, RI
β-caryophyllene	1,52	0,05	1420	MS, RI, STD
α–humulene	0,15	0,01	1453	MS, RI, STD

Results are expressed as mean of Fid peak area normalization ± standard deviation; RI: calculated linear Retention Index, ID: Identification Method.

Table 1: Percentage composition of *Rosmarinus officinalis* EO.

The radial growth dynamics, recorded for seven days from the treatment are reported in Figure 2. In particular, when the treatment was performed right after the inoculum (t 0), rosemary EO reduced the colony diameter by 50% when the fumigation was carried out at the concentration of 50 ppm and by 57% at 100 ppm. As regard the treatment performed after 24 h from the inoculation, a completely inhibition was obtained with a fumigation of 50 ppm. The treatment carried out after 48 hours, at a concentration of 50 ppm, was able to delay the pathogen growth for 5 days and by the end of the experiment the colony diameter was reduced by 70%, whereas a completely inhibition was observed at 100 ppm (Figure 2a).

The myrtle EO was less effective against the development of *P. digitatum* (Figure 2b). The ability to completely inhibit the fungal growth is confirmed when the vapor contact is applied after 24 h from the inoculation, while a reduction of the colony diameter was recorded for the treatments carried out at 0 and 48 hours, with a control percentage of only 30% for both EO concentrations.

The effect observed with fumigations carried out with α-pinene (Figure 2c), in general was similar to that of myrtle EO, even if a better control of the pathogen development was observed with 100 ppm 0 hours after the inoculum. Finally as regard treatments carried out with β-pinene (Figure 2d) a poor control activity was observed. Only in the treatment carried out after 24 h at 100 ppm a completely inhibition of the pathogen development was recorded.

The antifungal activity, on the growth of *P. digitatum*, among different EOs used is presented in Table 3. Statistical differences in the radial growth ($p<0.05$) were found among them at t 0 with a concentration of 100 ppm, and at t 48 with 50 and 100 ppm.

The data obtained with α- and β-pinene highlight the synergism that occurs between the pools of compounds of each plant EO with respect to the antimicrobial activity of the single component. Indeed, the major components of EOs are very important for their biological activity, but the minor components also play a significant role, as they can strengthen the effect of major components [21].

The effectiveness of EOs fumigations against *P. digitatum* appears dose dependent and related to the elapsed time between fungal inoculation and treatment, probably related to a variable sensitivity of the pathogen during the different physiological phase.

The results obtained are in general consistent with previous literature data where the antifungal effectiveness of plant EOs was assessed [22-27]. Pintore et al. [18] and Angioni et al. [28] reported the weak antifungal activity of Sardinian rosemary, on the contrary in our

Compound	Composition (%)	SD	RI	ID
propyl butanoate	2,97	0,67	918	MS, RI
α-thujene	0,22	0,07	928	MS, RI
α-pinene	35,01	2,81	938	MS, RI, STD
β-pinene	0,41	0,12	974	MS, RI, STD
myrcene	0,38	0,10	993	MS, RI, STD
propanoic acid 2 methyl 2 methylpropyl	0,62	0,03	1005	MS, RI
butanoic acid 2 methyl 2 methylpropyl	0,36	0,03	1019	MS, RI
p-cymene	0,13	0,09	1026	MS, RI, STD
cineole 1,8	28,37	2,68	1040	MS, RI, STD
ocimene beta (z)	0,23	0,23	1050	MS, RI
ocimene beta (E)	0,33	0,07	1058	MS, RI
γ-terpinene	0,18	0,11	1087	MS, RI, STD
terpinolene	0,23	0,01	1108	MS, RI, STD
linalool	15,81	2,78	1138	MS, RI, STD
pinocarveol cis	0,08	0,03	1177	MS, RI
terpinen 4-ol	0,20	0,04	1192	MS, RI, STD
α-terpineol	2,57	0,51	1197	MS, RI, STD
linalyl acetate	2,38	0,58	1258	MS, RI, STD
pinocarvyl acetate	0,06	0,04	1300	MS, RI
myrtenil acetate	0,08	0,08	1343	MS, RI
terpinil acetate	0,86	0,15	1350	MS, RI, STD
neryl acetate	1,08	1,46	1367	MS, RI, STD
geranyl acetate	2,59	1,87	1387	MS, RI, STD
methyl eugenol	1,16	0,46	1408	MS, RI
α-humulene	0,25	0,09	1453	MS, RI, STD
flavesone	0,41	0,19	1545	MS, RI
caryophyllene oxide	0,24	0,11	1584	MS, RI, STD
humulene epoxide	0,57	0,22	1610	MS, RI
Leptospremone (not ident isom)	0,11	0,07	1621	MS, RI

Results are expressed as mean of Fid peak area normalization ± standard deviation; RI: calculated linear Retention Index, ID: identification method.

Table 2: Percentage composition of *Myrtus communis* EO.

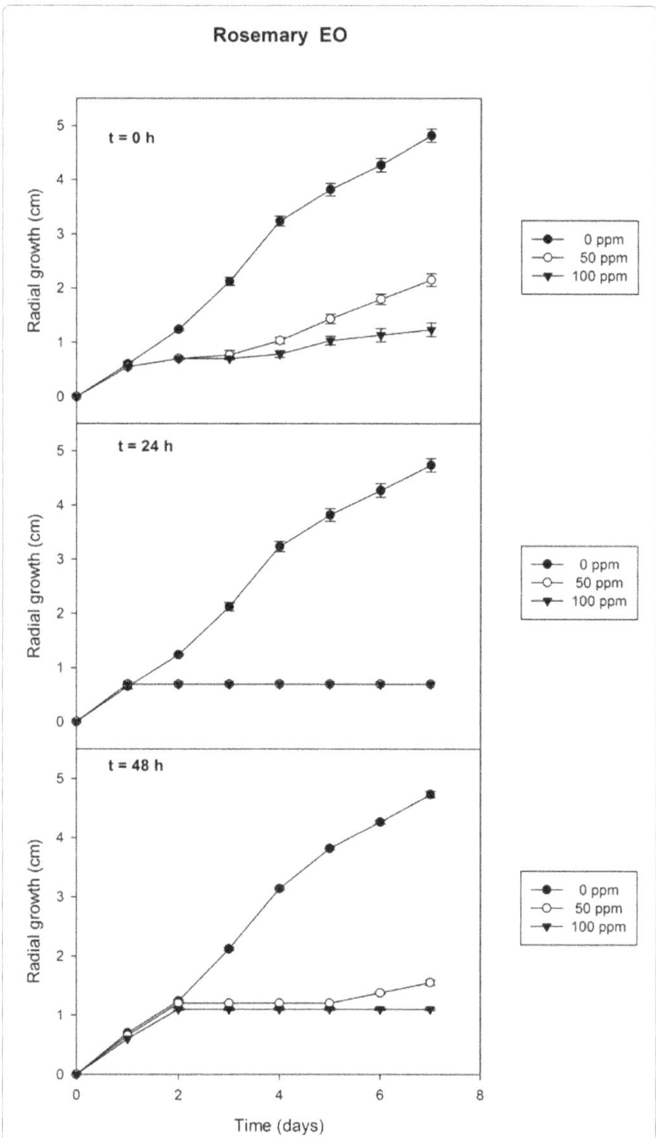

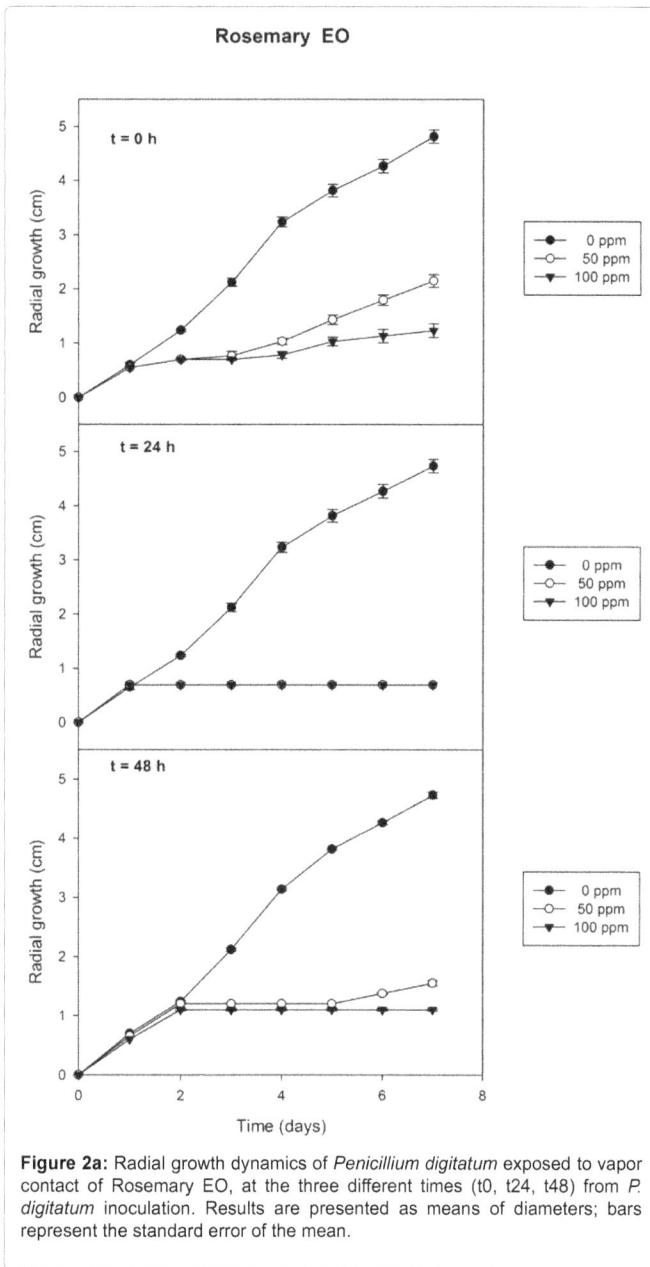

Figure 2a: Radial growth dynamics of *Penicillium digitatum* exposed to vapor contact of Rosemary EO, at the three different times (t0, t24, t48) from *P. digitatum* inoculation. Results are presented as means of diameters; bars represent the standard error of the mean.

work it show the greatest effectiveness, it is may be due to the treatment done by vapor contact and by the chemical composition of rosemary EO used in this study. Environmental factors were considered to play a key role in the chemical composition of plant EOs [29], that also depends on growing conditions season or vegetative period of plant [30-34].

Different modes of action are involved in the antimicrobial activity of EOs and to date they are not completely understood but there are some features on which some authors agree to explain the inhibitory activity: the hydrophobicity of these compounds leads them to cross cell membrane and interact with cell compounds [35,36]. Some hydrophobic compounds could change the microbial membrane permeability and the loss of differential permeability is generally considered the cause of cell death [2,37]. Phenolic compounds are known to alter microbial cell permeability, allowing the loss of macromolecules from the interior, they could interact with membrane proteins causing a deformation of structure and functionality [38]. All the actions culminate with the inhibition of germination, suppression of mycelia growth and germ tube elongation, therefore it has been suggested that the action of plant compounds might be related to the perception/transduction of signals involved in the switch from vegetative to reproductive development.

Scan electron microscopy observations were performed in order to obtain more information on the ultra-structural alterations of *P. digitatum* cells, treated with the plant EOs. SEM examinations revealed considerable morphological alterations (Figure 3), explained by the EO modification induced on the fungal morphogenesis and growth, through the interference of their components with the enzymes involved in cell wall synthesis, leading to changes in the hyphae integrity [39].

Rosemary oil vapor treated samples (Figures 3b-3d) showed the collapse of whole hyphae and complete inhibition of conidia production as compared with untreated control (Figure 3a), and

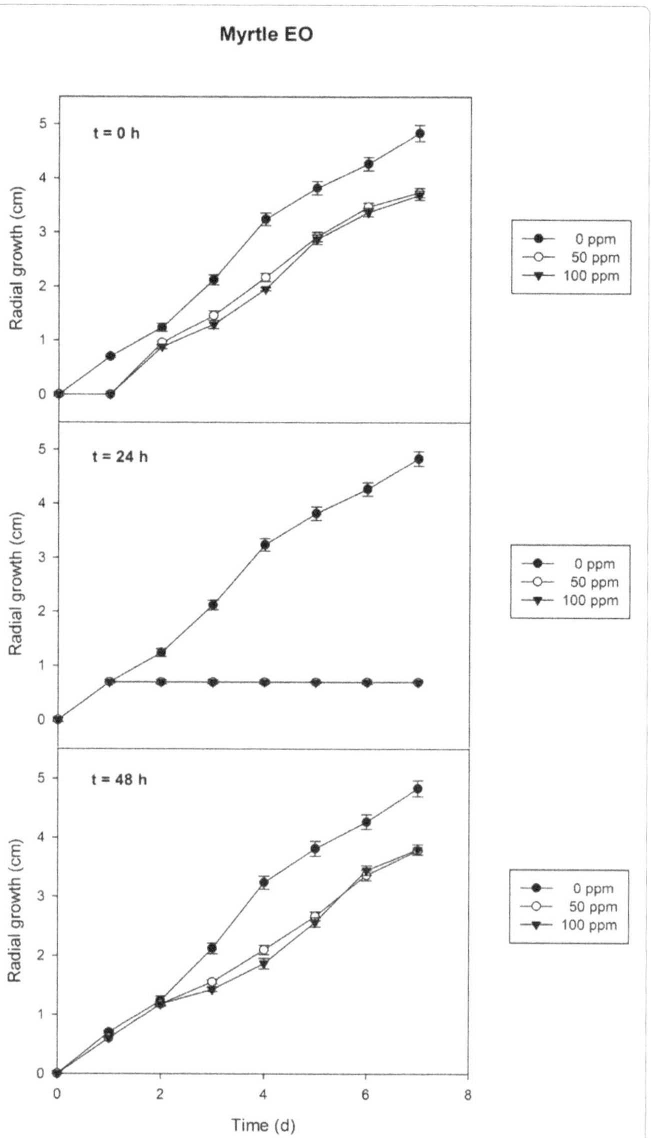

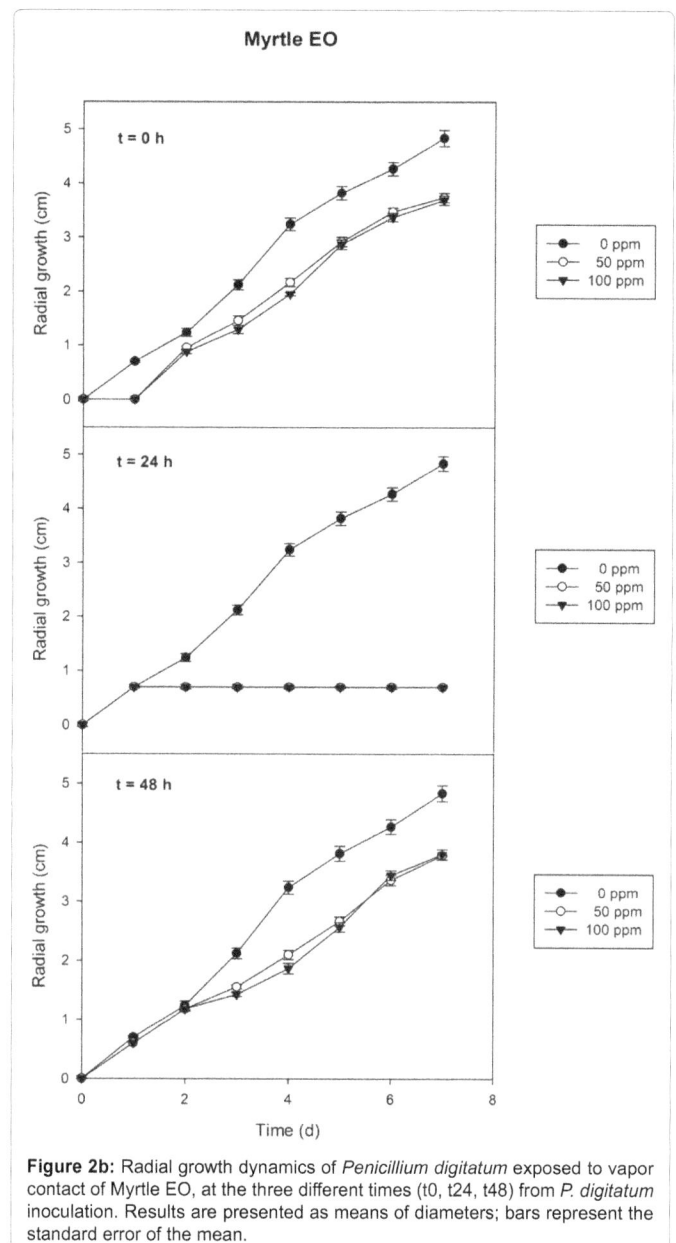

Figure 2b: Radial growth dynamics of *Penicillium digitatum* exposed to vapor contact of Myrtle EO, at the three different times (t0, t24, t48) from *P. digitatum* inoculation. Results are presented as means of diameters; bars represent the standard error of the mean.

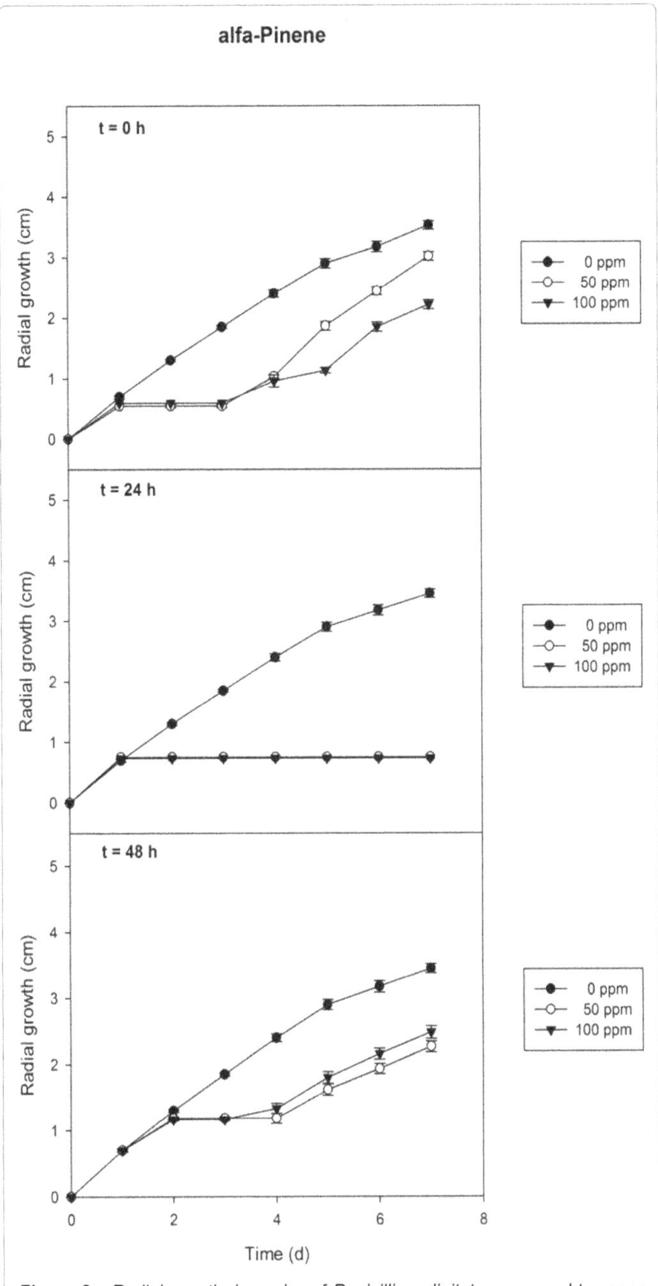

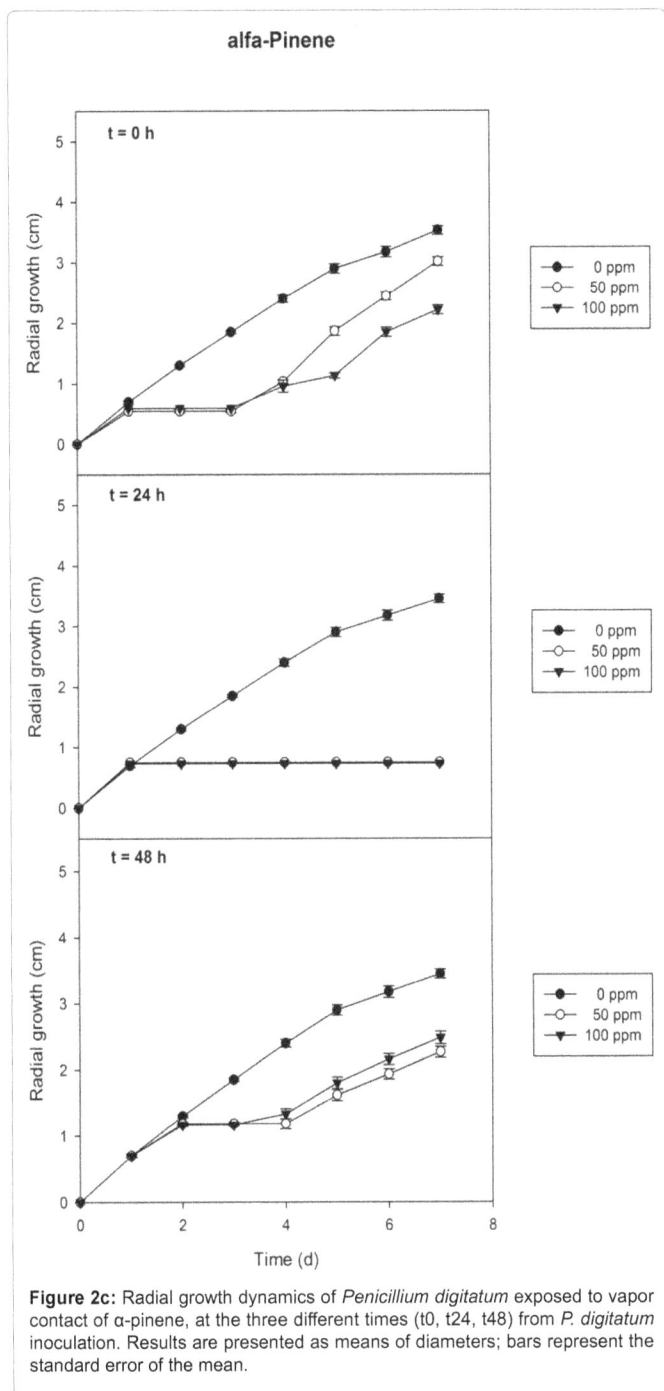

Figure 2c: Radial growth dynamics of *Penicillium digitatum* exposed to vapor contact of α-pinene, at the three different times (t0, t24, t48) from *P. digitatum* inoculation. Results are presented as means of diameters; bars represent the standard error of the mean.

they could be due to the loss of integrity of the cell wall and plasma membrane permeability [40,41].

There is evidence that plant compounds are able to affect several intracellular functions with consequent disruption of the normal mycelium development.

Conclusions

EOs from plants have antimicrobial activity against a variety of food borne fungi. In this study we tested the effect of *Rosmarinus officinalis* L. and *Myrtus communis* L. EOs on mycelia growth of green mold when applied as vapor contact in suitable box designed for this purpose. Results revealed that rosemary could be effective in controlling *P. digitatum*, while as light activity was observed with myrtle. Further researches are needed about treatment modalities and concentration because the effectiveness seems to be affected by these parameters.

Based on our findings, this work could be an important tool for the assessment of EOs inhibitory potential on the fungal growth, when the treatments are carried out by vapor contact, since that the most widely used methods, for determining the antimicrobial properties of EOs, are not ideals because of the complexity, volatility and water insolubility of these compounds.

Our results confirms that plants essential oils are one of the encouraging safe and environmentally-friendly candidates for future use as substitutes to conventional synthetic fungicides, for managing plants pathogens interactions and food contaminants and decays.

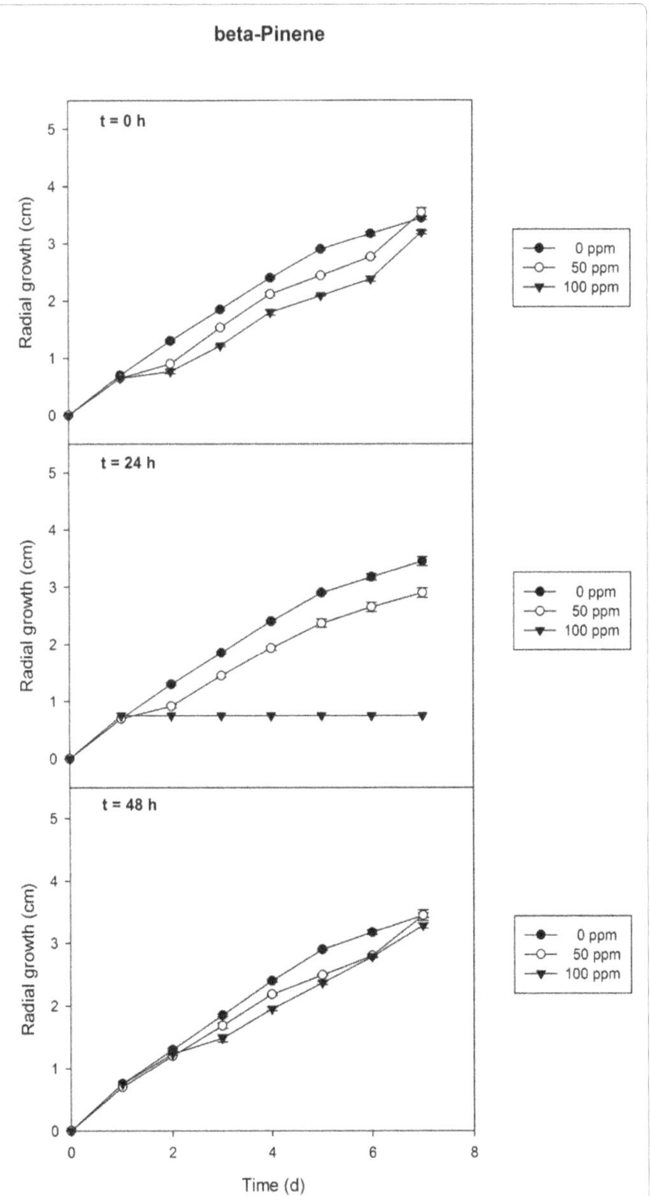

Figure 2d: Radial growth dynamics of *Penicillium digitatum* exposed to vapor contact of β-pinene, at the three different times (t0, t24, t48) from *P. digitatum* inoculation. Results are presented as means of diameters; bars represent the standard error of the mean.

Time (h)	Conc. (ppm)	EOs			
		Rosemary	Myrtle	α-pinene	β-pinene
t 0	0	4,73 a	4,87 a	3,53 b	3,45 c
	50	2,15 a	3,73 c	3,01 b	3,55 c
	100	**1,23 a**	**3,68 b**	**2,21 c**	**3,20 d**
t 24	0	4,73 a	4,87 a	3,53 b	3,45 b
	50	0,70 a	0,70 a	0,75 a	2,90 b
	100	0,70 a	0,70 a	0,73 a	0,75 a
t 48	0	4,73 a	4,87 a	3,53 b	3,45 b
	50	**1,55 a**	**3,78 b**	**2,26 c**	**3,45 d**
	100	**1,10 a**	**3,80 b**	**2,48 c**	**3,28 d**

Means followed by different letters in each row indicate significant differences according to Duncan's multiple range test at p<0.05

Table 3: Radial growth of *Penicillium digitatum* after seven days.

Figure 3a: Scanning Electron Micrographs of untreated *P. digitatum* inoculated PDA dishes (500x).

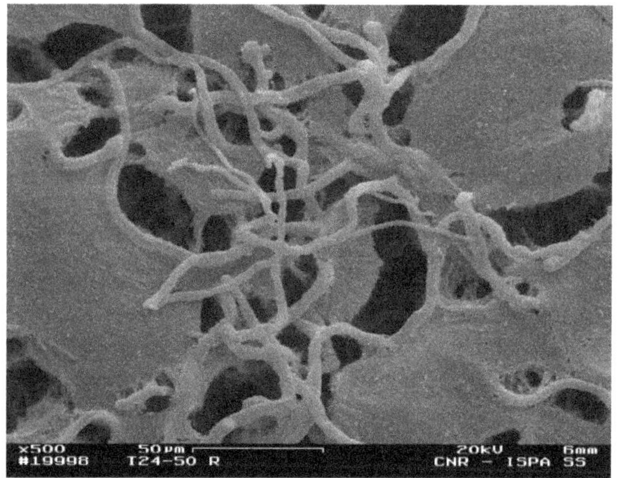

Figure 3b: Scanning Electron Micrographs of treated (Rosemary EO vapor contact) *P. digitatum* inoculated PDA dishes: treatment performed after 24 h from the inoculation with 50 ppm (500x).

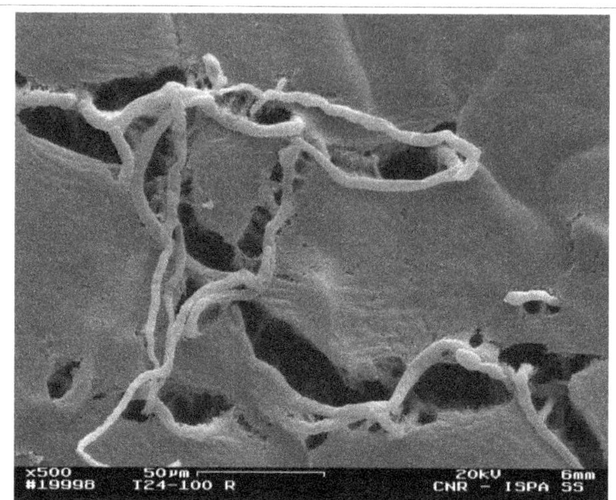

Figure 3c: Scanning Electron Micrographs of treated (Rosemary EO vapor contact) *P. digitatum* inoculated PDA dishes: treatment performed after 24 h from the inoculation with100 ppm (500x).

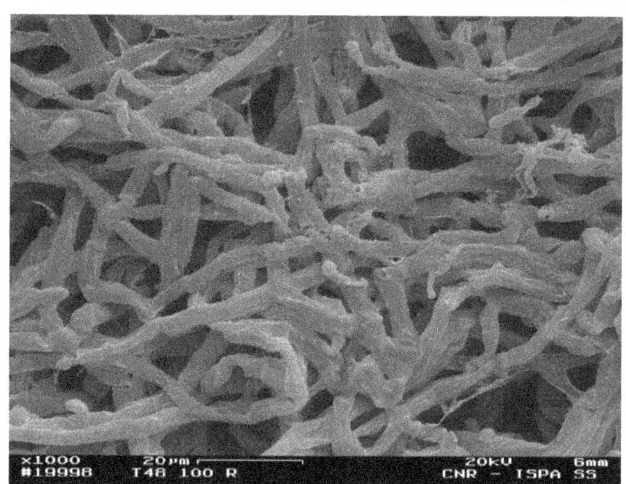

Figure 3d: Scanning Electron Micrographs of treated (Rosemary EO vapor contact) *P. digitatum* inoculated PDA dishes: treatment performed after 48 h with 100 ppm (1000x), were is well appreciable the collapsed mycelium.

Acknowledgement

Thanks to Salvatore Marceddu and Antonello Petretto for technical assistance.

References

1. Silva AS, Costa D, Albuquerque TG, Buonocore TG, Ramos F, et al. (2014) Trend in the use of natural antioxidants in active food packaging: A review. Food Addit Contam Part A Chem Anal Control Expo Risk Assess 31: 374-395.
2. Cabral LC, Pinto VF, Patriarca A (2013) Application of plant derived compounds to control fungal spoilage and mycotoxin production in foods. Int J Food Microbiol 166: 1-14.
3. Bata A, Lasztity R (1999) Detoxifcation of mycotoxin-contaminated food and feed by microorganisms. Trends in Food Sci and Technol 10: 223-228.
4. Magro A, Carolino M, Bastos M, Mexia A (2006) Efficacy of plant extracts against stored products fungi. R Iberoamericana de Micología 23: 176-178.
5. Etcheverry M, Torres A, Ramirez ML, Chulze S, Magan N (2002) In vitro control of growth and FUMO production by *Fusarium verticilloides* and *Fusarium proliferatum* using antioxidants under different water availability and temperature regimes. J Appl Microbiol 92: 624-632.

6. Torres AM, Ramirez ML, Arroyo M, Chulze SN, Magan N (2003) Potential use of antioxidants for control of growth and FUMO production by *Fusarium verticilloides* and *Fusarium proliferatum* on whole maize grain. Int J Food Microbiol 83: 319-324.

7. Marín S, Sanchis V, Ramos AJ (2004) Plant products in the control of *Mycotoxins* and *Mycotoxigenic fungi* on food commodities. In: Dubey NK (ed) Natural products in plant pest management. CAB Intern: 31-35.

8. Amorati R, Foti MC,Valgimigli L (2013) Antioxidant activity of essential oils. J Agric Food Chem 61: 10835-10847.

9. Bajpai VK, Baek KH, Kang SC (2012) Control of Salmonella in foods by using essential oils: A review. Food Res Int 45: 722-734.

10. Manso S, Pezo D, Gómez-Lus R,Nerín C (2014) Diminution of aflatoxin B1 production caused by an active packaging containing cinnamon essential oil. Food Control 45: 101-108.

11. Ferde M, Ungureanu C (2012) Antimicrobial activity of Essential oils against four food-borne fungal strains. UPB Sci Bull Series B 74.

12. Lis-Balchin M, Deans SG (1997) Bioactivity of selected plant essential oils against *Listeria monocytogenes*. J of Applied Microbiol 82: 759-762.

13. Davidson P, Taylor T (2007) Chemical Preservatives and Natural Antimicrobial Compounds. In: Doyle M, Beuchat L (3rdEdn), Food Microbiol: Fundamentals and frontiers. ASM Press, Washington DC, USA.

14. Omidbeygi M, Barzegar M, Esfahani ZH, Naghdibadi H (2007) Antifungal activity of thyme, summer savory and clove essential oils against *Aspergillus xavus* in liquid medium and tomato paste. Food Control 18: 1518-1523.

15. Dorman HJD, Deans SG (2000) Antimicrobial agents from plants: antibacterial activity of plant volatile oils. J of Applied Microbiol 88: 308-316.

16. Naeini A, Ziglari T, Shokri H, Khosravi AR (2010) Assessment of growth-inhibiting effect of some plant essential oils on different *Fusarium* isolates. J Medical Mycol 20: 174-178.

17. Van del Dool H, Kartz PD (1963) Generalization of the retention index system including linear temperature programmed gas-liquid partition chromatography. J Chromatogr A 11: 463-471.

18. Pintore G, Usai M, Bradesi P, Juliano C, Boatto G, et al. (2002) Chemical composition and antimicrobial activity of *Rosmarinus officinalis* L. oil from Sardinia and Corsica. Flavour and Fragrance J 17: 15-19.

19. Pintore G, Marchetti M, Chessa M, Sechi B, Scanu N, et al. (2009) *Rosmarinus officinalis* L.: chemical modification of the essential oil and evaluation of antioxidant and antimicrobial activity. Natural Product Communication 4: 1685-1690.

20. Petretto GL, Foddai M, Maldini MT, Chessa M, Venditti T, et al. (2013) A novel device for the study of antimicrobial activity by vapor-contact of volatile substances. CommAppl Biol 78: 65-72.

21. Bassolé IHN, Juliani HR (2012) Essential oils in combination and their antimicrobial properties. Molecules 17: 3989-4006.

22. Azerêdo GA, Stamford TLM, Nunes PC, Neto NJG, Oliveira MEG, et al. (2011) Combined application of essential oils from *Origanum vulgare* L. and *Rosmarinus officinalis* L. to inhibit bacteria and autochthonous microflora associated with minimally processed vegetables. Food Res Int 44: 1541-1548.

23. Nguefack J, Tamgue O, Dongmo JBL, Dakole CD, Leth V, et al. (2012) Synergistic action between fractions of essential oils from *Cymbopogon citratus*, *Ocimum gratissimum* and *Thymus vulgaris* against *Penicillium expansum*. Food Control 23: 377-83.

24. Freire MM, Jham GN, Dhingra OD, Jardim CM, Barcelos RC, et al. (2012) Composition, antifungal activity and main fungitoxic components of the essential oil *Mentha piperita* L. J Food Safety. 32:29-36.

25. De Sousa LL, De Andrade SCA, AguiarAthayde AGA, De Oliveira CEV, De Sales CV, et al. (2013) Efficacy of *Origanum vulgare* L. and *Rosmarinus officinalis* L. essential oils in combination to control postharvest pathogenic *Aspergilli* and autochthonous mycoflora in *Vitis labrusca* L. (table grapes). Int J Food Microbiol 165: 312-318.

26. Cannas S, Molicotti P, Ruggeri M, Cubeddu M, Sanguinetti M, et al. (2013) Antimycotic activity of *Myrtus communis* L. towards *Candida spp.* from clinical isolates. J Infect Dev Ctries 7: 295-298.

27. Felsociova S, Kacaniova M, Horska E, Vukovic N, Hleba L, et al. (2015) Antifungal activity of essential oils against selected terverticillate penicillia. Ann Agric Environ Med 22: 38-42.

28. Angioni A, Barra A, Cereti E, Barile D, Coisson JD, et al. (2004) Chemical Composition, plant genetic differences, antimicrobial and antifungal activity investigation of the essential oil of *Rosmarinus officinalis* L. J Agric Food Chem 52: 3530-3535.

29. Scora RW. (1973) Essential leaf oil variability in green, variegated and albino foliage of *Myrtus communis*. Phytochem 12: 153-155.

30. Sangwan NS, Farooqi AHA, Shabih F, Sangwan RS (2001) Regulation of essential oil production in plants. Plant Growth Regul 34: 3-21.

31. Flamini G, Cioni PL, Morelli I, Maccioni S, Baldini R (2004) Phytochemical typologies in some populations of *Myrtus communis* L. on Caprione Promontory (East Liguria Italy). Food Chem 85: 599-604.

32. Pengelly A (2004) Essential oils and resins. In: The constituents of medicinal plants. Australia. Allen and Unwin.

33. Gardeli C, Vassiliki P, Athanasios M, Kibouris T, Komaitis M (2008) Essential oil composition of *Pistacia lentiscus* L. and *Myrtus communis* L.: Evaluation of antioxidant capacity of methanolic extracts. Food Chem 107: 1120-1130.

34. Andrade EHA, Alves CN, Guimaraes EF, Carreira LMM, Maia JGS (2011) Variability in essential oil composition of *Piper dilatatum* L. C Rich Biochem Syst Ecol 39: 669-675.

35. Aleksic V, Knezevic P (2014) Antimicrobial and antioxidative activity of extracts and essential oils of *Myrtus communis* L. Microbiol Res 169: 240-254.

36. Amensour M, Sendra E, Abrini J, Bouhdid S, Pérez-Alvarez JA, et al. (2009) Total phenolic content and antioxidant activity of myrtle (*Myrtus communis*) extracts. Nat Prod Commun 4: 819-824.

37. Bouayed J, BohnT (2012) Nutrition, well-being and health. In Tech Publishers, India.

38. Fung DYC, Taylor SUE, Kahan J (1997) Effects of butylated hydroxyanisole (BHA) and butylated hydroxitoluene (BHT) on growth and aflatoxin production of *Aspergillus flavus*. J Food Safety 1: 39-51.

39. Rasooli I, Rezaei MB, Allameh A (2006) Growth inhibition and morphological alterations of *Aspergillus niger* by essential oils from *Thymus eriocalyx* and *Thymus x-porlock*. Food Control 17: 359-364.

40. Singh D, Kumar TRS, Gupta VK, Chaturvedi P (2012) Antimicrobial activity of antimicrobial activity of some promising plant oils, molecules and formulations. Indian J Exp Biol 50: 714-717.

41. Sumalan RM, Alexa E, Poiana AM (2013) Assessment of inhibitory potential of essential oils on natural mycoflora and *Fusarium* mycotoxins production in wheat. Chemistry Central J 7: 32.

Plugging of Parallel-Tube Devices with Fluid Foods

Steven Brandon C[1] and Paul Dawson L[2]*

[1]Membrane Application Services, SC, USA
[2]Department of Food Science and Human Nutrition, Clemson University, Clemson, SC, USA

Abstract

Plugging often occurs when fluid foods are processed in parallel-tube devices (e.g., tubular heat exchangers and membrane filtration modules, especially when fluids are concentrated). Plugging reduces productivity by reducing performance and causing downtime to remove the plug. The cause of plugging is explained here for the first time and a means of avoiding it is described. Plugging rarely occurs during normal operation while fluid food is being pumped through the tubes, but rather when the tubes are flushed with water, e.g., prior to routine cleaning. The study identified two prerequisites for plugging to occur: tubes must have unequal resistances to flow and the fluid food must possess a yield stress that exceeds a value which depends on the device geometry.

Keywords: Parallel; Plugging; Microfiltration; Membrane; Tubular; Heat exchanger; Yield stress; Non-Newtonian; Herschel-Bulkley; Tomato paste; Carbopol

List of Symbols

Symbol		Units
R^2	Coefficient of Determination	---
m	Consistency Index	---
ρ	Density	kg/m^3
D	Diameter	m
L_e	Entrance Length	m
n	Flow Behavior Index	---
n'	Flow Behavior Index; approximation	---
L	Length	m
l	Length of Column of Flushing Fluid	m
P_1	Pressure at Tube Inlet	Pa
P_2	Pressure at Tube Outlet	Pa
ΔP_i	Pressure Difference at Time Increment; i	Pa
ΔP_1	Pressure Difference; Tube 1	Pa
ΔP_2	Pressure Difference; Tube 2	Pa
ΔP_3	Pressure Difference; Tube 3	Pa
ΔP_4	Pressure Difference; Tube 4	Pa
K	Resistance Coefficient; Overall	---
$\dot{\gamma}$	Shear Rate	1/s
$\dot{\gamma}_w$	Shear Rate at the Wall	1/s
τ	Shear Stress	Pa
τ_w	Shear Stress at the Wall	Pa
V	Velocity; Average	m/s
μ	Viscosity	Pa-s
μ_{app}	Viscosity; Apparent	Pa-s
Q_1	Volumetric Flow Rate; Tube 1	m^3/s
Q_2	Volumetric Flow Rate; Tube 2	m^3/s
Q_3	Volumetric Flow Rate; Tube 3	m^3/s
Q_4	Volumetric Flow Rate; Tube 4	m^3/s
Q_{TOTAL}	Volumetric Flow Rate; Total	m^3/s
τ_0	Yield Stress	Pa
τ_B	Yield Stress; Bingham	Pa

Introduction

Problem statement

When fluid foods are concentrated in parallel-tube microfiltration modules the modules are frequently found to have one or more tubes plugged with solid paste-like material necessitating costly downtime to remove the plugs. In fact a whole industry has developed plans for plug removal [1]. The cause for this plugging phenomenon has not been well understood but has generally been believed to occur while very viscous fluids were being pumped through parallel channels. The thesis of the current study is that plugging of parallel-tube devices including both concentrating devices e.g. microfiltration membrane modules and evaporators and non-concentrating devices e.g. tubular heat exchangers occurs when yield stress fluids are flushed from the tubes with water. If this plugging phenomenon can be more clearly understood gradual dilution of viscous fluids during the end of processing runs could prevent the costly downtime and maintenance required when plugging occurs. If a pump is stopped and restarted while parallel tubes are filled with a yield-stress fluid the imposed pressure difference overcomes the yield stress and flow recommences. However if water rather than yield-stress fluid is pumped into flow channels that are initially filled with yield-stress fluid it is possible that one or more of the parallel channels will clear i.e. have all of the yield-stress fluid displaced with water before the other channels are cleared. The applied pressure difference across all parallel channels will then fall to that of the water-filled channel a level which may be insufficient to maintain flow of the yield-stress material remaining in the incompletely cleared channels.

If water is added to the yield-stress fluid moving through the parallel-tube flow device diluting the yield-stress fluid it is possible to displace the yield-stress fluid from the tubes without forming plug however the dilution must be gradual.

Plugging

In this paper the term "plugging" refers to blockage of a tube with yield-stress material which does not move even when flow is observed

***Corresponding author:** Paul Dawson L, Department of Food Science and Human Nutrition, Clemson University, Clemson, SC 29634, USA
E-mail: pdawson@clemson.edu

in neighboring tubes. Of course plugging may be caused by incorrect operation of equipment, for example the capture of solid objects in the tubes due to inadequate pre-filtration or in the case of filtration modules by failing to circulate the process fluid resulting in over concentration of process fluid in individual tubes.

Plugging of parallel-tubes devices results in isolation of the plugged tubes from the process reducing the performance and altering flow characteristics of the device. However since neither the loss of performance nor the change in flow characteristics are sensitive methods of plug detection the presence of plugged tubes is usually confirmed only by visual inspection.

Rheology background

Yield stress: Yield stress is defined as the minimum shear stress required initiating flow [2]. While the existence of yield stress has been challenged by arguing that all materials will flow when exposed to a shear stress given sufficient time from a practical standpoint there is little doubt that yield stress is an engineering reality [3,4].

Apparent yield stress: The term apparent yield stress has been introduced to cover cases where there is very little flow at low shear stress. Frequently it is not possible to decide whether a true yield stress exists or not. Therefore some workers prefer to refer to an 'apparent yield stress' which is an operational parameter and its evaluation involves extrapolation of data to zero shear rate often the value depending upon the range of data being used to evaluate it [5].

Bingham yield stress: Rao [6] cites Michaels and Bolger [7] who introduces the concept of the Bingham yield stress τ_B which is obtained by extrapolation of only the linear portion of the shear stress versus shear rate data to zero shear rates. The Bingham yield stress is part of the total yield stress at higher shear rates. For apple sauce the Bingham yield stress was found to be very close to the static yield stresses measured directly using the vane method [8].

Apparent slip velocity: Many inhomogeneous mixtures which contain suspended particles exhibit slip velocity at the wall due to migration of these particles away from the region of highest shear near the wall and toward regions of lower shear which for flow in tubes is at the center line of the tube [9]. The relative depletion of particles in the region near the wall referred to as the lubrication layer [10] results in lower apparent viscosity and reduced yield stress in the fluid near the wall. The consequence of this flow behavior is that bulk motion of the mixture occurs at shear stress levels below the apparent yield stress of the mixture. Since even in cases in which slip occurs the velocity of fluid in contact with the wall still approaches zero at the wall interface this phenomenon is here referred to as apparent slip velocity.

Meeker et al. [10] state that the magnitude of the apparent slip velocity depends on the shear stress. They describe three regimes of slip. First when the shear stress is at least 50% larger than the yield stress i.e. $\tau_{w/}\tau_0 \geq 1.5$ slip is negligible compared to the bulk flow. Secondly when the shear stress exceeds the yield stress by less than 50% i.e. when $1 < \tau_w/\tau_0 < 1.5$ slip becomes significant and the total deformation of the paste results from a combination of bulk flow and slip. Thirdly when the shear stress is less than or equal to the yield stress i.e. $\tau_{w/}\tau_0 \leq 1$ the bulk flow is negligible and the apparent motion is entirely due to the slipping of the paste. Apparent slip velocity has been measured using techniques such as magnetic resonance imaging [11] and ultrasonic flow meters [12,13] which do not disturb the velocity profiles of these fluids.

Herschel-Bulkley model: The Herschel-Bulkley model Equation (1) below may be used to describe the rheological behavior of a broad range of fluid types including Newtonian pseudo plastic dilatant and Bingham plastic fluids.

$$\tau = \tau_0 + m\dot{\gamma}^n$$

The Herschel-Bulkley model states that the shear stress that must be applied to a fluid to impart a given resulting shear rate in the fluid equals the sum of the fluid's yield stress and the product of the consistency index m and the shear rate raised to the power of the flow behavior index n [14].

The applicability of the Herschel-Bulkley model to a variety of fluid foods is supported by Aguilera and Stanley [15] who stated that "In many food liquids and suspensions τ versus $\dot{Y}$ data do not fit a straight line (i.e. the viscosity does not remain constant) but rather τ varies with $\dot{Y}$ so that an apparent viscosity μ_{app} has to be defined at each $\dot{Y}$. Moreover the relationship may not start at the origin which entails that some materials will not flow unless a critical stress or yield value τ_0 is reached after which they will flow like liquids."

Tubular viscometry: The shear rate at the wall for non-Newtonian fluids flowing in tubes is calculated using Equation (2) the Rabinowitsch-Mooney equation

$$\dot{\gamma}_w = \left[\frac{3n'+1}{4n'}\right]\frac{8V}{D}$$

Where n' is the point slope of the log (τ) vs. log (8 V/D) plot [16]. Since the average velocity in the tube is easily calculated from the measured volumetric flow rates determination of shear rate is not complex.

The formula for calculating shear stress at the wall for fluids flowing in tubes is derived from a force balance on a cylindrical element of fluid in the tube. This formula is given in Equation (3) below.

$$\tau = \frac{(P_1 - P_2)}{4(L/D)} \quad (3)$$

Here the relevant pressure difference is only that portion of the pressure difference imparted by frictional losses due to viscous flow rather than the measured total pressure.

Fluid mechanics background

Pressure differences: As shown in Equation (4) below the measured total pressure difference across each tube results in part from frictional pressure losses due to viscous flow and in part from momentum-change pressure losses due to changes in velocity and direction of the fluid at the entrance and exit of each parallel tube.

$$\Delta P_{TOTAL} = \Delta P_{FRICTION} + \Delta P_{MC}$$

Since the shear stress at the wall results only from the frictional pressure difference it is necessary to subtract the momentum-change pressure difference from the measured total pressure difference to obtain the frictional pressure difference.

The momentum-change pressure difference is given by the overall resistance coefficient K for each parallel tube using Equation (5).

$$\Delta P_{MC} = K\frac{\rho V^2}{2}$$

Since momentum-change pressure difference is due only to changes in the flow velocity and direction it is independent of fluid properties other than density. This means that the overall resistance coefficient determined while pumping a Newtonian fluid such as water through the tubes can be considered to be identical to that when non-Newtonian fluids are flowing through the tubes. This is supported by Carrère and René [17] who cite Midoux [18] stating that "For non-Newtonian fluids in the case of power law fluids with $0.1 < n < 2$ the resistance coefficient can be considered as not differing from the Newtonian value with $n = 1$." Having determined the overall resistance coefficient for each parallel tube in the test rig with water it is a simple matter to calculate the frictional pressure difference by subtracting the momentum-change pressure difference from the measured total pressure difference.

Flow in parallel tubes: For flow through parallel tubes the total volumetric flow rate equals the sum of the volumetric flow rates in each tube i.e.

$$Q_{TOTAL} = Q_1 + Q_2 + Q_3 + Q_4 .$$

Also the pressure difference across each parallel tube is the same i.e.

$$\Delta P_{TOTAL} = \Delta P_1 = \Delta P_2 = \Delta P_3 = \Delta P_4$$

This thesis was tested using an experimental approach in which flow of fluid foods was observed in four parallel transparent tubes [19]. Four fluids were studied: water glycerin Carbopol gel and tomato paste. These fluids were selected for reference characterization high viscosity without yield stress a well-defined yield stress model fluid and a representative yield stress food material respectively.

Materials and Methods

Experimental test rig

The test rig shown in Figure 1 was constructed from PVC and stainless steel (SS) piping. The four parallel tubes were transparent PVC Schedule 40 pipes with 21.2 mm inside diameter and 2.03 m length. The test fluids were moved through the tubes using a sanitary pump (Model 130 Waukesha Universal series positive-displacement rotary pump with a 5-to-1 variable speed controller). Steady temperature was maintained during the tests with a single-tube heat exchanger of 34.9 mm inside diameter 3.05 m length and 0.335 m^2 of heat transfer area.

To enable data to be collected at low velocities a diaphragm valve was used to control the bypass of some of the flow directly to the suction of the pump. A sanitary plug valve (labeled "MCV" in Figure 1) allowed selection between a once-through mode used when flushing the test rig with water and recirculation mode used to allow test fluids to be pumped continuously through the test rig.

Flow rates from 2.37 litres to 189 liters per minute were measured using a magnetic flow meter (Hersey® 25.4 mm Balanced Electrode Plane magnetic flow meter with a Model 100 Indicator ± 0.5% accuracy). Lower flow rates were measured by timing over a pre-measured distance the movement of small bubbles in the central plug region of the Carbopol gel or of particles near the surface of the tomato paste (when moving in slip flow at stresses below the yield stress of the paste). Pressure was measured using two ENFM® Bourdon-tube pressure gauges (0-413 kPa. ± 1.0% accuracy glycerin filled diaphragm protected) at headers upstream and downstream of the four parallel tubes. To improve accuracy of the upstream pressure measurement an additional pressure gauge was installed at the upstream header (sanitary Chicago Stainless Equipment® Bourdon-tube pressure gauge (0-690 kPa ± 1.5% accuracy glycerin filled diaphragm protected). Temperature was measured using an inline bi-metallic thermometer (-18°C to 121°C). The pH of the test fluids was measured using an inline double-junction glass pH electrode sensor with a CPVC housing connected to a digital pH meter (Fisher® Model No. 107). Test fluid added to or removed from the test rig was weighed using Ohaus® 0-50 kg electronic scales with Model No. I10 digital indicator.

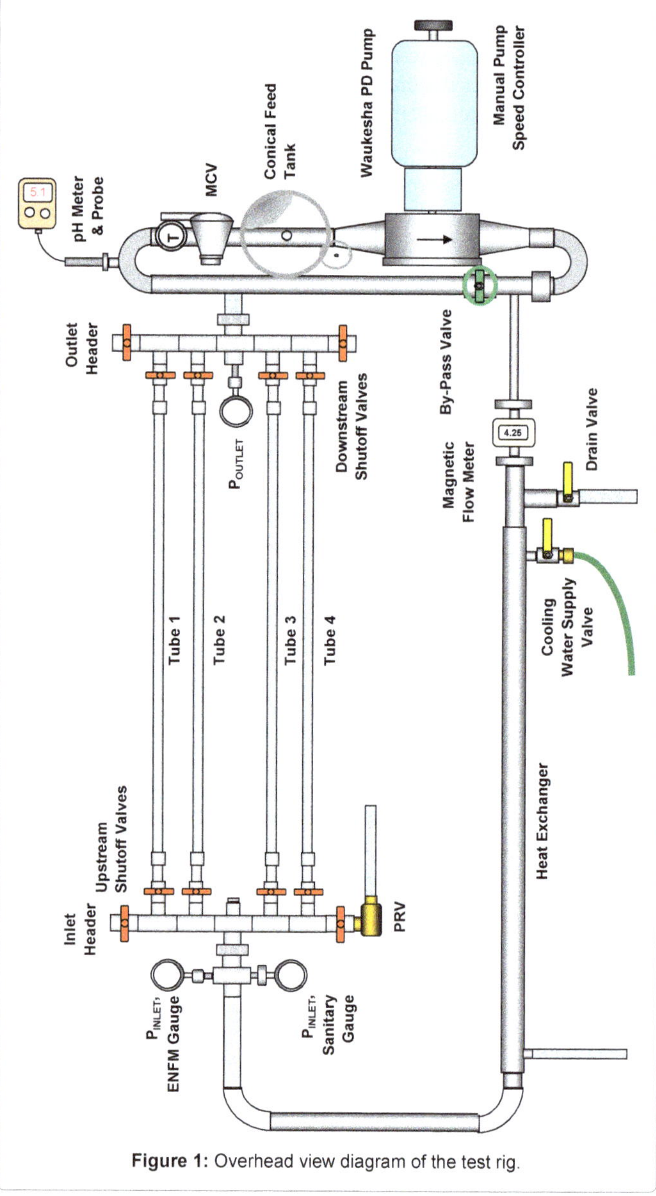

Figure 1: Overhead view diagram of the test rig.

Viscous test fluids were supplied to the test rig via a 19-L conical tank. Water was supplied from a 95-L tank. The pH of the 1.5% Carbopol gel was raised from pH 2.8 to pH 5 using 25% NaOH introduced into the system via a 500-mL SS funnel installed in the pump suction line.

Test fluids

Flushing tests were conducted on each of two fluids with non-zero yield stresses and one viscous Newtonian fluid as a control. At least three replicate tests for performed for each fluid.

a) Cold-break tomato paste was supplied by Campbell Soup Research and Development Dixon California USA in aseptically packaged bags collected from their processing line on 25 July 2003. Each bag contained approximately 2 kg of paste with solids content ranging from 41 to 43%. Several bags were required to fill the test rig. Distilled water was mixed with the paste diluting the solids content to about 32% solids. Dilution was necessary to fill the test rig with the paste. To conserve the limited supply of tomato paste paste was recovered from the test rig when the test rig piping was cleared prior to each plugging test. In subsequent tests the recovered paste which had been stored cold was mixed with previously unused paste to refill the test rig.

b) Canned hot-break Dei Fratelli Italian Style tomato paste was used in one additional test Test T7.

c) Carbopol® 940 (Noveon Inc. Cleveland Ohio USA) cross-linked polyacrylic acid food thickener was mixed dissolved in distilled water to prepare a 1.5% w/w aqueous solution. This fluid was used to fill the test rig and then adjusted from pH 2.8 to pH 5.0 ± 0.5 by the addition of 25% NaOH solution. The homogeneity of pH in the Carbopol gel was insured by pumping the solution through the test rig for several minutes while monitoring the pH and confirming that the lack of measurable changes in pH in the test fluid being pumped past the pH probe.

d) Glycerin was selected as the control fluid because it has a high viscosity but lacks a yield stress.

Test procedure

For each test fluid the following tasks were performed. First the rheological properties of each fluid were determined by the following procedure.

I. The pressure difference from the upstream to the downstream header was recorded over a range of flow rates.

II. The flow rate in the pipes was controlled by varying the rotational speed of the pump and by adjusting the feed by-pass valve (BPV). Volumetric flow rates above 2.5 liters per minute were read with the magnetic flow meter while flow rates below this rate were measured by timing the motion of air bubbles in the uniform-velocity plug-flow region of the flow.

III. Shear stress at the wall and shear rate at the wall were calculated from this data.

IV. The consistency index m the flow behavior index n and the yield stress of the test fluid were also determined.

The pump and all lines except for the four parallel transparent PVC pipes were flushed with filtered water. The valves connecting the parallel pipes to the downstream header were closed and the pipe drain valves were opened to direct flow from the pipes into buckets placed under each pipe's drain valve. The pump was restarted at low speed with the BPV open. The BPV was then closed to force all flow from the positive-displacement pump into the upstream header and pipe(s) flushed clear with water were recorded. The interface position between the test fluid and water in each plugged pipe was then recorded. This procedure was repeated at least twice for each test fluid (at least 3 repetitions).

Results and Discussion

Glycerin

A linear plot of shear stress versus shear rate for the glycerin data Figure 2 reveals the expected proportional relationship characteristic of Newtonian fluids. The reported viscosity of glycerin at the same

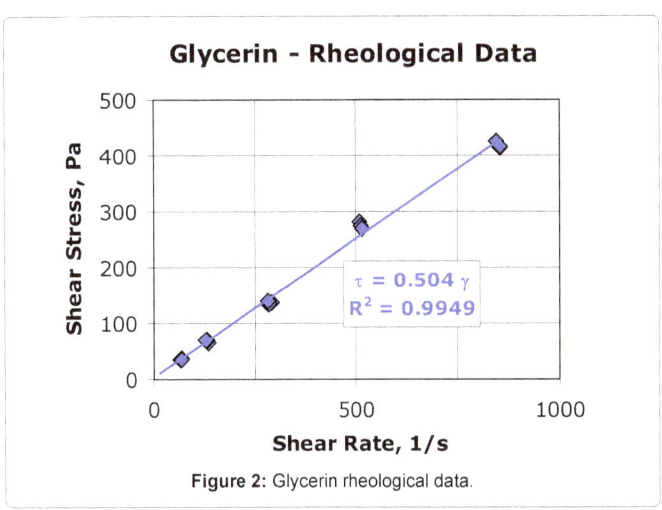

Figure 2: Glycerin rheological data.

temperature 30°C is 629 mPa-s [20] compared to 504 mPa-s in the present study.

The viscosity of glycerin is extremely sensitive to the presence of water [21]. Even though care was taken to minimize the presence of water as the test rig was filled with glycerin the lower than expected measured viscosity likely resulted from a small amount of residual water in the test rig. Nevertheless the glycerin was found to be a Newtonian fluid with no yield stress.

Carbopol gel

Figure 3 shows for each Carbopol 940 gel shear rate plotted against the difference between the shear stress at the wall and the yield stress of each test fluid. The formulas indicate the Herschel-Bulkley model formulas for each test. For both the Carbopol gel and the tomato paste a yield stress value was selected to provide the best curve fit of the data. This was achieved by selecting yield stress values that maximized the R^2 values of the power-law curve fit formulas. Combining this selected yield stress with the consistency index and the flow behavior index gives the complete Herschel-Bulkley formula for each of the test fluids. The values of the three Herschel-Bulkley parameters: yield stress consistency index and flow behavior index were determined for each Carbopol gel sample and are presented in Table 1.

The higher R^2 values obtained with the Herschel-Bulkley model than with the pseudo-plastic model as shown in Table 1 indicate that the Herschel-Bulkley model more accurately describes the flow behavior of this test fluid. This result agrees with Curran et al. [22] who report that for a 1.48% Carbopol gel "steady-shear data were shown to be fitted well by the Herschel-Bulkley model". For the three different tests run on Carbopol gel the close similarity in rheological parameters is believed to result from using gel prepared in a single large batch prior to running the series of tests on this fluid.

Tomato paste

As shown in Figure 4 greater variation in rheological properties was observed in the tomato paste. This variation is not surprising considering the fact that the batches of paste used in each test were prepared by mixing aseptically-packaged paste samples collected at different times in the tomato processing plant. These samples ranged from 41% to 43% solids.

Another factor that increased variation between the rheological properties of each batch of tomato paste was dilution of the paste

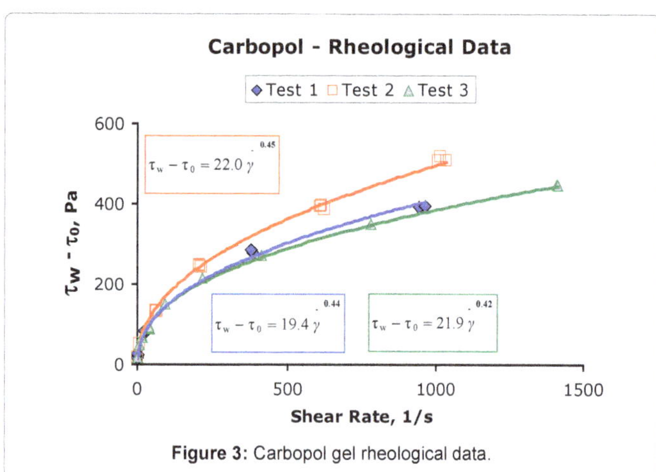

Figure 3: Carbopol gel rheological data.

Test	Pseudoplastic Model			Herschel-Bulkley Model			
	m	n	R^2	τ_0	m	n	R^2
	Pa-s^n	---	---	Pa	Pa-s^n	---	---
C1	116.8	0.199	0.967	103.8	19.47	0.442	0.998
C2	101.9	0.234	0.962	85.5	22	0.451	0.999
C3	115.2	0.189	0.926	106.4	21.91	0.415	0.994

Table 1: Carbopol gel rheological parameters.

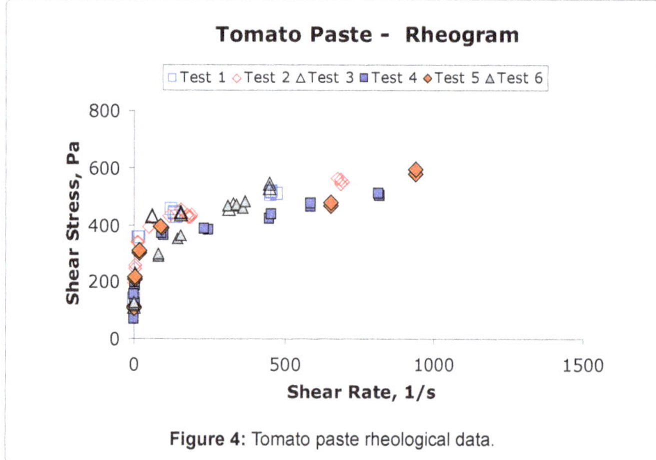

Figure 4: Tomato paste rheological data.

samples by adding distilled water equal to 30% of the mass of tomato paste. This dilution was necessary to reduce their apparent viscosity sufficiently to permit the paste to be drawn from the feed cone into the suction of the Waukesha PD pump. Despite the variation in rheological properties the objective of observing the behavior of a semi-solid food material as it is flushed from parallel tubes was achieved with the tomato paste.

As was done for the Carbopol rheological data a yield stress value was selected which gave the best fit (R^2 nearest 1.0) of the data using the Herschel-Bulkley model. The values of the three Herschel-Bulkley parameters: yield stress consistency index and flow behavior index were determined for each tomato paste sample and are presented in Table 2. Note that in all but two test cases no improvement in the fit of the data was obtained by applying a yield stress to the curve-fit formula. In other words these tomato paste samples appeared to behave as pseudo-plastic fluids lacking a dynamic yield stress. As reported in the literature tomato pastes are known to possess a static yield stress. This is supported by direct observation of the tomato paste in the current study (Figure 5).

Tomato ketchup a product based on tomato paste has been reported to possess a yield stress that is reduced from the initial static yield stress of an undisturbed sample to a lower dynamic yield stress after the paste has been exposed to shear stress resulting in breakdown of structure. In the rheological tests carried out in the current study only the apparent dynamic yield stress of each tomato paste sample was measured [23].

In the plugging tests where some motion is initiated as flushing water advances into the parallel tubes there may be some breakdown of structure but likely much less than during measurement of the rheological parameters when the paste is repeatedly pumped through the tubes for an extended time. During plugging tests the paste remained motionless in the tubes for a short time which may have allowed some recovery of structure and static yield stress which was not measured in the current study.

Since the static yield stress of the tomato paste samples used in each test were not measured the Bingham yield stress was determined by extrapolation of the linear portion of the shear stress versus shear rate data to zero shear rates, as an approximation of the apparent static yield stress. Table 3 presents the Bingham yield stresses for each tomato paste sample determined using shear rates above 50 s^{-1} which corresponds to the range of shear rates in the plugging tests.

As reported in the literature apparent slip is a significant factor affecting the flow behavior of tomato concentrates [12]. If the tomato pastes are regarded as being Bingham plastics in which the apparent dynamic yield stress equals the Bingham yield stress the observation of creeping flow at shear stresses below the yield stress would be entirely due to apparent slip. This agrees with the statement that when the shear stress is less than the yield stress i.e. $\tau_w / \tau_0 \leq 1$ the bulk flow is negligible and the apparent motion is entirely due to the slipping of the paste [10].

Test	Pseudoplastic Model			Herschel-Bulkley Model			
	m	n	R^2	τ_0	m	n	R^2
	Pa-s^n	---	---	Pa	Pa-s^n	---	---
T1	171.2	0.138	0.954	150	48	0.264	0.969
T2	160.2	0.192	0.97	0	160.2	0.192	0.97
T3	195.7	0.146	0.969	0	195.7	0.146	0.969
T4	136.6	0.208	0.936	0	136.6	0.208	0.936
T5	121.1	0.192	0.959	0	121.1	0.192	0.959
T6	142.7	0.201	0.939	0	142.7	0.201	0.939
T7	100.6	0.26	0.994	69.5	38.4	0.404	0.999

Table 2: Tomato paste rheological parameters.

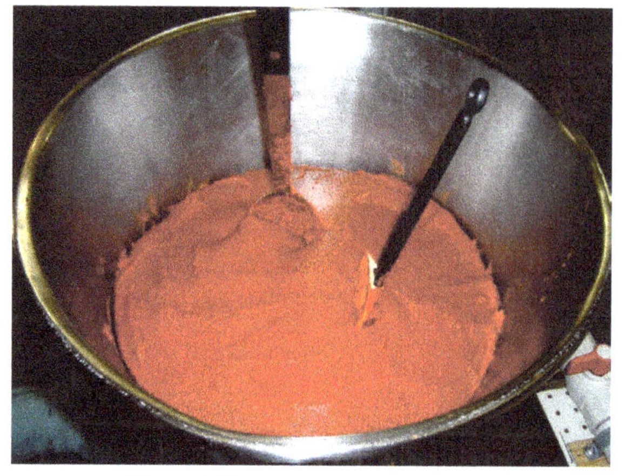

Figure 5: Tomato paste in the feed cone.

Test	Bingham Plastic Model			
	τ_B	m	n	R^2
	Pa	Pa-s^n	---	---
T1	323	0.122	1	0.902
T2	420	0.145	1	0.932
T3	397	0.139	1	0.967
T4	405	0.135	1	0.842
T5	343	0.1	1	0.982
T6	368	0.162	1	0.928
T7	251	0.66	1	0.973

Table 3: Tomato paste–Bingham yield stress.

Test	Fluid	Yield Stress	Tube				Plugged Tubes
		Pa	1	2	3	4	
G1	Glycerin	0	0	0	0	0	0
G2	Glycerin	0	0	0	0	0	0
G3	Glycerin	0	0	0	0	0	0
C2	Carbopol Gel	86	1	1	0	0	2
C1	Carbopol Gel	104	1	1	0	0	2
C3	Carbopol Gel	106	1	1	0	1	3
T7	Tomato Paste	251	1	1	1	0	3
T1	Tomato Paste	323	1	0	1	1	3
T5	Tomato Paste	343	1	1	0	1	3
T6	Tomato Paste	368	1	1	0	1	3
T3	Tomato Paste	397	1	1	1	0	3
T4	Tomato Paste	405	1	1	1	0	3
T2	Tomato Paste	420	1	1	1	0	3

1 = plugged tube; 0 = cleared tube

Table 4: Summary of test results.

This would also explain the observation made on tomato paste but not on Carbopol gel that after one tube had been cleared with flushing water the paste in the remaining tubes continued to creep slowly forward.

In the case of fluids such as tomato pastes in which slip occurs at shear stresses below the yield stress applying low shear stress would be expected to result in creeping flow of the paste. This means that in parallel-flow devices in which some tubes have been cleared by flushing with water and other tubes remain partially filled with paste material an extended period of pumping water through the cleared tubes would eventually clear the remaining tubes. However in a parallel-tube flow device such as a tubular heat exchanger flushing with water to clear tomato paste or other similar material from the tubes would be expected to be accomplished quickly. Unless the need is understood such extended flushing is unlikely to be performed. In this case the flow devices would not be fully cleared and the un-cleared tubes would be effectively plugged.

Plugging experiment results

As it can be seen in Table 4 in all tests run on yield-stress fluids two or three of the four parallel tubes remained plugged after flushing with water. In this table the yield stresses listed for the tomato paste tests are the Bingham yield stresses which are approximations of the apparent static yield stresses.

The following observations are based on the empirical results obtained in the plugging tests performed with the parallel-tube test rig. In all three tests using glycerin a viscous fluid with no yield stress all tubes were cleared with water. In all tests using fluids with a yield stress at least 2 of the 4 tubes were left partially filled with the yield stress fluid while the remaining tubes were completely cleared with water.

Conclusions

The cause of the formation of plugs in parallel-tube flow devices is explained. It has been shown that semi-solid fluids possessing a sufficient yield stress are not completely cleared from some parallel tubes when the device is flushed with water. The remaining tubes are effectively plugged. A solution to the problem of plugging is to slowly dilute the yield-stress process fluid during flushing gradually reducing its yield stress.

The experimental results lead to the following conclusions:

A. For plugging to occur the parallel tubes cannot have identical flow resistance.

B. For plugging to occur the fluid initially filling the tubes of a parallel-tube flow device must have a yield stress.

During normal operation plugging occurs only when the yield stress fluid is flushed with water or overly diluted process fluid.

References

1. Paseman R, Griffith L (1987) Cleaning the tube side of heat exchangers. Proceedings of the Fourth U.S. Water Jet Conference, Berkeley, CA.
2. Steffe JS (1996a) Rheological methods in food process engineering (2ndedn). Freeman Press, East Lansing, MI.
3. Barnes HA, Walters K (1985) The yield stress myth. Rheologica Acta 24: 323-326.
4. Hartnett JP, Hu RZ (1989) The yield stress-an engineering reality. J Rheology 33: 671-679.
5. Chhabra RP, Richardson JF (1999) Non-Newtonian flow in the process industries. Butterworth-Heinemann, New York.
6. Rao MA (1999a) Rheology of fluid and semisolid foods; Principles and applications. Aspen Publishers, Gaithersburg.
7. Michaels YS, Bolger JC (1962) The plastic behavior of flocculated kaolin suspensions. Indus Eng Chem Fundamemtal 1: 152-162.
8. Qiu CG, Rao MA (1988) Role of pulp content and particle size in yield stress of apple sauce. J Food Sci 53: 1165-1170.
9. Lee Y, Bobroff S, Mc Carthy KL (2002) Rheological characterization of tomato concentrates and the effect on uniformity of processing. Chem Engi Communi 189: 339-351.
10. Meeker SP, Bonnecaze RT, Cloitre M (2004) Slip and flow in pastes of soft particles: direct observation and rheology. J Rheology 48: 1295-1320.
11. Mc Carthy KL, Kerr WL (1998) Rheological characterization of a model suspension during tube flow using MRI. J Food Engi 37: 11-23.
12. Dogan N, Mccarthy MJ, Powell RL (2002) In-line measurement of rheological parameters and modeling of apparent wall slip in diced tomato; Hydrodynamical behaviour of non- Newtonian flows in a cross-flow filtration tubular module. Exp Fluids 25: 243-253.
13. Ouriev B (2002) Investigation of the wall slip effect in highly concentrated disperse systems by means of non-invasive UVP-PD method in the pressure driven shear flow. J Colloid 64: 740-745.
14. Rao MA (1999b) Rheology of fluid and semisolid foods; principles and applications. Aspen Publishers, Gaithersburg.
15. Aguilera JM, Stanley DW (1999) Microstructural principles of food processing and engineering (2ndedn). Aspen Publishers, Gaithersburg.
16. Steffe JS (1996b) Rheological methods in food process engineering (2ndedn). Freeman Press, East Lansing, MI.
17. Carrere H, Rene F, Tanguy PA (2002) Properties of Carbopol solutions as models for yield-stress fluids. J Food Sci 67: 176-180.
18. Midoux N (1985) Mécanique et rhéologie des fluides en genie chimique. Éditions Tec and Doc Lavoisier, Paris.
19. Munson BR, Young DF, Okiishi TH (1998) Fundamentals of fluid mechanics (3rdedn). John Wiley & Sons, New York.

20. Weats RC (1972) Handbook of chemistry and physics (53rdedn). Chemical Rubber Publishing, Cleveland, OH.

21. Gilmount R (2002) Liquid viscosity correlations for flowmeter calculations. CEP Magazine 98: 37-41.

22. Curran SJ, Hayes RE, Afacan A, Williams MC (2002) In-line measurement of rheological parameters and modeling of apparent wall slip in diced tomato suspensions using ultrasonics. J Food Sci 67: 2235-2240.

23. Yoo B, Rao MA, Steffe JS (1995) Yield stress of food dispersions with vane method at controlled shear rate and shear stress. J Textur Studie 26: 1-10.

Sensorial Assessment of Beef Sausage Processed by Wheat Germ Flour

Elbakheet SI*, Elgasim EA and Algadi MZ[2]

Faculty of Agriculture, University of Khartoum, Sudan

Abstract

The term sausage is derived from the Latin word (salsus) meaning salt or literally translated, refers to chopped or minced meat preserved by salting. In this study beef sausage was processed by additions of different replacement levels of meat by wheat germ flour (WGF) replacement levels were: 0% (as control) 10% and 15%. The processed beef sausages were packaged in foam trays, over wrapped with polyvinyl chloride (PVC) and stored refrigerated at 4°C ± 1 for up to 7 days. Several variables were determined using subjective and objective measurements, to evaluate the effects of replacement levels and storage periods on the sensory attributes of the processed beef sausage. The evaluation was conducted immediately after processing, three and seven days post processing day. Results demonstrated that lower scores in over all acceptability, aroma and flavor; but higher score ($p< 0.05$) in deviation from meat aroma. Fifteen percent replacement level sample had the highest ($p< 0.05$) on overall acceptability, flavor and aroma scores. Overall acceptability score, flavor score and aroma score, were increase with the increased of replacement levels. WGF act as binder in beef sausage production and could be a good substitute to others plant binders which are used as meat binder or extenders.

Keywords: Sensory assessment; Sausage; Polyvinyl chloride; Processing

Introduction

Nutritionally, meat is a very good source of essential amino acids, to a lesser extent, of certain minerals. Although vitamins and essential fatty acids are also present, meat also provides calories from protein, fat and limited quantities of carbohydrate [1]. Germ constitutes about 2.5% of the grain weight and comprises minimal amount of protein, but greatest share of fat, vitamins especially tocopherols [2]. Wheat germ is a by-product of wheat milling, and recently, it has attracted much attention due to its unique nutritional value. It contains 42% to 45% carbohydrate, 25% to 30% protein, 16% simple sugars, 4% to 5% minerals (total ash) and 10% to 12% lipid [3]. The continues successes of marketing meat on the innovation and consistent production of high quality products. Consumers are looking for convenient food products with new/exciting flavor, textures etc. [4]. Meat and meat products are highly perishable materials so sanitation and cooling is essentials in handling, marketing and processing of meat. The sanitation in the Sudan, in general is very poor regarding slaughtering, handling, marketing and processing of meat, except for very few meat plant and slaughter houses. Generally, meat products are widely consumed throughout the world; but unfortunately, their cost is high. To reduce this cost there is increasing interest in use of various non-meat proteins especially plant protein. The objectives of this study to evaluate the effects of partial replacement of meat by wheat germ flour on the quality characteristics of beef sausage.

Material and Methods

Materials

Food materials: Meat loins and round were obtained from Animal Production Research Center Kuku. The beef meat was stored frozen at -11 ± 1°C in freezer at Regional Training Center for Meat Quality, Grading and Meat Technology, Elkadaro. Wheat germ was obtained from Seen flour mills stored frozen. Spices, salt and sugar were obtained from local market of Khartoum North. The additional fat needed in the formulation was obtained from the local market. Uniform rendered fat free of protein was used.

Chemicals and reagents: Chemicals and reagent used were brought from the central lab stores of Khartoum University, sodium nitrite and ascorbic acid, were obtained from Looly Company, Khartoum.

Casings: Cellulose casings 23 mm in diameter were obtained from Looly Company, Khartoum.

Methods

Raw materials preparation

Meat preparation: Stored beef was allowed to thaw and sliced then ground through a 0.75 In, plate using a meat grinder. Ground beef was stored refrigerated at 4°C ± 1°C, for about 20 hr, a sample was taken to be analyses for protein fat and moisture content following AOAC Method [5] (Table 1).

Wheat germ preparation: Stored wheat germ was ground, to form wheat germ flour (WGF). Then a sample was taken and analyzed for protein, fat and moisture content, following AOAC Method [5] (Table 1).

Calculation for sausage formulation: The experiment designed to produce sausage with the following specification, protein 15%, fat 20% moisture, 58.3% added starch 4.7%, salt 1.5%, and spices 0.5% (Tables 2 and 3). Three batches with three replacements of meat by wheat germ were used every batch weight 2000 g.

Therefore,

$$\textit{Protein required} = \frac{15 \times 2000}{100} = 300\,g$$

$$\textit{Fat required} = \frac{20 \times 2000}{100} = 400\,g$$

***Corresponding author:** Elbakheet SI, Faculty of Agriculture, University of Khartoum, Sudan, E-mail: shawgi1416@gmail.com

Material	Protein %	Fat %	Moisture content %	Ash %	Crude fibre %	Carbohydrate	pH
Beef meat	22.6	3.2	71	0.98	-	0.3	6.29
Wheat germ	27.2	9.3	10.35	2.17	2.53	48.3	6.17

Table 1: Proximate analysis and pH of beef meat and wheat germ.

Components	Percentage	Weight in gms
Proteins	15	300
Fat	20	400
Starch	4.7	94
Water	58.3	1160
Salt	1.5	30
Spices	0.5	10

Table 2: Sausage formula.

Ingredient	Replacement level of meat*		
	0% protein	10% protein	15% protein
Beef meat	1327.43	1194.7	1128.32
Wheat germ	-	110.3	165.44
Starch	94	94	94100
Fat	357.52	355.15	94
Water	217.53	300.34	341.3
Salt	30	30	30
Sugar	10	10	10
Black pepper	3	3	3
Nutmeg	2	2	2
Cinnamon	2	2	2
Garlic	2	2	2
Sodium nitrite	0.13	0.12	0.11
Vitamin C	0.62	0.55	0.52

Table 3: Sausage formulation for all treatments.

$$Water\ required\ = \frac{58.3 \times 2000}{100} = 1166g$$

$$Starch\ required\ = \frac{4.7 \times 2000}{100} = 94g$$

$$Salt\ required\ = \frac{1.5 \times 200}{100} = 30g$$

$$Spices\ required\ = \frac{0.5 \times 200}{100} = 10g$$

Sodium nitrite 100 ppm.

Vitamin C 0.466 g/kg

First replacement level: Wheat germ 0% so the required protein was 100% from meat beef therefore beef require = $\frac{300 \times 100}{22.6}$ =1327.43

$$Fat\ in\ 1327.43g\ beef = \frac{3.2 \times 1327.43}{100} = 42.48$$

$$Fat\ to\ be\ added\ = 400 - 42.48 = 357.52$$

$$Moisture\ in\ 1327.43g\ beef = \frac{71 \times 1327.43}{100} = 942.47$$

Moisture from starch = 6

Total moisture = 948.47

Moisture to be added = 1166 - 948.47 = 217.53

Required sodium nitrite = 0.13 g

Required Vitamin C = 0.62 g

Second replacement level: Wheat germ 10% so the required protein was 90% from beef and 10% from wheat germ

$$There\ fore\ beef\ required\ = \frac{300 \times 90}{22.6} = 1194.7$$

$$Wheat\ germ\ required\ = \frac{300 \times 10}{27.2} = 110.3$$

$$Fat\ in\ 1194.7g\ beef\ = \frac{3.2 \times 1194.7}{100} = 38.23$$

$$Fat\ in\ 110.3\ wheat\ germ\ = \frac{6 \times 110.3}{100} = 6.62$$

Total fat = 44.85

Fat to be added = 400 - 44.85 = 355.15

$$Moisture\ in\ 1194.7g\ beef = \frac{71 \times 1194.7}{100} = 848.24$$

$$Moisture\ in\ 110.3g\ wheat\ germ\ = \frac{10.35 \times 110.3}{100} = 11.42$$

Moisture in 100 g starch = 6

Total moisture = 865.66

Moisture to be added = 1166 - 865.66 = 300.34

Sodium nitrite to be added = 0.12 g

Vitamin C to be added= 0.56 g

Third replacement: Wheat germ 15% so the required protein in 85% from beef 15% from wheat germ.

$$Therefore\ beef\ required\ = \frac{300 \times 85}{22.6} = 112.32$$

$$Wheat\ germ\ required\ = \frac{300 \times 15}{27.2} = 165.44$$

$$Fat\ in\ 1128.32\ g\ beef\ = \frac{3.2 \times 1128.32}{100} = 36.11$$

$$Fat\ in\ 165.44g\ wheat\ germ\ = \frac{6 \times 165.44}{100} = 9.95$$

Fat to be added = 400 - 46.04 = 353.96

$$Moisture\ in\ 1128.32\ g\ beef\ = \frac{71 \times 1128.32}{100} = 801.11$$

$$Moisture\ in\ 165.44\ g\ wheat\ germ\ = \frac{10.35 \times 165.44}{100} = 17.12$$

Moisture in 100 g starch = 6.

Total moisture = 824.23

Moisture to be added =1166 -824.23 = 341.8

Sodium nitrate = 0.11 g

Samples	A		B		C		
	Extremity like	moderately like	Like	slightly like	slightly dislike	dislike	Extremity dislike
Aroma	--	--	--	--	--	--	--
Flavour	--	--	--	--	--	--	--
Deviation from meaty aroma	--	--	-	--	--	--	--
Juiciness	--	--	-	--	--	--	--
Tenderness	--	--	--	--	--	--	--
Overall acceptability	7	6	5	4	3	2	1

Table 4: Sensory evaluation form (Flavour, deviation from meat aroma, juiciness, tenderness, and overall acceptability. Using scores as follows: 7 = Extremity like; 6 = moderately like; 5 =Like; 4 = slightly like; 3 = slightly dislike; 2 = dislike; 1 = Extremity dislike).

Independent variables	Replacement levels of meat by WGF			S.E
	0%	10%	15%	
Taste	4.5	4.95	5.6	± (0.34)
Order	4.6	5.05	5.65	± (0.3)
Variation from meat taste	5.20	4.60	5.15	± (0.19)
Juiciness	5.15	4.85	5.37	± (0.15)
Tenderness	5.20	5.35	5.85	± (0.2)
Overall acceptability	4.85	5.00	6.10	± (0.39)

Table 5: Sensory characteristic of cooked beef.

Vitamin C to be added = 0.52 g

Sausage preparation: Minced meat, salt, sugar, minced fat, spices, vitamin C, sodium nitrate and half of calculated ice water were introduced to a Hobart Chopper; the Chopper was then started for about 4 min. The added materials were dispersed uniformly. Then the ground wheat germ, starch were added together with the remainder of the calculated water. The entire mass was chopped for about 5 min. then transferred to manual stuffer to be stuffed into cellulose casing of 23 mm in diameter and linked at lengths of 15 cm. The framed sausage were heated in water at 98°C for about 40 min, followed by immediate cooling in ice water, for 15 min. The cooled processed sausage was peeled and packed in foam trays over-wrapped with polyvinyl chloride (PVC) and stored refrigerated for up to 7 days. The WGF replacement levels in beef sausage formulation and processing were performed following the same procedures explained above.

Method of analysis

Sausage were assessed at 0 day (i.e., immediately after processing) after three and seven days post processing.

Sensory evaluation: Ten member sensory panel consisting of M.Sc. and B.Sc. student of food science and technology Department, Faculty of Agriculture, University of Khartoum, semi-trained according to the procedure of Cross et al. [6]. The panel evaluated the cooked sausage sample with the different treatment for juiciness, tenderness, test, odor, differential from meat taste and over all acceptability. By mean of the scale (7 = Extremely like, 1 = Extremely dislike), (Table 4). Panelists received samples which were randomly numbered. Water at room temperature was made available to panel for cleaning the palate between the tested samples.

Statistical analysis: The data collected from the different treatments was subjected to analysis of variance and whenever appropriate the mean separation procedure of Duncan was employed [7]. The SAS program [8] Was used to perform the general linear model (GLM) analysis.

Results and Discussion

Sensory evaluation

The sensory evaluation of beef sausage extended with three replacement level is shown in (Table 5), the control samples scored high values in deviation from meat aroma that could be due to flavor of wheat germ flour. According to Canasambandam and Zayas [9] aroma and flavor are probably the most important attributes that influence the sensory properties of comminuted meat product extended with non-meat protein additives. Fifteen percent replacement levels had the highest score in: aroma, flavor, Juiciness, tenderness and over all acceptability among the treatment. Generally, it was relatively like control sample in juiciness and deviation from meat aroma, and these agrees with the finding of Canasambanadam and zayas [10] a trained panel found suggested an effect due to increasing levels of wheat germ protein flour on aroma and flavor of frankfurters. 10% replacement level sample usually, scores higher than 0% and less than 15% in: aroma, flavor, deviation from meat aroma, Juiciness, tenderness and over all acceptability.

Conclusion

The relative high scores of tenderness and Juiciness in the sample with replacement levels 15% may be due to high water binding of these samples. Judge et al. [1] indicated that many of the physical property of meat include color, texture and firmness of raw meat, Juiciness and tenderness of cooked meat are partially dependent on WHC. And mention that the portion of water present in free form and the ability of meat to bind water and factors that increase this ability will increase juiciness.

References

1. Judge MD, Albert ED, Forrest JC, Hedrick MB, Merkel RA (1990) Principle of meat science. (2ndedn), Kendall Hunt, Owa, USA.
2. Anon (1987) Wheat facts. Official publications of National Association of Wheat Growers 10: 17.
3. Majzoobi M, Ghias F, Farahnaky A (2015) Physicochemical assessment of fresh chilled dairy dessert supplemented with wheat germ. Int J Food Sci Technol.
4. Barbut S (2016) Principles of meat processing. The science of poultry and meat processing.
5. AOAC (1995) Official method of analysis (4thedn). Association of Official Analytical Chemists, Washington DC.
6. Cross HR, Moen R, Stanfield MS (1978) Guidelines for training and testing judges for sensory analysis of meat quality. J Food Technol 32: 48-54.
7. Steel RG, Torrie JH (1980) Principles and procedures of statistics. MC Grow Hill, New York, USA.
8. SAS (1988) SAS/ STAT users guide.
9. Canasambanadam R, Zayas JF (1992) Functionality of wheat germ protein in comminuted meat products as compared with corn germ and soy proteins. J Food Sci 57: 829-833.
10. Canasambanadam R, Zayas JF (1996) Frankfurters extended with wheat germ protein: Sensory properties and consumers response. J Food quality 19: 423-435.

Potentials of Trifoliate Yam (*Dioscorea dumetorum*) in Noodles Production

Akinoso R[1], Olatoye KK[2*] and Ogunyele OO[1]

[1]*Department of Food Technology, Faculty of Technology, University of Ibadan, Nigeria*
[2]*Department of Food, Agriculture and Bio-engineering, College of Engineering and Technology, Kwara State University, Kwara State, Nigeria*

Abstract

Trifoliate yam (*Dioscorea dumetorum)* is a high yielding but under exploited yam species. Potential of its flour in noodle production was investigated in this study. Trifoliate yam flour was produced and substituted for wheat flour at 20 -70% levels. Noodles were produced from the composite flour and evaluated for their proximate composition, functional properties, colour and cooking properties, using standard methods. Sensory attributes of the noodle were also determined using panelists. Data were analyzed using ANOVA at ($p<0.05$). The results showed that incorporation of trifoliate yam flour into wheat flour increased the moisture, crude fibre, ash and fat content of dried noodles. The moisture content ranges from 7.16 - 12.93%, crude fiber (0.72 - 1.30%), Ash (1.20 - 2.88%), fat (18.26 - 28.54%), protein (5.88 - 7.79%) and carbohydrate (51.18 - 62.77%). The water and oil absorption capacities of the noodles increased from 1.60 g to 2.03 g and 0.72 g to 1.21 g respectively, with increase in levels of trifoliate yam flour. Significant differences ($p<0.05$) exist between the colour of raw sheet noodles and the cooked noodles. The optimum cooking time and cooking loss were in the ranges (5.43 min to 7.06 min) and (9.31 - 15.09%) respectively. Sensory evaluation showed that the acceptability of uncooked and cooked dried noodles were at the moderate level. The substitution of different levels of trifoliate yam flour did not give significant ($p<0.05$) influence on sensory attributes except for the taste and color of dried noodles. Trifoliate yam has a good utilization potential in noodle production and could be used as substitute to wheat in human nutrition and in the food industries since they compare favorably in their nutritional compositions.

Keywords: Trifoliate yam; Noodles production; Sensory evaluation; Chemical composition

Introduction

Yam (*Dioscorea sp.*) is one of the most important staple food crops in West Africa especially Nigeria [1]. Yams are annual or perennial tuber-bearing and climbing plant with over 600 species, in which only few are cultivated for food and medicine [2]. Some yam varieties are widely used for food while others are underutilized. Trifoliate yam (*Dioscorea dumetorum*) is an under-exploited but high yielding yam species [3]. It has also been reported to be nutritionally superior to the commonly consumed yams with high protein and minerals [4]. It has starch grains that are smaller, more digestible than those of other yam species. Trifoliate yam tuber however, contains some anti-nutrient contents as a result of which slight bitterness may be experienced [5]. In addition, this yam species hardens few days after harvest leading to reduction in moisture and starch content and increase in sugar as well as structural polysaccharides [6]. Intensive processing like prolonged soaking and blanching are expected to eliminate these defects. Transforming its tubers into edible flour constitutes a means of conferring a long-term value onto it. Today, people's preference towards convenience food products has cumulated in many adverse consequences including hike in price and increase in the demand for importation of wheat. Noodles, which are convenient pasta products, are basically prepared from unleavened dough of durum wheat semolina and are only second to bread in popularity as staple food, globally [7]. They are nutritious and delicious, containing complex carbohydrates, which can provide long lasting energy and help to feel full for long periods. They are consumed in all five continents, with increased awareness of its nutritional benefits [8]. Noodles preparation by supplementations of wheat flour with other food materials have been documented; cassava [9], matured green banana and oat beta glucan [10], green banana [11], unripe plantain [12]. Better alternative usefulness of these materials is many; therefore involvement of underutilized materials like trifoliate yam could be more economically viable. Incorporation of its flour into wheat flour will reduce pressure on the latter while concomitantly promoting the industrial utilization of the former in pastry. Thus, this research work aimed at reducing food insecurity bane by promoting the utilization of trifoliate yam through substituting its flour in the noodle formulation.

Materials and Methods

Materials

Trifoliate yam (*Dioscorea dumentorum*), clean and free from bruises, was obtained from Malete, Ilorin, Kwara State. Sodium carbonate, potassium carbonate, gelatin, Iodized salt, ascorbic acid, Sodium phosphate (analytical grades), Eggs and edible oil and all the equipment used were supplied by the Department of Food Technology, University of Ibadan, Oyo state, Nigeria.

Methods

Production of trifoliate yam flour (TYF): Trifoliate yam tubers were processed into flour according to procedure of Abiodun et al. [13] as shown in Figure 1.

Preparation of noodles: Noodles were prepared following the procedures of Nagao [14] as shown in Figure 2. Noodles produced from six different blends of trifoliate yam flour and wheat flour with their codes was shown in Table 1. Basic ingredients and combinations were presented in Table 2.

Functional properties

Determination of bulk density: Bulk density and oil absorption capacity were determined by adopting the method of Udensi and Okaka [15]. 3 g of each sample was weighed into 10 ml graduated cylinders and tapping ten times against the palm of hand. The volume

***Corresponding author:** Olatoye KK, Department of Food, Agriculture and Bio-engineering, College of Engineering and Technology, Kwara State University, Malete, P.M.B 1530, Ilorin, Kwara State, Nigeria, E-mail: luckykaykay@yahoo.com

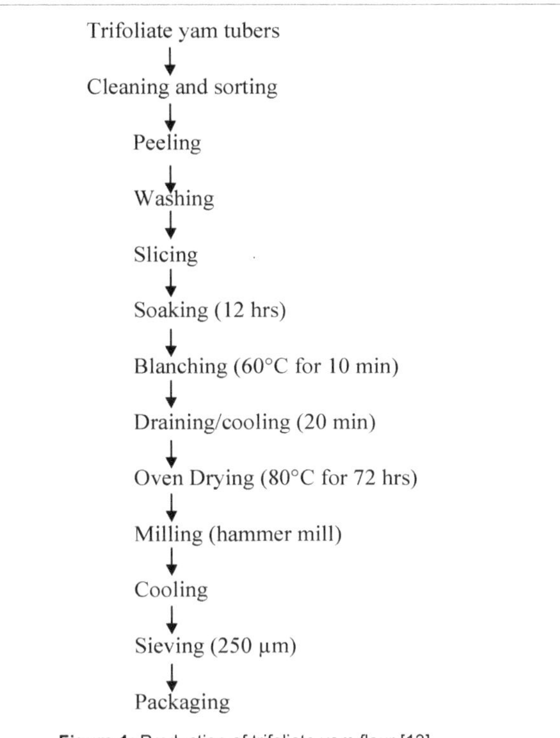

Figure 1: Production of trifoliate yam flour [13].

Code	TYF: WF
Y_0W_{100}	00:100
$Y_{20}W_{80}$	20:80
$Y_{30}W_{70}$	30:70
$Y_{50}W_{50}$	50:50
$Y_{70}W_{30}$	70:30
$Y_{100}W_0$	100:00

Table 1: Sample codes and blend ratios of trifoliate yam-wheat noodles.

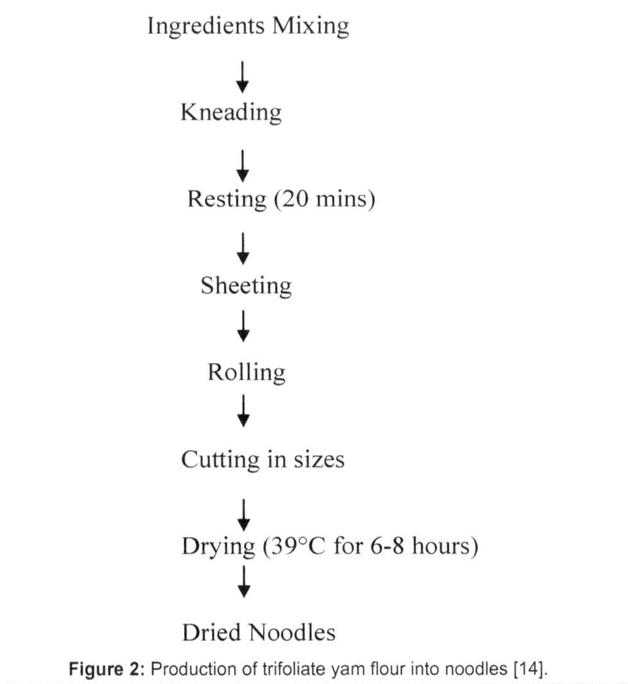

Figure 2: Production of trifoliate yam flour into noodles [14].

of sample was recorded after tapping was recorded and bulk density was expressed as g/ml.

Determination of water absorption capacity: Water binding capacity of noodles was determined according to the method of AACC [16]. Aqueous suspension of noodles was made by dissolving 2 g (dry weight) of noodle in 40 ml of distilled water. The suspension was agitated for 1 hour on a Griffin flask shaker and centrifuged at 2200 rpm for 10 minutes. The free water (supernatant) was decanted from the wet sample, drained for 10 minutes and the wet sample was then weighed. The water absorption capacity was calculated by difference using equation 1.

$$\%Water\ absorption\ capacity = \frac{Weight\ of\ bound\ water}{Weight\ of\ sample} \times 100 \quad (1)$$

Oil absorption capacity: The method of Kinsella JE and Melachouris [17] was used. One gram meal that was mixed with 10 ml refined vegetable oil (Gino) in a weighted 25 ml centrifuge tube was thoroughly stirred for 2 min and then centrifuge at 4000 rpm for 20 min. the supernatant was discarded, the adhering free oil was removed and the tube and content was re-weighed. Oil absorption capacity as expressed as weight of oil bound by 100g meal.

Determination of solubility index and swelling capacity: Solubility and Swelling power determinations were carried out based on a modification of the method of Iwuoha [18]. One gram of noodle was dissolved with distilled water to a total volume of 40 ml using a weighed 50 ml graduated centrifuge tube. The suspension was stirred just sufficiently and uniformly avoiding excessive speed since it might cause fragmentation of the starch granules. The slurry in the tube was heated at 85°C in a thermostatically regulated temperature water bath for 30 minutes with constant gentle stirring. The tube was then removed, wiped dry on the outside and cooled to room temperature. It was then centrifuged at 2200 rpm for 15 minutes. The supernatant was decanted into a pre-weighed moisture can. The solubility was determined by evaporating the supernatant in thermostatically controlled drying oven at 105°C and weighing the residue (Equation 2). The sediment paste was weighed and swelling capacity was calculated as the weight of sediment paste per gram of noodle used (Equation 3).

$$\%Solubility = \frac{Weight\ of\ soluble}{Weight\ of\ sample} \times 100 \quad (2)$$

$$Swelling\ capacity = \frac{Weight\ of\ sediment}{Sample\ weight - Weight\ of soluble} \quad (3)$$

Cooking time: Optimal cooking time was evaluated by observing the time of disappearance of the core of the noodle strand during cooking (every 30 s) by squeezing the noodles between two transparent glass slides. About 10g of noodles was cooked in 300 ml of distilled water in a covered 500 ml beaker. Cooking time was determined by the removal of a strand of noodle every 30 seconds and pressing the noodle between two pieces of watch glasses. Optimum cooking was achieved when the center of the noodles became transparent. Cooking was stopped by rinsing with distilled water.

Cooking loss: The cooking loss was determined by measuring the amount of solid substance lost to cooking water. A 10 g sample of noodles was placed into 300 ml of boiling distilled water in a 500 ml beaker. Cooking water was collected in an aluminum vessel which was placed in an air oven at 105°C and evaporated to dryness. The residue was weighed and reported as a percentage of the starting material. For each optimal cooking time and cooking loss value, five determinations were performed to obtain the mean values (Equation 4)

$$Cooking\ loss\,(\%) = \frac{Dried\ residue\ in\ cooking\ water}{Noodle\ weight\ before\ cooking} \times 100 \quad (4)$$

Water uptake: The water uptake was determined by the ratio of the weight of cooked noodle to the weight of noodle before cooking (Equation 5)

$$Water\ uptake\,(\%) = \frac{Weight\ of\ cooked\ noodle}{Weight\ of\ noodle\ before\ cooking} \quad (5)$$

Determination of colour: The color of the dried noodle sheets and the optimally cooked noodle samples were measured with a Chroma-meter (Minolta, Tokyo, Japan) equipped with a D65 illuminant using the CIE L*a*b* system. The L*, a* and b* readings were obtained directly from the instrument and provided measures of lightness, redness and yellowness, respectively. All measurements were performed in triplicate and mean value recorded.

Sensory evaluation: Cooked samples were evaluated for appearance, flavor, taste, texture, colour and overall acceptability by 30 untrained panelists using nine-point hedonic scales, where 9 and 1 were extremely like dislike respectively.

Statistical analysis: Means of triplicate data were recorded. There were subjected to one-way analysis of variance (ANOVA), means were separated using Duncan multiple range test at 0.05 significance level using SPSS software.

Results and Discussions

Proximate compositions

Substitution of TYF for Wheat flour significantly ($p<0.05$) influenced the proximate composition of noodles, (Table 3). The moisture content ranged between (9.09 - 12.16%) for 20 - 70% levels. Noodle sample $Y_{100}W_0$ had highest mineral content of (12.93%) db, while (control) Y_0W_{100} had the least (8.50%). The higher moisture content found in $Y_{100}W_0$ might be due to water holding capacity of fibers and polysaccharides in trifoliate yam flour during the dough formation. The results of moisture content were close to a safe moisture level (≤10%) for long term storage of flour. The capabilities of yam flours to hold more water are related to the higher fiber content compared with other flours [19]. Crude Protein ranged (5.88 - 7.79%) were significantly ($p<0.05$) reduced with substitution levels of TYF. This result was closer to (7 - 8%) minimum recommended for prevention of PEM (protein-energy malnutrition) by UNO/WHO [20]. Yap and Chen [21] similarly reported a decreased in protein content in the wet noodles when substitution proportion were raised. Crude Ash content ranged 1.20 - 2.88% and decreased when the trifoliate yam flour amount in some noodles increased and also increased in some noodles. Noodle prepared from $Y_{30}W_{70}$ has the highest ash content of 2.88% and was reduced to 2.57% at $Y_{70}W_{30}$. The ash content depends on the quality of the flour and thus corresponds to the higher mineral content, especially potassium. $Y_{100}W_0$ noodle has the lowest ash content of 1.20%. These values were still closer to 3% for maximum ash content in first quality dried noodle and higher than 0.012 - 0.030% ash content reported by Chang and Wu [22] in seaweeds powders substitution. However, $Y_{20}W_{80}$ and $Y_{70}W_{30}$ noodles showed no significant different ($p \le 0.05$) with Y_0W_{100} noodle. Ash content is a reflection of mineral status, though contamination can indicate a high concentration in a sample. Crude Fat content was at the range between 18.43 - 28.54%, and significantly higher ($p<0.05$) than 3.3% fat content recommended by the Ministry of Health on the wet noodle [23]. This might be due to higher oil absorption capacity of TYF than the WF in the control formulation. The results were in disagreement with Wang et al. [24] whose fat content of noodles decreases as the substitution level of green banana flour increases. This result was however similar to Senthil et al. [25] and lower than 13.37 - 21.50%, reported by Gabriel and Faith [26] in similar studies. The lipid components will contribute to the texture, flavor and aroma of foods, thereby prolonging satiety and facilitate the absorption of lipid-soluble vitamins [27]. Crude fibre contents of noodles incorporated with trifoliate yam flours were between 0.72 and 1.30%. $Y_{30}W_{70}$ noodle had higher fibre content than the Y_0W_{100} noodles i.e., control (0.99%). Fibre is important for the removal of waste from the body thereby preventing constipation and many health disorders [28]. Carbohydrate content, varies between (51.18 - 62.77%) and decreases with increased trifoliate yam flour addition. There were significant differences ($p<0.05$) between dried noodle control and noodle made from trifoliate yam flour substitution. Similar result was documented by Wirjatmadi et al. [29]. The Ministry of Health required not less than 14% of carbohydrate content on wet noodles [30].

Functional properties of trifoliate-wheat noodles

Functional properties are characteristics of a substance that affect its behavior and that of products to which it is added during food processing [31]. Results of functional Properties of Trifoliate-Wheat Noodles were presented in (Table 4). Control sample (Y_0W_{100}) had the highest degree of bulkiness (0.72) and bulkiness ranged between (0.61-0.72). Bulk density was significantly decreased with addition of TYF to Wheat flour. Bulk density is a measure of heaviness of a flour sample. Incorporated of TYF significantly increased ($p<0.05$) the WAC with increase in substitution level and ranged between (1.60 g to 2.03 g), with Y_0W_{100} the least and $Y_{100}W_0$ the highest. High water binding capacities

Ingredients (g)	Y_0W_{100}	$Y_{20}W_{80}$	$Y_{30}W_{70}$	$Y_{50}W_{50}$	$Y_{70}W_{30}$	$Y_{100}W_0$
WF	200	160	140	100	60	-
TYF	-	40	60	100	140	200
Sodium carbonates (g)	1.5	1.5	1.5	1.5	1.5	1.5
Potassium carbonates (g)	0.2	0.2	0.2	0.2	0.2	0.2
Sodium phosphate (g)	0.2	0.2	0.2	0.2	0.2	0.2
Gelatin (g)	6.0	6.0	6.0	6.0	6.0	6.0
Ascorbic acid	1.0	1.0	1.0	1.0	1.0	1.0
Iodized salt (g)	3.0	3.0	3.0	3.0	3.0	3.0
Egg (ml)	0.6	0.6	0.6	0.6	0.6	0.6
Edible oil (ml)	5.0	5.0	5.0	5.0	5.0	5.0
Water (ml)	130	130	130	130	130	130

WF: Wheat Flour; TYF: Trifoliate Yam Flour

Table 2: Recipe for trifoliate yam–wheat noodles.

are desirable as they increase the unit yield of products. It stabilizes starches against effects such as seneresis, which sometimes occurs during retorting and freezing [32-34]. Similar trend was observed on OAC, with $Y_{100}W_0$ being the highest (1.21 g) and control noodle, the least (0.72 g). This result disagrees with Siddaraju et al. [35]. Oil absorption is an important property in food formulations because fats improve the flavour and mouth feel of foods [17]. Water binding capacity is also an important functional characteristic in the development of ready-to-eat foods since high water absorption capacity may assure product cohesiveness [36]. Swelling capacity of noodles reduces with substitution levels and ranged between (3.33 g to 5.11 g). $Y_{100}W_0$ had the least value (3.33 g) and Y_0W_{100} the highest (5.11 g). Variation in the Swelling power can be traced to differences in their associative forces. This may explain its hard and waxy texture as revealed from sensory evaluation. Solubility indices of noodles varied significantly ($p<0.05$) and ranged (8.01 - 11.07%). There was no significant difference between the control sample and noodles from TYF/WF blends up to 30% substitution level. Riley et al. [30] reported that solubility increased with decreasing amylose content. The observation for the noodles is similar with the result obtained from of Soni et al. [37] who associated high solubility with high amylose content. The difference could be attributed to differences in granule sizes and their arrangement within their cells.

Colour of trifoliate yam-wheat noodles

Colour characteristics of raw sheet and optimally cooked noodles prepared from different levels of trifoliate yam flour substituted with wheat flour are shown in Table 5. The results indicated that, as the amount of trifoliate yam flour increased, the appearance of the raw sheet and the cooked noodles grew darker. The darkness is likely a product of the Maillard reaction between reducing sugars and proteins [38]. The redness (a*) also increased, while yellowness (b*) decreased with increased addition of TYF. These changes in colour can also be attributed to the higher amount of oxidized phenol compounds in TYF noodles. Similar results have been reported for crackers incorporated green banana flour [24] and yeast leavened banana-bread [38]. Color is a key quality trait [39] because of the visual impact at the point of sale. It provides some indication of the quality of the starting materials and, in some cases, the age of the product.

Cooking properties of trifoliate yam-wheat noodles

Inclusion of TYF in noodles production significantly ($p<0.05$) influenced the cooking properties. Cooking times of all noodle samples increased with increase in TYF and ranged from 5.43 min to 7.06 min, with Y_0W_{100} and $Y_{100}W_0$ being the least and highest respectively (Table 6). Similar trend was observed with cooking loss. The cooking loss is the amount of dry matter in the cooking water of optimally cooked noodles. An increase in the cooking loss with noodles substitute with TYF may have been due to weakening of the protein network by the presence of trifoliate yam flour. This may allow more solids to be leached out from the noodles into the cooking water [40]. These results are in the agreement with Ovando- Martinez et al. [41] who reported that partial or complete substitution of durum wheat semolina with fibre material can result in negative changes to pasta quality, including increased cooking loss. Water uptake and cooking weight also increases as the level of TFY in the noodle increases. $Y_{30}W_{70}$ has the lowest water uptake and cooking weight and is not significantly ($p<0.05$) different with that of control. The water uptake and cooking weight were observed to increase as TYF addition increases, with 100% having the highest in

Noodles Sample	Moisture (%)	Crude Ash (%)	Crude Fat (%)	Crude fibre (%)	Crude protein (%)	Carbohydrate (%)
Y_0W_{100}	8.50 ± 0.14^a	2.50 ± 0.03^a	18.43 ± 0.15^f	0.99 ± 0.00^a	7.79 ± 0.02^c	62.77 ± 0.34^a
$Y_{20}W_{80}$	9.09 ± 0.13^b	2.41 ± 0.07^{ab}	20.15 ± 0.12^e	0.97 ± 0.01^{ab}	7.21 ± 0.01^b	62.11 ± 0.06^c
$Y_{30}W_{70}$	9.77 ± 0.01^c	2.88 ± 0.00^c	23.73 ± 0.04^a	1.30 ± 0.01^d	6.94 ± 0.05^e	56.68 ± 0.10^b
$Y_{50}W_{50}$	11.69 ± 0.07^d	2.29 ± 0.08^b	28.54 ± 0.01^b	0.72 ± 0.01^c	6.30 ± 0.01^a	51.18 ± 0.14^d
$Y_{70}W_{30}$	12.16 ± 0.02^e	2.57 ± 0.07^a	21.89 ± 0.00^c	0.76 ± 0.03^c	6.72 ± 0.01^f	61.67 ± 0.09^c
$Y_{100}W_0$	12.93 ± 0.01^f	1.20 ± 0.00	26.26 ± 0.04^d	0.93 ± 0.00^b	5.88 ± 0.02^d	53.23 ± 0.07^e
Significant differences is indicated by different letters within the same rom. ($p \le 0.05$)						

Table 3: Proximate composition of trifoliate–wheat noodles.

Noodles Sample	Bulk Density g/ml	WAC (g)	OAC (g)	Swelling Capacity(g)	Solubility Index (%)
Y_0W_{100}	0.72 ± 0.00^a	1.60 ± 0.01^b	0.72 ± 0.01^e	5.11 ± 0.10^a	9.94 ± 0.05^a
$Y_{20}W_{80}$	0.67 ± 0.00^b	1.91 ± 0.01^a	0.75 ± 0.00^d	4.26 ± 0.01^c	9.96 ± 0.01^a
$Y_{30}W_{70}$	0.63 ± 0.00^d	1.61 ± 0.01^{bc}	0.87 ± 0.00^c	4.95 ± 0.02^b	9.92 ± 0.01^a
$Y_{50}W_{50}$	0.61 ± 0.00^e	1.71 ± 0.00^e	0.99 ± 0.00^a	3.54 ± 0.01^d	8.01 ± 0.02^b
$Y_{70}W_{30}$	0.66 ± 0.01^{bc}	1.63 ± 0.00^c	0.93 ± 0.01^b	3.50 ± 0.01^{de}	8.13 ± 0.02^c
$Y_{100}W_0$	0.63 ± 0.00^d	2.03 ± 0.01^d	1.21 ± 0.01^f	3.33 ± 0.03^e	11.07 ± 0.02^d
Significant differences is indicated by different letters within the same rom. ($p \le 0.05$)					

Table 4: Functional properties of trifoliate–wheat noodles.

Noodles Sample	L* Raw Sheet Noodle	a* Raw Sheet Noodle	b* Raw Sheet Noodle	L* Optimally Cooked Noodle	a* Optimally Cooked Noodle	b* Optimally Cooked Noodle
Y_0W_{100}	84.53 ± 0.01^f	-1.44 ± 0.01^c	19.77 ± 0.01^a	73.46 ± 0.01^f	-2.11 ± 0.01^a	20.07 ± 0.21^d
$Y_{20}W_{80}$	69.66 ± 0.01^e	2.24 ± 0.01^a	21.72 ± 0.03^b	53.03 ± 0.01^e	3.86 ± 0.01^b	15.53 ± 0.01^c
$Y_{30}W_{70}$	72.19 ± 0.01^d	2.26 ± 0.00^{ab}	19.42 ± 0.10^c	48.04 ± 0.01^d	3.24 ± 0.01^c	12.59 ± 0.01^a
$Y_{50}W_{50}$	68.03 ± 0.01^c	2.28 ± 0.01^b	18.53 ± 0.01^d	45.68 ± 0.01^c	4.77 ± 0.01^d	14.63 ± 0.01^b
$Y_{70}W_{30}$	71.95 ± 0.07^b	1.32 ± 0.01^d	18.31 ± 0.01^e	47.21 ± 0.01^b	3.88 ± 0.01^b	12.78 ± 0.01^a
$Y_{100}W_0$	66.82 ± 0.08^a	3.02 ± 0.01^e	18.83 ± 0.03^f	45.26 ± 0.01^a	4.86 ± 0.01^e	14.55 ± 0.01^b
Significant differences is indicated by different letters within the same rom. ($p \le 0.05$)						

Table 5: Colour characteristics of raw and optimally cooked noodles.

Noodles Sample	Cooking Time (mins)	Cooking loss (%)	Water uptake (g)	Cooking weight (%)
Y_0W_{100}	5.43 ± 0.02^a	9.36 ± 0.01^b	1.62 ± 0.01^c	162 ± 0.82^c
$Y_{20}W_{80}$	6.36 ± 0.01^b	9.31 ± 0.01^b	1.93 ± 0.02^a	193 ± 1.63^a
$Y_{30}W_{70}$	6.27 ± 0.01^b	11.19 ± 0.01^a	1.61 ± 0.01^c	161 ± 0.82^c
$Y_{50}W_{50}$	6.92 ± 0.10^c	13.06 ± 0.25^c	1.71 ± 0.01^b	171 ± 1.25^b
$Y_{70}W_{30}$	5.45 ± 0.04^a	15.09 ± 0.07^d	1.64 ± 0.01^c	164 ± 0.47^c
$Y_{100}W_0$	7.06 ± 0.08^c	12.46 ± 0.03^e	2.04 ± 0.01^d	204 ± 0.94^d
Significant difference is indicated by different letters within the same rom. ($p \leq 0.05$)				

Table 6: Cooking properties of trifoliate yam–wheat noodle.

Noodles Sample	Colour	Taste	Flavor	Texture	Over all acceptability
Y_0W_{100}	7.41 ± 2.12^b	6.81 ± 2.44^a	7.32 ± 1.64^a	6.92 ± 1.20^b	7.09 ± 1.83^a
$Y_{20}W_{80}$	4.61 ± 0.00^c	5.01 ± 1.70^b	5.11 ± 1.37^b	6.21 ± 2.30^a	6.26 ± 2.05^b
$Y_{30}W_{70}$	6.90 ± 2.47^a	5.31 ± 2.71^c	5.61 ± 1.71^c	6.20 ± 2.39^a	6.32 ± 2.28^b
$Y_{50}W_{50}$	5.30 ± 1.89^e	4.90 ± 2.23^d	5.90 ± 2.08^d	4.90 ± 2.23^c	5.92 ± 1.44^c
$Y_{70}W_{30}$	6.10 ± 2.81^f	4.63 ± 2.63^e	4.30 ± 2.36^e	4.20 ± 2.15^c	4.83 ± 2.82^d
$Y_{100}W_0$	4.49 ± 1.16^d	4.10 ± 2.25^f	4.80 ± 1.55^f	4.11 ± 1.73^d	5.52 ± 2.75^e
Significant difference is indicated by different letters within the same rom. ($p \leq 0.05$)					

Table 7: Sensory evaluation of trifoliate yam–wheat noodles.

both cases.

Sensory properties of trifoliate yam-wheat noodles

According to Alozie et al. [3] five on a scale of 9 is adjudged as acceptable. From Table 7 most panelists adjudged the optimally cooked noodles substituted with 20 - 50% trifoliate yam flour to be as acceptable as the control noodles, although there were significant differences ($p<0.05$) among some sensory attributes. Virtually all noodle samples were accepted, as no outright rejection was observed for any of the sensory parameters.

Conclusion

Noodles of comparable nutritional and sensory standards were produced from blends of trifoliate yam flour and wheat flour. Outcome of this study, if put into commercialization could reduce increase in the demand for importation of wheat for flour emanating from its exclusive utilization in bakery/pastry and also make it available for other useful purposes and thereby promoting food security. Anti-oxidant potential, microbial and shelf-life characteristics of these noodles are recommended for further studies.

References

1. Igyor MA, Ikyo SM, Gernah DI (2004) The food potential of yam (*Dioscorea bulbifera*). Niger Food J 22: 209-215.
2. IITA (2006) Yam: Research review. International Institute of Tropical Agriculture, Ibadan, Nigeria.
3. Alozie Y, Akpanabiatu MI, Eyong EU, Umoh IB, Alozie G, et al. (2009) Amino acid composition of *Dioscorea dumetorum* varieties. Pak J Nutri 8: 103-105.
4. Martin G, Treche S, Nuobi L, Agbor ET, Gwangwa S (1983) Introduction of flour from *Dioscorea dumetorum* in a rural area. Proceedings of the Second Triennial Symposium of the Society for Tropical Root Crops, African Branch, held in Douala Cameroon.
5. Eka OU (1998) Root and Tubers. In: Osagie AU, Eka OU (eds.) Nutritional quality of plant foods. Macmillian press, London, pp: 1-31.
6. Afoakwa EO, Sefa-Dedeh S (2001) Chemical composition and quality changes occurring in *Dioscorea dumetorum* pax tubers after harvest. Food Chem 75: 85-91.
7. Baiano A, Fares C, Peri G, Romaniello R, Taurino AM, et al. (2008) Use of toasted durum whole meal in the production of a traditional Italian pasta: chemical, mechanical, sensory and image analyses. Int J Food Sci Technol 43: 1610-1618.
8. Sissons M (2008) Role of durum wheat composition on the quality of pasta and bread: Invited Review Global Science Books. Food 2: 75-90.
9. Sanni LO, Bamgbose CA, Babajide JM, Sanni SA (2007) Production of instant cassava noodles. Proceedings of the 13th ISTRC Symposium.
10. Chong LC (2007) Utilization of matured green banana (*musa paradisiaca var. awak*) flour and oat beta glucan as fibre ingredients in noodles.
11. Saifullah R, Abbas FMA, Yeoh SY, Azhar ME (2009) Utilization of green banana flour as a functional ingredient in yellow noodle. Int Food Res J 16: 373-379.
12. Ojure MA, Quadri JA (2012) Quality evaluation of noodles produced from unripe plantain flour using xanthan gum. IJRRAS 13: 740-752.
13. Abiodun OA, Akinoso R, Oladapo AS, Adepeju AB (2013) Influence of soaking method on the chemical and fuctional properties of trifoliate yam (*Dioscorea dumetorum*) flours. J Root Crop 39: 81-87.
14. Nagao S (1996) Processing technology of noodle products in Japan. In: Kruger JE, Matsuo RB, Dick JW (eds.) Pasta and Noodle Technology. American Association of Cereal Chemists, St. Paul, MN, pp. 169-194.
15. Udensi EA, Okaka JC (2000) Predicting the effect of blanching, drying temperature and particle size profile on the dispersibilty of cowpea flour. Niger Food J 18: 25-31.
16. AACC (1999) Approved methods of the AACC, method 56-30.01. American Association of Cereal Chemists, St.Paul, Minnesota, USA.
17. Kinsella JE, Melachouris N (1976) Functional properties of proteins in foods: A survey and Critical Revision. Food Sci Nutri 7: 219-280.
18. Iwuoha CI (2004) Comparative evaluation of physicochemical qualities of flours from steam-processed yam tubers. J Food Chem 85: 541-551.
19. Tan LC (2003) Improving the application of gibberellic acid to prolong dormancy of yam tubers (*Dioscorea spp.*). J Sci Food Agri 83: 787-796.
20. FAO/WHO/UNU (1985) Energy and protein requirements. World Health Organisation Technology Report, Series no.724.
21. Yap CY, Chen Y (2001) Polyunsaturated fatty acid: Biological Significance, biosynthesis and production by macroalgae and microalgae like organism. In: Feng Chen, Yue Jiang (Eds) Algae and their biotechnological potential. Kluwer academic publisher pp: 1-32.
22. Chang HC, Wu LC (2008) Texture and quality properties of Chinese fresh egg noodle formulated with green seaweed (*Monostroma nitidum*). J Food Sci 73: 398-404.
23. Eko ND (2010) Quality evaluation of dried noodle with seaweeds purees substitution. Diponegoro University, Central Java, Indonesia.
24. Wang YQ, Zhang M, Mujumdar AS (2012) Influence of green banana flour substitution for cassava starch on the nutrition, color, texture and sensory quality in two types of snacks.
25. Senthil A, Ravi K, Bhat KK, Seethalakshmi MK (2002) Studies on the quality of fried snacks based on blends of wheat flour and saya flour. Food Qual Pref 13: 267-273.
26. Gabriel IO, Faith CU (2014) Production and evaluation of cold extruded and baked ready-to eat snacks from blends of breadfruit (*Treculia africana*),cashewnut (*Anacardium occidentale*) and coconut (*Cocos nucifera*). Food Sci Qual Manag 23: 2224-6088.
27. FAO (2010) Fats and fatty acids in human nutrition-Report of an expert consultation. FAO Food and Nutrition, Paper 91, Food and Agriculture Organization of the United Nations, Rome.
28. Hassan LG, Umar KJ (2004) Proximate and mineral composition of Seeds and pulp of *Parkia biglobosa*. Niger J Basic Appl Sci 13: 15-27.
29. Wirjatmadi B, Merryana A, Purwanti S (2002) Marketing strategy noodle dried seaweed (E*uchema cottonii*) and iodine rich fiber system with marketing mix. J Penelitian Media Eksata 3: 89-104.
30. Riley CK, Wheatley AO, Asemota HN (2006) Isolation and characterization of starches from eight *Dioscorea alata* cultivars grown in Jamaica. Afri J Biotechnol 5: 1528-1538.
31. IFIS (2005) Dictionary of food science and technology, International Food Information Service. Blackwell Publishing, Oxford, UK.

32. Ellis WO, Oduro I, Barimah J, Otoo JA (2003) Quality of starch from six Japanese sweet potato varieties in Ghana. Afri J Root Tuber Crop 5: 38-41.

33. Oduro I, Ellis WO, Nyarko L, Koomson G, Otoo JA, et al. (2001) Physicochemical and pasting properties of flour from four sweetpotato varieties in Ghana. Proceedings of the Eighth Triennial Symposium of the International Society for Tropical Root Crops (ISTRC-AB), Ibadan, Nigeria.

34. Baker RC, Wonghan P, Robbins KR (1994) Fundamentals of new food products developments. Science, Elsevier Science, Amsterdam.

35. Siddaraju NS, Ahmed F, Urooj A (2008) Effect of incorporation of *Dioscorea alata* flour on the quality and sensory attributes of indian dehydrated products. World J Dairy Food Sci 3: 34-38.

36. Kulkarni KD, Noel G, Kulkarni DN (1996) Sorghum malt-based weaning food formulations: preparation, functional properties and nutritive value. Food Nutri Bul 13: 322-327.

37. Soni RL, Sharma SS, Dun D, Gharia MM, Ahmeda J (1993) Physico chemical properties of *Quercus leucotrocophora* (oak). Starch/Starke 45: 127-130.

38. Mohamed A, Xu J, Singh M (2010) Yeast leavened banana-bread: Formulation, processing, colour and texture analysis. Food Chem 118: 620-626.

39. Mares DJ, Campbell AW (2001) Mapping components of flour and noodle color in Australian wheat. Aust J Agri Res 52: 1297-1309.

40. Rayas-Duarte P, Mock CM, Satterlee LD (1996) Quality of spaghetti containing buckwheat, amaranth, and lupin flours. Cereal Chem 73: 381-387.

41. Ovando-Martinez M, Sáyago-Ayerdi S, Agama-Acevedo E, Goñi I, Bello-Pérez LA, et al. (2009) Unripe banana flour as an ingredient to increase the indigestible carbohydrates of pasta. Food Chem 113: 121-126.

The Impact of Trade Facilitation on Agricultural Products Standard Compliance in Relation to Cameroon's Export Performance

Femshang M Charles*

Ministry of Secondary Education, Yaounde, Cameroon

Abstract

This is a research on the impact of trade facilitation on agricultural products standard compliance in relation to Cameroon's export performance. The background and literature review was based on concepts like the Trade facilitation, Standards, Residual levels and agreement like the Trade Facilitation Agreement. Secondary data was used for this study, precisely a time series data. The OLS technique was used in the data analysis and the null hypothesis was rejected implying that Standards compliance has a positive relationship with Cocoa Export. Even though some initiatives in relation to Trade Facilitation have been implemented to raise these export volumes, a significant effort by the government needs to be addressed. Looking at the advantages of Trade Facilitation in the long run, it is worth necessary to neglect the high cost of its initial implementation and strive towards compliance. There were recommendations made in the areas of Trade Facilitation promotion and Standards compliance.

Keywords: Standard compliance; Agriculture; Trade facilitation agreement; SPS and TBT agreement; PAHs

Introduction

Agriculture constitutes a significant portion of the world economies, it also contributes towards eradicating hunger and poverty, boosting intra-continental trade and investments, rapid industrialization, economic diversification, sustainable resource and environmental management, it further helps in creating jobs, human security and shared prosperity. In Cameroon for example, Agriculture accounts for 43% of her GDP, it employs about 70% of the active population and contributes to more than one third of total export earnings (2004 journal fresh produce). Main agricultural products include coffee, cattle meat, cocoa beans, bananas, maize, fresh vegetables and groundnuts. The majority of Cameroonian farmers are small-holders, generally obtaining low yields thus subsistence method of farming plays a dominant role in determining the characteristics of output after harvest.

The trend in Agricultural trade worldwide has been characterized by a shift in food particularly from unprocessed to processed food, diets have been shifted from staple food towards items such as fats, protein and vegetable. There has been an increase share of livestock products in the total share of export of agricultural products. The increase in composition of agricultural products is due to increasing population far more than domestic agricultural production, increase urbanization, increase in the income of individuals which in whole has increased the demand for agricultural goods thus making the sector more competitive and eventually lowering the prices of these goods [1]. One of this reason is because agricultural standards has been a key issue nowadays in determining exports due to its nature and characteristics of these products which includes its perishability, fragility, use of chemicals and fertilizers, seasonality in its production, price instability and fluctuation in its supply. For Trade facilitation to be possible, Agricultural product Standard compliance is therefore indispensable, since agricultural standards has as objectives of protect plant, animal and most of all human health. There is the need for standards compliance to be more scrutinized so as to meet food safety as well as timely delivery of agricultural goods in many developing and less developing nations and thus for trade facilitation [2,3]. Cameroon most especially whose agricultural trade sector has been experiencing a fall in export volume due to noncompliance standards set by her trading partners in her agricultural product sector especially the major exports like Banana, cocoa, coffee and Palm oil.

Problem statement

Agriculture being predominant in Cameroon, yet her traditional production and export sectors are almost static, or in decline. In addition, a majority of her traditional products are exported unprocessed, with little or no value added. This results to agricultural practices being small-scaled, with inadequate use of machines, poor understanding of methods of usage of chemicals and fertilizers. Consequently, Cameroon's domestic agricultural production is clearly suffering from standards as noticed from the persistent ban in some agricultural products, the NCCB identified some of these factors as: low yield, inconsistent production pattern, disease incidence, pest attack and use of simple farm tools. Although there has been the existence of Standardizing organizations like the ANOR (Agence de Regularisation De Produit Agriculture) but unfortunately, standard in Cameroon appears to be embryonic without any regulation, thus its application [4-7]. Without an appropriate body responsible for certifying that products conform to national standards and for granting the right to use the national emblem, certification is almost non-existent. Although there has been the existence of Standardizing organizations like the ANOR but unfortunately, standard in Cameroon appears to be embryonic without any regulation, thus its application. Without an appropriate body responsible for certifying that products conform to national standards and for granting the right to use the national emblem, certification is almost non-existent.

For the purpose of this research, I will concentrate on Cameroons chief export crops which are cocoa and coffee. In central Africa, Cameroon use to be the leading producer of these crops and the fourth largest exporter in Africa. Both crops face problems nowadays

***Corresponding author:** Femshang M Charles, Trade Analyst, Ministry of Secondary Education, Economics, BP 1130, Mendong, Yaounde, Cameroon
E-mail: femshang@gmail.com

which include; Fluctuation in output caused by seasonal fluctuation in production, poor weather, inadequate inputs, and periodic pest infections leading to poor quality outputs. The fall in quality and general production reduced export revenue by a quarter in 2012 to 578 million US dollars, meanwhile output of coffee dropped by more than 50% as noticed in 2013 due to same factors affecting cocoa. In addition, weak international prices have driven most farmers out of the market and this has resulted to Cameroon having a weak performance in terms of Trade Facilitation as seen in the graph below showing the overall performance of Cameroon in Trade Facilitation.

Research questions

This research therefore is aimed at providing possible answers to the following research questions:

I. What are the standard requirements by the EU on Cocoa export?

II. Does Standard compliance to EU Cocoa import requirements have an impact of Cameroon's Export volume of Cocoa?

III. What are the Trade Facilitation Initiatives put in place by Cameroon to comply with the EU standards?

IV. What are the challenges faced by Cameroon to comply with Trade Facilitation Agreement?

V. What are the impact of Trade Facilitation on Agricultural Products Standards?

The purpose of this study is to answer the above research questions. The study will be useful to the Cameroonian government in taking decisions on the appropriate allocation of her resources in the agricultural sector which serves as backbone of the economy, It will create awareness to most farmers and entire population on the advantages of standard compliance in terms of health, animal, plant and environmental protection, since standards gives assurance and certification of a product, its compliance will increases information on food safety and product quality which can lead to increased consumer confidence, reduce transaction costs and thus facilitate trade [8]. TFA will ensure the Streamlining of export processing procedures in the agriculture and agro-processing sectors to improve access to markets. Nevertheless, with the aim to increase the competitiveness of the agricultural sector by building export capacity, reducing trade transaction costs and deepening regional integration, government officials need to adhere to the Trade Facilitation Agreement. If Cameroonian agricultural producers can comply with these wide range of requirements, including technical regulations, product standards and customs procedures. Then the agricultural sector will be a highly beneficial sector to trade.

Background

The following conceptual framework will be applied to this work so as to provide a conceptual background to my problem statement; Trade facilitation, Standards, technical regulations, SPS and TBT agreements, Codex Alimentarius, WTO and TFA agreements, Agreement on Agriculture.

Trade facilitation

Trade facilitation can be defined as policies and measures aimed at easing trade costs by improving efficiency at each stage of the international trade chain. According to the WTO definition, trade facilitation is the "simplification of trade procedures", It involves collecting, presenting, communicating and processing data required for the movement of goods in international trade. The TFA as a wider part of the Bali Package aims at harmonizing and standardizing procedures and requirements for the movement of goods at the point of entry or exit of a country. Trade facilitation covers all the various steps that is taken in view of smoothing the flow of trade from the producer to the final consumer [9-11]. It relates to a wide range of activities at the border and within the border (procedures on exports and imports, formalities on transport, payment settlements, insurance and financial requirements). Therefore, Trade facilitation has as aim to develop a consistent, coherent, transparent, predictable and non-discriminatory environment for international trade procedures based on internationally accepted practices and norms resulting from: simplification of procedures and formalities, standardization and improvement of physical infrastructure and other facilities, harmonization of laws and regulations.

Need for trade facilitation: Given Cameroon's vision by 2035 being an emerging country with international trade being the main contributing variable, Cameroon has therefore taken into consideration to improve on her trade facilitation measures especially as she ranks among the lowest in trade facilitation performance (Table 1 and Figure 1).

For the case of Cameroon and any other country involve in trade, the main goal of trade facilitation is to reduce the transaction costs and complexity of international trade for businesses and improve the trading environment in Cameroon, while maintaining efficient and effective levels of government control.

According to statistics drawn from OECD Trade Facilitation Indicators index as shown on the graph above: Cameroon need to focus on the following trade facilitation area for which they are very behind compared to the world

The main area of focus in Cameroon (Order of priority) includes:

a. Infrastructure investment

b. Customs modernization and border crossing-environment

c. Streamlining of documentary requirements and information flows

d. Automation and EDI

e. Ports efficiency

f. Logistics and transport services: regulation and competitiveness

g. Transit and multimode transport

h. Transport security

Table 1 and Figure 1 show Cameroon's ranking in the level of implementation and performance of trade facilitation indicators compared to the best in the world (Singapore) and the best in Africa (Ghana) in the year 2014.

Country (2014)	LPI Rank	Customs	Infrastructure	International shipments	Logistics competence	Tracking & tracing	Timeliness
Singapore	5	3	2	6	8	11	9
Ghana	100	130	70	93	121	73	113
Cameroon	142	156	154	147	104	111	120
Source: OECD TF Logistics index 2014.							

Table 1: Trade facilitation indicators ranking Cameroon-2014.

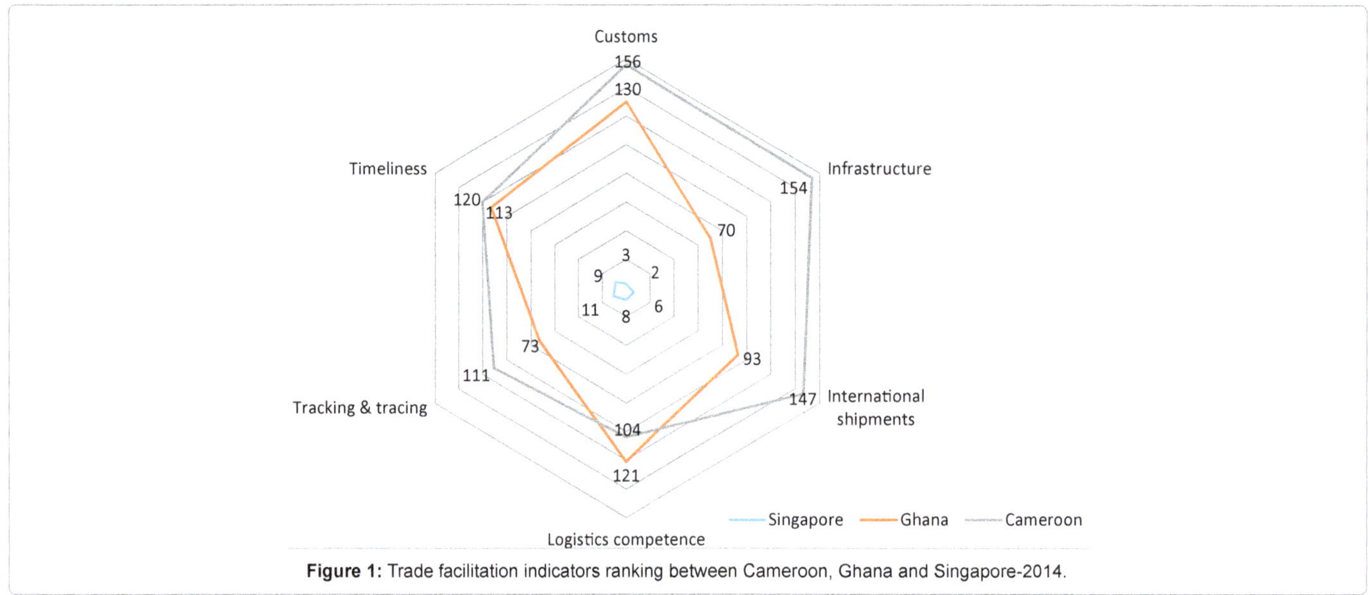

Figure 1: Trade facilitation indicators ranking between Cameroon, Ghana and Singapore-2014.

Government benefits	Trader benefits
Increased effectiveness of control methods	Cutting costs
Enhanced revenue collection	Reducing delays in delivery of goods
More effective and efficient use of resources	Faster Customs clearance
Improved trader compliance	Simple commercial framework
Accelerated economic development	Enhanced competitiveness
Encouragement of foreign investments	-----

Table 2: Various benefits of trade policies.

It is from this analysis where we see areas in trade facilitation for which Cameroon need to place more emphasis on, in terms of initiatives and investment projects to improve on her trade, we notice that Cameroon is ranked last implying that, there is much to be done by Cameroon to improve on her trade facilitation measures [12-15]. Therefore, for Cameroon to improve on her trade performance; she is required to invest much her trade facilitation measures.

Benefits of trade facilitation: Efficient and effective trade facilitation in Agricultural and food products can be of great benefits including improved food security, more export income, increased access to productivity advancements, and nutritionally varied diets. Because many foods are highly perishable, they require efficient trade regimes and border crossings. Food security can be effective when cross-border flows of food cargo are "facilitated" to minimize time spent in trade and thereby reduce physical losses and costs. Moreover, there is enormous potential for growth in regionally produced agriculture and food products in ASEAN to help achieve harmonization objectives for trade in goods [16]. Reasons why Trade Facilitation has become a call for concern are due to: increased trade volumes as a result of progressive trade liberalization, increased exchange of goods and services due to High technologies. Eventually, Trade facilitation results is beneficial to both the business community and the Government: Traders benefit is seen from quicker delivery of goods and reduced cost of transactions while the Government gain in terms of revenue collection from tax, increased economic efficiency, increased security and societal protection. Traders gain through faster delivery of goods and reduced transaction costs (Table 2).

Trade facilitation is a continuous process. Once trade facilitation mechanisms have been designed and launched by Government and business community, they should continue to jointly monitor trade related developments, both locally and internationally, in order to adapt the existing measures and design new ones to fit the changing environment. Trade facilitation appears to be a continuous process so once trade facilitation mechanisms is designed and put forward by the Government and the business community, it requires continuous join efforts of both in monitoring trade developments, both at home and internationally, so as to adapt the current measures and create new ones to adapt to the changing environment.

What is a standard?

Standard can be defined by ISO as a documented agreement that contains technical requirements to be used with consistency as rules or guidelines to ensure that products, materials, services and processes are fit for their purpose. The Codex Alimentarius of FAO is the organization that develops standards, guidelines and codes of good practice related to food. The purpose is to contribute to the safety, quality and fairness of international food trade [1]. Consumers can therefore trust the safety and quality of the food products they buy and importers can trust that the food they ordered will be in accordance with their specifications. Countries can adopt these standards or some of their partners as part of their national standards. In this way, Codex contributes to standards harmonization. It further recommends International Code of Practice including: general principles of food hygiene in respect of these products.

The agreement on trade facilitation and its main provisions

The WTO TFA was concluded during the Bali Ministerial Conference in December 2013. The agreement aims at expediting the movement of goods across borders, release and clearance of goods, including goods in transit, and at improving customs cooperation. In addition, it aims at enhancing support and assistance for capacity building in for least developed countries and enhance close cooperation among Members states on customs compliance issues and trade facilitation [17,18].

The main provisions in the TFA: Agreement contains two key provisions: Provision on facilitating customs and other border procedures; and Provision on special and Differential Treatment (S and DT) flexibilities for Least Developed Countries (LDCs) and developing countries.

Agricultural specific TFA provisions: The main provisions from the Trade Facilitation Agreement aiming at facilitating trade in agricultural products are as follows:

Article 1: Publication: This provision will enable trade activists to acquire and enquire relevant and up to date information on the procedures for trade of agricultural products. Thus predictability and reduce transaction costs that would otherwise occur from delays due to lack of information and facilitate trade.

Article 2: Opportunity to comment, information before entry into force, and consultations: Here, Agriculturalists have an opportunity to comment and consult on information such as proposed laws and regulations before entry into force. Such a measure will be useful in accommodating for the interests and views of stakeholders across the agriculture value chain.

Article 3: Advance rulings: Provide advance ruling in a reasonable, timely manner for applicants meeting the necessary requirements. Under this provision, traders in agricultural products will be able to clear products through customs in a minimum time to avoid delays and losses.

Article 4: Procedures for appeal and review: Under this provision a trader in agricultural products has the right to appeal or review on an administrative decision to an authority that issued the decision. This will promote transparency in the review claims in a manner that satisfies traders.

Article 5: Other measures to enhance partiality, non-discrimination and transparency: Countries are required to issue notification for inspections, testing procedures and detention with respect to feedstuffs covered under the notification for protection of human, animal and plant life or health. Such information is important to traders in agricultural products by saving them time and promoting compliance to requirements.

Article 6: Disciplines on fees and charges imposed on or in connection with importation and exportation and penalties: Traders are provided with relevant information on fees and charges connected with importation and exportation other than import and export duties and other charges. This is useful for predictability, compliance and procedures on agricultural products at the border for traders.

Article 7: Release and clearance of goods: The provision provides for expedited shipments for goods under customs control with the aim of providing a timely release for perishable goods. This measure will save traders in agricultural costs and losses that would otherwise be incurred in lengthy entry procedure.

Article 8: Border agency cooperation: Aligning working hours, procedures and formalities, common facilities and establishment of One Stop Border Post control will provide an efficient and timely service delivery to traders and thus facilitate trade in agriculture products.

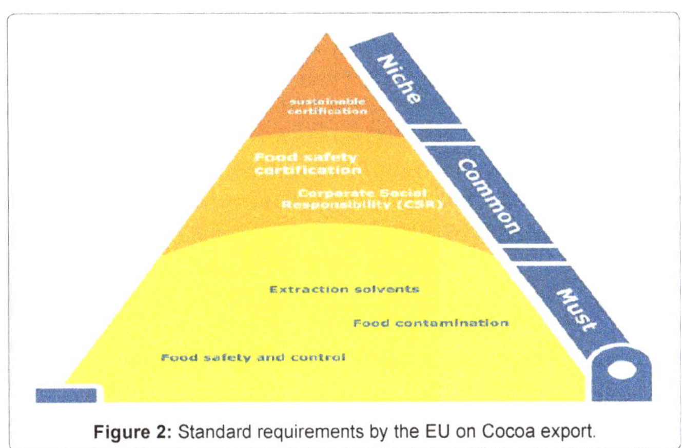

Figure 2: Standard requirements by the EU on Cocoa export.

Article 9: Movement of goods intended for import under customs control: By allowing movement goods under customs control within a country's territory to another country, traders in agricultural products will be able to save time that would otherwise be used on complying with regulatory requirements at the border.

Article 10: Formalities connecting with importation, exportation and transit: This provision aims to rapidly release and clear goods with minimum incidence and complexity of import, export, and transit particularly perishable goods, reducing time and cost compliance for traders of agricultural products.

Under the Agreement on Technical Barriers to Trade and the Agreement on the Application of Sanitary and Phytosanitary Measures included in the GATT 1994, countries have the right to implement health protection measures and regulations concerning human, animal, and plant health, as well as accompanying technical requirements, restrictions, and voluntary standards and procedures.

The EU requirements for coffee and cocoa can be divided into: (1) Musts: There are requirements that must be met in order to access the EU market, such as the legal requirements, (2) Common requirements: These are requirements which most of your competitors have already implemented like in the Case of Cameroon her competitor Nigeria already implemented, in other words, these the standards requirements you have to comply with so as to keep up with the market and be competitive, and lastly; (3) Niche market requirements for specific segments (Figure 2).

Food safety: Its hygeine, traceability and control: The key issue of EU food legislation is Food safety. This implies that, Food products throughout the entire supply chain so that the guarantee of food safety must be traceable, and also that appropriate action can be tahen in cases when food is tested and found unsafe this limits the risks of contamination [19]. A very important aspect in controling hazards of food safety is by defining the critical control points (HACCP) and by implementing food principles of food management (*Common requirements*). Subjecting Agricultural and food products under controls is an important aspect. Therefore Products that are not considered safe for consumption will be refused access to the EU markets.

Control of food imported to the EU

In the case of a repeated non-compliance of a specific products which originates from a particular countries, stricter meausres, such as a analytical test report and health certificate, will be required for before import. These Products that have shown repeated non-compliance will then be kept on a list included in the Annex of Regulation (EC)

669/2009. Due to problems with PAH residues found in Cameroon's Cocoa, it is subject to increased level of controls. In April 2013, the EU strengthened its regulation on PAH in cocoa beans.

Avoid contamination to ensure food safety

Contaminants: These are substances that may be present in a product as a result of the several stages of growing, packaging, processing, transport and storage. The exceeding limits for several contaminants force to place strict control so as to avoid negative impact on the quality of food and risks to human health. There are different forms of contamination which include:

i. Pesticides: The presence of pesticides is the most common reason for border authorities to reject products. This implies that most farmers from Cameroon has to be aware that products containing pesticides above the allowed level will be withdrawn from the EU market.

ii. Mycotoxic: These are moulds and fungi which consist of another reason why there exits many border rejections. For coffee and cocoa beans, Ochratoxin A (OTA) is set at 5 ug/kg.

iii. Poly-Aromatic Hydrocarbons (PAH) which can result from beans coming into direct contact with smoke as in the case of Cameroon, sometimes caused by artificial and poorly maintained driers, the limit for Benzo (a) Pyrene is 5.0 ug/kg of fat and 30 ug/kg approximately 0.03 mg/kg for the total sum of PAHs. The limit for the PAH was said to become stricter as of 1st April 2015. The total acceptable level of PAH by set by the EU In April 2013 was 0.2 mg/kg. This greatly affected Cameroons export to the EU.

iv. Samonella, which is a serious form of contaminiation that occurs ocassionally as a result of incorrect harvesting and drying techniques can be eliminated by irradiation but yet not allowed by the EU legislation.

v. Contamination by foreign matter like plastic and insects are a threat when food safety producers are not carefully managed.

Current situation in Cameroon

In the Center South and Eastern regions of Cameroon, the climatic condition favors the sun drying of cocoa. The rainfall in this area is not much. Unlike in the South West Province where the rain fall is very high and heavy, sun drying during the rainy season is farfetched. Hence cocoa drying during the rainy season is by the used of artificial dryers. These artificial dryers originally, were not properly designed and maintained [20,21]. Hence fume and smoke are bound to appear on the cocoa beans. Thanks to the European Union that provided 2500 Samoan ovens to the South West Farmers within a period of eight years. At least 20% of the demand was satisfied. Farmers were very glad. About 70% of the farmers still used local materials, which are very poorly designed. This greatly affected the sustainability of the cocoa sector in Cameroon and has resulted to a fall in her export volume in this sector. With the maximum level of PAH 0.2 ug/mg placed by the chief cocoa exporter, The Netherlands, Cameroon's cocoa production still faces lots of challenges.

Initiatives by Cameroon Government on Trade Facilitation

In the area of trade

• Initiatives for the Elimination of quantitative restrictions on imports, licences and other import and export authorizations; such as the introduction of the single window system (eGUCE) in 2005, allowing a single submission point for trading documents. Therefore, reducing delays in import and export procedures.

• Initiatives on up-dating the legislation on main trade institutions in Cameroon such as the Ministry of Trade, Products standards regulatory body (ARNO), Cameroon's customs Authority, as sole public-private bodies specialized on trade facilitation.

• Organization of specific training/awareness programs for companies in the field of trade facilitation such as the Cameroon trade Forum created in 2014.

• Introduction of measures designed to guarantee fairness in commercial transactions, in particular through metrology controls, a crackdown on discriminatory sales, refusal to sell, holding of stocks for speculative purposes and conditional sales.

• Implementation of legislation on dumping and competition; such as the signing of the structural adjustment program II in 1994.

• Reorganization of tax and customs regimes to make them more effective and consistent with the sub regional programme adopted within the Central African Economic and Monetary Community (CEMAC).

• Implementation, as of January 2007, of the computerized system of customs administration (ASYCUDA) and the installation of a container scanner at the port of Douala, the main port through which goods enter the country.

Legal initiatives pertaining to trade facilitation

Cameroon has signed agreements on legal basis which enables them, through a mutually beneficial manner, to strengthen and diversify their trade relations in a fairly manner in accordance with the provisions of WTO agreements such as the:

• Government secured adoption of the Investment Charter by the National Assembly. A sectorial machinery for its implementation, in particular the Investment Promotion Agency, the Export Promotion Agency and the National Agency for Standards and Quality have still to be put in place.

• 1994 Marrakech Agreement Establishing the World Trade Organization ratified by Decree No. 95/194 of 26 September 1995. Hence the Government has committed itself to implementing the agreement on goods (GATT), the Agreement on Services (GATS), the Agreement on Intellectual Property (TRIPS).

• EU-ACP Partnership Agreement (2000 Cotonou Agreement), the African Intellectual Property Organization (OAPI), the African Export - Import (AFREXIM Bank), the Organization for the Harmonization of Business Law in Africa (OHADA), United Nations Conventions on International Trade Law (UNCITRAL), the World Trade Organization (WTO), the Islamic Conference Organization (agreement ratified on 11 July 1983), and finally the international agreements on products and the Agreement on Services (GATS), the Agreement on Intellectual Property (TRIPS).

Regarding financial initiatives

The Government has taken measures to strengthen the financial sector such as the Douala Stock Exchange began operations in 2006.

Telecommunications initiatives

A Special Telecommunications Fund has been set up, designed to support investment policy in international trade. This is to enable the Government to provide effective funding for the universal telecommunications service, and to develop telecommunications in a consistent manner throughout the national territory.

Transport initiative

Air Transport Safety and Security Project have been put in place to Liberalize this sector, and to modernize civil aviation and bring it into line with international safety standards. And in the port subsector, measures have been taken to liberalize and simplify procedures at the Port of Douala.

Effects of trade facilitation on standards

Considering the fact that SPS measure is a component of standards, it has a direct link to Trade facilitation. For example, when Codex Alimentarius, the International Plant Protection Convention and the Office International Des Epizooties OIE), introducing the quality and safety management system and conducting equivalence assessments, such as the Hazard Analysis and Critical Control Point (HACCP), this will obviously facilitate access to international markets [17]. Therefore, to benefit from the Trade Facilitation Agreement developing countries and LDCs has to comply with standards.

TFA enables reduction in compliance costs of implementation of theses standard measures. Some of these costs are unnecessary measures, documentations, inefficient and complicated procedures, long time taken for inspections, little or no transparency, consistency and predictability in the implementation of SPS and TBT controls. These costs are sometimes direct (for example; document submission, fees and charges, cost of inspection) or indirect (e.g. delays at border, uncertainty about requirements and procedures, sometimes inadequate and contradictory documentation). Costs from export and import-related documentation and procedures vary from 1% to 15% estimates of cost of production of a product. These make an agricultural product's cost very high. Considering the fact that agricultural products constitute most exports of developing countries especially Cameroon and other LDCs, combined with poor bureaucracy and infrastructure, faced with these challenges increase her transaction costs and delays to clear exports, imports, and goods on transit. In most African countries, loss of revenue resulting from inefficient and complicated border procedures can be estimated at exceeding 5% of their GDP. Therefore, Trade facilitation stands as a possible means to augment revenues from trade and assist developing countries to part take in international trade. According to the OECD reports, simplification and harmonization of documents results in the reduction of trade costs by 3% for low-income countries and 2.7% for lower middle-income countries; streamlining procedures as one of the objectives of Trade Facilitation would reduce cost by 2.2% for lower middle-income countries and by 2.3% for low-income countries. When trade related information are available, such as the SPS standards, this would go a long way in generating cost savings of 1.6 percent for low-income countries 1.4% for lower middle-income countries [7].

In addition, implementing the TFA will help minimize some hidden form of protectionism that relates to SPS measures which some LDCs and developing countries face. Complying with the Agreement on SPS is related with border administration, such as rapid alert mechanisms, import alerts and, fees and charges, transparency of procedures, etc. All These elements are relevant for facilitation of trade, some of these elements may become SPS barriers. Eventually, the TFA helps to decrease SPS barriers by adopting e-documents, simplifying and reducing the required documents, cutting down multiple procedures, formalities and inspections, enabling cooperation between producers in the supply chain and SPS authorities, expediting clearance of goods including goods on transit.

Finally, The TFA Agreement provides provisions of S&D which enables developing countries and LDCs to implement the agreement on Trade Facilitation in a manner with conformity to their available, and where needed, they are liable to and Capacity Building and Technical Assistance (TACB). Although, in contrary to what LDCs and developing countries expected, this agreement did not include some important SPS provisions with guaranteed technical assistance. However, The TFA links the extent and the timing of implementation upon entry into force by LDCs and developing countries, providing to them more freedom to determine between the various categories A, B, and C, and making notifications to the committee in case of difficulties in implementing Categories B and C.

Literature Review

Trade facilitation is therefore now at the forefront of recent trade policy dialogue as developing countries look for ways to booster their economic growth through increased trade performance. This includes the transparency of trade policy and regulation, as well as product standards, infrastructure to support trade, and technology as it applies to lowering trade costs. Trade facilitation therefore include the improvement of transport infrastructure, the removal of government corruption, the modernization of customs administration, the removal of other non-tariff trade barriers, export marketing and promotion as well as improving the wider environment in which trade transactions take place. Lowering trade-related transaction costs is expected to result in significant improvement in Cameroon's ability to compete effectively in the global economy [6,9]. Further range of benefits to be realized include: improved revenue collection, improved border controls and security, lower administrative costs, encouragement of more trade and investment, and enhancement of the competitiveness of the domestic business in the home market as well as in the export market.

Economic benefits of standards from the Cameroon-Chococam, Compliance to standards will help producers to produce more efficiently and become competitive in the market. Standards can contribute XAF 9,991,750,000 representing over 5% of the Cocoa Company's annual turnover in the previous years.

The literature reporting empirical results for trade facilitation measures and agricultural trade was reviewed in order to advise future work on trade facilitation processes focused on more agriculture-specific constraints. While this exercise aims to identify evaluations using proxies of trade facilitation, it does not discuss measures such as governance and access to finance which fall outside of the scope of this paper but have been identified in the literature as potentially strong impediments to trade in agricultural products. Indeed, surveys conducted under the aid-for-trade initiative report that more than half of respondents from developing country agro-food suppliers listed access to finance as the main obstacle to their participation in value chains. Similarly, Moïsé et al. [7] point to issues of governance, and limited market information and access to financial services as significant bottlenecks to agricultural trade in developing countries. His study focusses predominantly on time delays, logistics and infrastructure, and customs efficiency or sanitary and phytosanitary (SPS) standards and technical barriers to trade (TBT), with very few studies overlapping into both categories.

Overall, evaluations which take indicators of trade times, logistics performance and infrastructure quality tend to find a significant relationship between trade facilitation proxies and trade flows or trade costs, though results vary significantly across product sectors and regions. These suggest that the length of export and import times and customs clearance delays can present important impediments to trade for time-sensitive agricultural products. Logistics performance and maritime connectivity also appear to have significant effects on trade costs. Nearly all of these studies rely on more general indicators such as the number of days needed to export or connectivity indices which may be more representative of aggregate trade flows and stylized shipments than sector specific constraints. This suggests that data on agriculture-specific measures may be limited [7].

Whether or not these barriers have different implications for agricultural products than those captured by the standard indicators for overall trade is worth investigating. Most empirical studies which attempt to evaluate the effects of customs performance on trade rely on broad measures such as customs times or the number of documents needed to trade as proxies for the customs environment, reporting somewhat limited results. This suggests a gap in available cross-country measures of customs efficiency pertaining not just to agricultural goods but to all trade flows. Here, initiatives to develop measures of customs and border efficiency, such as the OECD Trade Facilitation Indicators or ongoing efforts by the World Customs Organization could present promising avenues to evaluate more targeted at-the-border trade facilitation measures [11,18].

Studies evaluating the impacts of sanitary and phytosanitary standards and technical barriers to trade tend to report a negative effect on agricultural exports from developing countries to high income countries, but little effect on trade between developed countries, though again significant variations are observed across sectors. These studies also tend to highlight the importance of compliance capacity in determining whether standards form impeding barriers to trade.

While numerous studies evaluate the impacts of SPS and TBT measures on agricultural trade flows, relatively few have assessed the impacts of standards harmonization and compliance capacities largely because of data issues for quantitative assessments. The reviewed evaluations of standard harmonization are limited to either country specific, intra-EU or intra-OECD trade.

Analyzing harmonization measures in other regional groupings or the adoption of international standards across a wider sample of countries could provide valuable insight on the impacts of harmonizing standards on agricultural trade, though limited information on national standards may be an issue. Expanding measures of institutional compliance capacity and infrastructure such as Relative Rejection Rate Indicator, which summarizes compliance performance and at the border rejections of countries, and the Standards Compliance Capacity Index [18], which aims to provide a systemic and consistent framework to evaluate standards compliance capacity, could also present valuable openings to further assess the impact of compliance capacity and national compliance infrastructure.

Although gravity model type estimations such as the ones covered in this review do not directly identify the source of trade friction or trade costs at the country or country-pair level they can, as Noula et al. [12] suggest, form part of a comprehensive diagnosis and help advise policy packages, particularly when combined with country level assessments of trade facilitation, logistics and trade policy. Noula et al. [12] suggest that trade facilitation policy should pay special attention to improving transport and logistics performance, particularly in low income countries and in Sub-Saharan Africa where these could have highly significant impacts on trade costs. Improvements in such "hard" infrastructure have a greater impact on export growth compared to "soft" infrastructure, but they are expensive.

Similarly, Moïsé et al. [7] propose that a significant expansion of agricultural trade in low income countries could be achieved by easing constraints related to infrastructure quality and efficiently upgrading standards implementation, monitoring and certification capacities. The OECD and the WTO further identified the removal of obstacles to trade, the reduction of customs delays and border procedures, and the reduction of transport costs as key priorities for future aid-for-trade initiatives in the agro-food sector. Echoing these policy recommendations, the agricultural trade facilitation plan for the Greater Mekong Sub region outlined by the Asian Development Bank sets out short-term development strategies to develop and enhance the capacities of the region's quarantine agencies in improving permit insurance systems, product certification and inspection procedures through cooperation and information exchange. The plan further set out to examine and review regulatory frameworks and assess the adequacy of agricultural bureaucracies, in addition to a long-term strategy of liberalizing trade and improving cooperation among customs, border agencies and the private sector. Building upon research and policy lessons from past empirical studies, future quantitative and qualitative evaluations of trade facilitation focused on more agriculture-specific constraints could provide valuable insights on agricultural trade constraints and help advice more targeted policy initiatives. Trade patterns are evolving with traded intermediate inputs increasing in importance as firms increasingly engage in global value chains and consumers demand more food variety including year-round supplies of seasonal products. Consequently, trade facilitation measures that efficiently and speedily enable goods to cross borders while assuring food safety become paramount.

Based on OECD's work on the relationship between trade facilitation indicators and agricultural trade, the initial purpose of this study was to compliment that information by collecting data on the various procedures, timeliness and cost required for agricultural goods to cross a border either to enter as imports or to leave as exports. Among the variables identified in the literature and are of interest for this study, number of documents, time and cost, come from the World Bank's Doing Business database. But, those metrics are mostly based on trading manufactured goods that do not reflect the peculiarities of agricultural trade. There is a need therefore to collect such metrics specifically for agricultural products, especially perishable goods as they require special handling, they need to pass SPS and other health and safety standards and they require clearance not only from customs but also health authorities. Information on how long it takes to get agricultural goods to the border (inland transport), because of the widely dispersed farms and in developing countries the relatively small scale of production from small holders implies greater time and possibly cost than merchandise goods that are produced in a few plants that are probably located near major cities and transport hubs. There may be additional time required to clear customs because inspections may occur more than once for the same consignment by customs and by health authorities, if the various agencies do not co-ordinate. In addition to special logistics to assure continued cold storage as goods transit to or depart the border, information on whether cold storage is available and the procedures for getting goods in and out of those facilities and where they are Located (within or outside the clearance could affect time and cost.

The number of documents required may also be different for many agricultural products and all of these may necessitate longer time.

Timeliness has been identified in the literature as important to trade. Time in motion information as agricultural, especially perishable products move through the border would help identify bottlenecks and suggest procedures to speed the process without impeding countries abilities to provide the necessary food safety for their consumers or to protect their plants and animals from pests and other risks. Improvements in procedures that can speed up the process would facilitate trade. In general, to ascertain whether trade of agricultural goods, because of their particularities, is different from trade in general merchandise, needs agricultural specific metrics.

The original intent was to contact relevant agencies in OECD Member and developing countries to collect such information. The International Finance Corporation of the World Bank is in the process of collecting such information in its "Benchmarking the Business of Agriculture" project, an undertaking comparable to its "Doing Business" endeavor. A pilot to test the project in ten countries was initiated in 2014 with plans to eventually cover some 80 countries. Among other information, this project will provide time and motion indicators for the procedures and costs associated with the process of complying with food safety and health standards, along with describing the process for cross-border trade for specific agricultural produce. The aim is to provide policy makers with laws and regulations affecting the business of agriculture that are comparable across countries. Once these data become available, one would be able to revisit the question of whether and to what extent, trade facilitation of agricultural; particularly perishable products, differs from other goods and how they influence agricultural trade. This approach looks more promising than the one whereby OECD would directly collect questionnaire based information, given the difficulty of establishing the necessary contacts, especially in a number of developing countries. For this study, the next section describes developments in a few trade facilitating indicators from the Trading Across Borders component of the Doing Business database and follows with a description of developments in agricultural trade since the beginning of the 21st century especially for low and lower middle income countries. An innovative approach by Hummels and his co-authors to estimate the premium firms and consumers are willing to pay for timely delivery of goods is also discussed. SPS impact on LDCs agricultural exports sector the strict SPS standards requirements by developed countries, its implication to developing countries coupled with the lack of economic and technical resources of in standard setting process, has limited their market access to most developed countries. Many these developing countries have, as result, negative effects on their economies as a result of their failure to comply with these SPS standards requirements. This has resulted in a considerable loss of revenue from export, income and employment.

Research Methodology

John Beghin [14] defines Research design as a term that covers the aims of the research, research philosophy, the final selection of appropriate methodology, data collection techniques, the chosen method of data analysis and interpretation and an elaboration on how this combination blend into literature. A research design is a plan, structure and strategy of investigation that is employed to answer the research problem. Therefore, a research design is a plan that is adopted to answer questions in the most accurate, valid and objective way.

The research project was conducted basing on the guidelines advocated by Niang [4], that is, a case study method will involve at least two stages of work via:

- Gathering more specific data: the researcher achieved this through reading relevant journals and press publications.
- Presenting an analysis of findings and recommendations for action: in this study it was achieved through the feedback provided on interim and final bases to the case study organization, as well as the production of the final research report.

Since Cameroon has made no notifications on standard with the absence of proper standardisation organisation on Agricultural product sector precisely Cocoa sector which is my point of interest, the research was based on analysing the available data on the trend of Cocoa to bring out reasons for the fall in her export and recommend possible solutions.

Limitations

This research study has been out to show the impact of standard compliance but lacked data on Cameroon's Cocoa compliance. In addition, due to time constraint to contact directly farmers and get the realities of what they face with regards to compliance. I propose a further research on this topic for better results.

Results and Findings

Graphical techniques are more advantageous in that they convey and display findings clearly and in an easy-to-understand format. Buys et al. [6] stated that there is much truth in the adage: a picture is worth a thousand words. The next chapter will focus on data analysis and findings of the research study. The Gravity based Approach is used to quantify standards impacts on trade. It was also stated that the model is the most successful framework for empirical analysis of trade flow between countries (Table 3).

Model specification

$\ln ExpCcij = \mu i + \gamma \ln GDP\ i + \varphi \ln GDP\ j + \beta \ln Popi + \delta \ln Popj + \lambda\ PAH + \varepsilon ij$

Our dependent variables are Cocoa export which illustrates the outflow of the volume of trade from Cameroon to her major Cocoa importer, the Netherlands. The independent variables shall be explained in Table 4.

There is the use of a time series data for this research ranging from 2004 to 2014, the null and the alternative hypothesis of the F-test are as follows:

H_0: There exist no relationship between standards compliance and Export volume.

H_1: There is a significant relationship between standards compliance and Export volume.

The OLS regression method was used to analyse the data and the following results were obtained (Figure 3).

This is a time series since we study data in a long run trend and Co-integration can be used to improve long-run forecast accuracy for policy making. We are going to use the OLS technique to minimize the effect of errors arising from omission of important variables in the data so εij will capture the differences observed and value that would have been obtained if the relationship between dependent and independent variable was deterministic.

$ExpCc = 186770.2 + 1.86GDPi - 1.96GDPj + 0.56PoPi - 36.30Popj + 1.36PAH + \varepsilon ij$

The coefficient of the GDPi (Cameroon's GDP is positive (1.86), implying, everything being equal, a unit increase in GDP will increase the amount of Cocoa Export volume by 1.86 unit. There is strong

Year	PAH.ug/kg	Expsij	GDPi	GDPj	POPi	POPj	ExpCc
2004	0.5	230141	15775.4	646070	17.1653	16.2251	141,204,416
2005	0.5	209587	16587.9	672357	17.5536	16.3055	126,834,720
2006	0.5	221863	17953.1	719376	17.9484	16.378	114,300,040
2007	0.5	215837	20431.8	833148	18.35	16.4435	94,018,226
2008	0.4	400325	23322.3	931328	18.7588	16.5033	73,489,451
2009	0.4	543363	23381.1	858034	19.175	16.5593	70,552,227
2010	0.4	610990	23622.5	836390	19.5989	16.613	69,682,723
2011	0.4	512344	26587.3	893757	20.0304	16.6647	76,530,475
2012	0.3	394829	26472.1	823139	21.7	16.7567	62,602,021
2013	0.2	0	29567.5	853539	22.1599	16.8237	60,132,943
2014	0.2	563632	31602.4	865001	23.1099	16.8774	54,114,602

Table 3: Data presentation and analysis plan.

Independent variables	Abreviations
Natural log of Cameroons population	InPopi
Natural log of Netherlands (Importing Country) Population	InPopj
Natural log of Cameroon's GDP	InGDPi
Natural log of Importing Country's GDP	InGDPj
Natural log of total PAH standard flow of importing country	LnPAH
εij	Error term

Table 4: Variables used in the model.

```
. reg lnExpCc lnPOPj lnPOPi lnGDPj lnGDPi PAH

      Source |       SS       df       MS              Number of obs =      11
-------------+------------------------------           F(  5,     5) =   41.34
       Model |  .998012826     5  .199602565           Prob > F      =  0.0005
    Residual |  .024139929     5  .004827986           R-squared     =  0.9764
-------------+------------------------------           Adj R-squared =  0.9528
       Total |  1.02215275    10  .102215275           Root MSE      =  .06948

------------------------------------------------------------------------------
     lnExpCc |      Coef.   Std. Err.      t    P>|t|     [95% Conf. Interval]
-------------+----------------------------------------------------------------
      lnPOPj |  -36.30329   23.05112    -1.57   0.176    -95.55809    22.95151
      lnPOPi |   .5591581   2.605144     0.21   0.839    -6.137578    7.255894
      lnGDPj |  -1.964819   .8513339    -2.31   0.069    -4.153242    .2236049
      lnGDPi |   1.860117   1.300106     1.43   0.212    -1.481911    5.202145
         PAH |   1.358399    .876867     1.55   0.182    -.8956585    3.612457
       _cons |   125.9936    53.1311     2.37   0.064    -10.58426    262.5714
------------------------------------------------------------------------------
```

3 (a)

```
Linear regression                                      Number of obs =      11
                                                       F(  5,     5) =   92.46
                                                       Prob > F      =  0.0001
                                                       R-squared     =  0.9764
                                                       Root MSE      =  .06948

------------------------------------------------------------------------------
             |               Robust
     lnExpCc |      Coef.   Std. Err.      t    P>|t|     [95% Conf. Interval]
-------------+----------------------------------------------------------------
      lnPOPj |  -36.30329   24.56138    -1.48   0.199    -99.44032    26.83374
      lnPOPi |   .5591581   2.893943     0.19   0.854     -6.87996    7.998276
      lnGDPj |  -1.964819   .6121128    -3.21   0.024    -3.538305   -.3913325
      lnGDPi |   1.860117   1.090881     1.71   0.149    -.9440815    4.664316
         PAH |   1.358399   1.015396     1.34   0.239     -1.25176    3.968558
       _cons |   125.9936   55.36253     2.28   0.072    -16.32034    268.3075
------------------------------------------------------------------------------
```

3 (b)

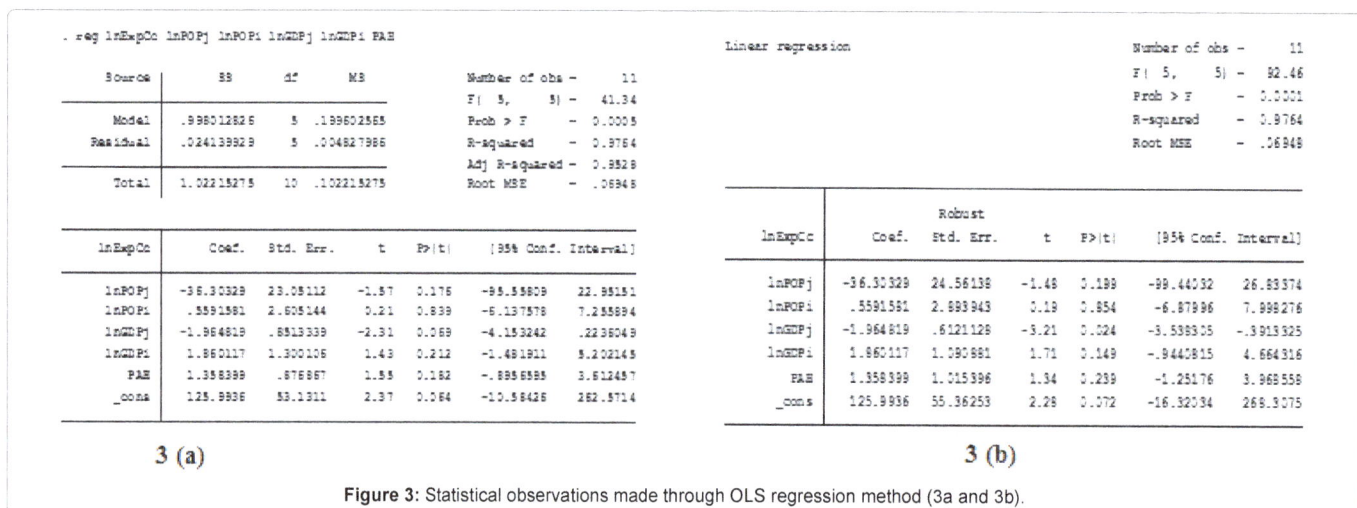

Figure 3: Statistical observations made through OLS regression method (3a and 3b).

relationship between Cameroon's GDP and CoCoa Export and is positive but insignificant since the P-value < 0.1 that is (0.149). This can be explained by the fact that most Cocoa farmers in Cameroon practice small scaled farming. An increase in a farmers GDP will provide resources for the purchase of farm inputs like fertilizers and seeds which thus increase her output and eventually increase the volume of exports.

The GDPj (Netherlands GDP) coefficient is negative (-1.96) implying a unit increase in Netherland's GDP will reduce the amount of exports by -1.96 unit, considering all other factors constant. There is a negative relationship between export and GDP and it is insignificant since the P-value < 0.1 (0.024). This value is valid at 95% significant level. The increase in Netherlands GDP can lead to the farmers look for other measures of raw material (Cocoa) to substitute import from Cameroon. In Addition, the PoPj is at -36.30 which shows a negative relationship with the cocoa export implying that, an increase population will lead too increase in food security. Increase in food security will make the EU to place a more stringent requirement for exports from Cameroon. This will then affect Cameroon's export negatively since compliance is not automatic. Increase in population can also lead to increase in the labour for in Netherland since population affects white-collar jobs, most people without jobs will seek for other means of survival by supplying labour to farms and other unskilled labour jobs.

The coefficient of the PoPi (Cameroon's Population is positive (0.56), implying, everything being equal, a unit increase in Population will increase the amount of Cocoa Export volume by 0.56 unit. There is positive relationship between Cameroon's Population increase and Cocoa Export. This can be explained by the fact that increase in population will increase the labour supply for the cocoa farmers coupled with the fact that cocoa farming is labour intensive.

The coefficient of PAH (Polycyclic Aromatic Hydrocarbon) is 1.36, implying, everything being equal, a unit increase in Compliance to the Maximum Residual level of PAH required by the Netherland (EU) for the import of Cocoa will increase the amount of Cocoa Export volume of Cameroon by 1.36 unit. There is strong relationship between Cameroon's PAH and Cocoa Export and is positive but insignificant since the P-value < 0.1 that is (0.24).

Conventionally, Cocoa Export and standard compliance are positively related, this makes us reject the null hypothesis H_0 which states there is no relationship between standard compliance and Export volume.

The value of R^2 is 0.97, implying 97% of Cocoa export is influenced by variables included in our model, it implies the results has a strong relationship between the dependent and independent variable since R^2 is close to one. Adjusted R-squared is 0.95, also close to one implying many factors affecting the dependent variable (Export) has been included in the regression analysis this makes the result.

From the above remarks, Cameroon is gradually becoming a marginal producer of low quality cocoa and coffee as compared to her rival in the export of these products, Nigeria who is recently taking her place in the EU market. In whole, the major problems that led to the fall in these export are the ageing cocoa and coffee farms which affects the yield per hectare, poor techniques of drying which negatively affects the bean quality (most of the cocoa and coffee bean seeds are dried in ovens which gets contaminated by smoke). In some areas, farmer's dry beans on tarmac roads which end up being contaminated with Polycyclic Aromatic Hydrocarbon (PAH) not good for human consumption. Due to PAH in 2012, the EU rejected a shipment of 2000 MT of coffee beans and hence in April 2013, they re-enforced strict measures on PAH contamination. To this effect, by December 2013, it resulted in a rejection of 3000MT of beans due to non-compliance to the EU requirements. This low beans quality and its rejection pulled down beans prices, Cameroon now trade at a discount of XAF 400/kg at the international market. Other problems faced in her external markets are; poor packaging, irregularity of supply and poor quality in the production process in terms of Level of Residuals.

Summary and Conclusion

These trade facilitation Initiatives investments, if adopted and implemented will benefits directly and indirectly to both Governments and business community: Governments benefit in terms of revenue collection, increased economic efficiency, better security and protection of society. Traders gain through faster delivery of goods and reduced transaction costs. These gains are considerable, especially for small companies, for whom the costs of compliance with procedures are considerable higher: in this case, trade facilitation enhances the capacity of SMEs to participate in international trade.

There is therefore the need to comply with standards which can be done depending on circumstances, various steps may be taken in pursuit of compliance. They may include the following:

A. Legal and regulatory change

B. Reform of institutional structures and responsibilities

C. Restructuring supply chains and increasing control over primary production

D. Modifications in production, post- harvest, processing, and treatment technologies

E. Modifications in firm - and farm - level quality assurance and safety management systems

F. Strengthening of information and surveillance systems

G. Technical and scientific research

H. Investments in physical infrastructure

I. Strengthening of accreditation and certification systems.

Some proposed Strategic policy recommendations in order to enhance Africa's participation in standards setting include:

a. The government should develop, strengthen or support regional policies that will encourage linkage with universities and national research institutions. This effort should be in collaboration with National Agricultural Research Systems (NARS) to enable the empowerment of regulatory institutions to commission standards setting projects and researches.

b. SPS standards-setting committees are encouraged to be established to represent the national and regional positions at international standards setting fora.

c. Governments at all levels should allocate funds to the accreditation of SPS laboratories; create incentives for private sector participation in funding initiatives; promote public-private partnerships including global partnerships in strengthening the capacities of laboratories; strengthening existing laboratories and identify centers of excellence for supporting accreditation activities.

d. Strengthen capacities for risk analysis as a core component of the strategies for implementation of CAADP country and regional compacts implementation processes.

e. Government should allocate resources to the provision of creating awareness tools among stakeholders. Such tools include policy briefs, media, radio, leaflets, posters, et cetera. Along this recommendation is the need for the establishment, operationalization and promotion of SPS enquiry points or nodes at community levels.

f. Capacity building: Training of Trainers (TOT) on various fields relevant to trade and food safety, Building capacity (human resources and equipment) for an effective quality and safety assurance for an effective quality and safety assurance systems, negotiations skills, improvement of infrastructure, especially the cold Improvement of infrastructure, establishment of testing and referral laboratories, Building of good and credible scientific database through regular sampling and analysis of samples, through regular sampling and analysis of samples, to counter any false claims. Development of bankable project proposals to help in accessing financial resources accessing financial resources. Participation in standard setting committees.

g. Financial resources: Establish a sustainable funding mechanism, Mobilization of funds from governments and Mobilization of funds from governments and development partners.

h. Policies: Development of sound policies and strategies for food safety and quality. Development of an effective and relevant legal framework to facilitate good hygiene practices. Hygiene practices.

i. Coordination and harmonization of important requirements: Given the complexity of SPS issues and other given the complexity of SPS issues and other requirements by importing countries.

Proposals on trade related investment areas to the government of Cameroon based on the initiatives carried forward by the government

To gear on the challenges of that facing Trade facilitation, Cameroon requires a comprehensive structure and coordinated approach that should necessitate in removing all the barriers that results in reducing trading costs and improve the business environment with efficient regulation. A lot has been done especially in the port of Douala, however the endless process is to increase the chain supply and Logistic management, for enhancing trade facilitation it is proposed necessary to impinge the following ones:-

Investment in a single window project to enable speeding up customs and border crossing procedure: The issue of speed up customs and border crossing procedure involves a significant reduction of number of documents. UNECA, 2014 fulfilled that, in order to succeed on this the Customs and Border Protection (CBP) tools should be organized into modernized procedures by switching from paper forms to electronic forms in deed this will decrease the time for both goods to be exported and imported. The trade document should be harmonized in accordance to international standardized manner, the clear transparency procedure and customs administration are required to a significant level of qualification and reliability.

Promoting the use of new technology that match slightly with efforts: Various customs administrations in Africa are using the Automated System Customs Data called ASYCUDA World as introduced by UNCTAD since 1994 that build on the experiences of ASYCUDA^{++} that helps in Trade Facilitation by strengthening the Custom administration through implementation of modern and reliable system (UNCTAD, 2013). The Cameroon government should invest in implementation of this new technology and capacity building for workers to be able to work on this new software.

Removing the illegal roadblocks and preventing the diversion of goods on countries roads: Trade costs are generally low for the country with very good condition business climate; illegal roadblocks and diversion are serious problem in developing countries as they are influenced by the higher trade cost and lower customs and border procedures. There is no uncertainty that, the removal of roadblocks and diversion of goods in Africa (SS) is said to be of most important accomplishment. Roadblocks in Africa are mostly caused by poor state of infrastructure and services, out of the road network in Africa only 580,066 or 22.7% of roads are paved and the remaining part being made of either earth or gravel.

Investment in adequate and efficiency transport infrastructure and services: The effort needed here is just and re habilitating and construction of existing road, expanding road network to isolated areas. Increasing connectivity of railway section of various track gauges, quality of the road has a significant impact to the time and cost of trade across borders.

Facilitation of efficient and transparent flow of goods: Tracking the goods from the source to destination, In Belgium there is more than 250 accessible networks from Air freight, sea freight and Land transportation for cargo to enter Belgium. The tracking of the cargo becomes easy for a sender even if a receiver can trace through using the track number if this is done in Cameroon, it will go a long way to facilitate trade.

Integration of border management system: Border efficient management system should have a significant impact towards a trade facilitation as the efficiency of the system need to work under the certified legal regulatory principles and transparent so as to make the system to work in a remarkable manner. One Stop Border Posts is very helpful to this policy strategy as it reduces time and cost of trading because it harmonization of the customs and border procedures. The responsibility of protecting the interest should be coordinated by the several state agencies includes policies, security, customs, and immigration, Health and Sanitation and the Bureau for Standards.

The use of public–private partnerships on custom-border reforms: The approach of involving the private sector is said to be of imperatives and continue to be a growth grounded in trade facilitation. To be effective in Trade facilitation, government's responses to private sectors needs should be driven by sustained Public-private consultation. Various studies observed that private sector traders share their view with government by many means.

However, due to Trade Facilitation Agreement, the expected impact of standards on agricultural trade is not necessarily clear. The one hand compliance with standards may lead to increased production costs which reduces trade, while on the other, standards may also increase information on food safety and product quality which can lead to increased consumer confidence, reduce transaction costs and thus facilitate trade. TFA will ensure the Streamlining of export processing procedures in the agriculture and agro-processing sectors to improve access to markets. Nevertheless, with the aim to increase the competitiveness of the agricultural sector by building export capacity, reducing trade transaction costs and deepening regional integration, government officials need to adhere to Trade Facilitation Agreement. If Cameroonian agricultural producers can comply with these wide range of requirements, including technical regulations, product standards and customs procedures. Then the agricultural sector will be a highly beneficial sector to trade and resources will be allocated efficiently. Trade facilitation and their indicators are very important in enhancing trade in both developed and developing countries.

References

1. FAO (1996) Food and international trade. Technical background document, FAO, Rome.
2. Ministry of Agriculture (2003) Rural sector development strategy paper. Ministry of Agriculture, Yaoundé, Cameroon.
3. Dankers C (2007) Sanitary and phytosanitary requirements and developing country agri-food exports: An assessment of the Senegalese groundnut sub-sector.
4. Niang PN (2004) Costs of compliance with export standards in the senegalese fisheries industry.
5. AU-UNECA (2011) Status of integration in Africa: Recent developments and initiatives, difficulties encountered and perspectives.
6. Buys P, Deichmann U, Wheeler D (2006) Road network upgrading and overland trade expansion in sub-saharan africa. Policy Research Working Paper No. 4097, World Bank, Washington DC.
7. Moisé-Leeman E, Lesser C (2009) Informal cross-border trade and trade facilitation reform in sub- Saharan Africa. OECD, Paris.
8. Ikenson D (2008) While doha sleeps securing economic growth through trade facilitation. cato's center for trade policy studies, Washington DC.
9. Njinkeu D, Wilson JS, Powo Fosso B (2008) Intra-African trade constraints: The impact of trade facilitation. Africa Econom Res.
10. Anderson K (2010) Globalization's effects on world agricultural trade, 1960-2050. Phil Trans R Soc 365: 3007-3021.
11. The Royal Society (2010) Globalization's effects on world agricultural trade. 1960-2050.
12. Armand N, Sama G, Linyong G, Munchunga G (2013) Impact of agricultural export on economic growth in Cameroon: Case of Banana, coffee and cocoa. Int J Bus Manage Rev 1: 44-71.
13. Odularu G, Tambi E (2011) Establishment of standards for international agricultural trade: Promoting Africa's participation. Trade Negotiations Insights.
14. Beghin J (2014) The protectionism of food safety standards in international agricultural trade. Agri Polic Rev.
15. Roberts D, Josling T (2011) Tracking the implementation of internationally agreed standards in food and agricultural production. International Food & Agricultural Trade Policy Council.

16. UNCTAD (2004) International trade negotiations, regional integration and south-south trade, especially in commodities. UNCTAD secretariat.
17. Kriti Gupta B, Sayed Saghaian K (2005) An institutional framework for meeting international food-safety market standards from a developing-country perspective. J Food Distri Res.
18. UNIDO (2006) Role of standards: A guide for small and medium-sized enterprises. United Nations Industrial Development Organization, Vienna.
19. RBS (2014) Rwanda Bureau of Standards (RBS) in Rwanda Trade Policy and Strategies. Ministry of Commerce, Industry, Investment Promotion, Tourism and Cooperatives, Cameroon.
20. EU (2013) Working document on standards and trade of agricultural products.
21. PAH (2007) Contamination/clean-up efforts at the alameda point skeet range.

Utilization of By-Product from Tomato Processing Industry for the Development of New Product

Karthika Devi B, Kuriakose SP, Krishnan AVC, Choudhary P and Rawson A*

Indian Institute of Crop Processing Technology, Thanjavur, India

Abstract

Extrusion cooking is recognized as a smart technology for food processors. It is a low cost, high temperature, short-time process. In this the starchy ingredients are input to create a puffed snack. However it contains multiple parameters that need to be rigorously controlled to develop an optimal process. Present study investigated the blends of corn flour, rice flour and tomato pomace (peel and seed), processed in a co-rotating twin-screw extruder and examined the effect of incorporation of tomato by-product derivatives on final extruded product quality of the ready -to-eat expanded product. Furthermore, the physio-chemical properties, post cooking quality were analyzed for the extruded product. As tomato pomace, corn and rice flour are naturally gluten free, the extruded product would appeal to people who suffer from gluten intolerances, allergies and celiac disease. Dried and milled tomato peel and seed at levels of 0 - 30% and 0 - 5%, respectively were added to the formulation mix. D-optimal mixture design was chosen, which generated 17 combinations; within these combinations, the control formulation existed. The formulations were processed in a twin-screw extruder with a combination of parameters including: solid feed rate kept constant, water feed adjusted to 14%, screw speed of 300 rpm - 350 rpm and process temperatures 30°C to 140°C. It was observed that the addition of tomato pomace significantly increased the crude fiber content and level of protein content in the final product. The expansion ratio, hardness, colour, and overall acceptability varied significantly with respect to tomato pomace addition. Sensory test panel indicated that tomato pomace extrudate could be incorporated into ready-to-eat expanded products up to the level of 30% and it was acceptable. Optimization using D-optimal mixture design suggested that the best formulation extruded product with high desirability was the one consisting of 40% corn flour, 30% rice flour, 25% tomato peel and 5% tomato seed. The results suggest that tomato pomace can be extruded with corn and rice flour into an acceptable and highly nutritious fiber enriched snack food.

Keywords: Extrusion; Tomato peel; Seed; D-optimal design

Introduction

Effective utilization of food by-product/waste as secondary source for the new product development is an emerging area of research. Residues from food industry waste (solid as well as liquid) have some potential benefits on health aspects. So recent research has been focusing on these food wastes for utilization as nutraceuticals and pharmaceuticals, and also for energy generation in the form of production of biogas, hydrogen and bio-ethanol etc. [1].

In food processing industry, food wastes require further processing before being used in food products. This transformation from food waste to value products implies high costs in research and development. Hence, it is essential to obtain important and high value-added products in order to justify the investment.

During the production season large quantities of tomato waste are generated, this may be from the insufficient processing of agricultural products. In recent years due to increase in population food production have increased and have led to over-consumption of processed food. However, this over production and consumption produce different category of tomato waste and these are generally remains largely underutilized, waste produced may be from households, losses occur in the food manufacturing industry, during food sector (ready to eat food, catering and restaurants) lost along distribution chain the waste consist of lot of nutrients and can be a promising sources for food supplementation therefore these should be investigated for further beneficiation for the industrial processing of food waste. In developing countries about 15% of population is starving [2], this large amount of food waste implies an increasing great loss of valuable materials. It also raises management problem due to associated resource consumption and pollutant emission. It has been estimated that for each ton of food waste there is an emission of about 2 tons of CO_2 [3].

In recent years, there is an increasing demand for conversion of fruit and vegetable wastes into useful products. The primary motivation is to minimize environmental impact of these by-products from food industry to avoid environmental problems and to utilize valuable constituents that remain, such as lycopene and dietary fiber. One viable method for utilization of fruit and vegetable by-products into useful products is extrusion processing due to its versatility, high productivity, relative low cost, energy efficiency and lack of effluents. Successful incorporation of tomato pomace into extruded products that deliver physiologically active components represents a major opportunity for food processors providing the consumer a healthy tomato pomace-based product to choose from which is currently lacking in the marketplace. Extrusion cooking is a popular food-processing technique, especially for the production of fiber-rich products, such as breakfast cereals, flat breads, dextrinized or cooked flour. Due to its high content of Total dietary fiber (TDF) and a high proportion of soluble dietary fiber (SDF), food processors are taking steps for an investigation into the use of tomato peel in a variety of extruded products which may be of importance from a nutritional point of view for the consumers.

Therefore, the objective of this research was to investigate process ability of mixing the tomato pomace into the rice and corn flour as a major ingredient to produce snack food in a twin-screw extruder. And to optimize the effect of the independent variables such as tomato

***Corresponding author:** Rawson A, Indian Institute of Crop Processing Technology, Thanjavur, India, E-mail: ashishrawson@gmail.com

pomace (peel and seed) content, rice and corn flour on the functional properties and physical properties of extrudates by using Mixture design, D-optimal.

Materials and Methods

Ingredients used for ready- to-eat snack preparation were: Maize flour and Rice flour. They were obtained from the local market, and were sieved to a particle size of 0.355 µm (mesh 44). The flours were stored at 4°C until used. Tomato pomace (TP) a byproduct from the tomato processing industry was obtained from the Sakthi Fruits tomato paste processing plant located in Erode, TN, India. Sample preparation flow chart given in Figure 1.

D-optimal mixture design was employed by fixing lowest and highest percentage of each independent variable for the design. These ranges were considered as the lower and upper bounds resulting in the constrained representing formulations (1-17) prepared for the experiment. A quadratic model with one centre point was selected, resulting in 17 combinations generated via the Design Expert software. Within these 17 combinations, the control was created, and two combinations were repeated twice to assess error within the model. The model that best fitted the response was selected during analysis of measurements. Analysis of variance (ANOVA) was carried out on each response model to identify the coefficient (R^2), and significant difference ($p < 0.0001$).

In four mixture Corn flour (A), Rice flour (B), Peel (C), Seed (D) made upto total of 100% of the actual formulations. The lower and upper boundaries of ingredients were determined to be as corn flour (40 - 60%), rice flour (30 - 40%), peel (0 - 30%) and seed (0 - 5%) added to 100% of the mixture design. The raw ingredients were weighed separately according to the formulation made by mixture design and blended in a mixer for 10 min. Moisture content of samples were determined by IR moisture meter (Kett, FD-240). These blends were chosen according to preliminary study and for the acceptable product's physical characteristics.

These samples were conditioned to 14% moisture (w.b.). Moisture content was chosen based on preliminary trials, to ensure least variations in hardness of the final product. Water addition was conducted by spraying calculated amount of water prior to extrusion and mixing continuously in a mixer, this preconditioning procedure was employed to ensure uniform mixing and hydration and to minimize variability of the feed material [4].

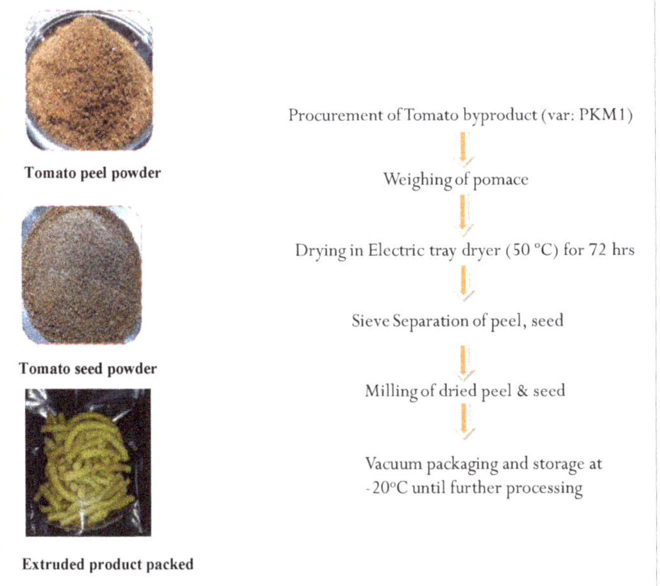

Tomato peel powder

Tomato seed powder

Extruded product packed

Procurement of Tomato byproduct (var: PKM1)

Weighing of pomace

Drying in Electric tray dryer (50 °C) for 72 hrs

Sieve Separation of peel, seed

Milling of dried peel & seed

Vacuum packaging and storage at -20°C until further processing

Figure 1: Sample preparation flow chart.

Constant variables were screw speed set at 300 rpm, die head temperature 130°C. The tomato pomace level was restricted to 0 - 30% range; this was based on the previous studies reported by Hsieh et al. [5]. Dependant variables (or responses) were selected based on descriptors that would best describe a high-quality puffed extruded snack on sensory quality. These were hardness, colour, protein, fat, crude fiber, expansion ratio (ER) and overall acceptance. The generation of response surface plot and statistical analysis were performed using Design Expert software. The main advantage of design is that it reduces the number of experimental runs needed to provide sufficient information for statistically acceptable results. ANOVA was adopted in the design of experimental combinations for the identification of significant difference in the formulations.

Extrusion cooking

Extrusion cooking trials were performed using a laboratory-scale Co-rotating twin-screw extruder (model SYSLG-IV) the barrel consisted of three independent temperature zones, zone 1 (entering zone), zone 2 (kneading zone), and at the barrel end, indicated as zone 3. Screw diameter was 30 mm. The exit diameter of the circular die used was 4.5 mm. The extrusion was carried out with following conditions; 300 rpm - 350 rpm screw speed and barrel temperature zone 30°C, 60°C, 100°C, 140°C respectively while feeder screw speed was set at 30 rpm throughout the experiment. The extrudates were dried in a hot air oven at 50°C. Final dried samples contained a maximum of 4%-5% (w.b.) moisture. Dried samples were stored in polythene bags at room temperature and used for further analysis.

Analysis of extruded products

Sectional Expansion index (SEI): The diameter of 20 pieces of extruded products was measured using a caliper (Mitutoyo, Tokyo, Japan), and the expansion ratio was calculated by dividing the average diameter of the products by the diameter of the die [6].

$$SEI = \frac{De^2}{Dd^2}$$

Where, De^2 is the diameter of the extrudate and Dd^2 is the diameter of the die.

Bulk density: Bulk density (BD) (g/cm^3) of 10 pieces of extrudates were calculated according to Alvarez-Martinez et al. [6] method, where m is mass (g) of a length L (cm) of extrudate, with diameter d (cm).

$$BD = \frac{4m}{\pi d^2 L}$$

Texture: The hardness of the sample was measured with a TA.XT plus, Texture Analyzer (Stable Micro System, Surrey, UK) equipped with 5 kg load cell. Hardness in N was determined by measuring the maximum force required to break the extruded samples (50 mm long) using three point bend test with a sharp-bladed probe (90 mm wide, 32 mm high and 9 mm thick). The test speed was 1 mm/s and the distance between two supports was 22mm. A force –time curve was recorded and analysed by Texture Exponent 32 software programme (version 5.0). Ten measurements were performed on each sample and averaged.

Colour: Hunter Lab Color Flex XE (Hunter Associates Laboratory

Inc., Reston, VA, USA) was used to determine colour values of the ground extruded in terms of *L*, *a* and *b* as measures of lightness, redness and yellowness respectively. The colorimeter was calibrated against a standard white tile ($L = 91.43$, $a = -0.74$, $b = -0.25$). The extrudates were ground in a mixer and passed through a 40 mesh sieve prior to colour analysis, in triplicate.

Water absorption and solubility indices: The water absorption index (WAI) is the weight of gel obtained per gram of dry ground sample. The WAI of extrudates was determined according to the AACC method 56-20. The ground extrudate was suspended in water at room temperature. After standing for 10 min, gently stirred during this period, samples were centrifuged for 15 min at 1000 g (REMI Centrifuge, REMI Elektrotechnik Ltd, Maharashtra, India). The supernatant was decanted into a tarred aluminum pan. The WAI was calculated as the weight of sediment obtained after removal of the supernatant per unit weight of original solids as dry basis. The water solubility index (WSI) is the percentage of dry matter recovered after the supernatant is evaporated from the water absorption determination. The supernatant was dried in a vacuum oven at 84.4°C and 20 - 24 mmHg gauge pressure for 24 h and weighed. The WSI was the weight of dry solids in the supernatant expressed as a percentage of the original weight of sample on dry basis [7,8].

Analytical methods

To characterize the chemical composition of the corn flour, rice flour, tomato peel and seed and the final extruded products, the protein content was measured using the Kjeldahl method [9], and fat content was determined using the Soxhlet method [9]. Ash content was obtained by drying the samples in a muffle furnace at 550°C for two hours. Crude fiber was determined by the acid sequence method using 1.25% H_2SO_4 and 1.25% NaOH for acid and alkaline hydrolysis, respectively. Carbohydrate content was calculated by the difference between the total and the contents of all other ingredients. Each analysis was carried in three replicates.

Sensory evaluation: All formulations were given for sensory evaluation to the panel members. Nine point hedonic scales were adopted and the categories were rated from 1 (absent/dislike extremely) to 9 (very high/like extremely) in order to evaluate the extrudate characteristics. The attributes examined were (Appearence, Colour, Flavour, Initial bite, Texture, Graininess, Tate, Umami, Tangy, Cohesivness, after taste). The panelists consisted of 10 semi-trained panellists (between 20-45 year old males and females) who are students and faculty members of the Department of Food Engineering IICPT, Thanjavur. Panelists were selected in preliminary sessions and experienced with the products and terminology.

Components	Corn Flour (w.b.)	Rice Flour (w.b)
Protein (%)[a]	10.03 ± 0.04	5.81 ± 0.12
Fat (%)	4.37 ± 0.60	0.79 ± 0.23
Ash (%)	0.50 ± 0.00	0.50 ± 0.00
Crude fiber (%)	0.72 ± 0.24	0.45 ± 0.42
Carbohydrate (%)[b]	77.26	82.95

[a]Protein = N × 5.95 for rice
[b]Carbohydrate by difference.
w.b: wet basis

Table 1: Proximate composition of rice and corn flour (Average ± SD).

Components	Tomato peel (w.b)	Tomato seed (w.b)
Protein(%)[a]	16.19 ± 0.00	26.39 ± 1.42
Fat (%)	6.50 ± 0.70	25.03 ± 1.45
Ash (%)	2.50 ± 0.00	3.5 ± 0.00
Crude fiber (%)	29.35 ± 5.79	13.37 ± 1.87
Carbohydrate (%)[b]	40.06	27.31

[a]Protein = N × 5.95
[b]Carbohydrate by difference.
w.b: wet basis

Table 2: Proximate composition of tomato peel and seed (Average ± SD).

Results and Discussion

Proximate analysis of extrudates

Chemical analysis was carried out for the raw material to know their composition (Corn flour, Rice flour, Peel and Seed) as given in Tables 1 and 2. It was observed that Crude fiber content was highest in tomato peel (29%) followed by seed (13%) then corn flour (0.72%) and rice flour (0.5%). Interestingly it was observed that tomato seed contained highest fat content (25%) followed by peel (6.5%) then corn flour (4.4%) and rice flour (0.8%). Similarly protein content was found to be highest in seed (26.4%) followed by peel (16.2%) corn (10%) and lastly rice flour (6%) carbohydrate responsible for expansion was highest in rice flour (82%) followed by corn flour (77.3%) were in peel it was highest about (40%) and lastly seed (27.3%). As illustrated, variation in the initial formulation will affect the properties of the final extruded product depending on changes in fat, protein, crude fiber, and carbohydrate.

Effect of crude fiber on the extruded products

Consumption of dietary fiber has been linked with various health promoting effects and thus has been recommended to be added in the food consumed. In the present study it was observed that crude fiber content was much higher in tomato pomace especially in tomato peel followed by tomato seed (Table 2). Hence it was observed that total crude fiber increased following addition of peel and seed in the extruded product formulation. The statistical analysis confirmed significant effect of the independent variable used in the formulation for extruded product.

The ANOVA showed significant effect ($p < 0.05$) for the mixtures combinations of AD, BC, CD, ABC, ACD (A-corn flour, B-rice flour, C-peel, D-seed) leading to increase in the crude fiber content as tomato peel and seed were added to the product, this effect is not only due to presence of tomato pomace but also due to some fiber content from the corn and rice flours (Table 3).

From Figure 2, it can be seen that the crude fiber content increased significantly following the increase in percentage of peel. To enhance the amount of fiber in food products it is obvious to add higher percentage of peel in the formulations as a source of fiber. Crude fiber, as defined by the association of Official Analytical Chemists is the residue of a feeding material after treatment with boiling sulphuric acid, sodium hydroxide, water, alcohol, and ether. It is a measure of the cellulose and lignin content mainly. The addition of fiber rich components like tomato peel affects the texture of the product such as hardness, moreover this effect is further intensified following thermal treatment during extrusion cooking if the components (corn flour, rice flour, tomato peel, seed) contains high amount of protein. This may also be due to the fact that at higher temperatures protein present in the components may get degraded and become insoluble. Similar results has been reported by several authors, while increasing fiber in the extruded product ultimately increases the hardness of extruded products as a result of its effect on cell thickness [10-12]. Cereal fibers

Run	Corn flour A (%)	Rice flour B (%)	Peel C (%)	Seed D (%)	Crude Fiber (%)	Protein (%)	Fat (%)	ER (%)	Hardness (N)	Colour (a)	OA
1	60	30	5	5	1.18	11.82	2.85	4.00	4.54	10.68	6
2	60	40	0	0	0.24	9.41	1.27	4.78	5.29	4.33	8
3	55.27	36.25	5.83	2.63	1.20	12.04	2.2	3.89	2.48	10.47	6
4	60	30	10	0	1.70	11.83	2.04	4.11	2.38	13.38	6
5	60	30	10	0	1.70	11.83	2.04	4.11	2.38	13.38	6
6	50	35	15	0	1.82	10.75	3.1	3.81	3.00	15.79	8
7	40	40	15	5	1.58	10.03	2.62	3.58	3.97	18.39	5
8	60	37.5	0	2.5	1.60	11.60	2.05	3.39	4.96	12.55	5
9	60	30	5	5	1.18	11.82	2.85	4.00	4.54	10.68	6
10	40	35	25	0	2.05	11.54	2.1	3.79	3.69	16.58	5.4
11	40	35	20	5	1.93	11.60	2.56	3.41	3.39	15.38	5
12	40	30	30	0	3.20	11.83	1.41	3.80	4.72	20.91	6
13	40	40	20	0	2.90	11.16	2.37	3.41	2.79	16.10	4.4
14	50	30	17.5	2.5	2.93	12.76	2.17	3.38	2.73	16.23	5.4
15	47.5	40	7.5	5	0.73	9.60	1.72	3.59	1.84	7.03	5.2
16	40	30	25	5	4.75	11.60	2.5	3.67	2.40	18.27	8
17	55	40	0	5	0.43	10.28	2.2	3.61	4.39	6.37	6

Table 3: Experimental design and results of response variable.

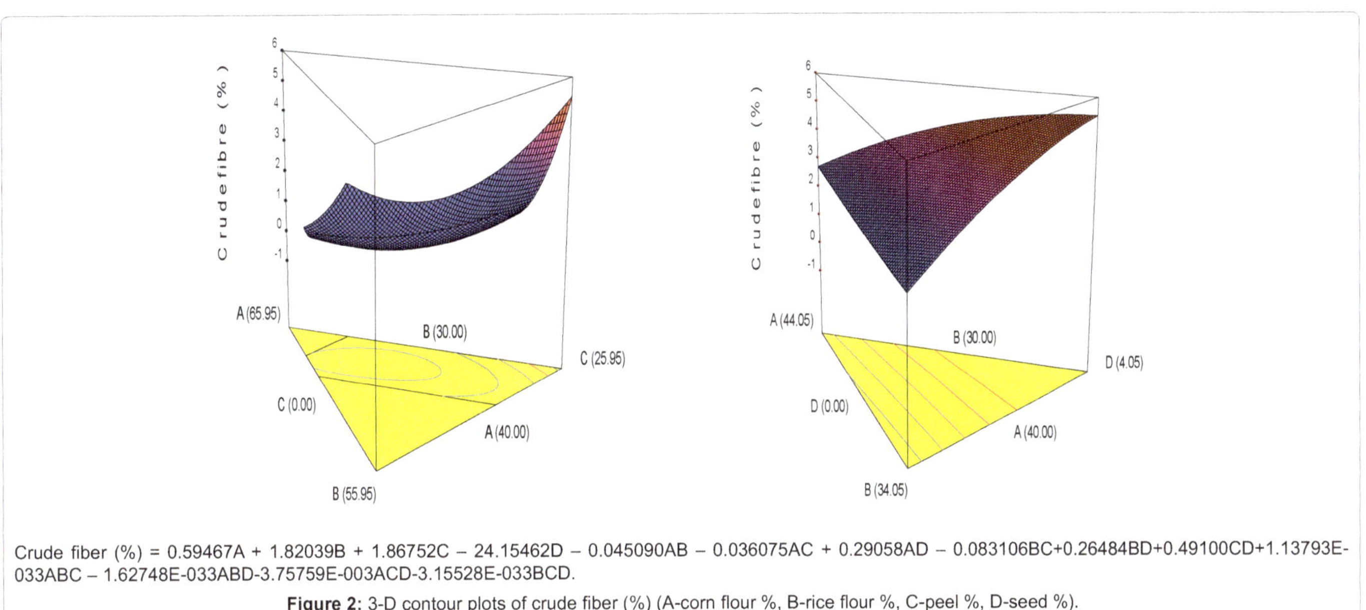

Crude fiber (%) = 0.59467A + 1.82039B + 1.86752C – 24.15462D – 0.045090AB – 0.036075AC + 0.29058AD – 0.083106BC+0.26484BD+0.49100CD+1.13793E-033ABC – 1.62748E-033ABD-3.75759E-003ACD-3.15528E-033BCD.

Figure 2: 3-D contour plots of crude fiber (%) (A-corn flour %, B-rice flour %, C-peel %, D-seed %).

are added to the extruded product due to the fact that the fiber rich foods produces many beneficial effects on the digestive tract, such as the regulation of the intestinal function, improvement of the tolerance to glucose in diabetics, or prevention of chronic diseases as colon cancer. Codex Alimentarius recommends that any product claiming to be a "source" of fiber should contain 3 g of fiber per 100 g of serving or 1.5 g of fiber per 100 kcal of serving or 10% of daily reference value per serving. The extruded product with high fiber is preferable in such a way that it should possess consumer acceptability in terms of hardness. Due to the nature of crude fiber having both insoluble and soluble properties, it can be included in food products with consumer preferences.

Effect of protein on extrudate

The effect of Protein content of the extruded product is studied and all data are significantly different with addition of tomato pomace into the product. The protein content in our study, found to be in range of 9.405 - 12.76% (wb) for all 17 combinations (Table 3 and Figure 3) including replication. The raw material protein content also had a significant impact on product qualities in terms of hardness.

From proximate analysis of raw ingredients it was observed that protein content of corn, rice, peel, seed was observed to be (10.03%, 5.81%,16.19% 26.39%) and Statistical analysis ANOVA showed that there is significance difference ($p \leq 0.01$) between the combinations AD, CD, ABC, ABD, BCD (A-corn flour, B-rice flour, C-peel, D-seed) when compared to control formulations. In treatment without pomace, there was no significant difference ($p > 0.05$) in the protein content between the extrudates. This hardness may be due to the extruders, with their shearing screws operating at high speeds, imparts significant structural changes to food components including proteins. However due to high protein content in (seed > peel > corn > rice) the product obtained also have the effect by the structural changes during extrusion which will increase the hardness and leads to lower expansion.

There are many reports suggested that as higher protein added to starch based- extruded snack, density and hardness are expressed to

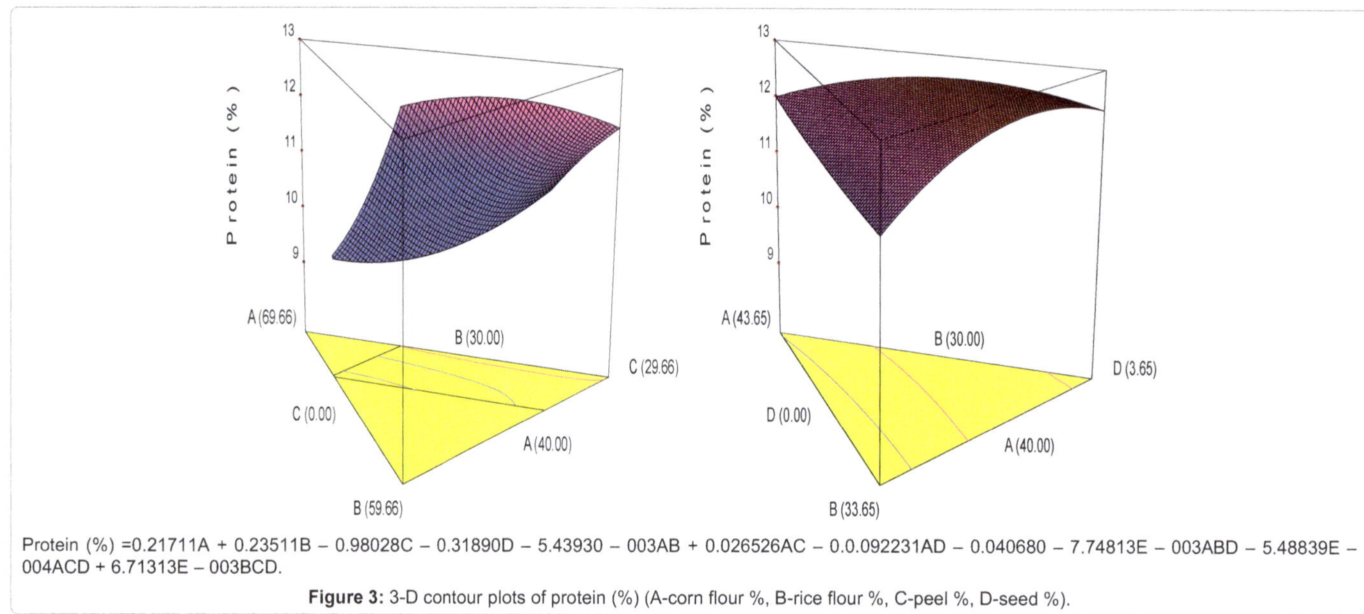

Protein (%) =0.21711A + 0.23511B – 0.98028C – 0.31890D – 5.43930 – 003AB + 0.026526AC – 0.0.092231AD – 0.040680 – 7.74813E – 003ABD – 5.48839E – 004ACD + 6.71313E – 003BCD.

Figure 3: 3-D contour plots of protein (%) (A-corn flour %, B-rice flour %, C-peel %, D-seed %).

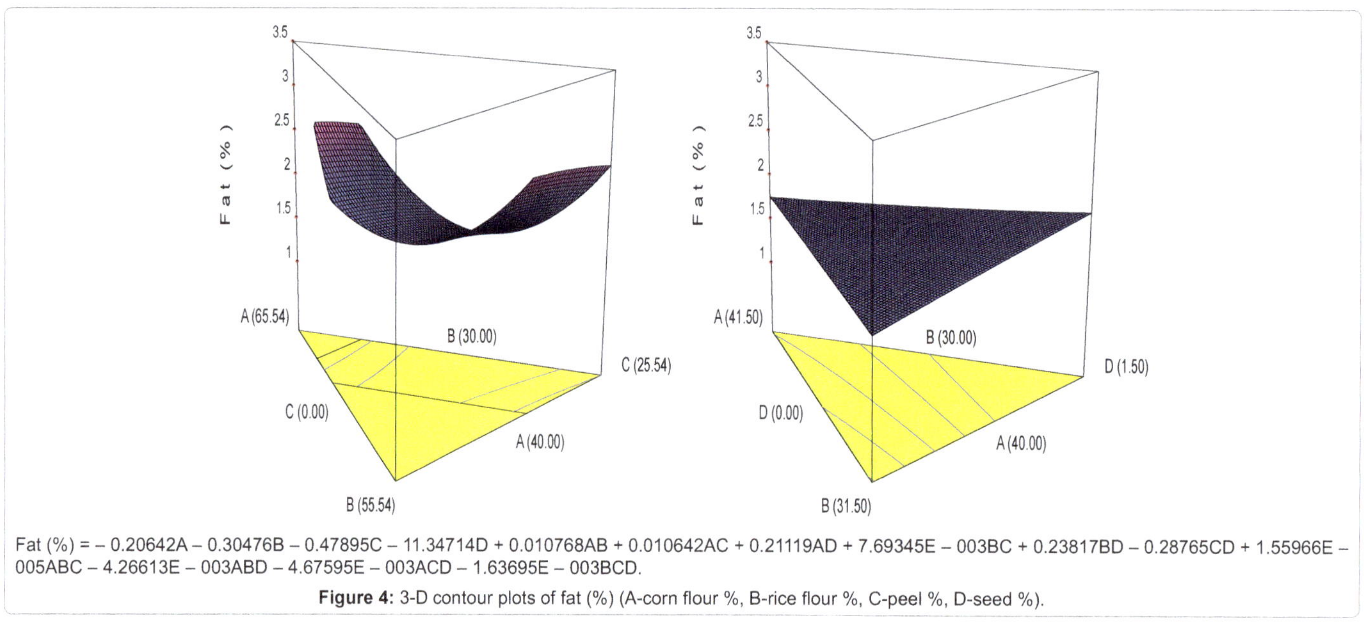

Fat (%) = – 0.20642A – 0.30476B – 0.47895C – 11.34714D + 0.010768AB + 0.010642AC + 0.21119AD + 7.69345E – 003BC + 0.23817BD – 0.28765CD + 1.55966E – 005ABC – 4.26613E – 003ABD – 4.67595E – 003ACD – 1.63695E – 003BCD.

Figure 4: 3-D contour plots of fat (%) (A-corn flour %, B-rice flour %, C-peel %, D-seed %).

increases. Adding proteins to extruded starch-based snacks increases the number of sites for crosslinking, but reduces the starch matrix, resulting in tough, non-expanded crusts [4,13-15]. On the other hand protein is essential for health and it can be suggested that 0%-5% seed is appropriate amount to be included in the extruded product for the consumer acceptance in terms of hardness, expansion and density.

Effect of fat on extrudate

Tomato seed is the important component in the extruded product which also affects the extrudates in a complex manner. The high amount of fat content was observed in seed of tomato as shown in (Table 2). Generally for extruded snack foods, it is desirable to have low levels of fat content [16]. In the present study it was observed that fat content in all extrudates did not exceed more than 3% and hence the product developed could be a low fat and high calorie snack food. However the fat content of extruded product enhances the taste of the product, which is directly proportional to the seed added in the extruded product formulation.

The fat content in the final extruded product varied significantly ($p < 0.001$) and was directly propositional to the percentage of seed present in the initial formulation. Similarly in the case of formulations which included only corn, rice and/or peel flour, no significant difference was observed in terms of fat content. This observation may be explained by the fact that less amount of fat were present in those formulations where tomato seed was absent as is evident by (Figure 4). Fahim Danesh et al. [17] found that tomato seed consisted of essential fatty acids which are useful for the substitution in new product as a source from by-product. They further stated that palmitic acid (12.26%) was the major saturated fatty acid, followed by stearic acid (5.15%) whereas Linoleic acid (56.12%) was the major unsaturated fatty acid followed by oleic acid (22.17%) in those tomato seeds. Substitution of fat in extrudates more than 10% leads to an increase in hardness of the final

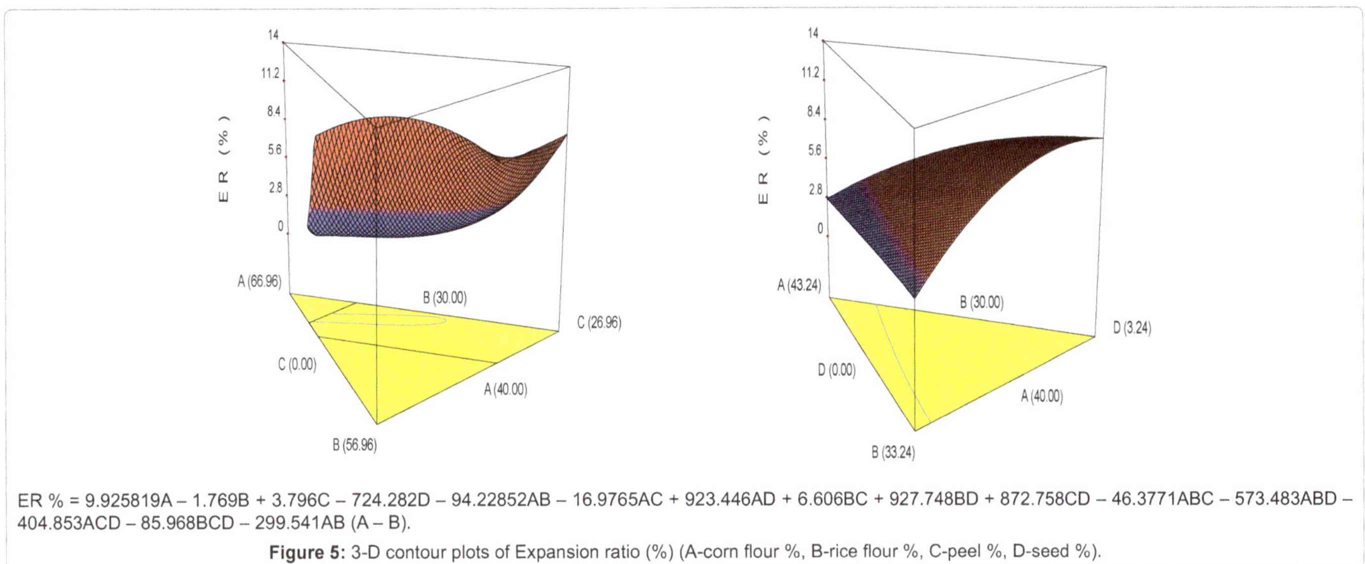

$$ER\ \% = 9.925819A - 1.769B + 3.796C - 724.282D - 94.22852AB - 16.9765AC + 923.446AD + 6.606BC + 927.748BD + 872.758CD - 46.3771ABC - 573.483ABD - 404.853ACD - 85.968BCD - 299.541AB(A - B).$$

Figure 5: 3-D contour plots of Expansion ratio (%) (A-corn flour %, B-rice flour %, C-peel %, D-seed %).

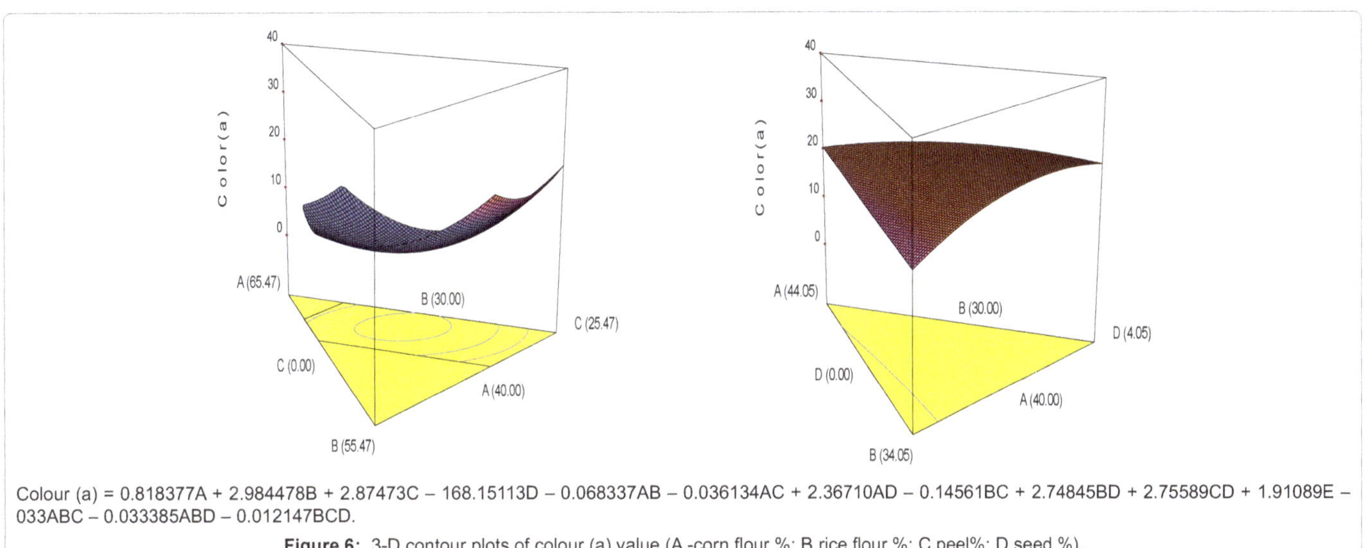

$$Colour\ (a) = 0.818377A + 2.984478B + 2.87473C - 168.15113D - 0.068337AB - 0.036134AC + 2.36710AD - 0.14561BC + 2.74845BD + 2.75589CD + 1.91089E-033ABC - 0.033385ABD - 0.012147BCD.$$

Figure 6: 3-D contour plots of colour (a) value (A -corn flour %; B rice flour %; C peel%; D seed %).

product (citation), considering these factors, in present study it was restricted to 5% (seed) as a substitute for producing low fat snack foods. Hence increasing the amount of seed to 5% which in turn increased the fat% to 3% did not significantly affect the expansion of the extrudate.

Effect of expansion ratio on extruded product

The expansion ratio of the extrudates seeks to describe the degree of puffing undergone by the dough as it exits the extruder. The stored energy was released in the expansion process, increasing the expansion ratio [18]. The extrusion of all combinations of corn flour, rice flour, peel and seed produced expanded snack at most conditions; however blends having higher peel % usually had lower expansion ratio. The expansion ratio measured for all the extruded samples ranged between (3.806 - 4.779%). The expansion ratios of the extruded snacks were similar to the published values of rice-based extrudates [19-21]. The response surface plot for the expansion ratio as a function of components is shown in (Figure 5)

The analytical results indicated that expansion indices were linearly affected ($p \leq 0.0001$) by the tomato pomace. As expected incorporation of tomato derivative reduced the expansion values when compared to the control (without by-product) (Figure 5), similar finding were observed by Dehghan-Shoar et al. [22]. Tomato pomace which is rich in fibers, in the formulation tends to rupture the cell walls before the gas bubbles may expand to their full potential during the process of extrusion cooking. Decrease in the expansion ratio by the addition of products rich in fiber was also observed in studies done by Atlan et al. [7] and Dehghan-Shoar et al. [22]. The decrease in expansion ratio may further be aided by the relative reductions in the amount of starch and protein, mainly responsible for the puffiness of the final product.

Colour

Colour is an important quality factor directly related to the acceptability of food products, and is an important physical property to report for extruded products. Results of colour recorded for the extrudates containing different concentrations of tomato pomace obtained using different mixture designs are presented in Figure 6. Among the colour parameters, the redness *a* values showed marked changes due to addition of tomato pomace only.

An increase in tomato peel level in product increased the '*a*' value of samples as expected due to the lycopene pigment in the tomato peel.

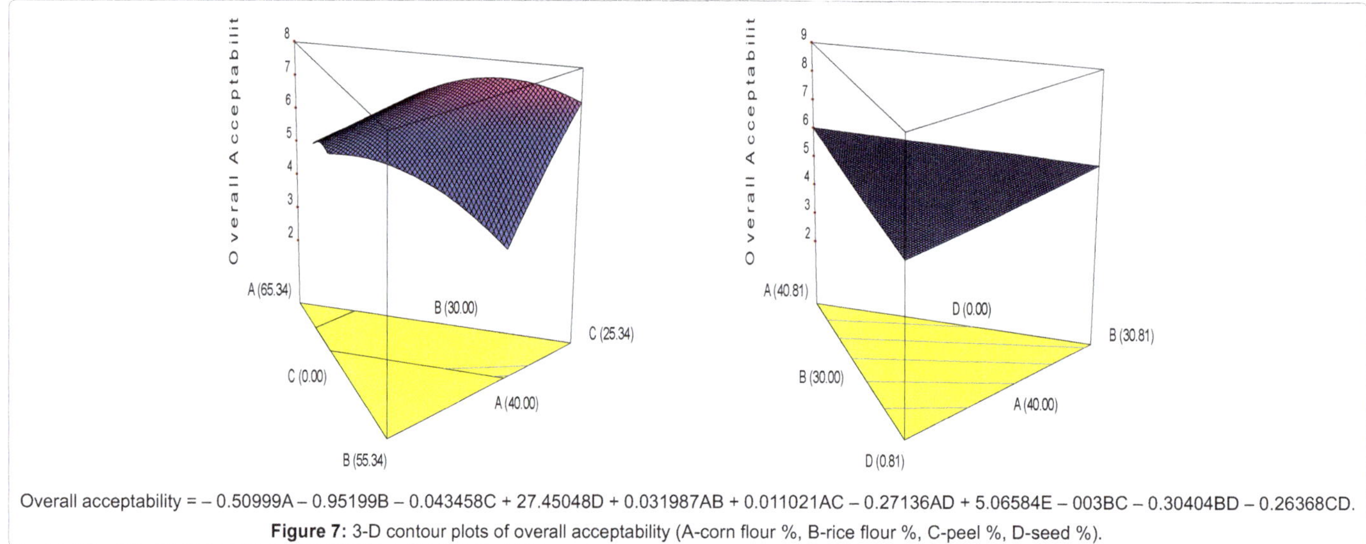

Overall acceptability = – 0.50999A – 0.95199B – 0.043458C + 27.45048D + 0.031987AB + 0.011021AC – 0.27136AD + 5.06584E – 003BC – 0.30404BD – 0.26368CD.

Figure 7: 3-D contour plots of overall acceptability (A-corn flour %, B-rice flour %, C-peel %, D-seed %).

Solutions					
Number	Corn flour	Rice flour	Peel	Seed	Desirability
1	40	30	25.24	4.76	0.84
2	54.82	30	13.42	1.75	0.58
3	46.78	40	12.04	1.18	0.55
4	60	30	5.72	4.28	0.48

Table 4: Solutions obtained from D-optimal mixture design.

Similar results were reported by Ilo and Berghofer [20], Altan et al. [7], Dehghan-Shoar et al. [22]. Results of ANOVA showed significant difference among the linear mixtures by the addition of pomace in the extrudates ($p < 0.05$).

There were however, some formulations like ABC and BCD which did not show significant difference among them with regards to colour. This phenomenon may be due to the yellowness colour which was not significant in most of the formulations and which are due to the carotenoid content in seed and corn flour may have some effect on the *a* value of the extrudate samples [22].

Overall acceptability

A 9 point hedonic scale was used for the sensory evaluation of the extruded products. The mean scores of overall acceptability showed that all products with tomato pomace were within the acceptable range. Statistical analysis showed that the model was ($P < 0.05$) significant with addition of pomace.

It was observed that tomato pomace had significant effect on final product quality in terms of appearance, colour, flavour and taste. Overall acceptability of the products ranged from (4.4-8), Extrudates with different levels of tomato pomace had higher scores with respect to control sample, combinations which had the Overall acceptability was the one with peel and seed ratio, 15: 0% and 25: 5% respectively (Figure 7).

Furthermore it could be explained that combination with seed also ranged high due to the fact that seed contained fat content which increases the taste of the sample over consumption. Although due to increase in peel and seed percentage some variation of textural properties were observed in all combinations, especially hardness may increase due to increase in fiber and protein content as discussed earlier this certainly affects the crispness of the product. Some product showed Umami flavour after taste which increased when addition of pomace (>20%) was done. Higher the peel percentage eventually increased the colour of the product.

Above all results indicated that extrusion of tomato pomace, in combination with rice and corn flour can produce acceptable extrudate snack food. Moreover the tomato pomace content was the most important parameter affecting the sensory properties of extruded product.

Optimization of formulation based on desirability by using mixture (D-Optimal) design

The optimization tool from the design expert software was utilized to derive the optimal formulations based on response results evaluated. This was implemented by choosing the criteria of maximum limits for both the favorable factors and desirable responses, i.e., tomato peel and seed (factor), corn flour and rice flour were kept in range and response with maximum limits were expansion ratio, protein, fat, crude fiber, colour (a) and overall acceptability. Undesirable responses (e.g. hardness) were minimized to produce formulations with a more crispy texture but less hardness. 17 formulations with predicted quality response values were generated based on these entered parameters. Depending on the best desirability ratings, one formulation was chosen. The details of these formulation can be seen in (Table 4) the combinations were F1, F2, F3, F4 (desirability 0.845 i.e., 84.5%) tomato pomace, corn flour and rice flour, die head temperature of 130°C and screw speed of 300 rpm.

Furthermore the model was validated by comparing the predicted data from the Mixture design with the experimental study, which was observed to be close and the % error was very low.

Conclusion

Using D-optimal mixture design as a tool, the effects of four factors on the physico-chemical characteristics, texture, post cooking quality (WAI & WSI) and chemical properties of extruded snack are highlighted in this study. The design demonstrated the ability of tomato pomace to affect all the responses. In all formulation, it was observed to be significant variation with the addition of tomato pomace on to the extruded product. Tomato pomace affected the expansion ratio, texture and hardness. This is attributed to the high fiber content present in tomato pomace and protein content in seed.

As discussed, fiber is well documented as having hydrophilic properties; beyond a certain limit, it can have a negative impact on the responses due to its ability to absorb excess water and damage aerated bubble structures. Optimization of these four factors was crucial for the development of a high-quality puffed snack. The optimal extrusion conditions and Tomato pomace inclusion to produce a high quality snack were calculated with the aid of the optimization tool to be; corn flour, corn flour and tomato pomace within the range and tomato pomace addition with desirability of 0.84%. The study successfully demonstrates how tomato pomace could be utilized in an extruded puffed snack. It also highlighted how an under-utilized by-product may be substituted for corn, rice flour and incorporated into an expanded puffed snack. Sensory acceptability of the product was showed to be most dependent with the pomace content. However high level of tomato pomace was included into the extruded snack without compromising the expansion characteristics of the snack while potentially improving its nutritive properties with high fiber and protein enriched snack product. These kinds of products could help in utilizing the by-product as a source of fiber for the production of valuable products in the future.

References

1. Mirabella N, Castellani V, Sala S (2014) Current options for the valorization of food manufacturing waste: A review. J Clean Prod 65: 28-41.
2. FAO (2012) The state of food insecurity in the world. Food and Agricultural Organization of the United Nations.
3. European Commission (2010) Preparatory study on food waste across EU 27. Technical Report-054, EC.
4. Stojceska V, Ainsworth P, Plunkett A, Ibanoglu S (2008) The effect of extrusion cooking using different water feed rates on the quality of ready-to- eat snacks made from food by-products. Food Chem 114: 226-232.
5. Hsieh F, Mulvaney SJ, Huff HE, Lue S, Brent JJ, et al. (1989) Effect of dietary fiber and screw speed on some extrusion processing and product variables. Lebens Wiss Technol 22: 204-207.
6. Alvarez-Martinez L, Kondury KP, Harper JM (1988) A general model for expansion of extruded products. J Food Sci 53: 609-615.
7. Atlan A, Kathryn McCarthy L, Maskan M (2008) Evaluation of snack foods from barley-tomato pomace belends by extrusion processing. J Food Engi 84: 231-242.
8. Jin Z, Hsieh F, Huff HE (1995) Effects of soy fiber, salt, sugar, and screw speed on physical properties and microstructure of corn meal extrudate. J Cereal Sci 22: 185-194.
9. AOAC (2005) Official methods of analysis of AOAC International (18thedn.) AOAC, Gaithersburg.
10. Mendonca S, Grossmann MVE, Verhé R (2000) Corn bran as a fibre source in expanded snacks. Lebensm Wiss Technol 33: 2-8.
11. Yanniotis S, Petraki A, Soumpasi E (2007) Effect of pectin and wheat fibers on quality attributes of extruded corn-snack. J Food Eng 80: 594-599.
12. Ainsworth P, Ibanoglu S, Plunkett A, Ibanoglu E, Stojceska V, et al. (2007) Effect of brewers spent grain addition and screw speed on the selected physical and nutritional properties of an extruded snack. J Food Eng 81: 702-709.
13. Martinez-Serna MD, Villota R (1992) Reactivity, functionality, and extrusion performance of native and chemically modified whey proteins. Food Extr Sci Technol 86: 387-414.
14. Onwulata CI, Smith PW, Konstance RP, Holsinger VH (2001) Incorporation of whey products in extruded corn, potato, or rice snacks. Food Res Int 34: 679-687.
15. Veronica AO, Olusola OO, Adebowale EA (2006) Qualities of extruded puffed snacks from maize/soybean mixture. J Food Eng 29: 149-161.
16. Yagci S, Gogus F (2009) Development of extruded snack from food by-products: a response surface analysis. J Food Process Eng 32: 565-586.
17. Fahimdanes M, Bahrami ME (2013) Evaluation of physicochemical properties of iranian tomato seed oil. J Nutri Food Sci 3: 1-4.
18. Thymi S, Krokida MK, Pappa A, Maroulis ZB (2005) Structural properties of extruded corn starch. J Food Eng 68: 519-526.
19. Asare EK, Sefa-Dedeh S, Sakyi-Dawson E, Afoakwa EO (2004) Application of response surface methodology for studying the product characteristics of extruded rice–cowpea–groundnut blends. Int J Food Sci Nutri 55: 431-439.
20. Ilo S, Berghofer E (1999) Kinetics of colour changes during extrusion cooking of maize gritz. J Food Eng 39: 73-80.
21. Ding QB, Ainsworth P, Tucker G, Marson H (2005). The effect of extrusion conditions on the physicochemical properties and sensory characteristics of rice based expanded snacks. J Food Eng 66: 283-289.
22. Dehghan-Shoar Z, Hardacre AK, Brennan CS (2010) The physico-chemical characteristics of extruded snacks enriched with tomato lycopene. Food Chem 123: 1117-1122.

The Study of Analytical Identification on Main Monomer Compounds of Spoiled Grass Carp by High Performance Liquid Chromatography of Quadrupole Time of Flight Mass Spectrometry

Sheng ZY, Si HJ, Ming LX, Rong CJ, Yi CZ and Hui ZY*

Sericulture and Agri-food Research Institute, Guangdong Academy of Agricultural Sciences, Key Laboratory of Functional Foods, Ministry of Agriculture, Guangdong Key Laboratory of Agricultural Products Processing, Guangzhou, China

Abstract

Background: The change of monomer compounds of materials during preservation is becoming increasingly important in the background of food preservation. It can help to understand underlying causes of quality change of materials in preservation substantially. The objective of this study was to infer and analyze the change of main monomer compounds during spoiled Grass carp (*Ctenopharyngodon Idellus*), and in favor of understanding the quality change of Grass carp in preservation.

Methods and materials: The spoiled Grass carp was studied as materials by High Performance Liquid Chromatography of Quadrupole Time of Flight Mass Spectrometry (HPLC-Q-TOF-MS). HPLC-Q-TOF-MS has two scan modes, positive one and negative one. Special related substances and molecular components of this characteristic fragment ion were identified by quasi-molecular ion peaks and accurate molecular weight of fragment ion from high-resolution mass spectrometry.

Results: 46 kinds of monomer compounds were determined in spoiled Grass carp, which have 3 kinds of non-nitrogenous compounds and 43 nitrogenous compounds. 43 nitrogenous compounds including 6 kinds of amino acids (2 kinds of α-amino acids), 10 kinds of amines, 12 kinds of amide compounds, 2 kinds of nitro compounds, 12 kinds of heterocyclic nitrogenous compounds, and 1 kinds of nitriles compound.

Conclusion and suggestion: Structure of monomer compounds in fresh and perishable materials can be inferred and identificated by HPLC-Q-TOF-MS. They can be used to increase efficiency in identification and analysis of chemical component. It will be benefit for identification, evolution, and deduction of active ingredients and new compounds of fresh material in preservation.

Keywords: High performance liquid chromatography coupled with quadrupole time-of-flight mass spectrometry (HPLC-Q-TOF MS); Analytical identification; Spoiled; Grass carp; Chemical compounds; Fragment ions

Introduction

Grass carp (*Ctenopharyngodon idellus*) which is one of the four major Chinese carps has been introduced into more than 100 countries [1]. It is an important economic species in freshwater and its global production is more than 4.5 million tons annually [2], it has become the most important freshwater fish in consumption in worldwide [3]. Compared to other types of meats, Grass carp is beneficial to human's body due to its low fat, low cholesterol and high unsaturated fatty acids. However, it is most at risk of damage and perishable as aquatic products because it is easy to cause the spoilage quickly and the freshness loss [4].

Similar to other types of fresh aquatic products, storage conditions play an important role in grass carp quality. Improper storage time and storage environment will bring about decrease in freshness and deterioration of fish, decrease in the freshness of the fish may influence meat quality and taste, or even deterioration that will bring about great loss to fisheries production. The fish spoilage is a multifaceted process; it involves physical, chemical and microorganism mechanisms and is related to color changes, texture collapse, protein denaturant, lipid oxidation, ATP degradation and microbial spoilage [5,6].

At present, There are many indicators in studying fish preservation, such as sensory assessment, cooking quality and structure, salt-soluble protein, thiol, disulfide bond, total volatile base nitrogen (TVB-N), electrical conductivity (EC), total viable counts (TVC), biogenic amines (BAs). freshness quality index (K value), SOD, MDA and ATP enzyme, changes in metal ion content, etc. [7-14], some of these indicators are formulated relevant standards in china, for example, the determination of total volatile base nitrogen (TVB-N) in aquatic products (SC/T3032-2007), the freshness index for fish muscle (K-value)-high performance liquid chromatography (SC/T3048-2014) and so on, they have laid the foundation for the research of preservation of fish and other aquaculture products.

However, during the process of fish preservation, most of these indicators reflected change of the apparent indexes not the internal material compositions, so there is insufficient in describing the preservation process or revealing the change rules.

In fact, the actual changes in apparent indexes are reflected by the existence and quantity of the monomeric compounds which is material foundation. Studying the monomeric compounds is necessary to reveal the changes in apparent indexes. During the period of storage, compounds decomposition and polymerization of Grass carp are always occurred by enzymolysis, microbial metabolism and

***Corresponding author:** Hui ZY, Sericulture and Agri-food Research Institute, Guangdong Academy of Agricultural Sciences, Key Laboratory of Functional Foods, Ministry of Agriculture, Guangdong Key Laboratory of Agricultural Products Processing, Guangzhou 510610, China, E-mail: zhangyhgx@163.com

temperature and so on as well as other aquatic materials. Thus, low attainment of the high degree of purity of monomeric compounds from fish is main bottleneck in research and has been one of the hot points in the fish study in recent years.

In recent years, with the development of the inspection instruments, the technology which combines both efficient separation of liquid chromatography with high sensitivity of mass spectrometry especially combined high resolution mass spectrometry with multi-stage mass spectrometry has been widely used in plant and animal components analysis and identification, it has opened up a new way on the study of natural products, heavy metals, pesticide residues and so on [15-20].

High Performance Liquid Chromatography of Quadrupole Time of Flight Mass Spectrometry (HPLC-Q-TOF-MS) is a kind of typical technology which has qualitative and quantitative analyses in effective ingredients by the combination of Liquid Chromatography and Mass Spectrometry, this technology can be utilized in the structure analysis of trace components without reference substance, it has highly efficient and sensitive advantages.

In this study, HPLC-Q-TOF-MS was utilized to analyze the main monomeric compounds of Grass carp (*Ctenopharyngodon idellus*) in the process of maintaining the freshness qualitatively; accurate information of molecular weight and fragment was got. At the same time, the main monomeric compounds were inferred according to structure databases of a variety of compound and the cracking rules of mass spectrometry. The aim was to provide the scientific basis for active ingredients identification and preservation research of Grass carp, and to provide a reference for chemical composition analysis and identification of fresh material.

In this experiment, chemical composition of the species and its change rule of Grass carp in different stages like fresh raw materials, preservation process and deterioration were studied. Study of chemical composition in deterioration of Grass carp is as follows.

Materials and Methods

Materials and chemicals

In refrigerator, Grass carp was kept for 6 months during -10°C (fresh Grass carp was bred in Guangzhou Huadu district, weighted 1200 g ± 200 g, obtained from Guangzhou Grandview supermarket, Guangdong, China). It was removed from the refrigerator and placed at room temperature for three days until it smelled badly. At the same time, its content of TVB-N was 32.5 mg/100 g, and identified it as rotten fish according to the National Food Safety Standard of china for Fresh and Frozen Aquatic Animal Products (GB 2733-2015).

Methanol (CH_3OH) which was purchased from Sigma-Aldrich-(St. Louis, MO, USA) was used as a solvent in LC/MS/MS, it was used throughout the study. Ultra-pure water was produced by Milli-Q type 3 ultra-pure water machines (Millipore, USA).

Equipment

Ekspert™ ultraLC type 110-XL HPLC and AB SCIEX Triple TOF ™ type 5600 Q-TOF MS, Duo Spray ™ ion source, with AB SCIEX-Analyst ® TF Software, Multi Quant ™ Software quantitative analysis Software were used. All equipment were supplied by the SCIEX companies in the United States.

Integml Milli-Q type 3 ultrapure water machine was supplied by the Millipore companies in the United States.

Experiment methods

Preparation of sample solution: Firstly, 10 g of spoiled grass carp was obtained by a sterile knife respectively, they were extracted with 100 ml ultrapure water and 100 ml chromatographic pure methanol respectively, refluxing extraction (extraction temperature is 95°C in ultrapure water and 70°C for reflux in chromatographic pure methanol, respectively) for 2 hours. The extract was stand for 24 h during 0°C, then centrifuged it (12 000 r. min-1,5 min). At the last, supernatant were filtered through 0.35μm filters and analyzed by HPLC–MS. Blank is frequently prepared as described above.

Analysis of test conditions

HPLC analysis: The chromatographic separation was carried on ZORBA×RPHD Eclipse pluse C18 column (2.1 mm × 100 mm, 1.8 μm particle size). The mobile phase, which consisted of methanol (A) and 0.1% formic acid in water (B) was delivered at a flow rate of 0.5 mL/min under the following gradient program: 10% (A) from 0.1 to 5 min, 20% (A) from 5 to 10 min, 25% (A) from 10 to 15 min, 30% (A) from 15 to 20 min, and returning to the initial condition over 5 min. The sample injection volume was 10 μl. The column oven temperature was set at 40°C.

MS analysis: Each sample was analyzed in positive and negative ionization modes. Column effluent was directed to the ESI source. The curtain gas, nebulizer gas, and heater gas were set to 30, 50, and 55 psi. Source temperature was 550°C for both modes. In an ESI source, positive ion mode produced 4500V atomizing voltage (ISVF) and 100 V declustering potential; negative one produced -4500 V atomizing voltage (ISVF) and -100 V declustering potential.

The experiment was carried out to collect by using one TOF MS survey scan (250 ms) and 4 TOF MS/MS scans (100 ms each). The scan types of TOF MS was 100-1000 m/z, and the scan types of TOF MS/MS was 50-1000 m/z. Acquisition of MS/MS spectra was controlled by IDA function of the Analyst TF software (AB Sciex, Concord, Canada) with application of following parameters–dynamic background subtraction, charge monitoring to exclude multiply charged ions and isotopes, and dynamic exclusion of former target ions for 5 s. Rolling collision energy was set whereby the software calculated the CE value to be applied as a function of m/z. Data quality was corrected by CDS system （automated calibrant delivery system, SCIEX, Concord, Canada） under DuoSpray source.

Data processing: MarkerView 2.0 (AB Sciex, Concord, Canada) was used to generate a peak table of m/z and RT for samples in the individual study using the following parameters. For peak detection: noise threshold of 50 counts, minimum chromatographic peak width of 3 scans, minimum spectra width of 10 mDa, background subtraction offset of 20 scans and subtraction multiplication factor of 1.2. For peak alignment: RT window of 0.7-10 min, RT tolerance of 0.3 min, mass tolerance of 12 ppm.

Results and Discussion

HPLC-Q -TOF -MS analysis of spoiled grass carp

It is important to prevent monomeric compounds of spoiled Grass carp degrade and accumulate; it is the basis of the component identification. In this experiment, ultrapure water and methanol were used as solvent extractions at a relatively low temperature (reflux extraction temperature of ultra-pure water is 95°C, and one of chromatographic pure methanol is 70°C). Then, scanning test of extract solution is processed by positive and negative ion modes, respectively.

The results showed that the characteristic information of the total ion current (TIC) under positive ion modes was stronger and with higher sensitivity than negative one in the methanol extract, and information of compounds in the methanol extract was more than one in aqueous extraction. Therefore, methanol extract under the scanning mode of positive ion (Fig. 1) were compared, analyzed and identified.

The analytical identification of main monomer compounds in spoiled grass carp

In experiment, possible elements (error is less than ± 5×10^{-6}) was calculated by the combination of the precise molecular mass from test of ESI positive ion mode with the high resolution data given by peakview 2.0 workstation. Then possible molecular formula of the main monomeric compounds was determined. The characteristic fragment ions and accurate chemical elements of the main monomeric one were obtained by the secondary mass spectrometry analysis.

46 kinds of monomeric compounds included 43 nitrogenous compounds and 3 non-nitrogenous compounds were identified by Combination of Chemspider database and fragment cracking rules of mass spectrometry.43 nitrogenous compounds including 6 kinds of amino acids (2 kinds of α-amino acids), 10 kinds of amines, 12 kinds of amide compounds, 2 kinds of nitro compounds, 12 kinds of heterocyclic nitrogenous compounds and 1 kinds of nitriles compound. Cracking way of TOF MS/MS fragment ions for each compound were inferred the details were shown in Tables 1 and 2.

Inference and analysis of the cracking way from fragment ions

Amino acids: Amino acid which has great effect on fish quality is an important component in Grass carp. Duo to its strong polarity, the retention time is short and the peak was aroused earlier in the chromatographic column. Amino acid which has low ionization efficiency is an amphoteric compound. In general, in positive ion mode, intensity of molecular ion peak in free amino acids is an order of magnitude higher than in negative one. So the positive ion mode which has higher sensitivity is more suitable for fragmentation study of amino acids In the positive ion mode, the carboxyl group and amino group of α-C of the α-amino acids are easy to lose, its fragmentation path mainly include decarboxylate, dehydroxylation or deamination, that is $[M+H\ -HCOOH]^+$, $[M+H\ -H_2O]^+$ or $[M+H\ -NH_3]^+$. The amino acids lose NH_3 (17u) to form $[R\text{-}CHCOOH]^+$ or lose HCOOH (46 u) to form $[R\text{-}CHNH_2]^+$ [21-24]. 6 amino acids which included 2 α-amino acids in the methanol extract of spoiled Grass carp were identified. L-Leucyl-L-lysyl-L-proline was used as an example, it was produced the proton molecular ion peak [L-Leucyl-L-lysyl-L-proline+H]$^+$ (m/z 357.0121) by electrospray ionization. The MS^2 spectrum was shown in Figures 1 and 2.

In MS/MS spectrum, the [L-Leucyl-L-lysyl-L-proline+H]$^+$ ions have two possible fragment pathways in positive mode by collision induced dissociation, the first one was lost NCOOH (59 u, is actually

No.	RT /min	Molecular formula	Molecular weight/u	Precursor ions【M+H】(m/z)		The main fragment ions of TOF-MS/MS
				Observed /u	Theoretical/u	
1	0.96	C_5H_9NO	99.1311	100.07645	100.0764	82.0670, 56.0517
2	2.15	$C_5H_{14}N_2$	102.1781	103.12331	103.12317	103.1233, 86.0964
3	2.59	C_8H_8	104.1491	105.0704	105.07019	105.0708, 79.0556
4	0.56	$C_4H_7N_3O$	113.1179	114.06642	114.06618	86.0729
5	0.78	C_8H_9N	119.1638	120.07097	120.0808	120.0815, 93.0716
6	1.89	C_8H_8O	120.1485	121.06498	121.06488	121.0650, 93.0709, 77.0407
7	2.6	$C_8H_{11}N$	121.1796	122.09646	122.09621	122.0965, 105.0710
8	0.57	$C_5H_4N_4O$	136.1115	137.04588	137.04566	137.0460, 119.0354, 94.0407
9	1.79	C_6H_9NOS	143.2068	144.04776	144.04774	144.0478, 126.0371, 113.0300
10	2.71	$C_{10}H_9N$	143.1852	144.08068	144.08058	144.0807, 128.0497, 103.0546
11	1.34	$C_7H_{16}N_2O$	144.2147	145.13337	145.1333	145.1334, 128.1080, 86.0978
12	2.58	$C_8H_{20}N_2$	144.2578	145.16961	145.16963	145.1695, 86.979, 69.0719, 60.0827
13	0.52	$C_7H_{15}N_2O$	145.2063	146.11728	146.11741	146.1182, 87.0455, 60.0834
14	0.53	$C_8H_{18}N_2O$	158.2413	159.14878	159.14872	159.1488, 142.1227, 86.0615
15	0.77	$C_9H_{11}NO_2$	165.1891	166.08598	166.08579	166.0860, 120.0813, 103.0547
16	0.92	$C_8H_{13}N_3O$	167.2083	168.11287	168.11282	168.1127, 150.1028, 112.0872
17	2.27	$C_9H_{20}N_2O$	172.2679	173.16472	173.16462	173.1647, 156.1377, 100.0751
18	1.5	$C_9H_{15}N_3O$	181.2349	182.12861	182.12851	182.1286, 164.1193, 95.0607
19	8.14	$C_8H_{10}N_2O_3$	182.1766	183.07811	183.07755	155.0465, 127.0149, 97.9687
20	2.12	$C_8H_{13}NO_2S$	187.2593	188.07371	188.07319	188.0737, 131.1180, 117.1027
21	2.2	$C_9H_{21}N_3O$	187.2825	188.17524	188.17527	188.1752, 152.0522, 117.1014
22	0.54	$C_9H_{20}N_2O_2$	188.2673	189.15895	189.15929	189.1599, 144.1384, 130.0867
23	1.7	$C_{10}H_{20}N_2O_2$	200.278	201.15949	201.15953	201.1595, 183.1488, 159.1495
24	0.53	$C_9H_5NOS_2$	207.2721	207.98815	207.98813	207.9882, 189.9757, 165.9665
25	2.76	$C_{11}H_{22}N_2O_2$	214.3046	215.1749	215.1751	215.1749, 159.1490, 129.9939
26	10.13	$C_{10}H_{16}O_5$	216.231	217.10692	217.10663	217.1069, 173.0812, 156.0709, 111.0450, 83.0501
27	14.48	$C_{14}H_{29}NO$	227.3862	228.23206	228.23202	228.2321, 88.0757, 70.659
28	0.56	$C_{11}H_{20}N_2O_3$	228.2881	229.15464	229.15465	229.7556, 114.0553, 96.0823
29	5.71	$C_{12}H_{24}N_2O_2$	228.3312	229.19088	229.19093	173.1662, 156.1383, 142.1231, 129.9960, 100.762

30	1.79	$C_{11}H_{23}N_3O_2$	229.3192	230.18631	230.18616	230.1863, 213.1581, 114.091, 100.0760
31	7.76	$C_{12}H_{10}N_4O_2$	242.2334	243.08769	243.08743	243.0877, 226.0607, 200.0820, 172.0870, 157.0642
32	2.82	$C_{14}H_{22}N_2O_2$	250.3367	251.17513	251.17475	251.1751, 147.8527, 138.0893, 121.0654
33	13.99	$C_{16}H_{33}NO_2$	271.4387	272.25852	272.25805	272.1556, 255.2345, 237.2162, 213.0615
34	2.48	$C_{14}H_{30}N_4O_2$	286.4136	287.24417	287.24408	287.2442, 171.1490, 100.0764
35	1.36	$C_{13}H_{20}N_2O_5S$	316.3733	317.11489	317.11499	176.8766, 114.0939
36	3.35	$C_{18}H_{33}N_3O_3$	339.4729	340.25962	340.25933	340.2596, 322.2500, 209.1650, 114.0922
37	0.49	$C_{17}H_{32}N_4O_4$	356.46038	357.24992	357.24985	357.0121, 298.1776, 242.1508, 229.1556
38	8.37	$C_{18}H_{30}N_4O_2S$	366.5214	367.21932	367.21933	367.2193, 280.1109, 262.1007, 245.1305
39	0.74	$C_{17}H_{16}N_2O_4S_2$	376.4499	377.0635	377.06344	377.0635, 208.0316
40	10.86	$C_{16}H_{26}N_2O_6S_2$	406.5174	407.12941	407.13006	407.1294, 176.0596, 158.0441, 140.0336
41	11.05	$C_{24}H_{44}N_4O_4$	452.6306	453.34371	453.34378	453.3437, 322.2493, 228.1595, 209.1650
42	10.95	$C_{20}H_{34}N_6O_6$	454.52056	455.26312	455.26285	455.2631, 371.2428, 328.1299
43	2.28	$C_{23}H_{44}N_6O_5$	484.6327	485.34504	485.34489	357.2510, 142.0861, 124.0759, 84.0820
44	11.34	$C_{30}H_{55}N_5O_5$	565.7882	566.42725	566.42732	566.0121, 453.3455, 340.2613, 322.2505
45	10.87	$C_{26}H_{28}N_6O_7S$	568.6015	569.18166	569.18302	569.1817, 407.1460, 252.0857, 158.0444
46	13.52	$C_{30}H_{46}N_6O_8$	618.7216	619.34533	619.34492	619.34492, 488.2489, 389.1813, 276.1336, 261.1228

Notes: (Name of compound): 1: 1-Methyl-3-pyrrolidinone; 2: 2,2-Dimethyl-1,3-propanediamine; 3: 1,3,5,7-Cyclooctatetraene; 4: Creatinine; 5: Indoline; 6: dihydrobenzofuran; 7: Phenylethanamine; 8: 3H-[1,2,3] Triazolo [4,5-b]pyridin-3-ol; 9: 1-(3,4-Dihydro-2H-1,4-thiazin-5-yl)ethanone; 10: 8-Methylquinoline; 11: 3-(1-Piperazinyl)-1-propanol; 12: 1,8-Diaminooctane; 13: N,N-Dimethyl(2-oxo-1-pyrrolidinyl)methanaminium; 14: Octanehydrazide; 15: 3-Amino-phenylpropionic acid; 16: 3-(4-Methyl-1-piperazinyl)-3-oxopropanenitrile;17: 1-(3-Methoxypropyl)-4-piperidinamine; 18: 2-(Diethylamino)-6-methyl-4(1H)-pyrimidinone; 19: 4-Methoxy-N-methyl-2-nitroaniline; 20: Pentanoicacid, 2-isothiocyanato-4-methyl-, methyl-ester(S)-; 21: 2-[4-(3-Aminopropyl)-1-piperazinyl]ethanol; 22: Propamocarb; 23: 1-BOC-3-aminopiperidine; 24: Bisthieno [3,4-b: 3',4'-d]pyridine 4-oxide; 25: 2-Methyl-2-propanyl (4-piperidinylmethyl)carbamate; 26: 4-Acetyl-4-methylheptanedioic acid; 27: NN-Dimethyllauramide; 28: N-Boc-DL-2-piperidinecarboxamide; 29: 2-Methyl-2-propanyl (2S)-2-isopropyl-1-piperazinecarboxylate; 30: 1-Butyl-3-[2-(4-morpholinyl)ethyl]urea; 31: Amacel Orange GR; 32: 2-(4-ethoxyphenyl)-2-morpholin-4-ylethanamine; 33: 2-Aminohexadecanoic acid; 34: 1,1'-(1,6-Hexanediyl)bis(3-isopropylurea); 35: Ethyl1-[(3,5-dimethyl-4-isoxazolyl) sulfonyl]-4-piperidinecarboxylate; 36: 2-Methyl-2-propanyl 3-{[cyclopropyl(L-valyl)amino]methyl}-1-pyrrolidinecarboxylate; 37: L-Leucyl-L-lysyl-L-proline; 38: 8-(sec-Butylsulfanyl)-3-methyl-7-octyl-3,7-dihydro-1H-purine-2,6-dione; 39: 6-[(5Z)-4-Oxo-5-(2-oxo-1,2-dihydro-3H-indol-3-ylidene)-2-thioxo-1,3-thiazolidin-3-yl]hexanoic acid; 40: N,N'-Bis(1,1-dioxidotetrahydro-3-thiophenyl)-1,4-cyclohexanedicarboxamide; 41: Cyclo(L-isoleucyl-L-leucyl-L-isoleucyl-L-leucyl); 42: (6S,9S,12S,15S)-6-(4-Aminobutyl)-9-(hydroxymethyl)-12-isopropyl-1,4,7,10,13-pentaazabicyclo[13.2.0]heptadecane-2,5,8,11,14-pentone; 43: ile-pro-lys-lys; 44: 3-Cyclohexyl-N-(ethoxycarbonyl)-L-alanyl-N-[(4S,5E,7R)-7-carbamoyl-9-methyl-5-decen-4-yl]-L-lysinamide; 45: 2'-Deoxy-N-[4-(methylsulfonyl)benzyl]-2'-[(phenoxyacetyl)amino]adenosine; 46: (2S)-5-[(Diaminomethylene)amino]-2-[({(2S,3aS,6R,7aS)-6-hydroxy-1-[(2R)-2-{[(2S)-2-hydroxy-3-(4-hydroxyphenyl)propanoyl]amino}-4-methylpentanoyl]octahydro-1H-indol-2-yl}carbonyl)amino]pentanoic acid.

Table 1: Dates and the main fragment ions from HPLC-Q-TOF-MS.

No.	Derivation pathways of the main fragment ions	Classification
1	$[M+H-H_2O]^+82.0670$, $[M+H-H_2O-CN]^+56.0517$	Heterocyclic nitrogenous
2	$[M+H]^+103.1233$, $[M+H-NH_3]^+86.0964$	Amines
3	$[M+H]^+105.0708$, $[M+H- C_2H_2]^+79.0556$	Non-nitrogenous compounds
4	$[M+H-CO]^+86.0729$	Heterocyclic nitrogenous
5	$[M+H]^+120.0815$, $[M+H-HCN]^+93.0716$	Heterocyclic nitrogenous
6	$[M+H]^+121.0650$, $[M+H-C_2H_4]^+93.0709$, $[M+H-C_2H_4O]^+77.0407$	Non-nitrogenous compounds
7	$[M+H]^+122.0965$, $[M+H-NH_3]^+105.0710$	Amines
8	$[M+H]\ ^+137.0460$, $[M+H-H_2O]^+119.0354$, $[M+H-HN_3]^+94.0407$	Heterocyclic nitrogenous
9	$[M+H]^+144.0478$, $[M+H-H_2O]^+126.0371$, $[M+H-OCH_3]^+113.0300$	Heterocyclic nitrogenous
10	$[M+H]^+144.0807$, $[M+H-NH_2]^+128.0497$, $[M+H-C_2H_3N]^+103.0546$	Heterocyclic nitrogenous
11	$[M+H]^+145.1334$, $[M+H-OH]^+128.1080$, $[M+H-C_3H_6OH]^+86.0978$	Heterocyclic nitrogenous
12	$[M+H]^+145.1695$, $[M+H-C_3H_9N]^+86.979$, $[M+H-C_3H_9N-NH_3]^+69.0719$, $[M+H-C_5H_{11}N]^+60.0827$	Amines
13	$[M+H]^+146.1182$, $[M+H-C_3H_9N]^+87.0455$, $[M+H-C_4H_{10}N_2]^+\ 60.0834$	Amines
14	$[M+H]^+159.1488$, $[M+H-NH_3]^+142.1227$, $[M+H-C_2H_5N_2O]^+86.0615$	Amines
15	$[M+H]^+166.0860$, $[M+H-HCOOH]^+120.0813$, $[M+H-HCOOH-NH_3]^+103.0547$	Amino acids
16	$[M+H]^+168.1127$, $[M+H-H_2O]^+150.1028$, $[M+H-C_3H_6N]^+112.0872$	Nitriles compound
17	$[M+H]^+173.1647$, $[M+H-NH_3]^+156.1377$, $[M+H-C_4H_9O]^+100.0751$	Amines
18	$[M+H]^+182.1286$, $[M+H-H_2O]^+164.1193$, $[M+H-C_3H_9N_3]^+95.0607$	Amines
19	$[M+H-CNH_2]^+155.0465$, $[M+H-CNH_2-CO]^+127.0149$, $[M+H-CNH_2-CO-NO]^+97.9687$	Nitro compounds
20	$[M+H]^+188.0737$, $[M+H-C_4H_9]^+131.1180$, $[M+H-C_5H_{11}]^+117.1027$	Amines
21	$[M+H]^+188.1752$, $[M+H-H_2O-NH_3]^+152.0522$, $[M+H-C_2H_3O-CH_2N]^+117.1014$	Amines
22	$[M+H]^+189.1599$, $[M+H-C_2H_6NH]^+144.1384$, $[M+H-C_2H_6NH- CH_2]^+130.0867$	Amide compounds

23	$[M+H]^+$201.1595, $[M+H-H_2O]^+$183.1488, $[M+H-C_2H_4N]^+$159.1495	Heterocyclic nitrogenous
24	$[M+H]^+$207.9882, $[M+H-H_2O]^+$189.9757, $[M+H-CNO]^+$165.9665	Heterocyclic nitrogenous
25	$[M+H]^+$215.1749, $[M+H-C_4H_8]^+$159.1490, $[M+H-C_4H_8-CH_4N]^+$129.9939	Amide compounds
26	$[M+H]^+$217.1069, $[M+H-C_2H_4O]^+$173.0812, $[M+H-C_2H_5O-OH]^+$156.0709, $[M-2HCOOH-CH_3]^+$111.0450, $[M+H-2COOH-C_2H_4O]^+$83.0501	Non-nitrogenous compounds
27	$[M+H]^+$228.2321, $[M+H-C_{10}H_{20}]^+$88.0757, $[M+H-C_{10}H_{20}-H_2O]^+$70.659	Amide compounds
28	$[M+H]^+$229.7556, $[M+H-CONH_2-C_3H_6COH]^+$114.0553, $[M+H-CONH_2-C_3H_6COH-H_2O]^+$96.0823	Amide compounds
29	$[M+H-C_4H_8]^+$173.1662, $[M+H-C_4H_9O]^+$156.1383, $[M+H-C_4H_9O-CH_2]^+$142.1231, $[M+H-C_4H_9O-C_2H_3]^+$129.9960, $[M+H-C_2H_3NO_2C_4H_8]^+$100.762	Amide compounds
30	$[M+H]^+$230.1863, $[M+H-OH]^+$213.1581, $[M+H-OH-C_5H_{11}N_2]^+$114.0919, $[M+H-OH-C_5H_{11}N_2-CH_2]^+$100.0760	Amide compounds
31	$[M+H]^+$243.0877, $[H-NH_3]^+$226.0607, $[M+H-C_2H_6N]^+$200.0820, $[M+H-C_2H_2NO_2]^+$172.0870, $[M+H-C_3H_5NO_2]^+$157.0642	Nitro compounds
32	$[M+H]^+$251.1751, $[M+H-C_4H_9NO-NH_3]^+$147.8527, $[M+H-C_4H_9NO-C_2H_2]^+$138.0893, $[M+H-C_4H_9NO-C_2H_2-NH_3]^+$121.0654	Amines
33	$[M+H]^+$272.1556, $[M+H-NH_3]^+$255.2345, $[M+H-NH_3-H2O]^+$237.2162, $[M+H-COOH-CH_2]^+$213.0615	Amino acids
34	$[M+H]^+$287.2442, $[M+H-C_5H_{12}N_2O]^+$171.1490, $[M+H-C_{10}H_{23}N_2O]^+$100.0764	Amide compounds
35	$[M+H-C_8H_{13}O_2]^+$176.8766, $[M+H-C_7H_{11}N_2O_3S]^+$114.0939	Heterocyclic nitrogenous
36	$[M+H]^+$340.2596, $[M+H-H_2O]^+$322.2500, $[M+H-C_2H_5NO_2C_4H_8]^+$209.1650, $[M+H-C_6H_{11}NO_2C_4H_8-C_3H_5]^+$114.0922	Amide compounds
37	$[M+H]^+$357.0121, $[M+H-C_3H_9N]^+$298.1776, $[M+H-C_3H_9N-C_4H_8]^+$242.1508, $[M+H-C_3H_9N-C_5H_9]^+$229.1556	Amino acids
38	$[M+H]^+$367.2193, $[M+H-C_6H_{15}]^+$280.1109, $[M+H-C_6H_{15}-H_2O]^+$262.1007, $[M+H-C_6H_{14}-2H_2O]^+$245.1305	Heterocyclic nitrogenous
39	$[M+H]^+$377.0635, $[M+H-C_5H_{11}NCOOH-C_3H_3]^+$208.0316	Amino acids
40	$[M+H]^+$407.1294, $[M+H-C_6H_{10}NOC_4H_7O_2S]^+$176.0596, $[M+H-C_6H_{10}NOC_4H_7O_2S-H_2O]^+$158.0441, $[M+H-C_6H_{10}NOC_4H_7O_2S-2H_2O]^+$140.0336, $[M+H-2C_4H_6O_2S-C_2H_2NO]^+$115.0392	Amide compounds
41	$[M+H]^+$453.3437, $[M+H-2C_4H_9-OH]^+$322.2493, $[M+H-3C_4H_9-3H_2O]^+$228.1595, $[M+H-3C_4H_9-4H_2O]^+$209.1650	Amide compounds
42	$[M+H]^+$455.2631 , $[M+H-C_5H_{10}N]^+$371.2428, $[M+H-C_5H_{10}N-CHNO]^+$328.1299	Amide compounds
43	$[M+H-C_6H_{10}NO_2]^+$357.2510, $[M+H-C_{16}H_{31}N_4O_4]^+$142.0861, $[M+H-C_{16}H_{31}N_4O_4-H_2O]^+$124.0759, $[M+H-C_{16}H_{31}N_4O_4-H_20-C_2H_2N]^+$84.0820,	Amino acids
44	$[M+H]^+$566.0121, $[M+H- C_6H_{11}NO]^+$453.3455, $[M+H- C_6H_{11}NO-H_2O-C_5H_8-C_2H_5]^+$ 340.2613, $[M+H- C_6H_{11}NO-H_2O-C_5H_8-C_2H_5]^+$ 322.2505	Amide compounds
45	$[M+H]^+$569.1817, $[M-C_9H_8NO_2]^+$407.1460, $[M-C_9H_8NO_2-C_7H_7O_2S]^+$252.0857, $[M-C_{12}H_{15}NO_5-C_8H_7O_2S]^+$158.0444	Heterocyclic nitrogenous
46	$[M+H]^+$619.34492, $[M+H-COOH-C_3H_8N_3]^+$488.2489, $[M+H-C_9H_{18}N_4O_3]^+$389.1813, $[M+H-C_{15}H_{29}N_5O_4]^+$276.1336, $[M+H-C_{15}H_{29}N_5O_4-CH_3]^+$261.1228	Amino acids
Notes: The name of compound is the same as Table 1.		

Table 2: The N compounds mainly fractured fragment ions derivation and classification.

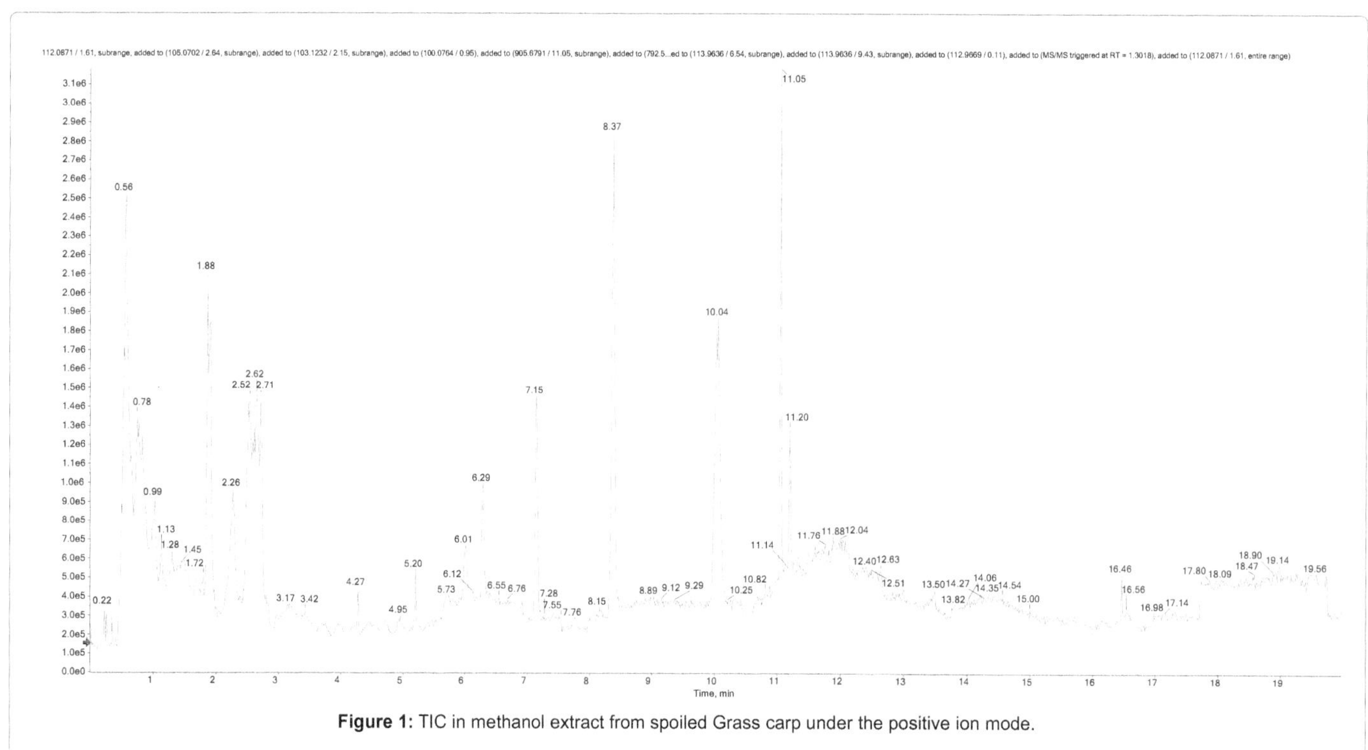

Figure 1: TIC in methanol extract from spoiled Grass carp under the positive ion mode.

took off NH_2 and COOH, redundant H reset to the other atoms), and to produce fragment ions at m/z 298.1756; the second one was to take off the peptide chain and lost $C_6H_{13}NO$ (115 u), then to create fragment ions at m/z 242.1501. The fragment ions continue to fractured, and lose $C_5H_{12}N_2$ (100 u), H_2O (18 u) and CO (28 u), respectively to form ions at m/z 142.0863, 124.0757 and 96.0814 m/z. Its possible fragment pathways were shown in Figure 3.

Amines: Amines are formally derivatives of ammonia, wherein one or more hydrogen atoms have been replaced by a substituent such as an alkyl or aryl group. Molecular ion of Amines is formed by loss of the electrons of N atom. β-fragment is the most fragment pathway in collision induced dissociation. R group (substituent group) of fat amine was lost easily, $CH_2 = CH_2$ (28u) group of ring amine was lost easily, too, and HCN (27u) of aromatic amine was lost easily [25]. 10 amine compounds were analyzed and identified in this experiment. Using 1,8-Diaminooctane as an example, it was produced the proton molecular ion peak $[1,8\text{-Diaminooctane} + H]^+$ (m/z145.1695) by electrospray ionization. The MS^2 spectrum shown in Figuer 4. In MS/MS spectrum, under collision induced dissociation, the $[1,8\text{-Diaminooctane} + H]^+$ lost C3H9N (59 u) firstly to produce ions at m/z 86.09, then the fragment ions fractured to lose NH_3 (17u) so as produced ions at m/z 69.07. Or $C_5H_{11}N$ (85 u) was lost to produce fragment at m/z 60.08 ions which is stable and highly abundant. Its possible fragment pathways were shown in Figures 4 and 5.

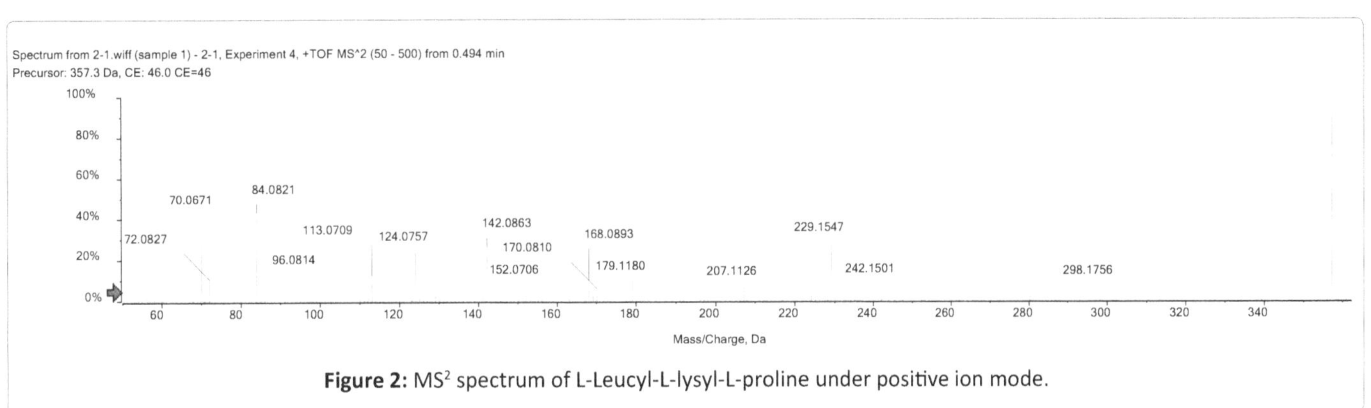

Figure 2: MS^2 spectrum of L-Leucyl-L-lysyl-L-proline under positive ion mode.

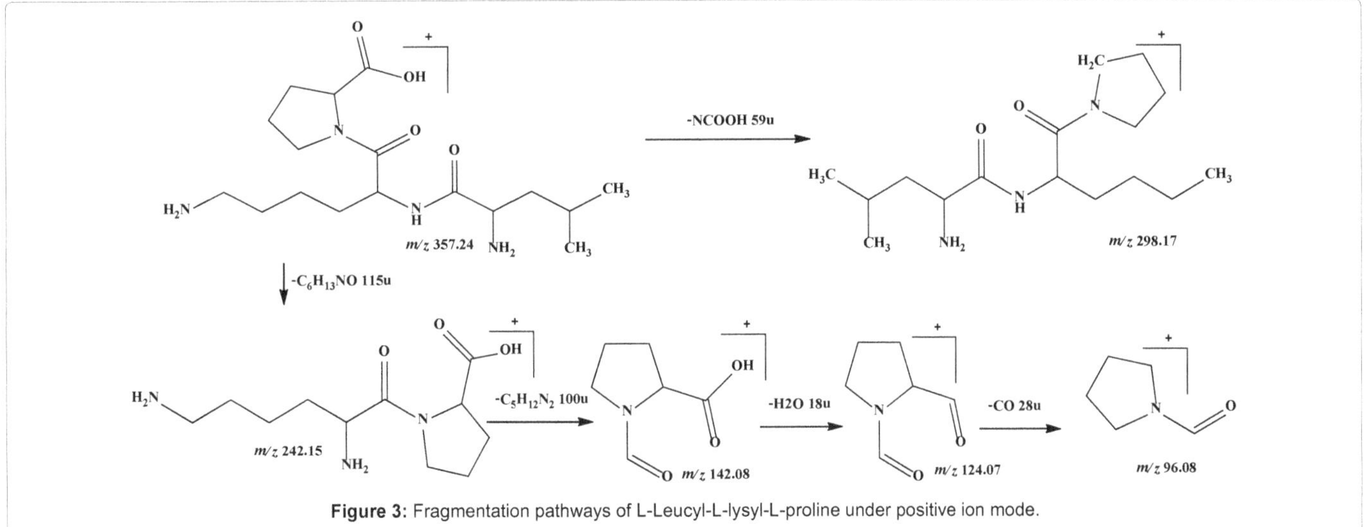

Figure 3: Fragmentation pathways of L-Leucyl-L-lysyl-L-proline under positive ion mode.

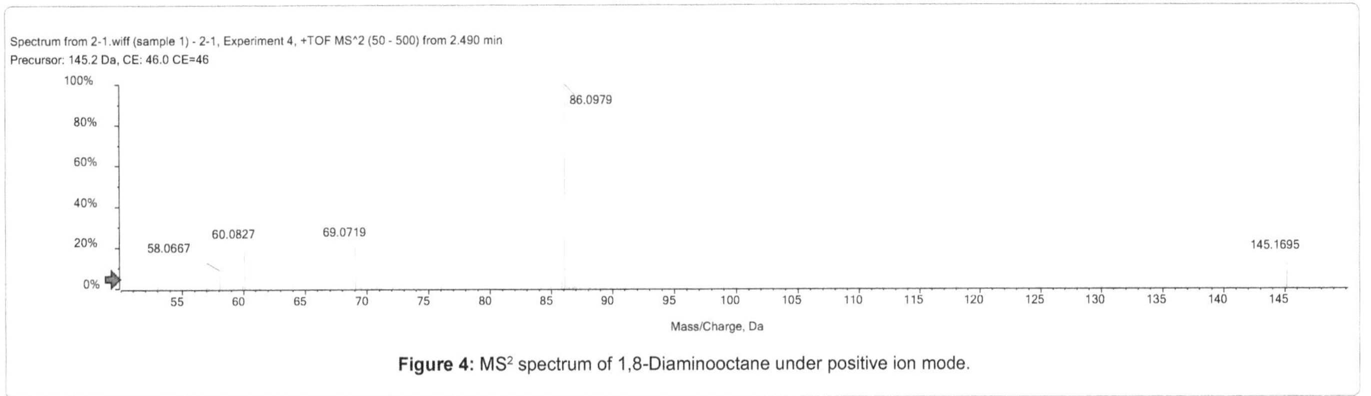

Figure 4: MS^2 spectrum of 1,8-Diaminooctane under positive ion mode.

Amide: Amide is a special form of amines; the simplest amides are derivatives of ammonia wherein one hydrogen atom has been replaced by an acyl group. Amide can also be considered as compounds that hydroxyl group of carboxylic acid molecules are replaced by amino or phenyl group. Its fragment pathway was similar to carboxylic acid, Mclafferty rearrangement is the important pathway among them [26]. 12 amide compounds were analyzed and indentified in this experiment. For example, Propamocarb produced the proton molecular ion peak [Propamocarb+H]$^+$ (m/z 189.1599) by electrospray ionization. The MS^2 spectrum is shown in Figure 6.

In MS/MS spectrum, under collision induced dissociation, the [Propamocarb+H]$^+$ ions lost C_2H_6NH (45u) firstly to produce ions at m/z144.1384, then the fragment ions continue to fractured to lose CH_2 (14u) and CH_6O (56u)and formed ions at m/z130.0867 and m/z84.0819. Its possible fragment pathways were shown in Figure 7.

Nitro compounds: Nitro compounds are organic compounds that contain one or more nitro functional groups ($-NO_2$), mainly including aliphatic and aromatic nitro compounds. It has a little or no proton molecular ion peak. The ion peaks of M-OH, M-CO and M-NO were easy produced in MS/MS spectrum by rearrangement of r-hydrogen atoms [27]. 2 nitro compounds were identified in this experiment. For example, the obvious proton molecular ion peak was not produced by 4-Methoxy-N-methyl-2-nitroaniline of electrospray ionization. The MS^2 spectrum is shown in Figure 8.

In MS/MS spectrum, under collision induced dissociation the [4-Methoxy-N-methyl-2-nitroaniline +H]$^+$ ions lost C_2H_4 (28u) firstly to produce ions at m/z155.05, then the unstable ions at m/z127.0157 was formed by CO (28u) lost and ions at m/z 98.9851 was formed by C_2H_4 (28u) lost. Its possible fragment pathways were shown in Figure 9.

Other compounds containing N: 13 other nitrogen compounds which include 12 N-heterocycles and 1 nitriles were identified in this experiment. Although the fragment pathways of other nitrogen compounds are complicated, there is a general rule in the secondary mass spectrometry, that [M+H]$^+$ was broken into different daughter ions, by the loss of neutral molecules such as H_2O, $2H_2O$, $3H_2O$, CH3OH, $2CH_3OH$, $3CH_3OH$, CH_2O, $2CH_2O$, $3CH_2O$, CO, CO_2, $CHNH_2$, $CHCH_2$, NO, NH_3 would be lost [28]. The details are shown in Table 2.

Non-nitrogenous compounds: 3 Non-nitrogenous compounds were identified in this experiment. Non-nitrogenous compounds were not identified in fresh fish and in the prophase of fish preservation processing, non-nitrogenous compounds which were not identified were produced of loss of N compounds in the process of preservation (Specific data published article). Thus, the quantity of non-nitrogenous compounds (type and quantity) can be used as an index of fresh fish.

For example, 4-Acetyl-4-methylheptanedioic acid proton molecular ion peak [4-Acetyl-4-methylheptanedioic acid +H]$^+$ (m/z 217.1071) was produced by electrospray ionization. The MS^2 spectrum is shown in Figure 10.

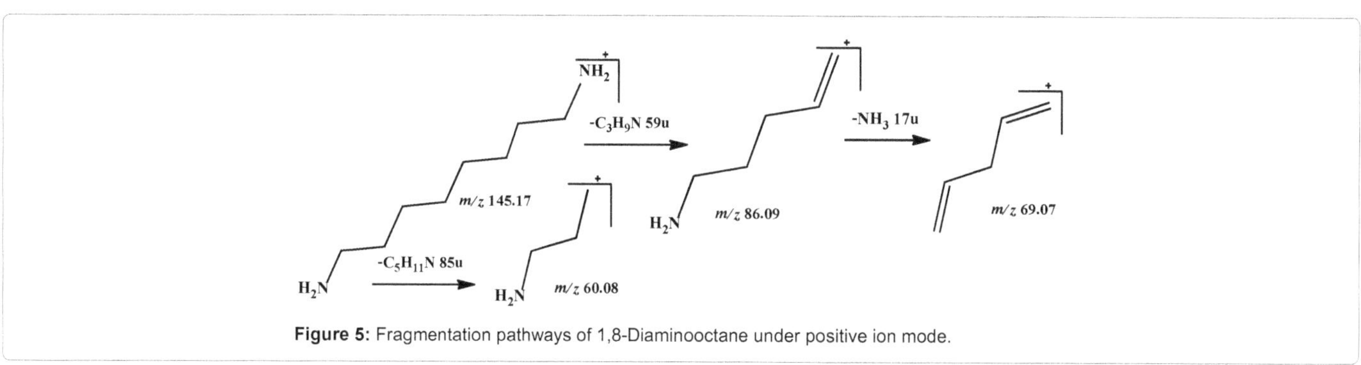

Figure 5: Fragmentation pathways of 1,8-Diaminooctane under positive ion mode.

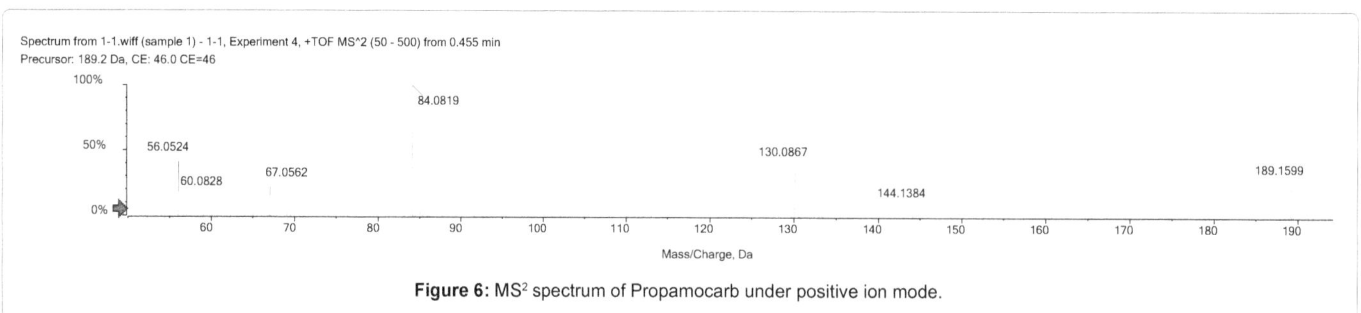

Figure 6: MS^2 spectrum of Propamocarb under positive ion mode.

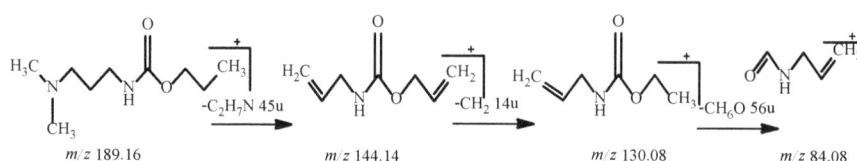

Figure 7: Fragmentation pathways of Propamocarb under positive ion mode.

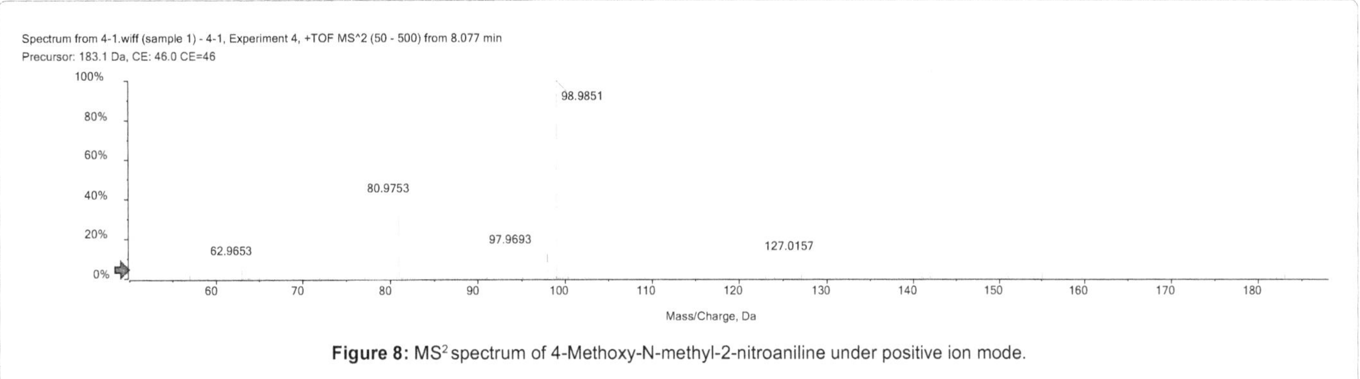

Figure 8: MS^2 spectrum of 4-Methoxy-N-methyl-2-nitroaniline under positive ion mode.

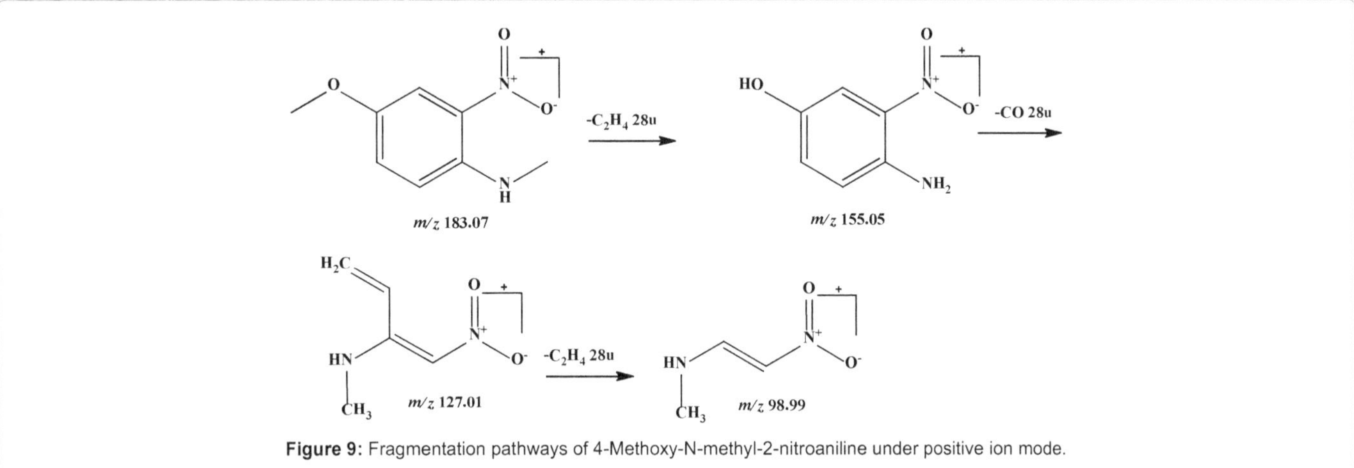

Figure 9: Fragmentation pathways of 4-Methoxy-N-methyl-2-nitroaniline under positive ion mode.

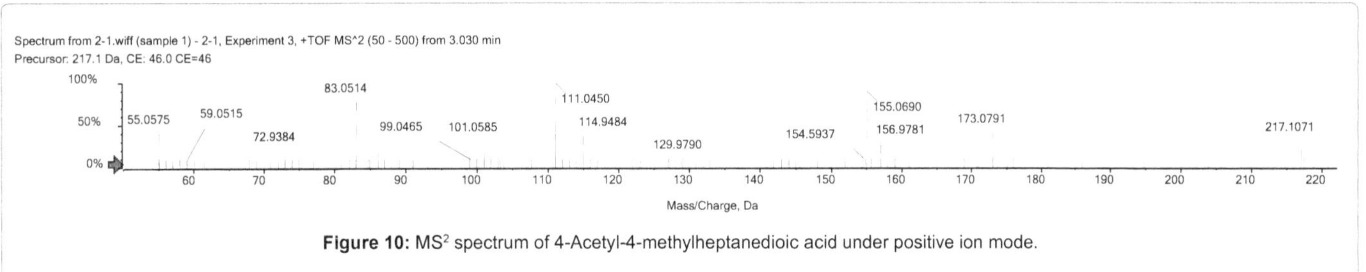

Figure 10: MS^2 spectrum of 4-Acetyl-4-methylheptanedioic acid under positive ion mode.

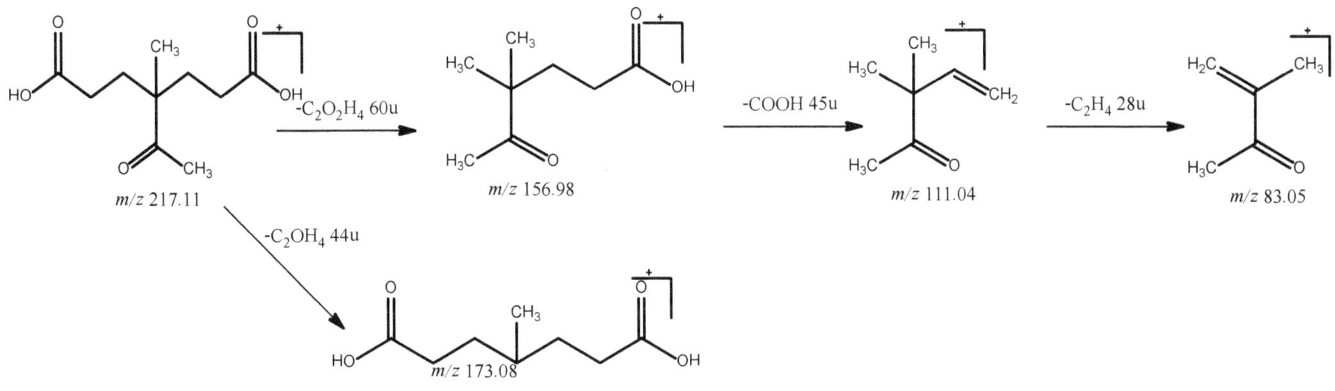

Figure 11: Fragmentation pathways of 4-Acetyl-4-methylheptanedioic acid under positive ion mode.

In MS/MS spectrum, the [4-Acetyl-4-methylheptanedioic acid $+H]^+$ ions first lost CH_3COOH (60u) to produce ions at m/z156.9781by collision induced dissociation, then the fragment ions lost COOH (45u) to produce ions at m/z 111.0450 and C_2H_4 (28 u) continued to fractured to produce ions at m/z 83.0514. Or it may lose C_2OH_4 (44u) to produce ions at m/z 173.0791. Its possible fragment pathways were shown in Figure 11.

Conclusion

46 kinds of monomeric compounds which has 3 kinds of non-nitrogenous compounds, 43 nitrogenous compounds in bad grass carp were determined by HPLC-Q-TOF-MS, 43 nitrogenous compounds includes 6 kinds of amino acids (2 kinds of α-amino acids), 10 kinds of amines, 12 kinds of amide compounds, 2 kinds of nitro compounds, 12 kinds of heterocyclic nitrogenous compounds and 1 kinds of nitriles compound.

Compared with the low-resolution MS methods such as quadrupole, triple quadrupole and ion trap mass spectrometry, HPLC-Q-TOF-MS has relative high resolution and the extraction functions of ion characteristics, it can measure mass of both parent ion and fragment ions accurately. And it can provide selectivity because it has ability to discriminate peak which has same nominal masses but different exact masses between interference and mass peaks having similar [29,30]. Consequently, the role of Q-TOF MS/MS instruments has high efficiency on identifying non-target compounds in complex matrices when the reference compounds were unavailable [29,30].

HPLC-Q-TOF-MS has less complex pretreatment such as excessive purification, derivatization and solid phase extraction to sample, it has lower requirement on Chromatographic separation and simplifies the research process, increases the efficiency of analysis [30]. For fresh materials like Grass carp, it is difficult to obtain monomer compounds of high purity by traditional extraction, separation and purification and is easy to decompose and polymerize the compounds in traditional one. Therefore, HPLC-Q-TOF-MS used have important significance for studying monomeric compounds.

Acknowledgement

This work is part of the research projects of Guangdong Science and technology (2015B020206001) and National High Technology Research and Development Program of China (2013AA102208). In addition, thanks to Huang haoshen (a student of Guangdong Pharmaceutical University in the class of 2012) for performing TOF-MS in the study.

References

1. Yang XX, Zhang MH, Xie JG (2012) Composing of fatty acid of *Ctenopharyngodon idellus* by GC-MS. Guangzhou Chem Industry 11:135-137.
2. Song X, Li SF, Wang CH, Xu JW, Yang QL (2009) Grass carp *(Ctenopharyngodon idellus)* genetic structure analysis among native populations in china and introduced populations in USA, Europe and Japan based on mitochondrial sequence. Acta Hydrobiol Sin 33: 709-716.
3. Stephan M, Hobsdawn P (2013) Australian fisheries and aquaculture statistics. Fisheries Research and 2014.
4. Cheng JH, Sun DW, Qu JH, Pu HB, Zhang XC, et al. (2016) Developing a multispectral imaging for simultaneous prediction of freshness indicators during chemical spoilage of grass carp fish fillet. Journal of Food Engineering 182: 9-17.
5. Alishahi A, Aider M (2012) Applications of chitosan in the seafood industry and aquaculture: a review. Food Bioprocess Technol 3:817-830.
6. Ghaly AE, Dave D, Budge S, Brooks M (2010) Fish spoilage mechanisms and preservation techniques: review. Am. J. Appl. 7: 859-877.
7. Qin N, Li DP, Hong H (2016) Effects of different stunning methods on the flesh quality of grass carp(Ctenopharyngodon idellus) fillets stored at 4°C. Food Chem 201: 131-138.
8. Ryder JM (2002) Determination of adenosine triphosphate and its breakdown products in fish muscle by high-performance liquid chromatography. J Agric Food Chem 33:678-680.
9. Özogul F, Gökbulut C, ÖzogulY, Özyurt G (2006) Biogenic amine production and nucleotide ratios in gutted wild sea bass *(Dicentrarchus labrax)* stored in ice, wrapped in aluminium foil and wrapped in cling film at 4°C. Food Chem 98:76-84.
10. ÖzogulY, Özogul F (2004) Effects of slaughtering methods on sensory, chemical and microbiological quality of rainbow trout *(Onchorynchus mykiss)* stored in ice and MAP. Eur Food Res Techno 219:211-216.
11. Song Y, Liu L, Shen H, You J, Luo Y (2011) Effect of sodium alginate-based edible coating containing different anti-oxidants on quality and shelf life of refrigerated bream *(Megalobrama amblycephala)*. Food Control 22: 608-615.
12. Hong H, Luo Y, Zhou Z, Shen H (2012) Effects of low concentration of salt and sucrose on the quality of bighead carp *(Aristichthys nobilis)* fillets stored at 4°C. Food Chem 133: 102-107.
13. Wang H, Luo Y, Huang H, Xu Q (2014) Microbial succession of grass carp *(Ctenopharyngodon idellus)* filets during storage at 4°C and its contribution to biogenic amines' formation. Int J Food Microbiol 190: 66-71.
14. Wang H, Luo Y, Yin X, Wu H, Bao Y, et al. (2014) Effects of salt concentration on biogenic amine formation and quality changes in grass carp *(Ctenopharyngodon idellus)* fillets stored at 4°C and 20°C. J Food Protect 77: 796-804.
15. Sele V, Sloth JJ, Holmelid B, Valdersnes S, Skov K, et al. (2014) Arsenic-containing fatty acids and hydrocarbons in marine oils - determination using reversed-phase HPLC-ICP-MS and HPLC-qTOF-MS. Talanta 121: 89-96.
16. Guo P, Qi YJ, Zhu CH, Wang Q (2015) Purification and identification of antioxidant peptides from Chinese cherry *(Prunus pseudocerasus Lindl.)* seeds. J Funct Foods 19: 394-403.
17. Zhang MX, Wang XC, Liu Y, Xu XL, Zhou GH (2012) Isolation and identification of flavour peptides from Puffer fish *(Takifugu obscurus)* muscle using an electronic tongue and MALDI-TOF/TOF MS/MS. Food Chem 135: 1463-1470.
18. Cheng XL, Wei F, Xiao XY, Zhao YY, Shi Y, et al. (2012) Identification of five gelatins by ultra performance liquid chromatography/time-of-flight mass spectrometry (UPLC/Q-TOF-MS) using principal component analysis. J Pharm Biomed Anal 62: 191-195.
19. Zhang XP, Jiang KZ, Lv HQ, Nie J, Li ZG (2015) Identification and characterization of major chemical compounds in the ethyl acetate extract from ficus pandurata hance aerial roots by HPLC-Q-TOF MS. Journal of Chinese Mass Spectrometry Society 4:310-320.
20. Yan Y, Chai CZ, Wang DW, Xin YY, Dan NZ, et al. (2013) HPLC-DAD-Q-TOF-MS / MS analysis and HPLC quantitation of chemical constituents in traditional Chinese medicinal formula Ge-Gen Decoction. J Pharm Biomed Anal 80: 192.
21. Zhou JL, Qi LW, Li P (2009) Herbal medicine analysis by liquid chromatography/time-of-flight mass spectrometry. J Chromatogr A 1216: 7582-7594.
22. Qu CL, Zhang HQ, Zhang HR, Bai YP, Wen H (2008) Studies on fragmentation pathways of amino acids and their interactions with ginsenoside Rb3 by spectrospray ionization mass spectrometry. Chem J Chinese U 9:1721-1726.
23. Huang YF, Hu J (2010) Simultaneous analysis of twenty free amino acids in tobacco using liquid chromatography-electrospray ionization/ion traps tandem mass spectrometry. Chin J Chromatogr 6: 615-622.
24. Wang Y, Li SM, He MW (2014) Fragmentation characteristics and utility of ammonium ions for peptide identification by MALDI TOF/TOF spectrometry. Chinese J Anal Chem 7: 1010-1016.
25. Daniel D, Dos Santos VB, Vidal DT, Do Lago CL (2015) Determination of biogenic amines in beer and wine by capillary electrophoresis-tandem mass spectrometry. J Chromatogr A 1416: 121-128.
26. Wu YL, Chen RX, Zhu Y, Zhao J, Yang T (2015) Simultaneous determination of sixteen amide fungicides in vegetables and fruits by dispersive solid phase extraction and liquid chromatography - tandem mass spectrometry. J Chromatogr B 989: 11-20.
27. Zoran K, Irena G, Reinhilde V, Magda C, Willy M (2012) Liquid chromatography tandem mass spectrometry method for characterization of mono-aromatic nitro-compounds in atmospheric particulate matter. J ChromatogrA 1268: 35-43.
28. Sun Y, Li H, Hu J, Li J, Fan YW, et al. (2013) Qualitative and quantitative

analysis of phenolics in *Tetrastigma hemsleyanum* and their antioxidant and anti-proliferative activities. J Agric Food Chem 61: 10507-10515.

29. Fu Y, Gao W, Yu JJ, Chen J, Li HJ, et al. (2012) Characterization and identification of baccharane glycosides in Impatientis Semen by rapid-resolution liquid chromatography with electrospray ionization quadrupole time-of-flight tandem mass spectrometry. J Pharm Biomed Anal 65: 64-71.

30. Chen XF, Wu HT, Tan GG, Zhu ZY, Chai YF (2011) Liquid chromatography coupled with time-of-flight and ion trap mass spectrometry for qualitative analysis of herbal medicines. J Pharmaceut Ana 4:235-245.

Translocation of *Klebsiella* from Enteral Diets: Study through Mouse Model System and Molecular Technique

Pereira SCL[1]* and Maria Cristina DV[2]

[1]*Federal University of Minas Gerais, School of Nursing, Department of Nutrition, Belo Horizonte-Minas Gerais, Brazil*

[2]*Federal University of Viçosa, Department of Microbiology, Viçosa-Minas Gerais, Brazil*

Abstract

Objective: This study was conducted to evaluate the translocation of *Klebsiella* strains to organs of normal and immunocompromised mice.

Methods: An enteral diet with 6.0 × 10^9 CFU/mL of *Klebsiella* was provided to immunodepressed and healthy animals. The presence of *Klebsiella* in animal organs was evaluated on MacConkey-inositol-carbenicillin agar and molecular typing was assessed by means of random amplified polymorphic DNA (RAPD).

Results: No typical *Klebsiella* colonies were recovered from liver, spleen, heart, kidney or lung samples when mice were supplied with the uncontaminated enteral formula. However, typical colonies were recovered from liver and lung samples from immunocompromised animals, regardless of whether they had received contaminated diet. Translocation was also detected in non-immunocompromised mice fed with contaminated diet. There were higher counts of typical *Klebsiella* colonies in samples collected from the guts of animals that received prednisone and/or carbenicillin. However, the highest colony count was obtained when both drugs were administered together with the enteral formula contaminated with *Klebsiella*. Translocation was confirmed through similarity to the DNA band patterns of orally administered *Klebsiella* strains.

Conclusion: *K. pneumoniae* was translocated into the lungs and liver of test mice fed with a *Klebsiella* pool. The presence of *Klebsiella* with a different DNA profile in liver samples from mice that only received medication suggests that autochthonous intestinal microbial strains can also become translocated when the immune system is depressed or when selective decontamination is promoted through the use of corticoids and antibiotics.

Keywords: Bacterial translocation; *Klebsiella*; Enteral diets; RAPD

Introduction

Klebsiella is an opportunistic pathogen that causes infections in newborns, elderly people and individuals suffering from chronic cardiac and neoplastic diseases, alcoholism, diabetes or chronic obstructive lung disease [1-3]. These infections are frequently associated with hospital environments and can result in significant mortality, especially due to exposure to several sources of contamination, or intensive therapy with immunosuppressive drugs such as corticoids and antibiotics [4].

Klebsiella pneumoniae is a capsulated Gram-negative bacterium found in the normal flora of the mouth, skin and intestine, and is also the third most commonly isolated microorganism in blood cultures from sepsis patients. Because of emerging antibiotic resistance, *K. pneumoniae* infection is becoming a major health threat [1].

Klebsiella is a common contaminant of enteral formulas [5,6], but the contribution of these diets to the epidemiology of hospital-acquired infections has so far received little attention.

In a noteworthy recent study, rapid and widespread dissemination of an epidemic clone of *K. pneumoniae* was found to cause a large nosocomial outbreak through the food chain. To our knowledge, this was the first report to provide insight on how transmission of multiresistant *Klebsiella* can occur through food, as the vehicle in the hospital setting [7]. To evaluate the pathogenic potential of clinical and environmental isolates of *Klebsiella*, virulence factors like the capsule, fimbriae, serum resistance and enterotoxin production should be taken into consideration [8]. However, bacterial pathogenesis is a complex event and varies as a function of the host-parasite relationship itself. The expression of a virulent phenotype is also influenced by environmental factors and, for this reason, can be better estimated by *in vivo* studies [8]. Bacterial translocation is defined as the passage of viable bacterial cells or their endotoxins from the intestinal tract into the mesenteric lymph nodes or other organs [9]. However, in critically ill patients, with various underlying diseases, this bacterial translocation may lead to infections and consequently to a further reduction in general health status. The mechanism of bacterial translocation has been widely and somewhat controversially investigated *in vitro* and in animal models. In human studies, several diseases have been correlated with bacterial translocation [9,10].

Translocation of orally inoculated bacteria is a proposed methodology for *in vivo* studies on pathogen virulence, especially regarding bacteria transmitted via contaminated food. This procedure uses the microbial ability to penetrate the intestinal epithelium and cause infections in vital organs as a virulence criterion [11].

Additionally, *Klebsiella* has recently been correlated with chronic intestinal diseases. The results from a large number of studies and reviews support the idea that *K. pneumoniae* is the most likely triggering factor involved in the initiation and development of ankylosing spondylitis

***Corresponding author:** Pereira SCL, Federal University of Minas Gerais, School of Nursing, Department of Nutrition, Belo Horizonte-Minas Gerais, Brazil
E-mail: simoneclpereira@gmail.com

and Crohn's disease [12]. Members of the family *Enterobacteriaceae*, particularly the species *Klebsiella oxytoca*, have been found to be more abundant in patients with active celiac disease than in controls [13].

Given this scenario, the present study aimed to evaluate the potential for *Klebsiella* strains isolated from enteral diets to infect vital organs and colonize the gut of experimental mice, using a molecular technique.

Materials and Methods

Bacterial strains

Five *K. pneumoniae* strains and one *K. oxytoca* strain isolated from hospital enteral diets were selected based on RAPD analysis (data submitted for publication). Serotypes, mucoid appearance present by colonies and antibiotypes were also taken into consideration during the selection (Table 1). Cultures were kept frozen at lower than -18°C in trypticasein-soy broth (TSB) and glycerol. They were thawed at room temperature and reactivated twice in TSB prior to use

Cell preparation

Activated cells of *Klebsiella* isolates were harvested by means of centrifugation at 5000 *g* for 15 min. Cells from all six cultures were suspended in 2 mL of enteral formula (Ensure, Abbott) of vanilla flavor, which had previously been prepared following the manufacturer's recommendations. Aliquots of the contaminated diets were diluted and plated onto trypticasein-soy agar in order to estimate the number of viable cells per milliliter of diet. The six cell concentrates were joined in the same flask and the volume was completed to 20 mL. The final concentration of each isolate in this mixture was approximately 10^{10} colony-forming units per milliliter (CFU/mL). All procedures were done under strict sterile conditions.

Animals

Five to six-week-old Swiss albino mice weighing between 22 and 26 g were randomized into groups of 30 animals that were housed in cages kept in temperature-controlled rooms (25°C).

Experimental design

The experiments were conducted under conditions of strict observance of ethical guidelines on animal experimentation and consisted of seven treatments (Groups I to VII) and one repetition. The baseline group (Group I) consisted of six mice that were fed standard chow and water *ad libitum*. The control group (Group II) consisted of 30 mice orally provided with enteral formula (Ensure®, Abbot) throughout the study period. The test groups (Groups III-VI), with 30 animals each, received the enteral diet inoculated with the *Klebsiella* pool. Mice in GIII were solely provided with this diet; those in GIV were also given prednisone (10 mg/kg/day); those in GV, carbenicillin (200 mg/kg/day); and those in GVI, corticoid plus antibiotic. Prednisone and carbenicillin were administered to the mice in GVII in the same way as in GVI. The animals in GVII were considered to be a second control group because they did not receive the contaminated diet. Four days after corticoid and antibiotic had been administered to the mice in GIII to GVI, all the animals received 100 μL of the enteral diet inoculated with the *Klebsiella* pool, by means of oral nutrition. Only carbenicillin (200 mg/kg/day) was continually administered until the end of the study period. One, two, four, six and nine days after oral administration of the bacterial pool, six animals in each group were sacrificed in order to detect the presence of *Klebsiella* in the liver, heart, spleen, kidney, lung and gut.

Translocation of Klebsiella to different organs

Immediately after the animals were killed, their spleen, heart, lung, liver, gut and kidney were collected, weighed and stored in sterile plastic bags (Whirl-Pak, Millipore). On day nine, typical *Klebsiella* colonies obtained from the intestinal tracts of the animals in GII-VII were also counted. The organs were macerated and dilutions were plated onto MacConkey-inositol-carbenicillin agar (Merck) [14]. Cultures were incubated at 37°C for 48 h; Petri dishes containing more than 20 colonies were selected and the number of CFU of *Klebsiella* per gram in each organ was estimated. Three to five typical *Klebsiella* colonies were picked from the same organ of each group on the sampling day, for RAPD analysis.

Genetic analysis on translocated Klebsiella

RAPD was used to identify which orally administered isolates became translocated from the intestinal tract into the sampled organs. For this reason, total DNA from the six *Klebsiella* cultures given to the animals and from the *Klebsiella* isolated from the different organs was extracted [15]. The 25 μl reaction mixture consisted of 10 mM of Tris-HCl (pH 8.0), 50 mM of KCl, 2.0 mM of $MgCl_2$, 0.1 mM of dNTP, 0.4 μM of RAPD primer, 1.0 U of *Taq* DNA polymerase (Promega) and 25 ng of total DNA. The primers OPD03, OPD07, OPD08, OPD12, OPD18, OPD20, OPF10 and OPF13 were purchased from Operon Technologies Inc., Alameda, CA, USA. Thermal cycling was performed in a PTC 100 machine (MJ Research Inc., Waterton, MA, USA), programmed for an initial denaturation step at 94°C for 5 min, followed by 30 cycles of 1 min at 94°C, 1 min at 40°C, and 1.30 min at 72°C. A final extension at 72°C for 7 min was performed. The amplification products were analyzed by means of electrophoresis on 1.5% agarose gels in order to facilitate the electrophoresis analysis. The gels were photographed under UV light using the Eagle-Eye II video system (Stratagene, La Jolla, CA, USA).

Results

No typical *Klebsiella* colonies were observed when samples of liver, spleen, heart, kidney and lung collected from mice in the baseline and control groups (GI and GII) were plated onto the selective medium.

Nº	Species	Isolate	Capsular serotype	Aspect mucoid	Antibiotic resistance
1	***K. oxytoca***	P4	SND	Absent	AMO e AMP
2	***K. pneumoniae***	P14	SNH	Absent	AMC, NAL, AMI, AMO, AMP, ATM, CFL, CTX, KN, NO, TET e TIC
3	***K. pneumoniae***	P15	SNH	Weak	AMO, AMP, ATM, KN e TET
4	***K. pneumoniae***	U4	SNH	Intense	NAL, AMO, AMP, ATM, CFL, CTX e TET
5	***K. pneumoniae***	U5	K4	Absent	AMO e AMP
6	***K. pneumoniae***	U8	K5	Absent	AMC, NAL, AMO, AMP, ATM, CFL, CTX, NO e TET

SND: serotype not determined. SNH: serotyping nonspecific but higher than K6. Amoxicillin + Clavulanic Acid-AMC, Nalidixic acid-NAL, Amikacin-AMI, Amoxicillin-AMO, Ampicillin -AMP, Aztreonam-ATM, Cephalothin-CFL, Cefotaxime CTX, Neomycin-NO, Tetracycline-TET, Ticarcillin + clavulanic acid-TIC.

Table 1: *Klebsiella* strains isolated from enteral formulas and employed for translocation studies.

However, they were detected in liver and lung samples from animals in GIII to GVII (Figure 1). *Klebsiella* could also be detected five and six days after administration of prednisone and carbenicillin, in liver samples from animals that had not been fed with a contaminated diet (GVII). RAPD analysis did not reveal any similarity between these isolates and the *Klebsiella* strains orally fed to the animals (Figure 2).

Klebsiella was found in the lungs of the animals provided with enteral formula contaminated with the six strains (GIII to GVI) (Figure 1). Four days after inoculation, *Klebsiella* was detected in the lungs of the animals that received prednisone and/or carbenicillin, while non-medicated animals showed *Klebsiella* two days after inoculation (Figure 1). RAPD analysis on mucoid colonies isolated from the lungs revealed

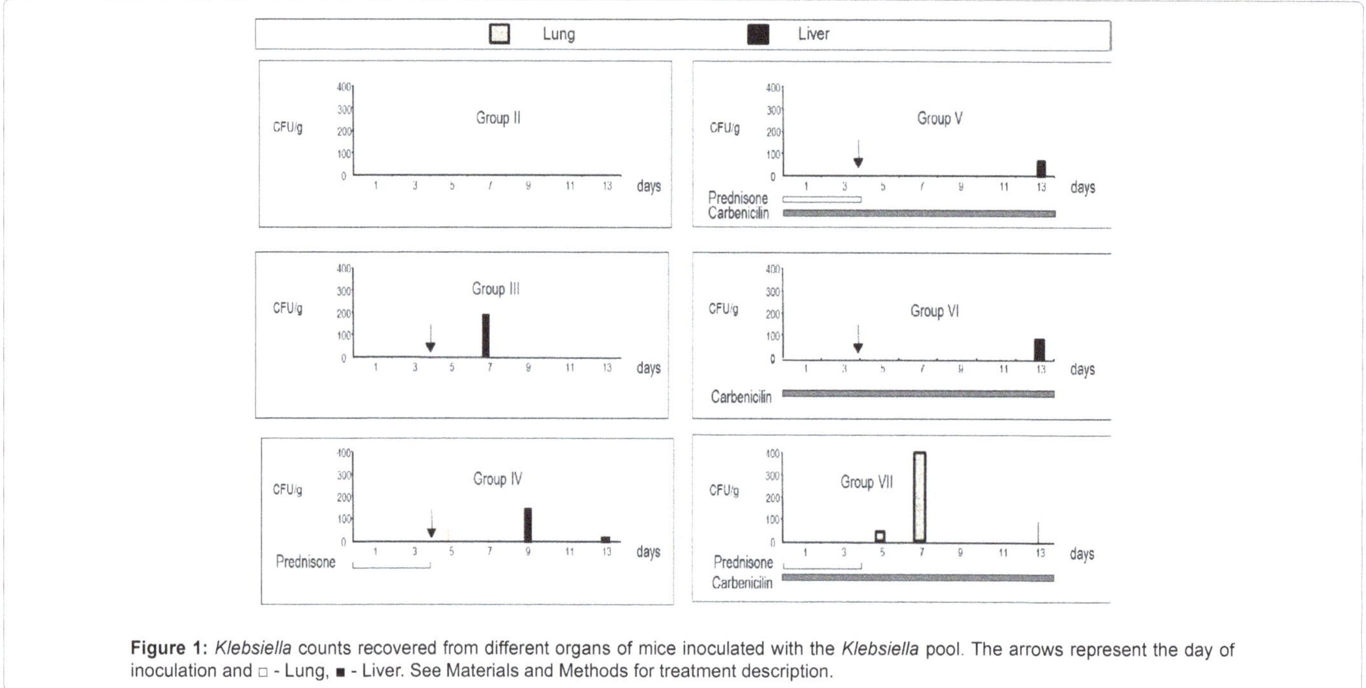

Figure 1: *Klebsiella* counts recovered from different organs of mice inoculated with the *Klebsiella* pool. The arrows represent the day of inoculation and □ - Lung, ■ - Liver. See Materials and Methods for treatment description.

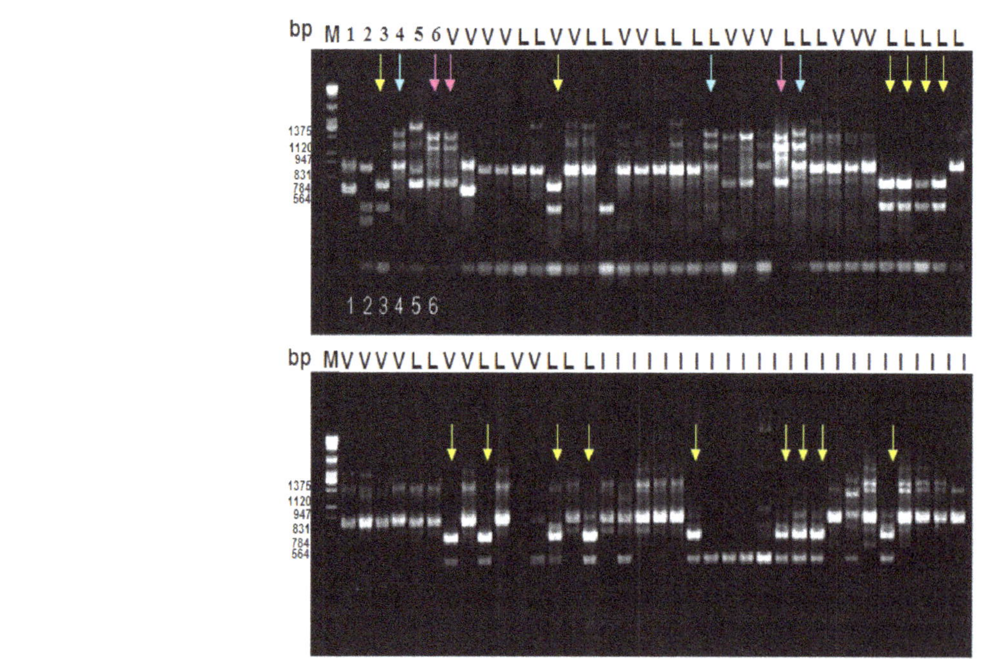

Figure 2: RAPD analysis of total DNA extracted from *Klebsiella* strains with the OPD20 primer. Numbers 1 to 6 refers to *Klebsiella* strains administered to mice. See Table 1 for strain description. Kidney, heart, lung, spleen, liver, and gut are represented by the letters K, H, L, S, V, and I, respectively. Isolates with identical RAPD profile are shown by the same arrows. DNA of phage Lambda restricted with the enzymes *Hind*III, *Eco*RI, and *Bam*HI, in base pairs (M).

that these bacteria had the same band profile as the *Klebsiella* strains fed to animals in Groups III and IV (Figure 2).

Higher counts of typical *Klebsiella* colonies in samples collected from the intestines of animals in Groups IV to VII were noted (Table 2). These mice all received prednisone and/or carbenicillin. However, the highest colony count was obtained when both drugs were administered together with the *Klebsiella* pool.

Discussion

Bacterial translocation is a phenomenon that seems to explain the origin of many infections caused by *Klebsiella* in hosts with dysregulated immune response. In this study, we confirmed that *Klebsiella* translocation does not occur in hosts with a healthy immune system. This was evident from the observation that no typical *Klebsiella* colonies were found in samples collected from the liver and lungs of animals that had not been fed with contaminated diet and had not been medicated with immunosuppressive drugs (Groups I and II, respectively). On the other hand, the results obtained from the control group (Group VII) showed that *Klebsiella* from the autochthonous microbial gut became translocated upon depression of the immune system and selective decontamination induced by antibiotics.

It has been well documented that antibiotic treatments cause aberrances in the host microbiota. Although it is generally believed that such changes become normalized within weeks of cessation of antibiotics, recent evidence challenges this notion. In the context of dysbiosis (an imbalance of protective and pathogenic bacteria in the gut), antibiotics can thus be viewed as a double-edged sword. They are effective in eradicating pathogens but also nonspecifically reduce microbial diversity, thereby making it possible for opportunistic bacteria to colonize the newly hospitable niches in the gut ecosystem [16]. Such is the case of *Klebsiella* Various disease states have been correlated with dysbiosis. Changes to the community structure of the intestinal microbiota are not without consequences, considering the wide effects that microbes have on both local and systemic immunity. Microbiotas associated with disease are defined by lower species diversity, fewer beneficial microbes and/or the presence of pathobionts [11].

In this study, the highest number of *Klebsiella* colonies was detected in the liver (Groups III-VI). This may have been because the gut and liver are connected by the portal venous system, which makes the liver more vulnerable to translocation of bacteria, bacterial products, endotoxins or secreted cytokines [17]. Thus, the presence of *Klebsiella* in the livers of the animals analyzed differed over the sampling period, which suggested that it is difficult for infectious processes to become established. This is because the mesenteric lymph node is the first line of defense, once the intestinal barrier has been breached [18].

The presence of *Klebsiella* in lung samples (Group VII) from non-medicated mice may be due to disequilibria in the commensal microbiota of the gut. Current evidence shows that dietary factors affect the microbial ecosystem in the gut. Contaminated diets and environmental changes may be responsible for this, though reducing the resistance to colonization by exogenous bacteria. This promotes bacterial growth and represents one of the major causes of infections by opportunistic bacteria [11].

Groups	*Klebsiella* count (CFU/g gut)	RAPD profile similarity with isolates administered to mice
II	3.5×10^2	ND*
III	5.7×10^2	ND
IV	4.5×10^3	Isolate 3 (P15)
V	2.3×10^3	ND
VI	6.6×10^4	Isolate 3 (P15)
VII	3.9×10^3	ND

*ND - not detect

Table 2: Colonization of the intestinal tract by the *Klebsiella* pool.

In this regard, the presence of *Klebsiella* in enteral diets needs to be considered as a possible source of contamination for hospitalized patients. This arises because enteral diets are especially prescribed for patients who are concentrated in intensive care units, where they are highly susceptible to nosocomial infections. These are patients who are subject to the intrinsic risks of nosocomial infections relating to complex underlying diseases, nutritional vulnerability, extreme ages (premature babies and elderly people) and the effects of immunosuppressant and/or immune depressive drugs and broad-spectrum antimicrobial agents. Furthermore, these patients are subject to extrinsic risks such as having a prolonged stay in the hospital environment, being subjected to routines of invasive procedures and lacking guaranteed quality in all hospital services, given the complex nature of hospital organizations [19].

Corroborating these findings, the study by Calbo et al [7] described the epidemiology of and control measures taken in relation to a foodborne nosocomial outbreak of *K. pneumoniae* in an acute-care hospital. Their report described rapid and widespread dissemination of an epidemic clone of *K. pneumoniae*, which caused a large nosocomial outbreak through the food chain. To our knowledge, this was the first report to provide insight on how transmission of *K. pneumoniae* can occur through food, as the vehicle in the hospital setting. The outbreak was stopped only after control measures were applied in the kitchen. The absence of new cases during the 14-month follow-up period suggests that these measures were effective.

Moreover, analysis on lung samples obtained from mice fed with prednisone (Group IV) detected exogenous bacteria. Use of corticosteroids significantly decreases the secretory IgA rate in the intestine and may be a plausible explanation for why these drugs ease bacterial translocation into extra-intestinal sites. Certain components of the microbiota are also likely to be critical for normal development of regulatory mechanisms that contribute towards mucosal homeostasis [10].

Our results show that colonization of the intestinal tract by *Klebsiella* strains isolated from enteral formulas may be favored when immunosuppressive and antimicrobial medications are used. Systemic and oral antibiotics cause overgrowth of bacteria in the gut through selective decontamination of some bacteria, such as antagonists, but also lead to colonization by resistant bacteria. In addition to eliminating pathogenic microorganisms, broad spectrum antibiotics also destroy the normal microbiota, thereby allowing an increase in the number of opportunistic bacteria, usually from the family *Enterobacteriaceae* [20].

This is because the microbiota plays a fundamental role in induction, training and functioning of the host immune system. In return, the immune system has largely evolved as a means of maintaining the symbiotic relationship between the host and these highly diverse and evolving microbes. When operating optimally, this immune system-microbiota alliance allows induction of protective responses to pathogens and maintenance of the regulatory pathways involved in continuation of tolerance to innocuous antigens. However, overuse of antibiotics, changes in diet and elimination of constitutive partners such as nematodes may have selected a microbiota that lacks the resilience and diversity required for establishing balanced immune responses. This phenomenon has been proposed to account for some of the dramatic rise in autoimmune and inflammatory disorders in parts of the world where the symbiotic relationship with the microbiota has been most affected [20].

Clinical studies on diseases have identified dysbiosis in patients with inflammatory bowel disease, including both Crohn's disease and ulcerative colitis. In addition to local diseases, intestinal dysbiosis has also been correlated with systemic diseases such as obesity, diabetes, atherosclerosis and nonalcoholic fatty liver disease [11].

We have provided evidence that *K. pneumoniae* becomes translocated into the lungs and liver of test mice fed with a *Klebsiella* pool, and these results emphasize the importance of strict control over microbial contamination during preparation and administration of these feeds.

This was made possible through using RAPD analysis in order to discriminating between autochthonous strains and those from the enteral diet. The amplified DNA fragments were specific for *Klebsiella* strains in liver, lung and intestine samples. In a recent study, the discriminatory power of the pulsed field gel electrophoresis (PFGE) and random amplified polymorphic DNA (RAPD) methods for typing 54 clinical isolates of *Klebsiella pneumoniae* were compared. All the isolates could be typed using RAPD, while 3.6% of them could not be typed by PFGE. PFGE is considered to be the gold standard for typing *K. pneumonia*. The repeatability of the two typing methods was 100%, with satisfactory reproducibility. It was concluded that an optimized RAPD protocol is technically less demanding and time consuming, which makes it a reliable typing method that is competitive with PFGE [21,22]. It should be emphasized that dissemination of data analyses such as that of the present study and other recent studies should be done routinely among all the healthcare and administrative professionals of hospital institutions. Many professionals will, through becoming aware of the results relating to their own specific services, start to rethink their practices and become more involved in prevention and control practices, with surveillance in its true sense.

References

1. Thornton MM, Chung-Esaki HM, Irvin CB, Bortz DM, Solomon MJ, et al. (2012) Multicellularity and antibiotic resistance in Klebsiella pneumoniae grown under bloodstream-mimicking fluid dynamic conditions. J Infect Dis 206: 588-595.
2. Meatherall BL, Gregson D, Ross T, Pitout JD, Laupland KB, et al. (2009) Incidence, risk factors, and outcomes of Klebsiella pneumoniae bacteremia. Am J Med 122: 866-873.
3. Fung CP, Lin YT, Lin JC, Chen TL, Yeh KM, et al. (2012) Klebsiella pneumoniae in gastrointestinal tract and pyogenic liver abscess. Emerg Infect Dis 18: 1322-1325.
4. Lin YT, Liu CJ, Fung CP, Tzeng CH (2011) Nosocomial Klebsiella pneumoniae bacteraemia in adult cancer patients - characteristics of neutropenic and non-neutropenic patients. Scand J Infect Dis 43: 603-608.
5. Hurrell E, Kucerova E, Loughlin M, Caubilla-Barron J, Hilton A, et al. (2009) Neonatal enteral feeding tubes as loci for colonisation by members of the Enterobacteriaceae. BMC Infect Dis 9: 146-154.
6. Okuma T, Nakamura M, Totake H, Fukunaga Y (2000) Microbial contamination of enteral feeding formulas and diarrhea. Nutrition 16: 719-722.
7. Calbo E1, Freixas N, Xercavins M, Riera M, Nicolás C, et al. (2011) Foodborne nosocomial outbreak of SHV1 and CTX-M-15-producing Klebsiella pneumoniae: epidemiology and control. Clin Infect Dis 52: 743-749.
8. Fertas-Aissani R, Messai Y, Alouache S, Bakour R (2012) Virulence profiles and antibiotic susceptibility patterns of Klebsiella pneumonia strains isolated from different clinical specimens. Pathologie Biologie 61: 209-216.
9. Wiest R, Garcia-Tsao G (2005) Bacterial translocation (BT) in cirrhosis. Hepatology 41: 422-433.
10. Peloquin JM, Nguyen DD (2013) The microbiota and inflammatory bowel disease: insights from animal models. Anaerobe 24: 102-106.
11. Chan YK, Estaki M, Gibson DL (2013) Clinical consequences of diet-induced dysbiosis. Ann Nutr Metab 63 Suppl 2: 28-40.
12. Rashid T, Wilson C, Ebringer A (2013) The Link between Ankylosing Spondylitis, Crohn's Disease, Klebsiella, and Starch Consumption. Clinical and Developmental Immunology.
13. Sánchez E, Donat E, Ribes-Koninckx C, Fernández-Murga ML, Sanz Y, et al. (2013) Duodenal-Mucosal Bacteria Associated with Celiac Disease in Children. Applied and Environmental Microbiology 79: 5472-5479.
14. Bagley ST, Seidler RJ (1978) Primary Klebsiella identification with MacConkey-inositol-carbenicillin agar. Appl Environ Microbiol 36: 536-538.
15. Martins ML, Araújo EF, Mantovani HC, Moraes CA, Vanetti MC, et al. (2005) Detection of the apr gene in proteolytic psychrotrophic bacteria isolated from refrigerated raw milk. Int J Food Microbiol 102: 203-211.
16. Jernberg C, Löfmark S, Edlund C, Jansson JK (2007) Long-term ecological impacts of antibiotic administration on the human intestinal microbiota. ISME J 1: 56-66.
17. Quigley EM, Stanton C, Murphy EF (2013) The gut microbiota and the liver. Pathophysiological and clinical implications. J Hepatol 58: 1020-1027.
18. Henao-Mejia J, Elinav E, Thaiss CA, Licona-Limon P, Flavell RA, et al. (2013) Role of the intestinal microbiome in liver disease. J Autoimmun 46: 66-73.
19. Viale P, Giannella M, Lewis R, Trecarichi EM, Petrosillo N, et al. (2013) Predictors of mortality in multidrug-resistant Klebsiella pneumoniae bloodstream infections. Expert Review of Anti-infective Therapy. 1: 1053-1063.
20. Belkaid Y, Hand TW (2014) Role of the Microbiotain Imm unity and Inflammation. Cell 27: 121-141.
21. Ashayeri-Panah M, Eftekhar F, Ghamsari MM, Parvin M, Feizabadi MM, et al. (2013) Genetic profiling of Klebsiella pneumoniae: comparison of pulsed field gel electrophoresis and random amplified polymorphic DNA. Braz J Microbiol 44: 823-828.
22. Li W, Raoult D, Fournier PE (2009) Bacterial strain typing in the genomic era. FEMS Microbiol Rev 33: 892-916.

Using Barley Beta Glucan, Citrus, and Carrot Fibers as a Meat Substitute in Turkey Meat Sausages and their Effects on Sensory Characteristics and Properties

Naourez Ktari[1]*, Imen Trabelsi[2], Intidhar Bkhairia[1], Mehdi Triki[3], Mohamed A Taktak[4], Hafedh Moussa[4], Moncef Nasri[1] and Riadh B Salah[2]

[1]*Laboratory of Enzyme Engineering and Microbiology, National School of Engineering of Sfax, University of Sfax, Tunisia*
[2]*Laboratory of Microorganisms and Biomolecules (LMB), Centre of Biotechnology of Sfax, Tunisia*
[3]*Laboratory of Food Analysis, National School of Engineering of Sfax, University of Sfax, Tunisia*
[4]*Chahia Company, Road of Sidi Salem, Sfax 3003, Tunisia*

Abstract

This study was aimed to evaluate the effect of addition of two levels (1% and 2%) of Barley Beta Glucan (BBG), citrus (Ceamfibre 7000), and carrot (ID809) fibers on chemical, sensorial properties, oxidation and microbial quality of turkey meat sausage, during 21 days of storage at 4°C. The findings indicated that the addition of the three fibers decreased fat and protein contents but increased moisture content. Color parameters were significantly ($p<0.05$) affected by the fiber type and content. The meat substitution resulted in a tendency toward lighter coloration. A significant progression in the textural hardness, elasticity and chewiness of fiber-added sausages, was observed. The addition of the three fibers, at 1% and 2% level, induced a decrease in hardness when compared to control. Furthermore, color, taste, flavor and overall acceptability attributes were similar to the control when the fibers were incorporated. According to sensorial evaluations, the three polysaccharides remained at 1% to the consumer acceptability of sausage. Besides, the addition of the three polysaccharides had a significant effect on sausage safety because of its reduction of the lipid oxidation degree monitored by TBARS and it had no effect on microbial proliferation. Overall, the results of this study indicate that the three polysaccharides can be applied at 1% in turkey meat sausage to increase their nutritional status whilst maintaining the quality and safety attributes of the product.

Keywords: Fibers; Turkey sausages; Texture; Sensory testing; Lipid oxidation; Microbial proliferation

Introduction

Development of new product is a constant challenge for both applied and scientific research. The design of new food products required the generation of best formulation by ingredients optimization [1]. The present trend of using food for health purposes rather than just for nutrition opens up whole new fields for the meat industry. Thus, meat processors, in particular, continue to search for the ideal fat replacer or substitute, which gives all the fat-related attributes without the accompanying health concerns. It is well documented that high dietary fat intake is related to obesity, hypertension, cardiovascular disease and coronary heart disease, because of the high amounts of saturated fatty acids and cholesterol [2]. However, with fat playing a decisive role in product properties and consumer acceptance, the reduction of fat in meat products poses a challenge to the food industry. Fat stabilizes the meat emulsion, reduces cooking loss, improves water holding capacity and provides juiciness and hardness [3-6]. Besides, fat plays a major role in affecting sensory characteristics (appearance, flavor and texture) and consumer acceptance [7,8]. Furthermore, from a physiological standpoint, fat is a source of vitamins and essential fatty acids. It also constitutes the most concentrated source of energy in the diet (9 kcal/g). Hence, the removal of fat from meat products affects their flavour, juiciness and mouthfeel. In addition to that, it reduces satiety value. In order to produce low fat sausages similar to those of their full-fat counterparts, it is necessary to maintain similar sensory and health profiles. Among the possible options, reformulations by adding or substituting meat ingredients by dietary fiber are being explored [9,10].

Dietary fiber is derived from various plants, fruits, and nuts. It has different functional characteristics depending on processing conditions such as grinding and drying [11]. The addition of dietary fiber to meat products, such as pates, salami and other sausages, has been shown to improve cooking yield, water binding, fat binding, and texture [12]. Several authors have studied the use of dietary fiber as a functional ingredient in dry fermented sausages, with good results. Fernández-López et al. [13] observed that the addition of 1% orange fiber (by-product of juice production) decreases the growth of micrococci and the amount of nitrite required to increase the microbial stability of raw meats. Yalinkiliç et al. [14] observed that the addition of orange fiber (4%) in sucuk (Turkish uncooked cured sausage) has an effect on bacterial growth. On the one hand, it favors the growth of the lactic acid bacteria. On the other hand, the *Micrococcus* and *Staphylococcus* counts decrease, and the Enterobacteriaceae count is below detectable level ($<10^2$ CFU/g).

Considering the importance of dietary fiber as a frequently used food ingredient in the design of functional foods, the aim of this study was to assess the effect of BBG, Ceamfibre 7000, and ID809 fibers as mechanically separated turkey meat replacers on the quality and safety characteristics of Tunisian turkey meat sausage, including chemical composition, color, texture profile, sensory properties, lipid peroxidation and microbial evaluation.

Materials and Methods

Materials

Mechanically separated turkey (MST) meat was obtained from local processors (Chahia, Tunisia). MST meat was produced from turkey

***Corresponding author:** Naourez Ktari, Laboratory of Enzyme Engineering and Microbiology, National School of Engineering of Sfax, University of Sfax, P.O. 1173-3038 Sfax, Tunisia, E-mail: naourez.ktari@yahoo.fr

carcass after meat cutting. Analytical grade NaCl, $NaNO_2$, ascorbic acid, and sodium tripolyphosphate were used. Modified starch (E1422) was from Sigma Chemical Co. (St. Louis, MD, USA). Cold distilled water (4°C) was used in all formulations.

Origin and appearance of dietary fibers: BBG, Ceamfibre 7000, and ID809 fibers were supplied in powder form by a local meat company (CHAHIA, Sfax, Tunisia). They were initially purchased from F.P.S. GROUPE MANE (Marne La Vallee, France).

BBG is an excellent functional ingredient in a wide range of food and beverage applications. It is one of the richest available sources of concentrated natural beta glucan soluble fiber which is commonly used to make foods and beverages healthier. ID809 is a dietary fiber with strong functional properties. ID809 is manufactured from cooked mashed carrots from which carrot juice has been removed and then the pomace is dehydrated, milled and screened to ensure particle uniformity. The ingredient is GMO free, gluten free and cholesterol free. Ceamfiber 7000 is a natural fiber ingredient purified from citrus peel with high functional properties for a wide variety of applications. It is an all-natural food ingredient. Consequently, no E-number is needed, leading to a clean label. The Ceamfiber 7000 has proved successful in improving cooking yield, drip loss control, reducing fat/meat/solids content, maintaining form stability, preventing jelly formation and fat separation, improving texture of the end-product or replacing other gums and ingredients in the final food formulation. ID809 fibers were dark brown while BBG and Ceamfibre 7000 fibers were of white and yellow appearances, respectively (Figure 1).

Sausage preparation

The standard sausage formulation consisted of: MST meat, cold water, modified starch, NaCl, $NaNO_2$, Sodium tripolyphosphate, and ascorbic acid (Table 1). The three fibers were formulated into turkey meat sausage as described in Table 1. The percentages of spice additives were unchanged compared to the control sample. The main difference consisted in decreasing meat content and increasing fiber content.

Figure 1: (a) Ceamfiber 7000 fibers, (b) ID809 fibers, and (c) BBG fibers.

Cold water was added to frozen MST meat, which was then ground in a commercial food processor (Moulinex, Paris, France), equipped with a 5 cm blade for 5 min at the highest speed. Salts, sodium tripolyphosphate and other ingredients were slowly added to the ground MST meat while processing. After that, modified starch and fibers were incorporated until completely blended. Stuffing was carried out manually into 27-mm-diameter reconstituted collagen casings and hand-linked to form approximately 8-cm-long links. Then, sausages were heat-processed in a temperature controlled water-bath (Haake, Kalsruhe, Germany) maintained at 90°C until a final internal temperature of 74°C was reached. The temperature was measured using a Type-T (copper–constantan) thermocouple inserted into the center of a link. Afterwards, Sausages were cooled immediately using tap water and stored at 4°C until analysis. The procedure for preparation of turkey meat sausages is given in Figure 2.

Physico-chemical analysis

The moisture and ash contents of turkey meat sausages and fibers were determined according to the standard methods 930.15 and 942.05, respectively. Total nitrogen content was determined by using the Kjeldahl method according to the AOAC method number 984.13 [15]. Crude protein was estimated by multiplying total nitrogen content by the factor of 6.25. Crude fat was determined gravimetrically after Soxhlet extraction of samples with hexane. The pH values of fibers were measured in a homogenate prepared with 1 g of sample and distilled water (25 ml) using a pH meter (Model 340, Mettler-Toledo GmbH, Schwerzenbach, Switzerland). Oil binding capacity (OBC) of fibers was measured according to Lin et al. [16]. The three fibers were added to a concentration of 100 mg to 10 ml of corn oil in 50 ml centrifuge tubes. The mixtures were stirred and the tubes were then centrifuged at 2500 g for 30 min. The free oil was decanted and the absorbed oil determined. Water binding capacity (WBC) was measured according to Mac-Connel et al. [17]. Briefly, the three fibers were separately added to a concentration of 100 mg to 10 ml of distilled water in 50 ml centrifuge tubes and stirred overnight at 4°C. After that, the mixtures were centrifuged at 10,000 g for 30 min. The free water was decanted, and the absorbed water determined. All analytical determinations were performed at least in triplicate. Values of different parameters were expressed as the mean ± standard deviation.

Color evaluation of sausages

Color measurements of sausage samples were evaluated at different storage times using a Color Flex spectrocolorimeter (Hunter Associates Laboratory Inc., Reston, VA, USA) and reported as L*, a* and b* values, where L* refers to the measure of lightness, a* to the chromatic scale from green to red, and b* to the chromatic scale from blue to yellow [18]. All determinations were carried out in triplicate.

Ingredients (%)	Control	BBG (1%)	BBG (2%)	ID809 (1%)	ID809 (2%)	Ceamfibre 7000 (1%)	Ceamfibre 7000 (2%)
MST	62.4	52.4	42.4	43.4	24.4	53.4	44.4
Water	30	39	48	48	66	38	46
Modified starch (E1422)	6	6	6	6	6	6	6
NaCl	0.12	0.12	0.12	0.12	0.12	0.12	0.12
Sodium tripolyphosphate	0.28	0.28	0.28	0.28	0.28	0.28	0.28
$NaNO_2$	0.08	0.08	0.08	0.08	0.08	0.08	0.08
Ascorbic acid	0.045	0.045	0.045	0.045	0.045	0.045	0.045

Table 1: Formulations of turkey meat sausages with the three fibers.

Texture properties of sausages

Texture profile analysis (TPA) was done on cooked samples at different storage times using a texturometer (texture analyzer, Lloyd Instruments, Ltd., West Sussex, UK). The center cores of the sausage samples were cut (2 cm in diameter, 2 cm height) and placed between flat plates and a cylindrical probe (12 mm in diameter). Subsequently, samples were compressed to 50% of their original height in a double cycle at a constant rate of 40 mm/min. The texture profile parameters, i.e. hardness (N), elasticity (mm), and chewiness (Nmm) were computed from the resulting force-deformation curves [19].

Sensory evaluation

Five attributes namely taste, flavour, color, texture, and overall acceptability were evaluated on turkey meat sausage samples. Thirty-eight healthy subjects aged between 23 and 48 years old, with no history of taste disorders, were recruited from the staff at Chahia Company and also students at Food Analysis Laboratory of National School of Engineering of Sfax. Subjects were informed on the duration of the testing and gave consent in the same way. The majority of the subjects were familiar with sensory evaluation techniques. Subjects (codes 1 to 38) were asked to avoid drinking coffee and tea, eating, or smoking 1-hour prior testing. Experiments were conducted in an appropriately designed and lighted room. Water was served for the purpose of cleaning the mouth before the test and in-between the samples. Sausage slices of 3-mm thickness were placed in white polystyrene plates with lids coded with 3-digit random numbers. Panelists scored sausage samples ranging from 0 (dislike extremely) to 9 (like extremely).

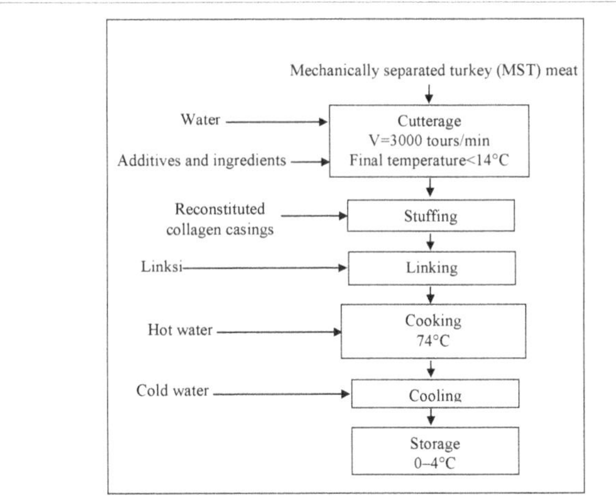

Figure 2: Procedure for preparation of Turkey meat sausage.

	BBG	ID809	Ceamfibre 7000
Protein	2.03% ± 0.04	1.74% ± 0.21	2.92% ± 0.04
Fat	0.88% ± 0.18	0.49% ± 0.13	0.79% ± 0.4
Ash	0.28% ± 0.04	1.59% ± 0.54	1.61% ± 0.1
Total fibers	91.78% ± 1.1	87.57% ± 0.81	86.95% ± 0.64
Moisture	5.4% ± 0.68	8.71% ± 0.5	7.62% ± 0.23
pH	6.07 ± 0.09	5.12 ± 0.17	7.68 ± 0.25
WBC (g/g)	9	18	8
OBC (g/g)	10.5	6.3	7.5

Physico-chemical composition was calculated based on the dry matter. Values are given as mean ± SD from triplicate determinations (n = 3).

Table 2: Physico-chemical compositions of the BBG, Ceamfibre 7000 and ID809 fibers.

Effect of the three dietary fibers on turkey meat sausage lipid oxidation

The extent of lipid oxidation in each turkey meat sausage was determined by the thiobarbituric acid reactive substances (TBARS) assay as described by Hogan et al. [20]. The final TBARS value was expressed as mg of malondialdehyde (MDA) equivalents per kg of sample.

Microbiological profile

Portions of 0.5 g of sausages were removed aseptically using a spoon, transferred to a stomacher bag (Seward Medical, Worthing, West Sussex, UK), containing 4.5 ml of sterile NaCl solution (0.9%), and homogenized using a stomacher (Lab Blender 400, Seward Medical) for 60 seconds at room temperature. For microbial enumeration, 0.1 ml samples of serial dilutions (1:10, NaCl 0.9%) were spread on the surface of agar plates. Total coliforms were determined using Brain Heart Infusion Broth (Pronadise, laboratorios CONDA, Madrid-Spain). Mesophilic germs were determined using Plate Count Agar (Pronadise, laboratorios CONDA, Madrid-Spain) and Yeast was determined using Potato Dextrose Agar (Lab M, United Kingdom). The plates containing 25-250 colonies were selected and counted. All microbial counts were converted to logarithms of colony-forming units per ml of sample (log CFU/ml).

Statistical analysis

All measurements were carried out in triplicate. Data were subjected to analysis of variance (ANOVA) using the General Linear Models procedure of the Statistical Analysis System software of SAS Institute (SAS, 1990). Differences among the mean values of the various treatments were determined by the least significant difference (LSD) test, and the significance was defined at $P<0.05$. The differences equal to or more than the identified LSD values were considered statistically significant.

Results and Discussion

Physico-chemical and techno-functional properties of the three dietary fibers

The physico-chemical and techno-functional properties of BBG, Ceamfibre 7000, and ID809 fibers were presented in Table 2. On a dry weight basis, the fibers containing protein ranged from 1.74% to 2.92%, fat ranged from 0.49% to 0.88%, ash ranged from 0.28% to 1.61% and total fibers ranged from 86.95% to 91.78%. All fibers had relatively low moisture content (5.4%-8.71%). No significant differences ($P>0.05$) in fat and protein were found among the three fibers. A significant difference ($P<0.05$) was observed between the WBC of the three fibers (9 g/g, 18 g/g, and 8 g/g for BBG, ID809, and Ceamfibre 7000, respectively). The high WBC of the three fibers suggests that they could be used as functional ingredients in food formulations to modify texture and viscosity, reduce dehydration during storage, and reduce energetic value.

Chemical properties of turkey sausages formulated with the three fibers

The chemical composition of turkey sausages formulated with the three fibers at two levels (1% and 2%) was determined and compared to that of control Turkey sausage (Table 3). Proximate analysis proved that the control sausage, recorded significantly ($p \leq 0.05$) lower moisture content (76.35%) than fibers added sausages (77.81%-84.05%). The moisture content of formulated sausages increased concomitantly with

Composition (%)	Control	BBG (1%)	BBG (2%)	ID809 (1%)	ID809 (2%)	Ceamfibre 7000 (1%)	Ceamfibre 7000 (2%)
Dry matter	23.65 ± 0.24	22.10 ± 1.52	19.19 ± 0.12	20.8 ± 0.5	15.95 ± 1.08	22.19 ± 0.00	19.7 ± 0.35
Ash	2.62 ± 0.3	2.55 ± 0.12	2 ± 0.68	2.68 ± 0.05	2.84 ± 0.14	2.6 ± 0.08	2.41 ± 0.33
Protein	15.84 ± 0.22	14.11 ± 0.19	12.38 ± 0.53	13.19 ± 0.3	10.01 ± 0.52	14.45 ± 2.26	12.94 ± 0.93
Fat	4.57 ± 0.09	4.33 ± 0.1	3.88 ± 0.15	3.08 ± 0.12	1.94 ± 0.08	3.9 ± 0.15	2.43 ± 0.28

Table 3: Chemical properties of turkey meat sausages formulated with the three fibers.

Storage (days)	Color parameters	Control	BBG (1%)	BBG (2%)	ID809 (1%)	ID809 (2%)	Ceamfibre 7000 (1%)	Ceamfibre 7000 (2%)
0	a*	13.23 ± 0.05	13.05 ± 0.08	13.52 ± 0.09	13.74 ± 0.34	12.89 ± 0.05	14.4 ± 1.74	15.1 ± 0.91
	b*	7.83 ± 0.17	7.64 ± 0.02	7.1 ± 0.03	7.73 ± 0.05	7.91 ± 0.01	7.93 ± 0.21	7.81 ± 0.04
	L*	50.84 ± 1.17	51.84 ± 0.01	52.38 ± 0.22	50.66 ± 0.46	51.49 ± 0.17	52.1 ± 0.2	51.3 ± 1.26
7	a*	14.81 ± 0.03	14.01 ± 0.22	14.89 ± 0.02	14.90 ± 0.05	15.08 ± 0.07	15.13 ± 0.8	14.16 ± 0.13
	b*	7.55 ± 0.08	7.49 ± 0.01	7.30 ± 0.03	7.33 ± 0.02	8.16 ± 0.01	8.02 ± 0.15	8.2 ± 0.11
	L*	50.69 ± 0.36	53.26 ± 0.47	50.48 ± 0.11	51.47 ± 0.09	51.59 ± 0.11	50.61 ± 1.2	51.9 ± 0.52
14	a*	14.39 ± 39	14.47 ± 0.41	14.72 ± 0.05	16.49 ± 0.05	13.89 ± 0.24	14.14 ± 0.18	13.42 ± 0.19
	b*	7.49 ± 0.39	7.14 ± 0.20	6.98 ± 0.08	7.89 ± 0.03	7.75 ± 0.01	8.15 ± 0.03	8.1 ± 0.4
	L*	50.95 ± 0.74	52.9 ± 0.31	51.87 ± 0.03	50.52 ± 0.08	51.92 ± 0.56	51.3 ± 0.28	52.6 ± 0.68
21	a*	15.49 ± 0.01	14.32 ± 0.41	13.90 ± 0.07	14.03 ± 0.05	14.28 ± 0.1	12.15 ± 0.06	11.54 ± 0.04
	b*	7.27 ± 0.05	6.48 ± 0.08	6.61 ± 0.02	7.37 ± 0.3	7.06 ± 0.34	8.2 ± 0.03	8.44 ± 0.15
	L*	50.97 ± 0.10	52.74 ± 0.03	52.02 ± 0.03	52.73 ± 0.01	53.29 ± 0.18	53.9 ± 0.54	53.3 ± 0.12

Table 4: Experimental results relative to instrumental color parameter for turkey sausages formulated with the three fibers during storage at 4°C.

the level of dietary fibers. Among all formulations, sausage formulated with ID809 (2%) recorded significantly the highest ($p \leq 0.05$) moisture retention (84.05%). The increase in moisture percent among sausage formulations with the increasing level of dietary fibers could be due to their high water binding capacity (WBC). This is in accordance with the results obtained by Szczepaniak et al. [21] who found a significantly increased moisture content in finely comminuted thick wiener type sausages with the increasing of potato fiber level. Another study by Choi et al. [22] reported that the moisture content increased in reduced-fat emulsion sausages with added brown rice fiber. Likewise, Fernández-Ginés et al. [23] indicated that lemon albedo as a source of dietary fiber increased the moisture content of bologna sausage. Similar trends in moisture content were observed by García et al. [24] for the moisture content of dry fermented sausage supplemented with wheat, oat and fruit fiber, which were higher than that of the control owing to the high water retention of the fibers.

The protein and fat contents of control sausage were found to be significantly ($p \leq 0.05$) higher than those of sausages formulated with BBG, Ceamfibre 7000, and ID809 fibers. In fact, protein and fat contents decreased significantly ($p \leq 0.05$) with the increasing levels of dietary fibers. This could be due to the replacement of MST meat, the only source of proteins and fats, by dietary fibers. Our findings are in agreement with several other previously reported works. Huda et al. [25] have, for instance, reported that the mean fat and protein values of mutton nuggets formulated with apple pomace at different levels were significantly higher than the control formulated without fiber. Additionally, Sánchez-Zapata et al. [26] have reported that fat percent of dry-cured sausages formulated with tiger nut fiber decreased with the increasing dietary fibers concentrations.

The ash contents of partial meat substituted sausages formulated with BBG and Ceamfibre 7000 fibers were found to decrease with the increasing levels of the two fibers, whereas that of sausage formulated with ID809 fiber was found to increase with the increasing level of the fiber. This may be due to the higher ash content of ID809 fiber. The results presented above are in agreement with several other previously reported findings. Verma et al. [27] showed that the ash content of low fat chicken nuggets decreased with the increasing levels of apple pulp, while Lee et al. [28] indicated that the ash content of sausage supplemented with kimchi fiber was higher than that in the control formulated without fiber. Similar results were previously reported by Turhan et al. [29] and Choi et al. [22] showing that low-fat beef burgers with hazelnut pellicle and reduced-fat emulsion sausages with brown rice fiber have increased ash content compared to that in a control.

Color analysis

The first characteristic which consumers use to evaluate meat product qualities is color [30]. Table 4 presents the development of lightness coordinate (L*), redness coordinate (a*), and yellowness coordinate (b*) during the storage of sausages added with 1% and 2% fibers. L* seems to be the most informative parameter for color changes in meat and meat products [31]. In the control samples, L* values keep constant during the storage. Turkey sausage with added fibers had a similar behaviour, but with higher values. The increase in L* could be attributed to fibers water retention that allows a high light reflection. Similar results were previously reported by Sánchez-Zapata et al. [26], showing that the addition of tiger nut fiber to Spanish dry-cured sausage increase L* value. At the beginning of storage, sausages containing Ceamfiber 7000 (1% and 2%) showed higher redness ($p<0.05$) than control. Thereby, no significant difference was observed in the fiber BBG, ID809 when compared to the control sausage. Redness increased during the storage for all formulation till the 7th day. This increase has been attributed to moisture loss which provokes an increase in pigment concentration [32]. At the end of storage, a* values decreased in all formulations. This decrease could be related to lipid oxidation [33]. There were no differences ($p>0.05$) in b* values between control and samples with fibers added at the beginning (until the 7th day of storage). After that, yellowness was affected slightly during the storage. The observed changes in b*, during the dry-curing process, are probably due to the oxygen consumption by microorganisms during their exponential growth phase. Consequently, the decrease in oxymyoglobin greatly contributes to the value of this color coordinate, because microorganisms produce metabolites that induce the oxidation of meat and fat present in the sausage and, by so doing, contribute to

Storage (days)	Texture parameters	Control	BBG -1%	BBG (2%)	ID809 -1%	ID809 -2%	Ceamfibre 70000 (1%)	Ceamfibre 7000 (2%)
0	Hardness (N)	7.74 ± 0.38	6.42 ± 0.32	4.3 ± 0.01	3.78 ± 0.11	1.91 ± 0.01	6.22 ± 0.18	4.09 ± 0.38
	Elasticity (mm)	12.27 ± 0.12	14.52 ± 0.28	12.44 ± 0.1	12.08 ± 0.21	12.05 ± 0.11	13.52 ± 0.46	11.80 ± 0.32
	Chewiness (Nmm)	29.64 ± 0.8	30.87 ± 0.14	20.22 ± 0.4	19.12 ± 0.31	9.47 ± 0.21	30.29 ± 0.35	16.14 ± 0.29
7	Hardness (N)	7.73 ± 0.3	7.24 ± 0.22	4.61 ± 0.45	5.59 ± 0.32	2.39 ± 0.15	4.62 ± 0.09	4.16 ± 0.22
	Elasticity (mm)	13.07 ± 0.06	12.55 ± 0.13	12.00 ± 0.2	12.32 ± 0.25	13.69 ± 0.25	12.82 ± 0.23	14.07 ± 0.06
	Chewiness (Nmm)	33.91 ± 0.91	32.87 ± 0.23	24.87 ± 0.3	24.59 ± 0.24	12 ± 0.23	22.87 ± 1.63	26.94 ± 1.66
14	Hardness (N)	7.88 ± 0.36	15.77 ± 0.45	11.25 ± 0.2	16.42 ± 0.12	5.14 ± 0.22	5.14 ± 0.24	4.01 ± 0.08
	Elasticity (mm)	14.35 ± 0.09	12.68 ± 0.28	10.62 ± 0.3	10.64 ± 0.11	12.81 ± 0.24	13.87 ± 0.11	11.97 ± 0.06
	Chewiness (Nmm)	53.93 ± 2.68	53.72 ± 0.36	34.17 ± 0.2	53.67 ± 0.41	15.87 ± 0.32	27.41 ± 0.55	24.29 ± 0.4
21	Hardness (N)	4.98 ± 0.03	8.67 ± 0.34	7.0 ± 0.14	7.84 ± 0.11	3.64 ± 0.12	4.66 ± 0.12	2.93 ± 0.14
	Elasticity (mm)	12.22 ± 0.27	11.60 ± 0.36	11.88 ± 0.2	11.39 ± 0.21	12.13 ± 0.11	13.97 ± 0.22	11.30 ± 0.54
	Chewiness (Nmm)	15.92 ± 0.44	38.89 ± 0.26	35.42 ± 0.3	39.98 ± 0.31	15.77 ± 0.31	19.65 ± 1.13	8.41 ± 2.68

Table 5: Experimental results relative to instrumental texture parameter for turkey sausages formulated with the three fibers during storage at 4°C.

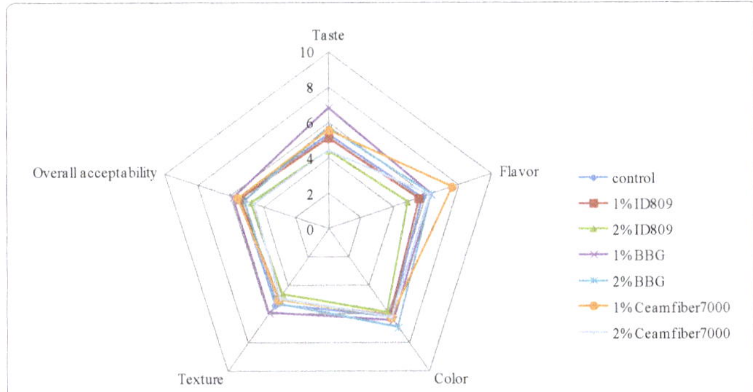

Figure 3: Sensorial analysis of turkey meat sausage formulated with BBG, Ceamfiber 7000, and ID809 fibers.

the decrease of this value [34]. Other authors have explained this color evolution by the presence of yellow compounds in fibers [35].

Textural properties of turkey sausages

The evolution of textural parameters (hardness, elasticity and chewiness) of turkey sausages as a function of the level of dietary added fibers and storage time is shown in Table 5. The findings revealed that the addition of the three fibers, at different levels, to turkey sausage induced a decrease in hardness when compared to control turkey sausage. The significant changes in hardness that were recorded could presumably be attributed to their higher water content and lower fat content. Similar results were obtained by Vural et al. [36] who indicated that the addition of sugar beet fiber decreases the hardness values of meat batters. However, it was observed that the addition of tiger nut fiber to dry-cured sausages increases the hardness [26]. García et al. [24] studied the addition of cereal and fruit fibers in low fat dry fermented sausages, and determined that the addition of cereal fibers (wheat and oat) at 3% to fat reduced sausages produces a significant increase in hardness, while the addition of fruit fibers (peach, apples and orange) decreases the hardness of products. Depending on the amount and type of fibers, the elasticity of formulated sausages was similar to control sausage. However, the chewiness of the new products increases when compared to control sausage. Indeed, it has been reported that owing to their water binding ability and swelling properties, insoluble fibers can influence food texture [23]. Insoluble fiber can increase the consistency of meat products through the formation of an insoluble 3-dimensional network capable of modifying rheological properties of sausages [37].

Sensory evaluation

Sensory analysis of prepared sausages was carried out by checking color, flavor, texture, taste and overall acceptability of the formulated sausages. The scores obtained by each attribute are shown in Figure 3. Attributes such as color and overall acceptability were similar to the control sausage when BBG, Ceamfibre 7000 and ID809 fibers were incorporated at a level of 1%. In addition, the substitution of meat with 1% BBG and 1% Ceamfibre 7000 fibers allows obtaining a product with a better taste and flavor, respectively. Besides, BBG fiber at 2% improves taste, flavor, and color of sausage when compared to control sausage. Unfortunately, the incorporation of Ceamfibre 7000 and ID809 fibers at 2% remarkably decreases the sensory quality of sausages.

Microbial evaluation

The bacterial count of total coliforms, mesophilic germs and yeast in turkey sausage formulated with the three fibers (BBG, ID809 and Ceamfiber 7000) during 21 days of storage at 4°C is shown in Table 6. At the beginning of storage (First week), no microbial germs were observed in all formulated turkey sausages. Initially, the addition of the three types of dietary fiber did not cause microbial change in turkey meat sausages. This outcome is consistent with the findings of the research literature which suggest that the addition of dietary fiber does not influence microbial growth in dry-fermented sausages with reduced fat content [24]. Only during the second week that the samples, with ID809 and Ceamfiber 7000 fibers added, showed the developments of the counts of coliforms, mesophilic germs and yeast. During storage at 4°C, microbial germs increased with time ($p<0.05$) in all sausages, to reach at the end of storage 16.47 log CFU/ml, 15.5 log CFU/ml and 22.29 log CFU/ml in sausages formulated with 2% BBG, ID809 and Ceamfiber 7000 fibers, respectively. Furthermore, none of the microbial groups was affected by fiber concentration during storage.

Effect of the three dietary fibers on lipid peroxidation in turkey meat sausage

Lipid oxidation is a major problem in high-fat food. During storage,

Storage (days)	Microorganism log (cfu/ml)	Control	BBG (1%)	BBG (2%)	ID809 (1%)	ID809 (2%)	Ceamfibre 70000 (1%)	Ceamfibre 7000 (2%)
0	Total coliforms	0	0	0	0	0	0	0
	Mesophilic germs	0	0	0	0	0	0	0
	Yeast	0	0	0	0	0	0	0
7	Total coliforms	0	0	0	0	0	6.68	7.21
	Mesophilic germs	0	0	0	0	0	10.53	10.8
	Yeast	0	0	0	5.24	5.3	12.3	12.81
14	Total coliforms	12.2	14.5	15.45	13.9	12.46	9.77	9.77
	Mesophilic germs	0	12.6	12.7	13.4	11.6	13.04	11.91
	Yeast	14.4	12.2	11	13.7	13.5	17.18	19.55
21	Total coliforms	16.86	17.03	16.5	16.45	17	18.67	18.9
	Mesophilic germs	16.3	16.3	15.13	15.18	15.7	14.58	12.58
	Yeast	16.6	16.48	16.47	15.49	15.5	25.46	22.29

Table 6: Development of total coliforms and mesophilic germs present in turkey sausages formulated with the three fibers during storage at 4°C.

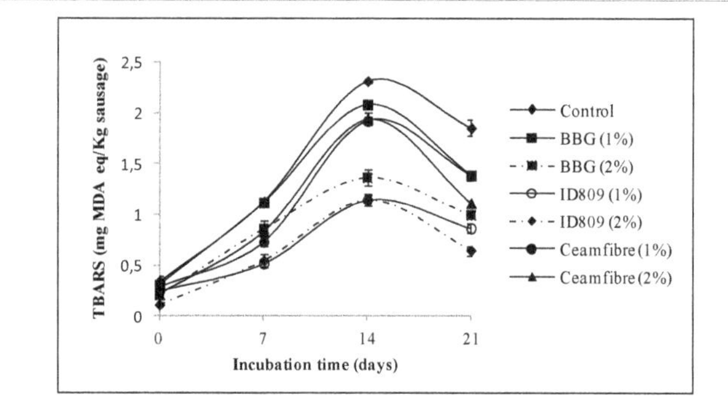

Figure 4: Lipid oxidation of turkey meat sausage formulated with BBG, Ceamfiber 7000, and ID809 fibers. (Bars represent the SD from triplicate determinations).

secondary oxidation products formed may react with nucleic acid, protein, lipid and polysaccharides and exert cytotoxic and genotoxic effects. In the present study, malondialdehydes (MDA) were chosen as markers for lipids oxidative deterioration (Figure 4). MDA reacts with TBA to give the TBA reactive substances (TBARS) detectable by spectrophotometry at 532 nm. As expected, the results showed that TBARS values of the control increased as a function of storage time. The value was 1.13 mg MDA/kg of turkey meat sausage after 7 days of storage. A decrease in TBARS values was observed after 14-days of storage. This was probably due to the loss of oxidation products formed, particularly low molecular weight volatile compounds. Indeed, MDA and other short-chain products of lipid oxidation are not stable for a long period of storage. Oxidation of these products yields alcohols and acids, which are not determined by the TBA test [38].

The incorporation of BBG, ID809 and Ceamfibre 7000 fibers in turkey meat sausage reduced the TBARS formation during storage. Indeed, on storage day 14, ID809 fiber (1% and 2%) reduced the meat lipid oxidation by more than 50% as compared to the control. After 14 days of storage at 4 °C, gradual decreases in TBARS values were observed. This would be due to the replacements of the amount of fat which are the causing agents of oxidation by fibers.

Conclusion

Economic, sensory and health aspects of sausages are the main reasons that processors of formulated foods increase non-meat fibers innovative sausages. The addition of BBG, ID809 and Ceamfibre 7000 fibers at 1% content to Tunisian turkey meat sausage provides a healthier product (lesser percentage of fat) and can help reduce costs associated with these sausages. Furthermore, the substitution of mechanically separated turkey by the three fibers at 1% content does not modify color, taste, flavor and overall acceptability attributes; being sometimes even better. Interestingly, the substitution of meat with 1% BBG fiber allows obtaining a product with a better taste. However, the addition of higher percentages of dietary fibers (2%) resulted in a remarkable decrease in sensory and textural profile. Furthermore, dietary fibers substitutions do not modify microbial growth and improve oxidative stability in formulated sausages. Overall, the dietary fibers described in this work could be a strong candidate for future applications in a wide range of meat products.

Acknowledgment

This research was supported by the Tunisian Ministry of Higher Education and Scientific Research.

References

1. Jousse F (2008) Modeling to improve the efficiency of product and process development. Compr Rev Food Sci F 7: 175-181.
2. Jiménez-Colmenero F (1996) Technologies for developing low-fat meat products. Trends Food Sci Tech 7: 41-48.
3. Miklosa R, Lametscha R, Nielsenb MS, Lauridsenb T, Einarsdottirc H, et al. (2013) Effect of fat type and heat treatment on the microstructure of meat emulsions. Inside Food Symposium 1-5.
4. Choi YS, Choi JH, Han DJ, Kim HY, Lee MA, et al. (2010a) Effects of replacing pork back fat with vegetable oils and rice bran fiber on the quality of reduced-fat frankfurters. Meat Sci 84: 557-563.
5. Choi YS, Park KS, Choi JH, Kim HW, Song DH, et al. (2010b) Physico-chemical properties of chicken meat emulsion systems with dietary fiber extracted from makgeolli lees. Korean J Food Sci Ani Resour 30: 910-917.
6. Yoo SS, Kook SH, Park SY, Shim JH, Chin KB, et al. (2007) Physicochemical characteristics, textural properties and volatile compounds in comminuted sausages as affected by various fat levels and fat replacers. Int J Food Sci Technol 42: 1114-1122.
7. Tokusoglu O, Unal K (2003) Fat replacers in meat products. PJN 2: 196-203.
8. Weiss J, Gibis M, Schuh V, Salminen H (2010) Advances in ingredient and processing systems for meat and meat products. Meat Sci 86: 196-213.
9. Elleuch M, Bedigian D, Roiseux O, Besbes S, Blecker C, et al. (2011) Dietary fibre and fibre-rich by-products of food processing: Characterisation, technological functionality and commercial applications: A review. Food Chem 124: 411-421.
10. Ktari N, Smaoui S, Trabelsi I, Ben-Salah R (2014) Chemical composition, techno-functional, sensory properties and effects of three dietary fibers on the quality characteristics of Tunisian beef sausage. Meat Sci 96: 521-525.

11. Femenia A, Bestard MJ, Sanjuan N, Rosselló C, Mulet A, et al. (2000) Effect of rehydration temperature on thecell wall components of broccoli (*Brassica oleracea* L. Var. *italica*) plant tissues. J Food Eng 46: 157-163.

12. Cofrades S, Guerra MA, Carballo J, Fernandez-Martin F, Colmenero FJ, et al. (2000) Plasma protein and soy fiber content effect on bologna sausage properties as influenced by fat level. J Food Sci 65: 281-287.

13. Fernández-López J, Sendra E, Sayas-Barberá E, Navarro C, Pérez-Alvarez JA, et al. (2008) Physico-chemical and microbiological profiles of "*salchichón*" (Spanish dry fermented sausage) enriched with orange fiber. Meat Sci 80: 410-417.

14. Yalinkiliç B, Kaban G, Kaya M (2012) The effects of different levels of orange fiber and fat on microbiological, physical, chemical and sensorial properties of sucuk. Food Microbiol 29: 255-259.

15. AOAC (2000) Official methods of analysis (17thedn.). Washington, DC: Association of Official Analytical Chemists.

16. Lin MJY, Humbert ES, Sosulski FW (1974) Certain functional properties of sunflower meal products. J Food Sci 39: 368-370.

17. Mac-Connel AA, Eastwood A, Mitchell WD (1974) Physical characterization of vegetable foodstuffs that could influence bowel function. J Sci Food Agric 25: 1457-1464.

18. Jamilah B, Harvinder KG (2002) Properties of gelatins from skins of fish black tilapia (*Oreochromis mossambicus*) and red tilapia (*Oreochromis nilotica*). Food Chem 77: 81-84.

19. Bourne BW (1978) Texture profile analysis. Food Technol 32: 62-65.

20. Hogan S, Zhang L, Li J, Wang H, Zhou K, et al. (2009) Development of antioxidant rich peptides from milk protein by microbial proteases and analysis of their effects on lipid peroxidation in cooked beef. Food Chem 117: 438-443.

21. Szczepaniak B, Piotrowska E, Dolata W (2007) Effect of partial fat substitution with dietary fiber on sensory attributes of finely comminuted sausages. part II. potato fiber and bran preparation. Pol J Food Nutr Sci 57: 421-425.

22. Choi YS, Kim HW, Song DH, Choi JH, Park J, et al. (2011) Quality characteristics and sensory properties of reduced-fat emulsion sausages with brown rice fiber. Korean J Food Sci Ani Resour 31: 521-529.

23. Fernández-Ginés JM, Fernandez-López J, Sayas-Barberá E, Sendra E, Pérez-Álvarez JA, et al. (2004) Lemon albedo as a new source of dietary fiber: Application to bologna sausages. Meat Sci 67: 7-13.

24. García ML, Dominguez R, Galvez MD, Casas C, Selgas MD, et al. (2002) Utilization of cereal and fruit fibers in low fat dry fermented sausages. Meat Sci 60: 227-236.

25. Huda AB, Parveen S, Rather SA, Akhter R, Hassan M (2014) Effect of incorporation of apple pomace on the physico-chemical, sensory and textural properties of mutton nuggets. Int J Adv Res 2: 974-983.

26. Sánchez-Zapata E, Zunino V, Pérez-Alvarez JA, Fernández-López J (2013) Effect of tiger nut fibre addition on the quality and safety of a dry-cured pork sausage ("*Chorizo*") during the dry-curing process. Meat Sci 95: 562-568.

27. Verma AK, Sharma BD, Banerjee R (2010) Effect of sodium chloride replacement and apple pulp inclusion on the physico-chemical, textural and sensory properties of low fat chicken nuggets. Food Sci Technol 43: 715-719.

28. Lee MA, Han DJ, Jeong JY, Choi JH, Choi YS, et al. (2008) Effect of kimchi powder level and drying methods on quality characteristics of breakfast sausage. Meat Sci 80: 708-714.

29. Turhan S, Sagir I, Ustun NS (2005) Utilization of hazelnut pellicle in low-fat beef burgers. Meat Sci 71: 312-316.

30. McKee L, Cobb E, Padilla S (2012) Quality indicators in poultry products. In: Leo ML Nollet, (ed), Handbook of meat, poultry and seafood quality. Wiley-Blackwell.

31. Gimeno O, Ansorena D, Astiasaran I, Bello J (2000) Characterization of chorizo de Pamplona: Instrumental measurements of colour and texture. Food Chem 69: 195-200.

32. Alesón-Carbonell L, Fernández-López J, Sayas E, Sendra E, Pérez-Alvarez JA, et al. (2003) Utilization of lemon albedo in dry-cured sausages. J Food Sci 68: 1826-1830.

33. Pérez-Alvarez JA, Fernández-López J (2006) Chemistry and biochemistry of color in muscle foods. In: Hui YH, (ed.) Food Biochemistry and Food Processing. Blackwell Publishing, Iowa.

34. Pérez-Alvarez JA, Sayas-Barberá ME, Fernández-López J, Aranda-Catalá V (1999) Physicochemical characteristics of Spanish-type dry-cured sausage. Food Res Int 32: 599-607.

35. Sayas-Barberá E, Viuda-Martos M, Fernández-López F, Pérez-Alvarez JA, Sendra E, et al. (2012) Combined use of a probiotic culture and citrus fiber in a traditional sausage 'Longaniza de Pascua'. Food control 27: 343-350.

36. Vural H, Javidipour I, Ozbas OO (2004) Effects of interesterified vegetable oils and sugarbeet fiber on the quality of frankfurters. Meat Sci 67: 65-72.

37. Backers T, Noli B (1997) Dietary fibres meat processing. Int Food Market Technol 1: 4-8.

38. Fernández J, Pérez-Álvarez JA, Fernández-López JA (1997) Thiobarbituric acid test for monitoring lipid oxidation in meat. Food Chem 59: 345-353.

Permissions

All chapters in this book were first published in JFPT, by OMICS International; hereby published with permission under the Creative Commons Attribution License or equivalent. Every chapter published in this book has been scrutinized by our experts. Their significance has been extensively debated. The topics covered herein carry significant findings which will fuel the growth of the discipline. They may even be implemented as practical applications or may be referred to as a beginning point for another development.

The contributors of this book come from diverse backgrounds, making this book a truly international effort. This book will bring forth new frontiers with its revolutionizing research information and detailed analysis of the nascent developments around the world.

We would like to thank all the contributing authors for lending their expertise to make the book truly unique. They have played a crucial role in the development of this book. Without their invaluable contributions this book wouldn't have been possible. They have made vital efforts to compile up to date information on the varied aspects of this subject to make this book a valuable addition to the collection of many professionals and students.

This book was conceptualized with the vision of imparting up-to-date information and advanced data in this field. To ensure the same, a matchless editorial board was set up. Every individual on the board went through rigorous rounds of assessment to prove their worth. After which they invested a large part of their time researching and compiling the most relevant data for our readers.

The editorial board has been involved in producing this book since its inception. They have spent rigorous hours researching and exploring the diverse topics which have resulted in the successful publishing of this book. They have passed on their knowledge of decades through this book. To expedite this challenging task, the publisher supported the team at every step. A small team of assistant editors was also appointed to further simplify the editing procedure and attain best results for the readers.

Apart from the editorial board, the designing team has also invested a significant amount of their time in understanding the subject and creating the most relevant covers. They scrutinized every image to scout for the most suitable representation of the subject and create an appropriate cover for the book.

The publishing team has been an ardent support to the editorial, designing and production team. Their endless efforts to recruit the best for this project, has resulted in the accomplishment of this book. They are a veteran in the field of academics and their pool of knowledge is as vast as their experience in printing. Their expertise and guidance has proved useful at every step. Their uncompromising quality standards have made this book an exceptional effort. Their encouragement from time to time has been an inspiration for everyone.

The publisher and the editorial board hope that this book will prove to be a valuable piece of knowledge for researchers, students, practitioners and scholars across the globe.

List of Contributors

Kobia Joyce, Asare DA and Asenso TN
Department of Animal Science, Kwame Nkrumah University of Science and Technology, Kumasi, Ghana

Emikpe BO
Department of Pathobiology, School of Veterinary Medicine, Kwame Nkrumah University of Science and Technology, Kumasi, Ghana

Yeboah Richmond
Department of Biological Sciences, Kwame Nkrumah University of Science and Technology, Kumasi, Ghana

Jarikre TA and Jagun-Jubril Afusat
Department of Veterinary Pathology, Faculty of Veterinary Medicine, University of Ibadan, Nigeria

Hassan AH
Department of Agriculture Systems Engineering, College of Agricultural and Food Sciences, King Faisal University, Saudi Arabia

Azam MM and Eissa AHA
Department of Agriculture Systems Engineering, College of Agricultural and Food Sciences, King Faisal University, Saudi Arabia
Agriculture Engineering Department, Faculty of Agriculture, Minoufiya University, Shibin El-Kom, Egypt

Birhanu Tesema Areda
Faculty of Agricultural Sciences, Department of Animal and Range Science, Blue Hora University, Ethiopia

Iheagwara MC
Department of Food Science and Technology, Federal University of Technology, Owerri, Imo State, Nigeria

Okonkwo TM
Department of Food Science and Technology, University of Nigeria, Nsukka, Enugu State, Nigeria

Naveen Kumar M and Das SK
Department of Agricultural and Food Engineering, Indian Institute of Technology, Kharagpur, India

Ryo Inagaki, Chikako Hirai, Yuko Shimamura and Shuichi Masuda
Laboratory of Food hygiene, Graduate School of Nutritional and Environmental Sciences, University of Shizuoka, Shizuoka, Japan

Hammad HHM, Meihu Ma and Guofeng Jin
National R&D Center for Egg Processing, College of Food Science and Technology, Huazhong Agricultural University, Wuhan, Hubei, PR China

Lichao He
College of Food and Biotechnology, Wuhan Institute of Design and Science, Wuhan, Hubei, PR China

Ayeloja AA
Fisheries Technology Department, Federal College of Animal Health and Production Technology, Moor Plantation, Ibadan, Nigeria

George FOA
Department of Aquaculture and Fisheries Management, Federal University of Agriculture, Abeokuta (FUNAAB), Abeokuta, Nigeria

Peixoto RRA, Villa JEL and Cadore S
Institute of Chemistry, University of Campinas, Campinas, SP, Brazil

Silva FF
Agilent Technologies Brasil, Av. Dr. Marcos Penteado Ulhoa, Barueri, SP, Brazil

Abker AM and Elkhedir AE
Industrial Research and Consultancy Center (IRCC), Khartoum, Sudan

Madwi HA
Department of Food Science and Technology, Faculty of Agriculture, University of Khartoum, Sudan

Dawood SY
Agriculture Ministry, Khartoum, Sudan

Pokhum C and Chawengkijwanich C
National Nanotechnology Center, National Science and Technology Development Agency, Pathumthani, Thailand

Kobayashi F
Nippon Veterinary and Life Science University, Musashino, Tokyo, Japan

Adeyanju JA and Olajide JO
Department of Food Science and Engineering, Ladoke Akintola University of Technology, Ogbomoso, Oyo State, Nigeria

Adedeji AA
Department of Biosystems and Agricultural Engineering, University of Kentucky, Lexington, KY, USA

Fiza Nazir
Division of Post-Harvest Technology, SKUAST-Kashmir, India

Nayik GA
Department of Food Engineering and Technology, SLIET, Punjab, India

Hema K
Department of Fish Processing Technology, Fisheries College and Research Institute, Thoothukudi, Tamil Nadu, India

Shakila RJ
Department of Harvest and Post Harvest Management, Fisheries College and Research Institute, Thoothukudi, Tamil Nadu, India

Shanmugam SA
Tamil Nadu Fisheries University (TNFU), Nagapattinam, Tamil Nadu, India

Jawahar P
Department of Fisheries Biology and Resource Management, Fisheries College and Research Institute, Thoothukudi, Tamil Nadu, India

Pereira SCL
Universidade Federal de Minas Gerais, Escola de Enfermagem, Departamento de Nutrição, Belo Horizonte-MG, Brazil

Vanetti MCD
Universidade Federal de Viçosa, Departamento de Microbiologia, Viçosa-MG, Brazil

Priscila Alonso DS
Institute Federal Goiano Campus Rio Verde, Rodovia Sul Goiana - Rio Verde - Goiás, Brazil

Marcio Caliari and Manoel Soares SJ
Federal University of Goiás, Food Technology Department, Goiânia, Goiás, Brazil

Kesava Reddy C, Sivapriya TVS, Arun Kumar U and Ramalingam C
School of Biosciences and Technology, VIT University, Vellore, India

Shobana Devi R and Nazni P
Department of Food Science and Nutrition, Periyar University, Salem, Tamil Nadu, India

Sanaullah Noonari, Irfana NM, Raiz AB and Muhammad IK
Assistant Professor, Faculty of Agricultural Social Sciences, Sindh Agriculture University, Tandojam, Pakistan

Shahbaz Ali
Research Assistant, Department of Agricultural Economics, Faculty of Agricultural Social Sciences

Hafez NE and Awad AM
Department of Food Science and Technology, Faculty of Agriculture, El-Fayoum University, Egypt

Ibrahim SM and Mohamed HR
Fish Processing and Technology Laboratory, Fisheries Division, National Institute of Oceanography and Fisheries, Egypt

Nakano T and Ozimek L
Department of Agricultural, Food and Nutritional Science, University of Alberta, Canada

Shehzad A and Sadiq Butt M
National Institute of Food Science and Technology, Faculty of Food, Nutrition and Home Sciences, University of Agriculture, Faisalabad, Pakistan

Tanweer S
National Institute of Food Science and Technology, Faculty of Food, Nutrition and Home Sciences, University of Agriculture, Faisalabad, Pakistan
Bioactive Natural Product Laboratory, Plant Soil and Science Building, Michigan State University, MI, USA

Shahid M
Department of Biochemistry, Faculty of Basic Sciences, University of Agriculture, Faisalabad, Pakistan

Podpora B, Świderski F, Sadowska A, Piotrowska A and Rakowska R
Department of Functional and Organic Food and Commodities, Faculty of Human Nutrition and Consumer Sciences, Warsaw University of Life Sciences, Warsaw, Poland

Guizani SEO and Moujahed N
National Institute of Agricultural Sciences of Tunisia, 43 Av. Charles Nicolle 1082 Tunis Mahrajene, Tunisia
Laboratory of animal and food resources, Tunisia

Pakhare KN and Dagadkhair AC
Department of Technology, Food Technology Division, Shivaji University, Kolhapur, India

Udachan IS
MIT College of Food Technology, Rajbaugh, Loni Kalbhor, Pune, India

Andhale RA
College of Food Technology, VNMKV, Parbhani, India

Oltiev AT and Majidov KH
Bukhara Engineering and Technological Institute, Bukhara City, Uzbekistan

Mudasir Ahmad Bhat and Hafiza Ahsan
Department of Post-Harvest Technology, SKUAST-K, Shalimar, Srinagar

Bimal Bibhuti and Yadav AK
Department of Food Process Engineering, Vaugh School of Agricultural Engineering and Technology, Sam Higginbottom Institute of Agriculture, Technology and Sciences, Allahabad, India

Ladu G, Cubaiu L, d'Hallewin G and Venditti T
CNR - ISPA, Trav. La Crucca, 3 - 07040 Sassari, Italy

Pintore G and Petretto GL
Department of Chemistry and Pharmacy, University of Sassari, via Muroni, 23 / A - 07100 Sassari, Italy

Steven Brandon C
Membrane Application Services, SC, USA

Paul Dawson L
Department of Food Science and Human Nutrition, Clemson University, Clemson, SC, USA

Elbakheet SI, Elgasim EA and Algadi MZ
Faculty of Agriculture, University of Khartoum, Sudan

Akinoso R and Ogunyele OO
Department of Food Technology, Faculty of Technology, University of Ibadan, Nigeria

Olatoye KK
Department of Food, Agriculture and Bio-engineering, College of Engineering and Technology, Kwara State University, Kwara State, Nigeria

Femshang M Charles
Ministry of Secondary Education, Yaounde, Cameroon

Karthika Devi B, Kuriakose SP, Krishnan AVC, Choudhary P and Rawson A
Indian Institute of Crop Processing Technology, Thanjavur, India

Sheng ZY, Si HJ, Ming LX, Rong CJ, Yi CZ and Hui ZY
Sericulture and Agri-food Research Institute, Guangdong Academy of Agricultural Sciences, Key Laboratory of Functional Foods, Ministry of Agriculture, Guangdong Key Laboratory of Agricultural Products Processing, Guangzhou, China

Pereira SCL
Federal University of Minas Gerais, School of Nursing, Department of Nutrition, Belo Horizonte-Minas Gerais, Brazil

Maria Cristina DV
Federal University of Viçosa, Department of Microbiology, Viçosa-Minas Gerais, Brazil

Naourez Ktari, Intidhar Bkhairia and Moncef Nasri
Laboratory of Enzyme Engineering and Microbiology, National School of Engineering of Sfax, University of Sfax, Tunisia

Imen Trabelsi and Riadh B Salah
Laboratory of Microorganisms and Biomolecules (LMB), Centre of Biotechnology of Sfax, Tunisia

Mehdi Triki
Laboratory of Food Analysis, National School of Engineering of Sfax, University of Sfax, Tunisia

Mohamed A Taktak and Hafedh Moussa
Chahia Company, Road of Sidi Salem, Sfax 3003, Tunisia

Index

P

R

S

T

W

Y

www.ingramcontent.com/pod-product-compliance
Lightning Source LLC
Chambersburg PA
CBHW080633200326
41458CB00013B/4618
* 9 7 8 1 6 4 1 1 6 1 7 0 1 *